Active Volcanoes of the World

Series editors

Corrado Cimarelli, München, Germany
Sebastian Müller, Mainz, Germany

More information about this series at http://www.springer.com/series/10081

Nick Varley · Charles B. Connor
Jean-Christophe Komorowski
Editors

Volcán de Colima

Portrait of a Persistently Hazardous Volcano

Springer

Editors
Nick Varley
Facultad de Ciencias
Universidad de Colima
Colima, Mexico

Jean-Christophe Komorowski
Institut de Physique du Globe de Paris
Paris, France

Charles B. Connor
School of Geosciences
University of South Florida
Tampa, FL, USA

ISSN 2195-3589 ISSN 2195-7029 (electronic)
Active Volcanoes of the World
ISBN 978-3-642-25910-4 ISBN 978-3-642-25911-1 (eBook)
https://doi.org/10.1007/978-3-642-25911-1

Library of Congress Control Number: 2018959259

This Springer imprint is published by the registered company Springer-Verlag GmbH, DE part of Springer Nature
The registered company address is: Heidelberger Platz 3, 14197 Berlin, Germany

Preface

This volume summarizes some of the recent scientific works to characterize the eruptive history, magmatic system and hazards associated with Volcán de Colima (Mexico), one of the most continuously active composite volcanoes on Earth during the historic era. Indeed, the history of scientific investigation of Volcán de Colima parallels the history of volcano science as a whole. This is not surprising, because from the very beginning of volcano science, Volcán de Colima was recognized as a laboratory for observing and understanding fantastic and rare eruptive phenomena. The volcano has been the subject of an impressive range of studies, which together, have increased our knowledge of this beautiful stratovolcano, and thus provided a better chance of an effective risk reduction strategy for the future. Many of the observations and fresh theories that have been derived can easily be applied to similar systems worldwide. Each chapter in this volume builds upon this record of research, conducted by generations of scientists who have studied the volcano. It provides an overview of the state of our knowledge, and will serve as a launching point for new initiatives in the forthcoming years.

Medina (1983) reports that the earliest records of Volcán de Colima eruptions were compiled in the 'Archivo de Indias', a library commissioned in 1572 by Philip II in Seville (Spain). These records include reports on eruptions in 1560 and large, possibly Plinian, eruptions in 1576 C.E. and 1585 C.E. The record of eruptive activity throughout the sixteenth–early nineteenth centuries was compiled from scattered reports by Arreloa (1903, 1915), Bárcena (1887) and Ordonez (1897), and was later summarized concisely by Mooser (1958, 1961). Charles Lyell mentions the 1818 C.E. eruption of Volcán de Colima in his famous text, Principles of Geology (Lyell 1830), but erroneously attributes the eruption to the reactivation of Volcán Jorullo. He writes that 6 inches (15 cm) of tephra fell in the city of Guadalajara as a result of the 1818 eruption, a figure that is possibly exaggerated! Nevertheless, it is clear from his description that Volcán de Colima figured prominently in the earliest discourse about physical processes of active volcanoes in the scientific literature.

Arguably, the first scientific monitoring of Volcán de Colima was undertaken in 1893–1905 by Diaz (1906), with the support of José María Arreola. During this time period, regular observations of the volcano were made by the Observatorio Meteorológico y Vulcanológico de Zapotlán. Diaz (1906) summarizes these observations in monthly notes and sketches of activity. At the time, these were among the best scientific records of eruptive activity at any volcano in the world, certainly rivalling the contemporary efforts of Thomas Jaggar (Kilauea, HI, USA) and Giuseppe Mercalli (Vesuvius, Italy). The efforts of Arreola, Diaz and colleagues brought Volcán de Colima more fully to the attention of the international volcano science community and established the volcano as a dominant feature of the Trans-Mexican Volcanic Belt. Their observations also provided important context for understanding of the January 1913 eruption, which impacted a large population then living in the region, especially due to tephra fallout in Ciudad Guzman and nearby areas (Waitz 1936). It clearly demonstrated the potential risks to those living near this massive volcano.

An understanding of the geology of Volcán de Colima and surrounding terrain began to emerge in the mid-twentieth century. Herrera (1967) produced a geological map of the region, followed by regional maps and descriptions by Luhr and Carmichael (1980) and Robin et al. (1987). After the 1980 eruption of Mount St. Helens (WA, USA), a number of groups quickly recognized the prevalence of volcano debris avalanches at Volcán de Colima and Nevado de Colima, identifying debris flow deposits on the south flank of Volcán de Colima (e.g. Luhr and Prestegaard 1988). Further studies have located debris avalanche deposits from the older edifice, Nevado de Colima (Robin et al. 1987; Capra and Macías 2002), and multiple deposits (e.g. Roverato et al. 2011). These studies represent a significant shift in our understanding of the volcanic system, which had been partially biased by the limited range of volcanic activity observed historically. The recognition of volcanic debris avalanches in the stratigraphic record of Volcán de Colima (e.g. Stoopes and Sheridan 1992; Sheridan and Macías 1995; Cortes et al. 2010a) as well as state-of-the-art physical models of the process of flank instability (Borselli et al. 2011) has resulted in significant revision of hazard assessments.

Most recently, mapping of the Colima volcano complex has resulted in a consistent set of chronostratigraphic units and a map that can be readily applied to hazard assessments on a variety of spatial and temporal scales (Cortes et al. 2010b). Important insights into the generation and emplacement of pyroclastic density currents generated at Volcán de Colima from dome and lava flow collapse have been made during detailed field studies, starting with the 1991 eruption (Rodriguez-Elizarraras et al. 1991) up until the recent eruptions (Saucedo et al. 2002; Reyes-Dávila et al. 2016). The hazard maps have recently been updated with the consideration of different eruptive scenarios, ranging from common Vulcanian eruptions and dome-collapse events, through to large VEI 6 Plinian eruptions with possible debris avalanches. Numerical modelling resulted in probabilistic maps, providing the authorities with a range of versatile tools for emergency planning and the construction of effective risk mitigation strategies (Varley et al. 2017). Finally, in the first comprehensive ambient seismic noise tomography analysis, Spica et al. (2017) produce a 3D model of the Volcán de Colima plumbing system providing thought-provoking clues to understand how and where magmas are produced and stored at depth below Volcan de Colima. Their new model remarkably shows a deep, large and well-delineated elliptic-shape magmatic reservoir below the Colima volcano complex at a depth of about 15 km, in which magma is mainly stored in conduits or inter-fingered dykes as opposed to a horizontally stratified magma reservoir.

Chapters in this volume continue this work of discerning the full range of eruptive activity in this marvellously complex volcanic system. Crummy et al. (Chap. 1) provide new insights into the Holocene tephra stratigraphic record of the volcano, which is key to documenting its history of explosive volcanic eruptions. Norini et al. (Chap. 2) tackle the complex topic of the basement geology of the volcano and the impact of basement faults on the stability of volcano edifices of the Colima Volcano Complex. Cortes et al. (Chap. 3) clarify the stratigraphic record of volcano debris avalanche deposits since the Pliocene, for both Volcán de Colima and Nevado de Colima. Chapters 4, 5 and 6 investigate some of the most profound hazards of the volcano produced by both eruptive and non-eruptive activity: tephra fallout, pyroclastic density currents and lahars. Cumulatively, these chapters document the state of knowledge for a broad range of activities.

In addition to modelling variations in the eruptive style at Volcán de Colima and its eruptive products, a huge effort has been devoted to monitoring of the volcano and to forecasting potential future activity. Geophysical monitoring began in the mid-1970s with the installation of portable seismic networks. This network was made permanent early in the eruption crisis of 1991 (Núñez-Cornú et al. 1994). Varley (Chap. 7) and Arámbula-Mendoza et al. (Chap. 8) describe a dramatic range of integrated monitoring efforts at Volcán de Colima, including seismic, thermal and gas monitoring. These two chapters document a substantial change in monitoring of the volcano over the last 30 years, and demonstrate that it is one of the best monitored volcanoes in North America, largely through the efforts of the Universidad de Colima and its students, along with national and international collaborations. In addition, Varley (Chap. 7) presents a detailed account of the 1998–2015 period.

At the same time as mapping and geophysical studies were underway in the late 1970s, James Luhr and Ian Carmichael, together with colleagues, began to unravel the magmatic history of the volcanic system in detail (e.g. Luhr and Carmichael 1980, 1981, 1990). Fundamentally, they found that the cyclicity in eruptive activity inferred from the eruption record was matched by geochemical cycles, driven by crystal fractionation and magma mixing in the volcano's shallow reservoir. Reubi et al. (Chap. 9) and Lavallée et al. (Chap. 10) expand upon this earlier work, and through the use of new and innovative techniques, provide a 4D view of the magmatic system that is otherwise inaccessible.

In the final chapters of the volume, de la Cruz et al. (Chap. 11) and Rodríguez (Chap. 12) write extremely valuable perspectives on mitigation of risk for those living near the volcano, both by providing long-term probabilistic hazard assessment, and by investigating the social consequences and response to this information. Risk mitigation at the volcano has provided various challenges during recent years, such as the issue of it being located on the border of two states, each with their differences, both in approach and resources.

As with any persistently erupting volcano, it is often difficult to draw the line when reporting details of eruptions. In the case of this volume, the limit is the July 2015 eruption, which proved to be the most important since the last cataclysmic event in 1913. Details of this eruption appear in Chaps. 7 and 8. Between this eruption and the final production of this book, further important activity occurred in 2016 and early 2017. At the end of September 2016, a large batch of magma filled the crater and produced another lava flow, which was emplaced on top of the 2015 flow to the south. The event led to a crisis response, with meetings at national level, increased observational overflights and preparations for evacuations. Notably, the eruption was purely effusive despite a fast ascent rate and evidence of a high volatile content. Large volumes of sulphur dioxide were emitted, which coincided with some heavy rainfall. This produced a considerable fallout of acid rain, producing crop failure or damage at unprecedented levels at this volcano.

Early 2017 featured a series of larger Vulcanian explosions, with the emplacement of a pyroclastic density current in La Arena ravine reaching 4.2 km. Smaller emplacements were observed to the north. Although smaller in magnitude than previous explosions, most notably, those of 2005, they produced an impressive shower of volcanic bombs and in one case, an extraordinary shock wave that rattled windows in the capital Colima as well as cities further afield.

Those of us who have worked in volcano science for many decades recognize the profound changes in our understanding of volcanoes represented by the contributions in this volume. Tremendous advances have occurred, both because of innovative scientific advances and through the hard work of many individuals in the field, laboratory and observatory. We are not finished, as major challenges remain to understand the nature of Volcán de Colima, its activity and potential impacts, while improving our preparedness for the potential future large eruptions. We hope this volume will be useful to scientists engaged in these efforts for many years to come, as well as to the general reader with a profound interest in volcanoes and what we can learn about them. The compilation of state-of-the-art volcano science in this volume will undoubtedly contribute to improve risk perception and societal resilience in the face of volcanic eruptions.

Colima, Mexico — Nick Varley
Tampa, USA — Charles B. Connor
Paris, France — Jean-Christophe Komorowski

References

Arreola, J.M.: Brief notice of observations of Colima. J. Geo. **11**:751–761 (1903)

Arreola, J.M.: Catalogo de las erupciones antiguas del Volcán de Colima, Memorias Soc. Cient A Alzate. **32**:443–481 (1915)

Barcena, M.: Inform sobre el estado actual del Volcán de Colima. Anales Minst. de Formento, Mexico, 385–365 (1887)

Borselli, L., Capra, L., Sarocchi, D., De la Cruz-Reyna, S.: Flank collapse scenarios at Volcán de Colima, Mexico: a relative instability analysis. J. Volcanol. Geotherm. Res. **208**, 51–65 (2011)

Capra, L., Macıas, J.L.: The cohesive Naranjo debris-flow deposit (10 km^3): a dam breakout flow derived from the Pleistocene debris-avalanche deposit of Nevado de Colima Volcano (México). J. Volcanol. Geotherm. Res. **117**(1–2), 213–235 (2002)

Cortes, A., Macias, J.L., Capra, L., Garduño, V.H.: Sector collapse of the SW flank of Volcán de Colima, México. The 3600 yr BP La Lumbre-Los Ganchos debris avalanche and associated debris flows. J. Volcanol. Geotherm. Res. **197**, 52–66 (2010a)

Cortés, A., Garduño, V.H., Macías, J.L., Navarro-Ochoa, C., Komorowski, J.C., Saucedo, R., Gavilanes, J.C.: Geologic mapping of the Colima volcanic complex (Mexico) and implications for hazard assessment. Geol. Soc. Am. Spec. Pap. **464**, 249–264 (2010b)

Diaz, S.: Efemerides del Volcán de Colima (1893–1905). Sria. de Formento, Mexico, 168 pp (1906)

Herrera, C.: Geolofia de las Volcanes de Colima. Ph.D. Dissertation, Faculty of Engineering, UNAM (1967)

Luhr, J.F., Carmichael, I.S.: The Colima volcanic complex, Mexico. Contrib. Mineral Petrol. **71**(4), 343–372 (1980)

Luhr, J.F., Carmichael, I.S.: The Colima volcanic complex, Mexico: Part II. Late-quaternary cinder cones. Contrib. Mineral. Petrol. **76**(2), 127–147 (1981)

Luhr, J.F., Carmichael, I.S.:. Petrological monitoring of cyclical eruptive activity at Volcan Colima, Mexico. J. Volcanol. Geotherm. Res. **42**(3), 235–260 (1990)

Luhr, J.F., Prestegaard, K.L.: Caldera formation at Volcán Colima, Mexico, by a large large holocene volcanic debris avalanche. Mexico. J. Volcanol. Geotherm. Res. **35**(4), 335–348 (1988)

Medina Martínez, F.: Analysis of the eruptive history of the Volcán de Colima, Mexico (1560, 1980). Geofísica Internacional, **22**(2) (1983)

Mooser, F.: Catalogue of active Volcanoes of the World, Part XI, 1, "Central America". International Association of Volcanology, Rome 10–13 (1958)

Mooser, F.: Los Volcanes de Colima. Bol. No. 61 del Inst. de Geol. Mexico, 49–61 (1961)

Núñez-Cornú, F., Nava, F.A., De la Cruz-Reyna, S., Jiménez, Z., Valencia, C., García-Arthur, R.; Seismic activity related to the 1991 eruption of Colima Volcano, Mexico. Bull. Volcanol. **56**(3), 228–237 (1994)

Ordoñez, E.: Les Volcans Colima et Ceboruco. Memorias de la Soc. Cient. A Alzate, **11**, 325–333 (1897)

Reyes-Dávila, G., Arámbula-Mendoza, R., Espinasa-Pereña, R., Pankhurst, M.J., Navarro-Ochoa, C., Savov, I., Vargas-Bracamontes, D., Cortés-Cortés, A., Gutiérrez--Martínez, C., Valdés-González, C., Domínguez-Reyes, T., González-Amezcua, M., Martínez-Fierros, A., Ramírez-Vázquez, C.A., Cárdenas-González, L., Castañeda-Bastida, E., Vázquez Espinoza de los Monteros, D., Nieto-Torres, A., Campion, R., Courtois, L., Lee, P.D.: Volcán de Colima dome collapse of July, 2015 and associated pyroclastic density currents, J. Volcanol. Geotherm. Res. **320**, 100–106 (2016)

Robin, C., Mossand, P., Camus, G., Cantagrel, J.-M., Gourgaud, A., Vincent, P.M.: Eruptive history of the Colima volcanic complex (Mexico). J. Volcanol. Geotherm. Res. **31**:99–113 (1987)

Rodríguez-Elizarrarás, S., Siebe, C., Komorowski, J.C., Espíndola, J.M., Saucedo, R.: Field observations of pristine block and ash flow deposits emplaced April 16–17, at Volcán de Colima, Mexico. J. Volcanol. Geotherm. Res. **48**, 399–412 (1991)

Roverato, M., Capra, L., Sulpizio, R., Norini, G.: Stratigraphic reconstruction of two debris avalanche deposits at Colima Volcano (Mexico): insights into pre-failure conditions and climate influence. J. Volcanol. Geotherm. Res. **207**(1–2), 33–46 (2011)

Saucedo, R., Macas, J.L., Bursik, M., Mora, J.C., Gavilanes, J.C., Cortes, A.: Emplacement of pyroclastic flows during the 1998–1999 eruption of Volcan de Colima, Mexico. J. Volcanol. Geotherm. Res. **117**, 129–153 (2002)

Saucedo, R., Macias, J.L., Gavilanes, J.C., Arce, J.L., Komorowski, J.C., Gardner, J.E., Valdez- Moreno, G.: Eyewitness, stratigraphy, chemistry, and eruptive dynamics of the 1913 Plinian eruption of Volcán de Colima, México. J. Volcanol. Geotherm. Res. **191**, 149–166 (2010)

Sheridan, M.F., Macías, J.: Estimation of risk probability for gravity-driven pyroclastic flows at Volcan Colima, Mexico. J. Volcanol. Geotherm. Res. **66**(1–4), 251–256 (1995)

Spica, Z., Perton, M., Legrand, D.: Anatomy of the Colima volcano magmatic system, Mexico. Earth Planet. Sci. Lett. **459**, 1–13 (2017)

Stoopes, G.R., Sheridan, M.F.: Giant debris avalanches from the Colima Volcanic Complex, Mexico: implications for long-runout landslides (>100 km) and hazard assessment. Geology, **20**(4), 299–302 (1992)

Varley, N.R., Cernas, J.A., García, A., López, Z., Meza, M.I.: Riesgos Volcanológicos, Universidad de Colima, Protección Civil del Estado de Colima, Colima (2017)

Waitz, P.: Datos historicos y bibliograficos acera del Volcán de Colima. Memorias de la Soc. Cient. A Alzate, **53**, 349–386 (1936)

Contents

Holocene Eruption History and Magmatic Evolution of the Colima Volcanic Complex

Julia M. Crummy, Ivan P. Savov, Carlos Navarro-Ochoa and Dan J. Morgan

Abstract

The Colima Volcanic Complex (CVC) erupts in both highly explosive and effusive fashion. Detailed radiocarbon dating of tephra fallout deposits exposed in road-cuts on the flanks of Nevado de Colima reveal at least 25 major Plinian eruptions occurred during the last 30,000 years. Deposits from these eruptions are basaltic to andesitic in composition showing an arc affinity in their immobile trace element abundance patterns. Detailed studies of the mineralogy, major and trace element and isotope chemistry reveal two distinct magma types erupted at the CVC during the Holocene. Group I, which comprise the bulk of eruption deposits, are calc-alkaline basaltic-andesites to andesites, while Group II magmas show strong petrological and geochemical similarities to alkaline lamprophyric magmas that formed monogenetic cinder cones to the east and west of the northern CVC. The presence of alkaline and calc-alkaline magmas at the CVC has allowed us to investigate the nature between these magma types in an arc setting. Data presented here reveal that the Group II magmas formed from pulses of alkaline melts intercepting the magmatic storage region of the CVC stratovolcanoes, mixing with the calc-alkaline Group I magmas. These pulses appear to have recurrence rates on the timescales of a few thousand years.

Keywords

Volcan de Colima · Arc magmas · Magmatic evolution · Magma mixing

1 Introduction

Historically, activity at Volcán de Colima is dominantly effusive with rare explosive sub-Plinian to Plinian eruptions, the last of which occurred in January 1913 (Bretón González et al. 2002). However, throughout the late Pleistocene and Holocene, numerous highly explosive eruptions occurred at the CVC, as revealed by tephra deposits exposed on the flanks of Nevado de Colima (Fig. 1; Luhr et al. 2010).

J. M. Crummy (✉)
British Geological Survey, The Lyell Centre, Research Avenue South, Edinburgh EH14 4AP, UK
e-mail: juli@bgs.ac.uk

J. M. Crummy · I. P. Savov · D. J. Morgan
School of Earth and Environment, University of Leeds, Leeds LS2 9JT, UK

C. Navarro-Ochoa
Centro Universitario de Estudios E Investigación En Vulcanología, Universidad de Colima, Colima, Mexico

N. Varley et al. (eds.), *Volcán de Colima*, Active Volcanoes of the World,
https://doi.org/10.1007/978-3-642-25911-1_5

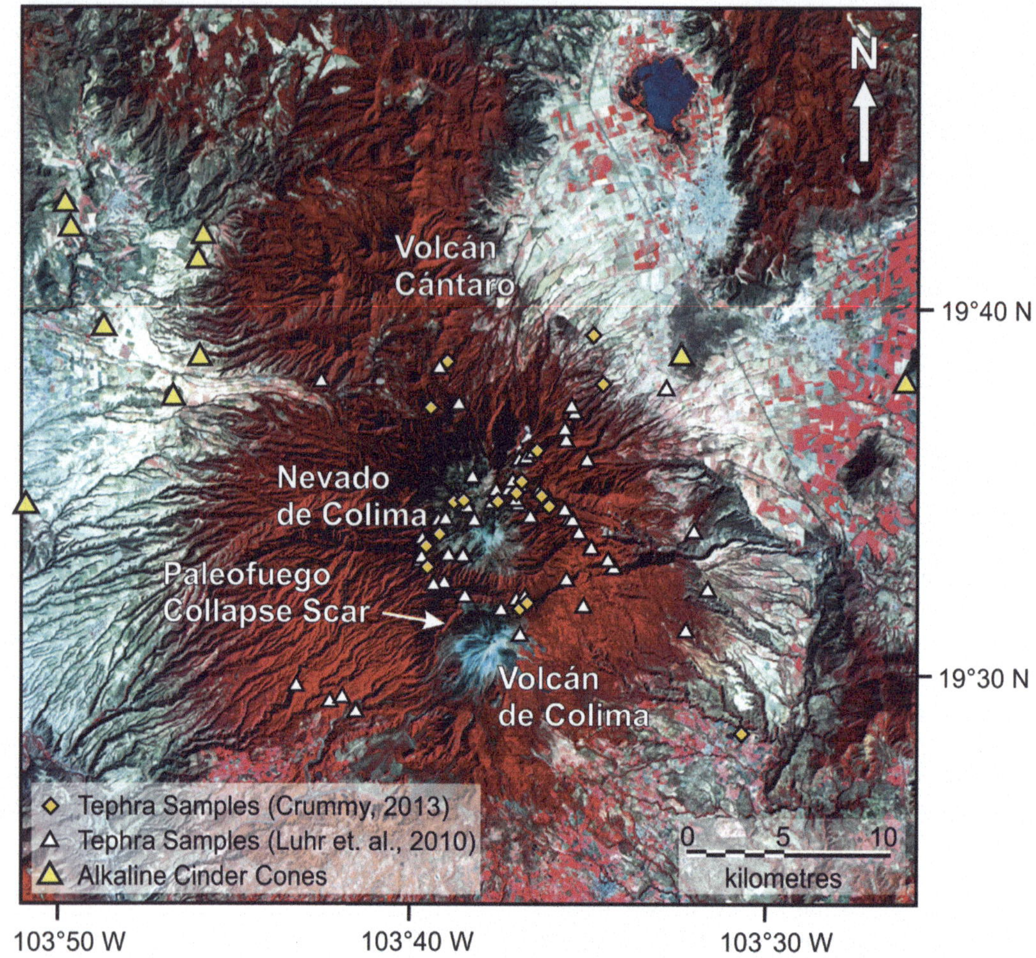

Fig. 1 Sample location map for the Holocene tephra fallout deposits of the CVC. Locations are from new sites sampled in 2010 and 2011 (yellow diamonds; Crummy 2013) and Luhr et al. (2010; white triangles). Also shown are monogenetic cinder cone locations after Luhr and Carmichael (1981) and Carmichael et al. (2006). The majority of samples are located in access roads on the upper slopes of Nevado de Colima. The image is a Landsat image from the United States Geological Survey (USGS). Red areas are forested, pink areas are farmland, and blue areas are unvegetated

Since the 1980s, Volcán de Colima has been the focus of many studies investigating the geochemistry and petrology of the erupted deposits, largely focusing on the current phase of activity and the 1913 eruption (Luhr and Carmichael 1980, 1982, 1990a, b; Medina-Martínez 1983; Robin et al. 1991; Luhr 1993, 2002; Robin and Potrel 1993; Komorowski et al. 1997; Righter 2000; Atlas et al. 2006; Luhr et al. 2006, 2010; Reubi and Blundy 2008; Savov et al. 2008; Saucedo et al. 2010; Verma and Luhr 2010). Based on historical records and petrological and geochemical studies, Luhr (2002) proposed a cycle of ~100 years for explosive Plinian activity at Volcán de Colima.

Much of the work carried out on Volcán de Colima is concerned with forecasting future activity and monitoring current activity (see Varley, this volume); however, few of these studies consider the volcano's explosive past, or patterns of activity which may emerge from a better understanding of the explosive Holocene (and late Pleistocene) eruption deposits. These deposits are largely unstudied and poorly understood, with few attempts to characterize them: the first by Luhr and Carmichael (1982), and two further more detailed studies by Luhr et al. (2010) and Crummy et al. (2014).

This work incorporates, and expands upon the work of Luhr et al. (2010), focusing on the stratigraphic relationships of the explosive Holocene eruption deposits, and the petrological and geochemical evolution of the erupted magmas during the past 13,000 years. Detailed petrological and geochemical investigations of the erupted tephra fallout deposits presented here reveal a link between the calc-alkaline volcanism responsible for the CVC stratovolcanoes, and alkaline volcanism that formed monogenetic cinder cones to the east and west of the northern CVC. New data, in combination with published and unpublished datasets, are used to investigate the nature of the CVC magmatic plumbing system and the link between alkaline and calc-alkaline volcanism in an arc setting.

2 Volcanic Activity of the CVC

Historical records of activity at Volcán de Colima extend back to 1519 and the arrival of the Spanish conquistadors (Bretón González et al. 2002). Prior to the establishment of volcano observatories in the cities of Colima and Zapotlan in 1898, records of activity are from eyewitness accounts. Bretón González et al. (2002) presented a detailed history of volcanic activity since 1519, summarized here in Fig. 2, in combination with reports from the Smithsonian Institute (Global Volcanism Program 2016). Since 1519, there have been nine highly explosive, VEI = 4 (Volcanic Explosivity Index after Newhall and Self 1982) events, the last of which occurred in 1913. Only the 1818 and 1913 eruptions are well-documented in historical records, with detailed descriptions of thick ash-fall and pyroclastic flows (Bretón González et al. 2002; Luhr 2002). Tephra fallout deposits from these two eruptions are the only historical fallout deposits preserved in the tephra record (Luhr et al. 2010). It is therefore assumed that the 1818 and 1913 events were larger-volume eruptions than the seven other reported VEI = 4 historical events (Fig. 2; Bretón González et al. 2002; Luhr et al. 2010).

The February 1818 Plinian eruption of Volcán de Colima produced a sustained ash column, which resulted in ash-fall in Mexico City ~470 km from the vent, and pyroclastic flows that travelled ~14 km along ravines (Bretón González et al. 2002). For 51 years after the 1818 eruption, activity at Volcán de Colima was restricted to the rise of the magma within the crater, and it was assumed that the volcano was inactive (Luhr and Carmichael 1990b; Bretón González et al. 2002). In January of 1913, Volcán de Colima once again erupted explosively producing a sustained ash column that reached an estimated height of 18–23 km above sea level (a.s.l.) and pyroclastic flows that travelled up to 15 km from the vent (Luhr and Carmichael 1980; Saucedo et al. 2010). For more details of the 1913 Plinian eruption, please refer to Connor et al., this volume. Following the 1913 explosive eruption, the volcano again entered an effusive phase of activity punctuated by small Vulcanian explosions (Robin et al. 1987; Bretón González et al. 2002; Luhr 2002). This type of activity is ongoing at the time of writing (September 2016), and is termed here as interplinian.

2.1 Interplinian Activity

Interplinian deposits are poorly exposed at the CVC having been buried or destroyed by subsequent eruptions; therefore, the nature of such activity is characterized by material erupted between 1869–1909 and 1957–present day. Luhr and Carmichael (1980) were the first to study the Colima andesites, describing the petrology and geochemistry of lava flows and ballistics erupted in 1869, 1913, 1961–1962 and 1975–1976. Further studies on interplinian activity focused on effusive and small explosive (VEI 2–3) deposits erupted between 1961 and 2005 (Robin et al. 1987, 1991; Luhr and Carmichael 1990b; Luhr 1993, 2002; Robin and Potrel 1993; Navarro-Ochoa et al. 2002; Savov et al. 2008).

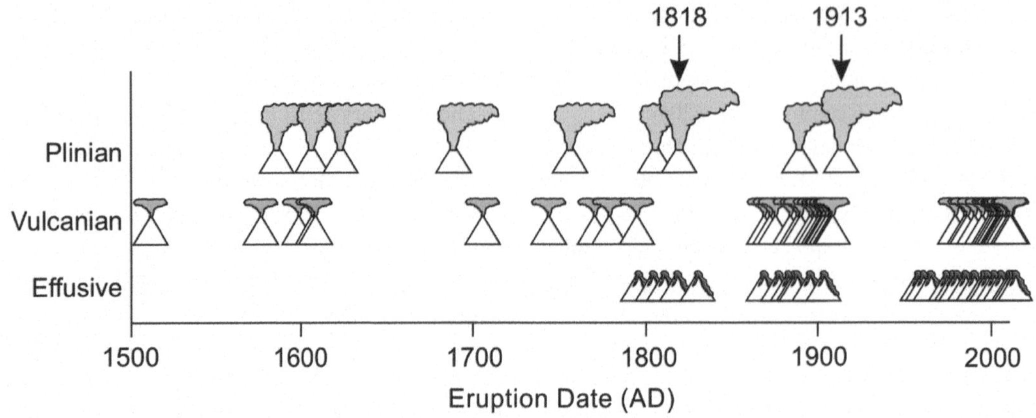

Fig. 2 Schematic diagram summarizing volcanic activity at Volcán de Colima since records began in 1519 using the data from Bretón González et al. (2002) in combination with reports from the Smithsonian Institute (Global Volcanism Program 2016). There is very little information in the records of lava flows from Volcán de Colima, and effusive deposits exposed on the flanks are all post-1818 lava flows (Bretón González et al. 2002). Numerous Vulcanian-type eruptions have occurred throughout historical times and up to nine VEI = 4 eruptions have occurred since the 1500s. Tephra is only preserved from the 1818 and 1913 eruptions and there are numerous eye-witness accounts of these eruptions; therefore, it is assumed that these were larger eruptions than the other seven

In 1869, 51 years after the large explosive 1818 eruption, lava erupted on the northeast side of the Volcán de Colima edifice forming the parasitic cone of Volcancito (Luhr and Carmichael 1990b). Resulting lava flows travelled into the northeast part of the caldera floor and along a ravine to the southeast. Activity continued at Volcancito until 1878 when activity in the main crater resumed (Bretón González et al. 2002). For the next 29 years, Volcán de Colima displayed effusive activity punctuated by small explosive eruptions that produced ash clouds above the vent and block-and-ash flows down the flanks of the volcano (Bretón González et al. 2002). Records of these eruptions are from eye-witness accounts, and no details of column height or flow length were given. From 1904 to 1909 explosive activity decreased and the volcano entered a more fumarolic phase with occasional small explosions (Bretón González et al. 2002).

In 1957, the first activity from the summit of Volcán de Colima since the 1913 eruption was recorded with the extrusion of a new lava dome, followed by the production of lava flows and weak explosions in 1960 and 1961 (Luhr and Carmichael 1980). For the next 30 years, activity was dominated by effusive lava dome extrusion and partial dome collapses, resulting in lava flows reaching up to 4 km in length (Luhr 2002).

A more explosive phase of activity commenced in 1999 with an explosion in February that sent ballistic projectiles up to 3.5 km away from the vent, and in July an eruption that produced an ash column 12 km a.s.l. and pyroclastic flows down the southern flank (Bretón González et al. 2002; Savov et al. 2008; Global Volcanism Program 2016). This explosive phase of activity is continuing at the time of writing (September 2016). For more details of recent activity the reader is referred to Luhr (2002) and Savov et al. (2008), and the chapters of Varley and Reubi et al. in this book.

2.2 Plinian Activity

The tephra fallout deposit of the 1913 explosive eruption blankets Nevado de Colima, varying in thickness from 60 cm ($\sim$4 km from the vent) to 6 cm ($\sim$10 km from the vent; Luhr et al. 2010; Saucedo et al. 2010; Connor et al., this volume). The 1818 eruption deposit is much thinner and less well-exposed (Luhr et al. 2010). These two units form the top of a stratigraphy comprising numerous tephra fallout and ash-rich surge deposits resulting from explosive sub-Plinian and Plinian eruptions throughout the Holocene and Late Pleistocene (Fig. 3; Luhr et al. 2010).

During the 1990s and 2000s, Jim Luhr, Carlos Navarro-Ochoa and Ivan Savov sampled and described CVC tephra fallout deposits preserved and exposed on the flanks of Nevado de Colima and in quarries and gullies on the rift floor (Luhr et al. 2010). Through detailed sampling and dating of charcoal found within ash horizons and tephra units, they were able to identify and date at least 25 deposits, erupted between ~30,000 yrs B.P. and the present day (Fig. 3; Luhr et al. 2010). Further field campaigns carried out in January 2010 and February 2011 have built upon the work of Luhr et al. (2010) on the tephrochronology and granulometry of the CVC deposits (Crummy 2013; Crummy et al. 2014). To date, eruption deposits have been described at 89 localities

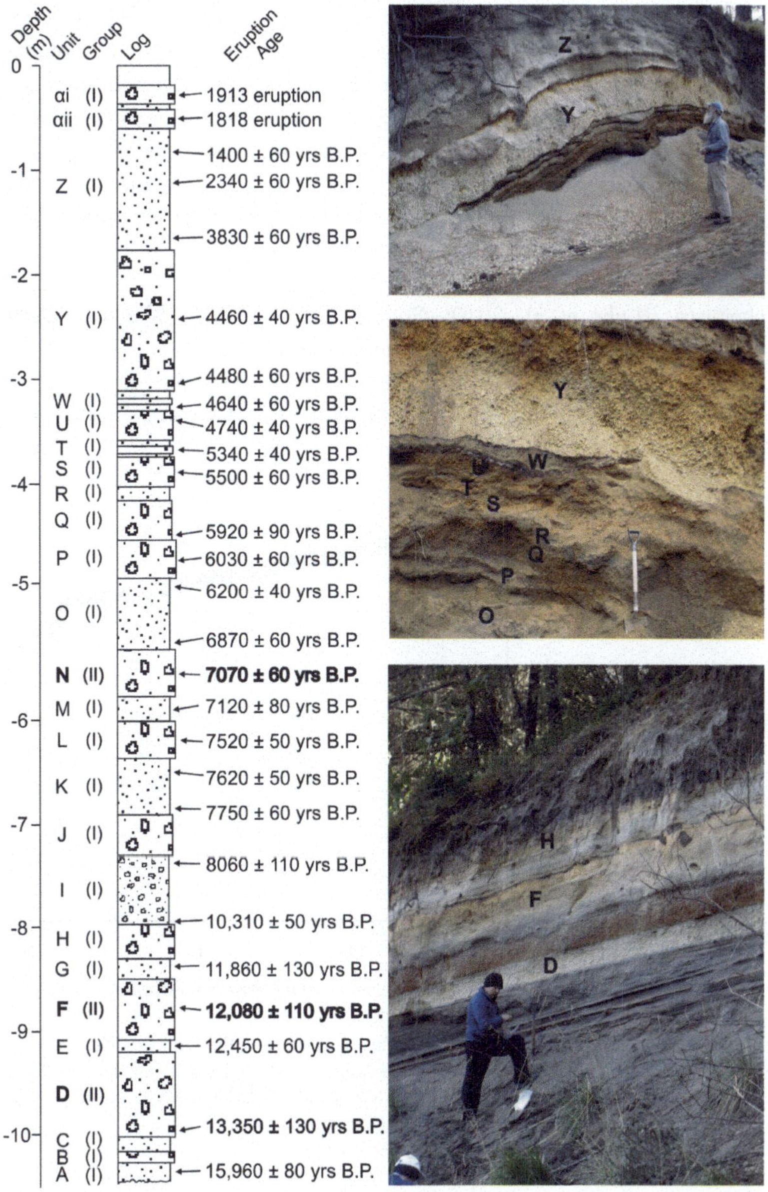

Fig. 3 Composite stratigraphic section and field photographs of deposits resulting from explosive eruptions exposed in road-cuts on the flanks of Nevado de Colima. The unit names and ages are as reported in Luhr et al. (2010). The Group II eruption deposits (units N, F and D) are highlighted in bold

across an area of ~500 km^2, including exposures in quarries on the rift floor (Fig. 1).

In total, 182 radiocarbon ages have been obtained from the CVC eruptive stratigraphy, yielding ages from 80 ± 50 to 29,930 ± 210 yrs B.P. (Fig. 4; Luhr et al. 2010). Of these, 143 ages represent the Holocene (0 to 9000 yrs B.P.). Nevado de Colima is densely vegetated; therefore, the eruption deposits on the volcano's flanks are only exposed in road cuts. Nevado de Colima is a national park which has one main access road on the northeastern flank. Once inside the national park (entrance at 3480 ma.s.l.), there are many more roads. Accordingly, the majority of the tephra deposits are exposed at high elevations, with the older units having been buried or eroded by subsequent eruptions. There are a few roads on the lower slopes; however, the area is prone to landslides and the roads are not maintained.

Using a combination of radiocarbon dating, together with field mapping, petrology and geochemistry (Luhr et al. 2010; Crummy 2013; Crummy et al. 2014), a detailed stratigraphy of the CVC Holocene explosive eruption deposits has been compiled up to 13,000 yrs B.P. (Fig. 3). The field characteristics of the eruption units are given in Table 1.

The stratigraphy of eruption deposits exposed on the flanks of Nevado de Colima comprises tephra fallout deposits varying in thickness from a few centimetres (~10 km from the vent) to over 1 m (~3 km from the vent). The fallout deposits comprise pumice or scoria horizons, with individual clasts varying from <1 to 17 cm. The largest clast sizes are typically found in the most proximal localities, ~3 km from the vent. Angular rock fragments are abundant in all the erupted units, measuring up to 2 cm across (along the long-axis). Interbedded with the tephra layers are ash-rich surge or pyroclastic flow horizons with thicknesses from 2 cm to over 1 m. These ash beds are typically rich in charcoal fragments, and commonly contain reworked pumice and scoria from underlying tephra fallout deposits.

2.3 Monogenetic Cinder Cones

To the east and west of the extinct Volcán Cántaro, located to the north of Nevado de

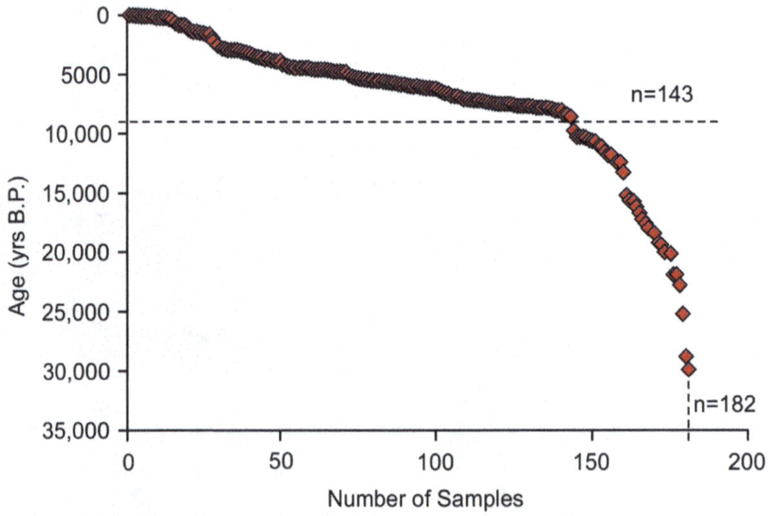

Fig. 4 Radiocarbon ^{14}C ages for the explosive Plinian CVC eruption deposits (Komorowski et al. 1997; Luhr et al. 2010). Out of a total of 182 dates, 143 of these were sampled between 9000 yrs B.P. and present. The younger eruption deposits are well exposed high on the flanks of Nevado de Colima, while the older eruption deposits are less well exposed, confined to lower elevations where the slopes are densely vegetated and access is difficult

Table 1 Field characteristics of CVC tephra fallout deposits discussed in the text

Unit	Eruption age (yrs B.P.)	Group	Color (dry)	Thickness (cm)	Max. pumice size (cm)	Max. lithic size (cm)	Sorting	Composition
Y	4460 ± 40	I	Cream	48–140	17	7	Inverse	Andesite
W	4480 ± 60–4540 ± 60	I	Dark brown	8–28	6	5	Normal	Basaltic-andesite
U	4740 ± 40–4760 ± 70	I	Dark orange	15–20	4	2	Normal	Basalt/basaltic-andesite
S	5430 ± 50–5500 ± 60	I	Dark brown-orange	12–46	4.5	2	Normal	Basaltic-andesite
P	5980 ± 50–6150 ± 40	I	Cream	6–40	9	6	Normal	Basaltic-andesite
N	6950 ± 50–7070 ± 60	II	Orange	12–60	12	2	Normal	Basaltic-andesite
L	7520 ± 50–7530 ± 80	I	Dark brown	4–50	6	3	Inverse	Andesite
J	7750 ± 60–7760 ± 50	I	Pale orange-cream	7–15	6	3	Inverse	Andesite
H	9770 ± 60–10,310 ± 50	I	Dark grey	15–42	5	2	Inverse	Basaltic-andesite
F	11,840 ± 70–12,080 ± 150	II	Grey	30–90	5	8	Normal	Basalt
D	12,460 ± 60–13,350 ± 130	II	Grey	32–180	4.5	3	Inverse	Basalt

Eruption ages are from radiocarbon ^{14}C dating of charcoal found within or between tephra fallout deposits (Komorowski et al. 1997; Luhr et al. 2010). Maximum lithic fragment and pumice clast sizes are found in sections 4–5 km from the currently active vent

Colima and where volcanic activity at the CVC began approximately 1.7 Ma (Luhr and Carmichael 1990a), eleven cinder cones erupted on the rift floor between 1.2 Ma and 62 ka (Fig. 1; Luhr and Carmichael 1981; Allan and Carmichael 1984; Carmichael et al. 2006). The two oldest cones erupted at 1.2 and 0.5 Ma, producing <0.003 km^3 of calc-alkaline basalt and basaltic-andesite (Luhr and Carmichael 1981; Allan and Carmichael 1984; Carmichael et al. 2006). Nine of the cinder cones are alkaline, producing ~1.3 km^3 magma between 450 and 62 ka (Luhr and Carmichael 1981; Allan and Carmichael 1984; Carmichael et al. 2006). The majority of the alkaline mafic magmas (>99%) erupted between 240 and 60 ka (Carmichael et al. 2006). During this time, calc-alkaline volcanism had migrated south from Volcán Cántaro and was ongoing at the Nevado de Colima volcano (Robin et al. 1987; Luhr and Carmichael 1990a; Cortés et al. 2005, 2010).

The alkaline cinder cones were first described by Luhr and Carmichael (1981), who classified them as primitive basanites to minette lamprophyres with 47.6–50.3 wt% SiO_2, 7.4–15.3 wt% MgO and 2.5–4.4 wt% K_2O. The occurrence and source of the alkaline magmas is discussed in detail in the geological literature (Luhr and Carmichael 1981; Wallace and Carmichael 1989; Lange and Carmichael 1990; Luhr 1997; Carmichael et al. 2006; Maria and Luhr 2008; Vigouroux et al. 2008; Cai 2009); however, the relationship between the calc-alkaline and the alkaline magmas remains poorly understood.

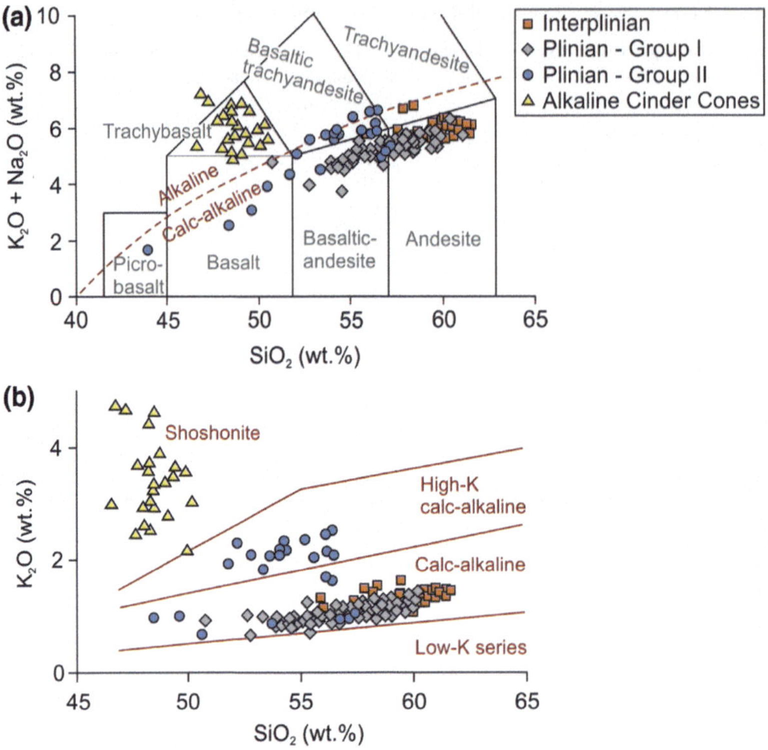

Fig. 5 Classification of the CVC eruption deposits. **a** Total alkalis-silica (TAS) classification diagram after Le Maitre et al. (2002) for CVC interplinian and Plinian eruption deposits and the alkaline cinder cone magmas. The dashed line separating the alkaline from calc-alkaline fields is after Irvine and Baragar (1971). **b** K_2O versus SiO_2 classification diagram after Le Maitre et al. (2002). Data sources for the interplinian and Plinian deposits are given in the text. Plinian tephra fall deposits range in composition from picro-basalt to andesite. Interplinian deposits are typically andesitic in composition. The alkaline cinder cone magmas are high-K shoshonites (Luhr and Carmichael 1981; Carmichael et al. 2006; Maria and Luhr 2008; Vigouroux et al. 2008; Cai 2009)

3 Mineralogy of Erupted Deposits

The majority of CVC eruptive products are basaltic-andesite to andesite in composition (Fig. 5) with the typical mineral assemblage of plagioclase + orthopyroxene + clinopyroxene + hornblende + Fe–Ti oxides ± olivine (Luhr and Carmichael 1980, 1982; Savov et al. 2008; Luhr et al. 2010). Olivine phenocrysts are typically present only in the most mafic units.

The interplinian lavas are typically highly crystalline and have up to 52 vol% crystals comprising the aforementioned mineral phases (Luhr and Carmichael 1980). The groundmass is also highly crystalline, comprising glass and a mesostasis of the same minerals, with the exception of hornblende, which is not found in the groundmass (Luhr and Carmichael 1980). Xenocrystic olivine displaying pyroxene and oxide overgrowths occur in many samples (Luhr and Carmichael 1980). Hornblende phenocrysts all display disequilibrium with breakdown rims of pyroxene and Fe–Ti oxides (Luhr and Carmichael 1980, 1990b; Luhr 2002; Savov et al. 2008).

The bulk of the tephra fallout deposits exposed in road-cuts comprise pumice and scoria clasts with 10–25 vol% phenocrysts and microphenocrysts of the above mineral phases. The groundmass comprises highly vesiculated intermediate to felsic glass (54–74 wt% SiO_2) with abundant microlites of the same mineral phases with the exception of hornblende. Plagioclase, orthopyroxene and clinopyroxene phenocrysts all display complex zoning patterns interpreted to represent multiple magma recharge events (Crummy et al. 2014).

Three of the tephra fallout deposits, units N, F and D of the Luhr et al. (2010) stratigraphy, which erupted c. 7000, c. 12,000 and c. 13,000 yrs B.P. (Fig. 3), are mineralogically distinct from the rest of the CVC tephra deposits. Samples from these units have low crystallinity (10–15 vol%) comprising phenocrysts and microphenocrysts of plagioclase + clinopyroxene + olivine + hornblende + phlogopite + Fe–Ti oxides + orthopyroxene. The groundmass comprises highly vesiculated, dark, mafic to felsic glass (49.5–66.9 wt% SiO_2) with abundant microlite crystal phases dominated by plagioclase and clinopyroxene.

Based on the mineralogy and whole-rock geochemistry (see below), the explosive eruption deposits have been divided into two groups. Group I comprises the bulk of the tephra units, while Group II comprises the three distinct (phlogopite-containing) tephra units N, F and D (see Table 2).

Each of the three units that make up the Group II eruption deposits have phlogopite phenocrysts and/or microphenocrysts, which are not present in scoria and pumice from the Group I deposits. The Group II units are, however, mineralogically distinct from each other (Table 2). Unit N scoria phenocrysts comprise predominantly plagioclase and clinopyroxene, with up to 1 vol% olivine, and trace abundances of hornblende and phlogopite. Scoria from unit F has a lower crystallinity when compared with units N and D, and the mineral assemblage is dominated by clinopyroxene, olivine and phlogopite with up to 1 vol% plagioclase and trace hornblende. Unit D scoria has the highest crystallinity, comprising predominantly hornblende and plagioclase, with clinopyroxene and up to 1 vol% olivine and phlogopite. The groundmass also appears to be more crystalline in unit D, with abundant microlites. Fe–Ti oxides are present in all samples from 1 to 3 vol%, and are most abundant in unit D. As observed within the Group I tephra samples, plagioclase and clinopyroxene phenocrysts from Group II scoria reveal crystallization histories involving complex magma recharge and mixing events (Crummy et al. 2014). For more detailed descriptions of the mineralogy of the Colima calc-alkaline tephra, the reader is referred to Crummy et al. (2014).

Table 2 Mineralogy of pumice and scoria from the CVC Plinian tephra fallout deposits

Unit	Group	Crystallinity (vol%)	Plag (vol%)	Cpx (vol%)	Opx (vol%)	Hbd (vol%)	Olivine (vol%)	Phlog (vol%)	Fe–Ti oxides (vol%)
Y	I	6–10	3–5	0–trace	0–trace	2–4	–	–	Trace–1
W	I	13–19	7–10	1–2	1–3	2–5	Trace–1	–	1
U	I	20–25	12–15	1–2	3–4	2–5	1–2	–	1
S	I	12–17	7–10	1	1–2	1–3	1–2	–	1
P	I	14–22	10–15	1–2	1–3	Trace	Trace	–	1
N	II	9–13	5–10	1–2	Trace	Trace	Trace–1	Trace	1
L	I	11–19	5–10	Trace	1	3–5	–	–	1–2
J	I	11–16	7–10	1	1–2	1	–	–	1–2
H	I	13–17	2–10	1–2	2–3	Trace–5	Trace–2	–	1
F	II	7–12	Trace–1	2–5	–	Trace	1–3	1–2	1–2
D	II	10–15	1–3	Trace–2	Trace	3–7	Trace–1	0–1	1–3

The three units that comprise Group II contain lower abundances of plagioclase and contain phlogopite. The mineral proportions are given as volume % of the whole sample, and were estimated from thin sections on the optical microscope

Plag plagioclase; *Opx* orthopyroxene; *Cpx* clinopyroxene; *Hbd* hornblende; *Phlog* phlogopite

The alkaline cinder cones were described by Luhr and Carmichael (1981) as forming a transitional series from basanite to minette based on the appearance and increasing abundance of phlogopite, sanidine, leucite and apatite. The typical mineral assemblage of the basanites, as defined by Luhr and Carmichael (1981), comprises phenocrysts and microphenocrysts of olivine + clinopyroxene + plagioclase (labradorite An_{50-70}) + microphenocrysts of titanomagnetite. Leucite-basanites have the characteristic mineral assemblage of olivine + clinopyroxene phenocrysts, and microphenocrysts of sanidine + leucite + titanomagnetite + apatite + phlogopite (Luhr and Carmichael 1981). The minettes comprise phenocrysts of olivine + clinopyroxene + phlogopite + apatite and microphenocrysts of sanidine, leucite, titanomagnetite, apatite and phlogopite (Luhr and Carmichael 1981).

4 Whole-Rock Geochemistry

A database of whole-rock major and trace elements for the CVC has been compiled from newly collected data (Crummy 2013), as well as from the CVC literature (Luhr and Carmichael 1980, 1981, 1982; Allan and Carmichael 1984; Robin et al. 1991; Luhr 1993, 2002; Robin and Potrel 1993; Komorowski et al. 1997; Carmichael et al. 2006; Maria and Luhr 2008; Savov et al. 2008; Vigouroux et al. 2008; Cai 2009; Luhr et al. 2010; Saucedo et al. 2010) and unpublished data obtained by James Luhr and Ivan Savov. The full database comprises 398 samples from the CVC, including the alkaline cinder cones; of these 252 are from the Holocene and late Pleistocene explosive tephra deposits. The ranges of major element composition for the Group I and Group II eruption deposits are listed in Table 3.

4.1 Major Elements

The CVC eruptive products show a wide range in composition from basalt to andesite based on the Total-Alkalis Silica (TAS) and K_2O versus SiO_2 diagrams of Le Maitre et al. (2002; Fig. 5). The majority of the deposits are calc-alkaline; however, some samples appear to be trending towards the high-K alkaline cinder cone compositions.

The geochemical evolution of the CVC eruption deposits is shown on time series plots

Table 3 Ranges of major element concentrations in weight % for the Group I and Group II calc-alkaline basalts to high silica andesites

Unit	SiO_2	TiO_2	Al_2O_3	FeOt	MnO	MgO	CaO	Na_2O	K_2O	P_2O_5
Y	53.9–60.4	0.4–0.7	17.9–21.5	4.7–5.8	0.10–0.11	1.8–2.8	5.0–5.9	3.8–4.9	1.0–1.5	0.17–0.31
W	54.7–57.1	0.6–0.8	17.7–18.9	5.7–6.6	0.11–0.12	3.9–5.9	6.4–7.5	3.8–4.4	0.8–1.1	0.20–0.21
U	50.7–57.8	0.5–0.8	17.4–21.4	5.6–7.0	0.10–0.12	2.3–6.1	5.5–7.5	3.3–4.4	0.7–1.3	0.19–0.35
S	54.6–58.6	0.5–0.8	17.1–19.1	4.8–6.5	0.09–0.12	3.6–5.8	5.7–7.5	4.0–4.4	0.9–1.3	0.17–0.15
P	55.4–59.7	0.5–0.8	17.2–19.6	4.7–5.8	0.08–0.11	2.6–6.2	5.6–7.1	4.0–4.7	1.0–1.3	0.16–0.22
N	**52.8–56.5**	**0.8–0.9**	**16.3–18.8**	**5.9–6.0**	**0.11–0.11**	**3.3–6.2**	**5.7–7.2**	**3.5–4.2**	**1.6–2.5**	**0.28–0.56**
L	58.0–59.9	0.6–0.7	16.9–17.8	4.9–5.4	0.09–0.10	3.9–5.1	5.9–6.3	4.2–4.7	1.0–1.2	0.15–0.18
J	53.4–59.5	0.6–1.0	16.4–17.5	5.2–7.0	0.10–0.12	4.6–6.0	5.9–8.1	3.6–4.7	1.1–1.9	0.15–0.39
H	54.5–58.7	0.7–0.9	15.7–18.0	5.8–7.2	0.11–0.13	4.0–7.3	6.7–8.0	2.9–4.4	0.8–1.1	0.13–0.50
F	**43.9–53.3**	**0.8–1.3**	**14.4–19.2**	**6.9–9.1**	**0.09–0.13**	**6.4–8.1**	**6.9–8.0**	**1.3–3.5**	**0.4–3.5**	**0.27–0.68**
D	**53.7–57.5**	**0.8–0.9**	**15.9–18.6**	**6.2–6.9**	**0.11–0.13**	**2.8–6.1**	**6.4–8.0**	**3.4–4.4**	**1.0–2.2**	**0.21–0.51**

Data are from Luhr and Carmichael (1981, 1982) and Luhr et al. (2010) and unpublished data collected by Ivan Savov and Jim Luhr related to the Holocene tephra fallout deposits, and by Julia Crummy as part of her Ph.D. thesis at the University of Leeds (Crummy 2013)
The Group II units are in bold

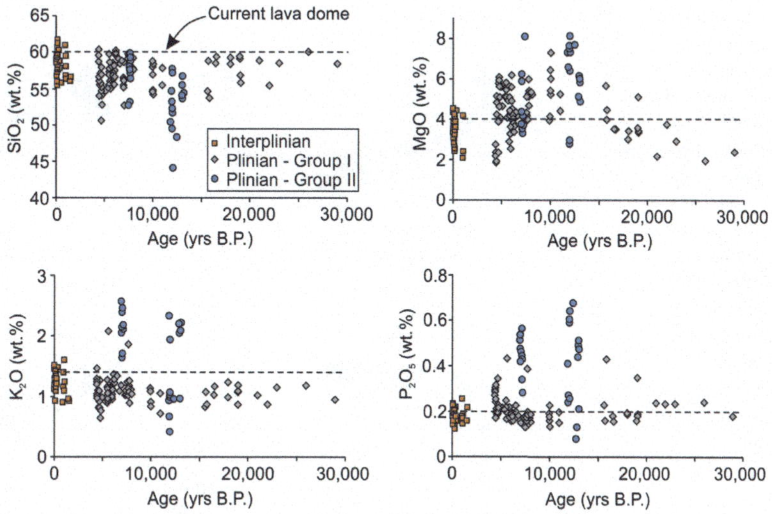

Fig. 6 Time series plots showing the variations in whole-rock major element geochemistry for the CVC eruption deposits over the past 30,000 years. Also shown is the composition of the current lava dome (2007–2010; unpublished dataset). The deposits are divided into two groups based on the geochemical distinctions; the deposits with high MgO, K_2O and P_2O_5 form Group II (blue circles)

(Fig. 6). The interplinian eruption deposits show little variation in whole-rock major element geochemistry, while the explosive (Plinian) eruption deposits show a wide variation in SiO_2, MgO, K_2O and P_2O_5. Overall the MgO content appears to increase from 30,000 yrs B.P. to 12,000 yrs B.P., and then decreases from ~10,000 yrs B.P. to the present. This pattern is somewhat reflected in the SiO_2 content, which shows increases in the same samples that show lower MgO (in the last 10,000 yrs). However, there are fewer samples from 10,000 to 30,000 yrs B.P., so it is difficult to draw concrete conclusions from the available dataset.

The variations of K_2O and P_2O_5 with time reveal the majority of the eruption deposits show little compositional variation; however, the Group II samples, which erupted c.7000, c.12,000 and c.13,000 yrs B.P., display distinctly high K_2O and P_2O_5 (Fig. 6).

The Group I tephra which, in terms of composition, incorporate the interplinian products, are calc-alkaline basaltic-andesites to high silica andesites with SiO_2 contents ranging from 50.7 to 60.3 wt% (Fig. 5). Overall the whole-rock major element systematics show an evolutionary trend toward increasing K_2O and Na_2O and decreasing MgO, CaO, FeOt, TiO_2, MnO and P_2O_5 with increasing SiO_2 (Fig. 7). These trends are consistent with evolution from a mafic magma parent, where the crystallization of the ferromagnesian minerals olivine, orthopyroxene, clinopyroxene and hornblende, and the Ca-rich plagioclase, leave residual melts that are more and more depleted in MgO, CaO, FeOt, TiO_2, MnO, while increasing their SiO_2 and alkalis.

The Group II units are more mafic than the Group I deposits, comprising basalt to basaltic-andesite with SiO_2 contents ranging from 43.9 to 57.5 wt% (Fig. 5). Overall, the Group II eruption deposits follow a similar differentiation trend to the Group I calc-alkaline series, with increasing K_2O and Na_2O and decreasing MgO, CaO, FeOt, MnO and P_2O_5 with increasing SiO_2 (Fig. 7). However, the Group II eruption deposits are geochemically distinct, with elevated K_2O (up to 3.5 wt%) and P_2O_5 (up to 0.68 wt%) relative to the Group I tephra (Figs. 6 and 7).

Figure 8 shows SiO_2 versus MgO and SiO_2 versus K_2O variation diagrams for Group II tephra with the main mineral phases determined via electron microprobe analyses (Crummy 2013). As minerals crystallize from a mafic parent melt, the composition of the melt is driven away from the

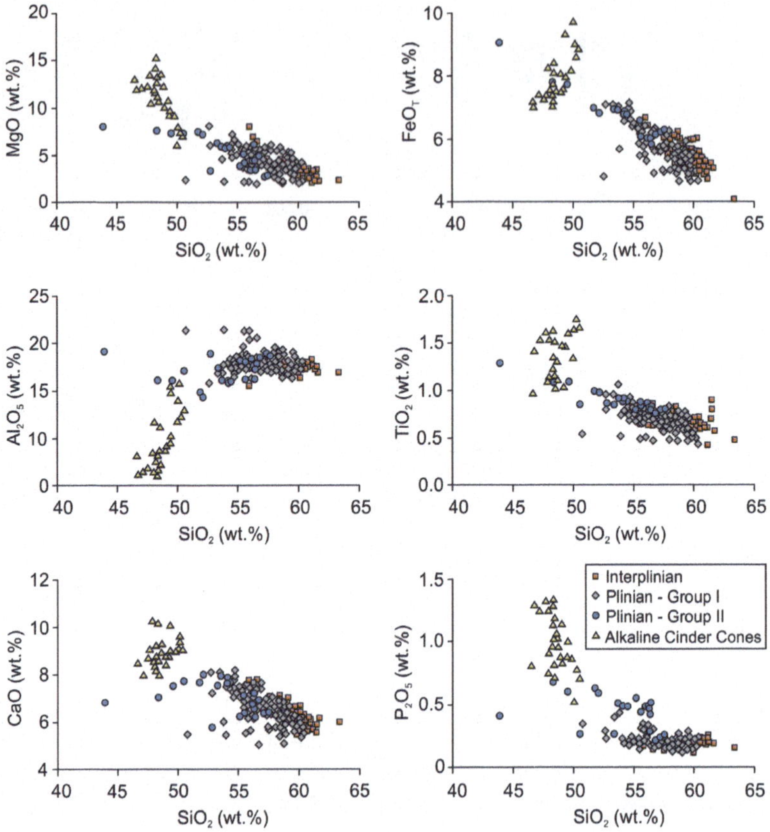

Fig. 7 Whole-rock major element variation diagrams for interplinian and Plinian eruption deposits of the CVC. Also plotted are whole-rock major element compositions for the alkaline cinder cones (Luhr and Carmichael 1981; Carmichael et al. 2006; Maria and Luhr 2008; Vigouroux et al. 2008; Cai 2009). The interplinian deposits are typically more evolved with higher SiO_2 and Al_2O_3, and lower MgO, FeOt, TiO_2, CaO and P_2O_5 contents relative to the Plinian deposits

composition of the extracted mineral assemblage, towards higher SiO_2 and K_2O and lower MgO concentrations. By projecting the differentiation trend back from the groundmass glass composition, through the parental melt composition, it is possible to determine the extracted mineral assemblage. The Group I magmas follow a trend defined by the extraction of plagioclase + hornblende + clinopyroxene + orthopyroxene (Fig. 8). The trend towards high K_2O and low MgO with increasing SiO_2 displayed by the Group II eruption deposits could be explained by the fractionation of plagioclase + phlogopite + clinopyroxene + orthopyroxene + olivine + hornblende from a mafic parent melt. However, not all Group II samples follow this trend. For example, some of the tephra samples from unit D follow the differentiation trend displayed by the Group I eruption deposits, while others lie along the Group II fractionation trend (Fig. 8). Similar variations are observed in the unit N and F samples, with many lying off the differentiation trends. These variation diagrams therefore reveal that not all the Group II samples can be linked to the Group I magmas through simple fractional crystallization from a common mafic parent. Accordingly, the wide range of K_2O (and P_2O_5, CaO, Al_2O_3) within the Group II units suggests a magma mixing process between two compositionally distinct magma batches.

The alkaline cinder cone deposits are characterized by low SiO_2 and Al_2O_3, and high TiO_2, MnO, MgO, CaO, K_2O and P_2O_5 (Fig. 7). These magmatic rocks show relatively little variation in SiO_2 content (47.6–50.3 wt%), but show large variations in MgO (7.4–15.3 wt%), K_2O (2.5–4.4 wt%) and P_2O_5 (0.7–1.3 wt%). The high MgO, K_2O and P_2O_5 contents are reflected in their mineralogy, with assemblages dominated by mafic minerals, and the presence of phlogopite, sanidine and apatite phenocrysts and microphenocrysts (Luhr and Carmichael 1981).

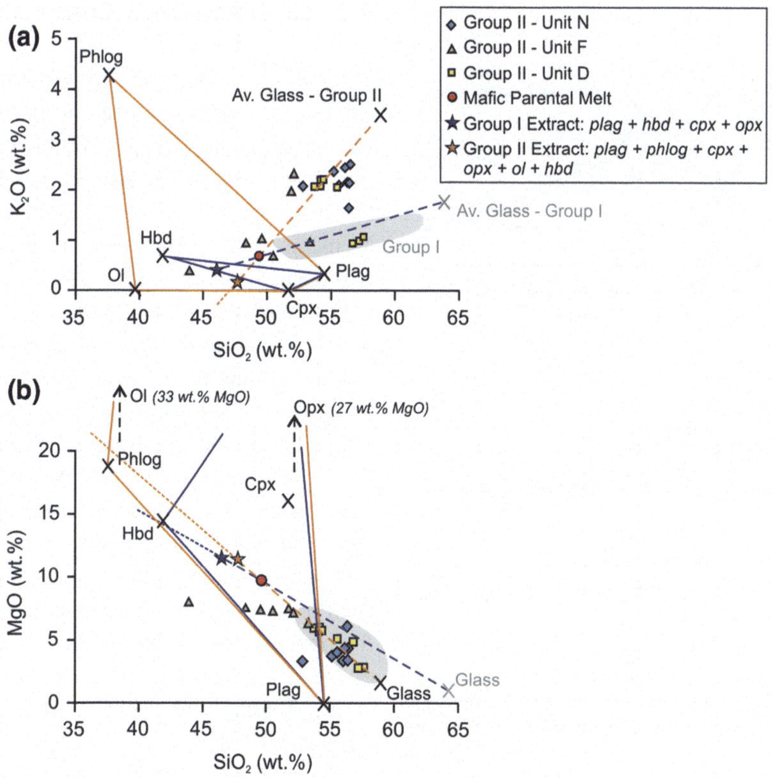

Fig. 8 K_2O and MgO variation with SiO_2 showing the control of the main mineral phases on the chemical trends. Projecting the melt composition from the groundmass glass back through the parental melt reveals the fractionating mineral assemblage driving magma differentiation. Sample SAY-22E from a basaltic cinder cone east of the CVC is used to represent the parental melt composition (Luhr and Carmichael 1981; Luhr 1997; Verma and Luhr 2010). The Group I samples follow a trend of increasing K_2O (**a**) and decreasing MgO (**b**) with increasing SiO_2 reflecting the extraction of plagioclase (plag) + hornblende (hbd) + orthopyroxene (opx) + clinopyroxene (cpx) [+olivine] from a mafic parental melt. The Group II eruption deposits reveal variations that cannot be explained by fractional crystallization alone. Some of the Group II samples follow the Group I trend; however, some samples from the same eruption unit follow a trend reflecting the extraction of plagioclase + phlogopite (phlog) + clinopyroxene + orthopyroxene + olivine (ol) + hornblende, and some do not appear to follow any fractionation trend

The alkaline cinder cones form a distinct group that lies off the calc-alkaline magma differentiation trend indicating a separate source composition and subsequent fractionation trends (Figs. 5 and 7).

4.2 Trace Elements

Incompatible trace element abundance patterns in the Group I and Group II CVC eruption deposits are consistent with those typical of subduction-related magmas (Fig. 9a; Pearce 1982; Straub et al. 2010 and references therein). All CVC tephra samples show enrichments in large ion lithophile elements (LILE: Rb, Ba, Th, and K) relative to high field strength elements (HFSE: Nb, Ta, Ti, Hf, Zr,) attributed to the addition of slab-derived fluids/melts to depleted mantle wedge sources (Vigouroux et al. 2008 and references therein). The relative depletions in the HFSE reflect the presence of stable mineral phases in the subducting slab or mantle wedge, such as ilmenite, titanite, rutile, zircon, apatite and garnet (Pearce 1982). The Group II samples display stronger enrichments in the incompatible

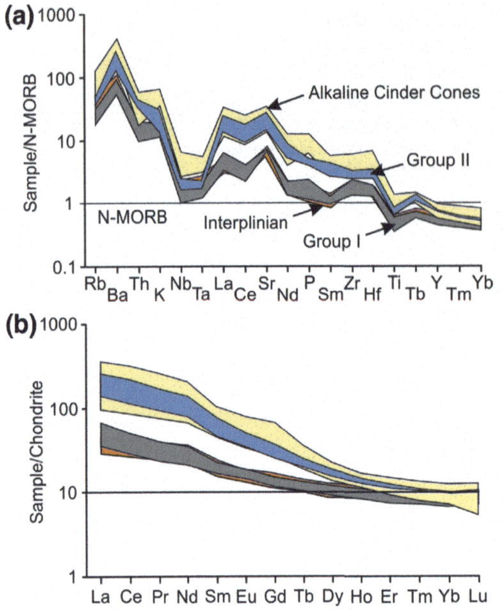

Fig. 9 Incompatible whole-rock trace element abundances normalized to N-MORB and REE normalized to Chondrite for the CVC eruption deposits. Normalizing values are from Sun and McDonough (1989) and Nakamura (1974), respectively. All the CVC eruption deposits, including the alkaline cinder cones, have subduction-related trace element abundance patterns characterized by depletions in the HFSE (Ta, Nb) relative to the LILE (Rb, Ba, K). The interplinian and Group I Plinian samples have overlapping compositions, while the Group II samples show stronger enrichments in all the incompatible trace elements, partially overlapping the alkaline cinder cone compositions

trace elements relative to the Group I units, and partially overlap the alkaline magma compositions (Fig. 9a).

Chondrite-normalized rare earth element (REE) abundance patterns in CVC magmas reveal that the light REE (La to Nd) are always enriched relative to the middle REE and heavy REE (Sm to Lu; Fig. 9b), reflecting a control of amphibole, clinopyroxene and garnet in the mantle source (Hanson 1980). Again, the abundances of the REE in the Group II eruption deposits overlap the abundances in the alkaline cinder cone magmas. These show very high light REE abundances (Fig. 9b), possibly reflecting lower degrees of partial melting (Hanson 1980).

4.3 Sr–Nd Isotopic Compositions

The radiogenic isotopic composition of a magma is inherited from its source during partial melting and remains unchanged as fractional crystallization progresses in the absence of contamination by an isotopically distinct component (Faure and Mensing 2005). However, isotopic variations will occur if, during ascent, the magma interacts with other batches of previously formed magma or crystalline mush, or assimilates wall-rocks or basement lithologies with isotopically distinct compositions (Hawkesworth and van Calsteren 1984; Faure and Mensing 2005). Accordingly, the $^{87}Sr/^{86}Sr$ and $^{143}Nd/^{144}Nd$ isotopic ratios have been widely used to fingerprint magma source(s) and identify open- and closed-system processes occurring in the magma storage region such as magma mixing, crustal assimilation and fractional crystallization (AFC; e.g. Kempton et al. 1991; Tatsumi et al. 1992; Straub et al. 2010; Verma and Luhr 2010; Schmidt and Grunder 2011).

Within Group I eruption deposits, $^{87}Sr/^{86}Sr$ and $^{143}Nd/^{144}Nd$ show a narrow range, varying from 0.70338 to 0.70371 and 0.51290 to 0.51295, respectively (Table 4; Fig. 10). The Group II units display slightly higher $^{87}Sr/^{86}Sr$ and lower $^{143}Nd/^{144}Nd$ than the majority of the Group I samples. Tephra units N and D display a narrow range in $^{87}Sr/^{86}Sr$ and $^{143}Nd/^{144}Nd$, varying from 0.70365 to 0.70372 and 0.51291 to 0.51294, respectively (Fig. 10). Unit F displays a much wider range in $^{87}Sr/^{86}Sr$ (0.70358–0.70408) and $^{143}Nd/^{144}Nd$ (0.51279–0.51293). However, there is a clear distinction within unit F samples; two of the samples have more radiogenic $^{87}Sr/^{86}Sr$ and less radiogenic $^{143}Nd/^{144}Nd$ corresponding with their lower SiO_2 content (Fig. 10a, b). The clear separation of these two samples from the rest of the Group II eruption deposits suggests either a separate source, or contamination from a magma or crystalline mush with high $^{87}Sr/^{86}Sr$ and low $^{143}Nd/^{144}Nd$ values. The relatively flat trend displayed by the Group I and the majority of the Group II samples suggest fractional crystallization from a homogenous melt.

Table 4 $^{87}Sr/^{86}Sr$ and $^{143}Nd/^{144}Nd$ isotope ratio compositions for the Group I and Group II eruption deposits

Sample	Unit	$^{87}Sr/^{86}Sr$	2σ error (×10^{-6})	$^{143}Nd/^{144}Nd$	2σ error (×10^{-6})
VF95-09X	Y	0.703627	±4	0.512940	±11
VF97-06D	Y	0.703633	±6	0.512912	±4
VF97-13B	Y	0.703623	±6	0.512916	±6
VF98-02W	W	0.703582	±4	0.512916	±6
VF95-06W	W	0.703574	±6	0.512941	±8
VF10-01U	U	0.703586	±4	0.512924	±6
VF10-02U	U	0.703593	±4	0.512932	±6
VF10-04U	U	0.703635	±5	0.512921	±7
VF10-07U	U	0.703594	±4	0.512934	±4
VF95-06P	S	0.703555	±4	0.512953	±3
VF95-09T	S	0.703592	±4	0.512932	±6
VF10-03Q	P	0.703577	±13	0.512948	±5
VF10-07M3	P	0.703604	±4	0.512927	±5
VF10-01P	P	0.703564	±4	0.512948	±7
VF10-03Pi	P	0.703557	±10	0.512902	±7
VF10-07D	**N**	**0.703723**	**±3**	**0.512908**	**±6**
VF95-09G	**N**	**0.703711**	**±4**	**0.512916**	**±4**
VF10-03N	**N**	**0.703710**	**±5**	**0.512922**	**±4**
VF95-09E	L	0.703597	±3	0.512909	±5
VF10-03M2	L	0.703604	±4	0.512932	±8
VF10-03L	J	0.703610	±4	0.512937	±6
VF00-06R	H	0.703384	±3	0.512952	±5
VF94-06E	H	0.703705	±5	0.512913	±5
VF97-01Di	H	0.703689	±5	0.512913	±5
VF95-06E	**F**	**0.703576**	**±4**	**0.512927**	**±10**
VF01-02Ps	**F**	**0.704084**	**±4**	**0.512791**	**±4**
VF01-05PA	**F**	**0.704044**	**±4**	**0.512787**	**±5**
VF10-04F	**F**	**0.703681**	**±4**	**0.512913**	**±5**
VF95-01B	**D**	**0.703645**	**±5**	**0.512936**	**±5**
VF01-02Ni	**D**	**0.703696**	**±4**	**0.512920**	**±10**
VF01-05N	**D**	**0.703695**	**±4**	**0.512908**	**±5**

Values are reported to 2σ error on the 6th decimal digit, and have been normalized to the NBS-987 and La Jolla standards for Sr and Nd, respectively

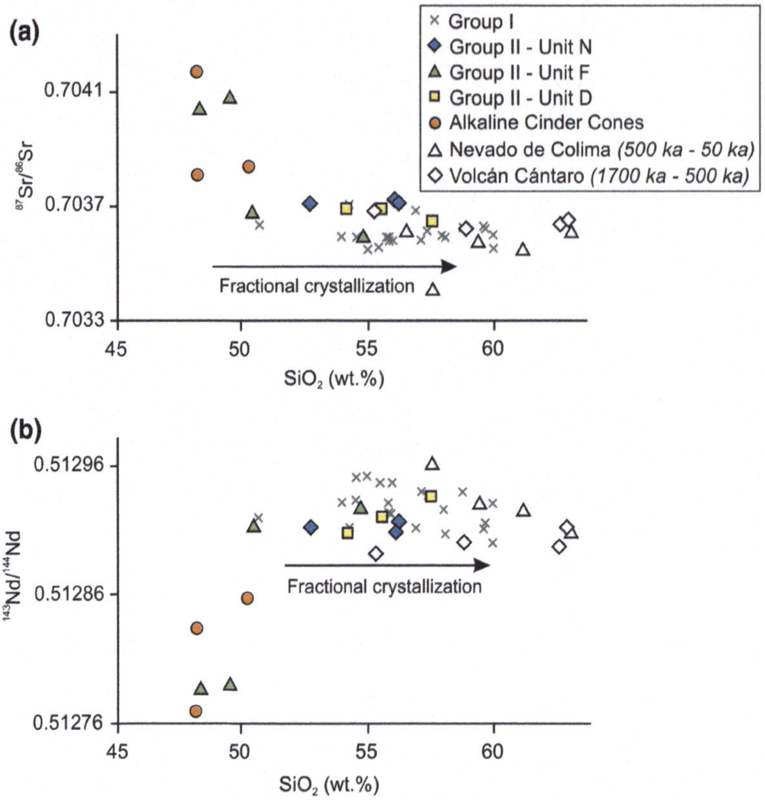

Fig. 10 $^{87}Sr/^{86}Sr$ and $^{143}Nd/^{144}Nd$ variation with SiO_2 for the CVC tephra deposits and the alkaline cinder cones. Data for the alkaline cinder cones are from Cai (2009). Isotopic data for the Group I and Group II samples were analyzed on Triton series TIMS instrument at Univ. Leeds as part of this study and analytical details can be found in Crummy (2013). Error bars are smaller than all the symbols used. Also shown are isotopic compositions for Nevado de Colima and Volcán Cántaro rocks after Verma and Luhr (2010). The Group I and the majority of the Group II eruption deposits show little variation in $^{87}Sr/^{86}Sr$ and $^{143}Nd/^{144}Nd$ with SiO_2 suggesting fractional crystallization from a relatively homogenous source. Two samples from unit F display isotopic compositions similar to those of the alkaline cinder cones, suggesting a common source, or mixing event

The cinder cone magmas have high $^{87}Sr/^{86}Sr$ (0.70381 to 0.70417) and low $^{143}Nd/^{144}Nd$ (0.51277 to 0.51287; Cai 2009) relative to the Group I eruption deposits and the majority of the Group II units (Fig. 10). Two samples from unit F of the Group II eruption deposits have similar isotopic compositions to two of the alkaline magmas consistent with mixing between isotopically distinct magmas.

The variation of $^{87}Sr/^{86}Sr$ versus $^{143}Nd/^{144}Nd$ for the calc-alkaline Group I and II eruption deposits, and the alkaline cinder cones reveals that all the CVC magmas fall within the mantle array (Fig. 11; Zindler and Hart 1986). Collectively, the CVC magmas appear to define a trend away from the estimated mantle wedge composition (Nimz et al. 1995; Gómez-Tuena et al. 2007) towards a more enriched component, with the alkaline cinder cone magmas having more enriched radiogenic isotopic signatures. Our findings, based on the new Sr and Nd isotope dataset, are in agreement with the Pb isotope ratios characteristic for CVC magmas reported by Verma and Luhr (2010).

5 Magmatic Evolution of the CVC

The geochemical and petrological variations shown by the CVC magmas reveal a heterogeneous source region in the mantle wedge under

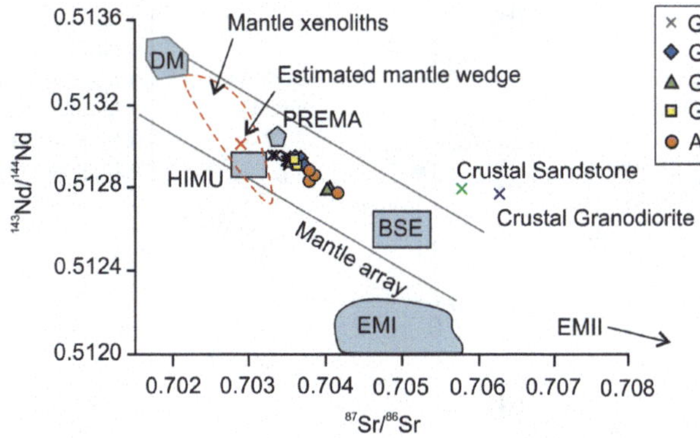

Fig. 11 $^{87}Sr/^{86}Sr$ versus $^{143}Nd/^{144}Nd$ diagram showing the relation of the CVC tephra deposits and the alkaline cinder cone magmas to the mantle reservoirs as defined by Zindler and Hart (1986). Data sources and analytical techniques and errors are the same as those on Fig. 10. Also shown are the mantle xenolith compositions from Northern Mexico (defined by dashed line; after Nimz et al. 1995), the estimated Mexican mantle wedge composition of Gómez-Tuena et al. (2007), and the compositions of granodiorite (Verma and Luhr 2010) and sandstone (Centeno-García et al. 1993) crustal rocks from the western Trans Mexican Volcanic Belt. The CVC magmas lie along the mantle array, trending towards a more enriched mantle source. The Group I and Group II eruption deposits appear to trend away from the mantle wedge composition towards the composition of the alkaline cinder cone samples

the CVC. The trace element abundance patterns displayed by CVC magmas are controlled by the influence of fluids from the subducting slab and the resulting partial melting of the overlying mantle. The fluid fluxing of the mantle wedge is responsible for fluid mobile element and LILE enrichments, while the relatively low HFSE abundances are controlled by residual mineral phases in the subducting slab (Tatsumi et al. 1992).

Incompatible trace element variation diagrams for the Group I and Group II calc-alkaline eruption deposits and the alkaline cinder cone magmas are shown in Fig. 12. Overall, the Group I eruption deposits form a cluster with no strong slab-derived fluid or sediment melt component (high Ba/Th, Ba/La, Ce/Pb and Th; Saunders et al. 1991; Schmidt and Poli 2004), and no strong slab-melt, or residual garnet signature (low La/Yb and Sr/Y coupled with Dy/Yb; Defant and Drummond 1990). The Group II eruption deposits display much stronger enrichments in Th suggesting a higher sediment component relative to the Group I samples (Fig. 12; Plank and Langmuir 1998). The opposite is displayed by the Ce/Pb ratio, with the Group II deposits containing higher concentrations of both Ce (70–130 ppm) and Pb (9–24 ppm) relative to the Group I samples (20–30 ppm Ce and 5–8 ppm Pb), indicating enrichment by both sediment melts and fluids. These deposits also show increasing Dy/Yb correlating with increasing La/Yb and Sr/Y reflecting contribution from a garnet-bearing source. The alkaline cinder cone magmas follow a similar trend to the Group II units, with a wide range of compositions reflecting fluid and sediment melt enrichment, and residual garnet in the source.

Several models have been proposed suggesting that the lamprophyric CVC magmas result from low degree melting of an enriched (veined) mantle source (Wallace and Carmichael 1989; Luhr 1997; Maria and Luhr 2008; Vigouroux et al. 2008). Such enriched metasomatic veins of phlogopite, clinopyroxene and apatite form as a result of the interaction of fluids from the subducting slab with depleted mantle wedge peridotites (Foley 1992). These veins have a lower melting point than the surrounding dry peridotite wall-rock, therefore low percentage melting is capable of producing the highly incompatible element-enrichments characteristic for the CVC lamprophyric melts (Foley 1992).

The elemental and isotopic data presented here supports models for alkaline mafic melt

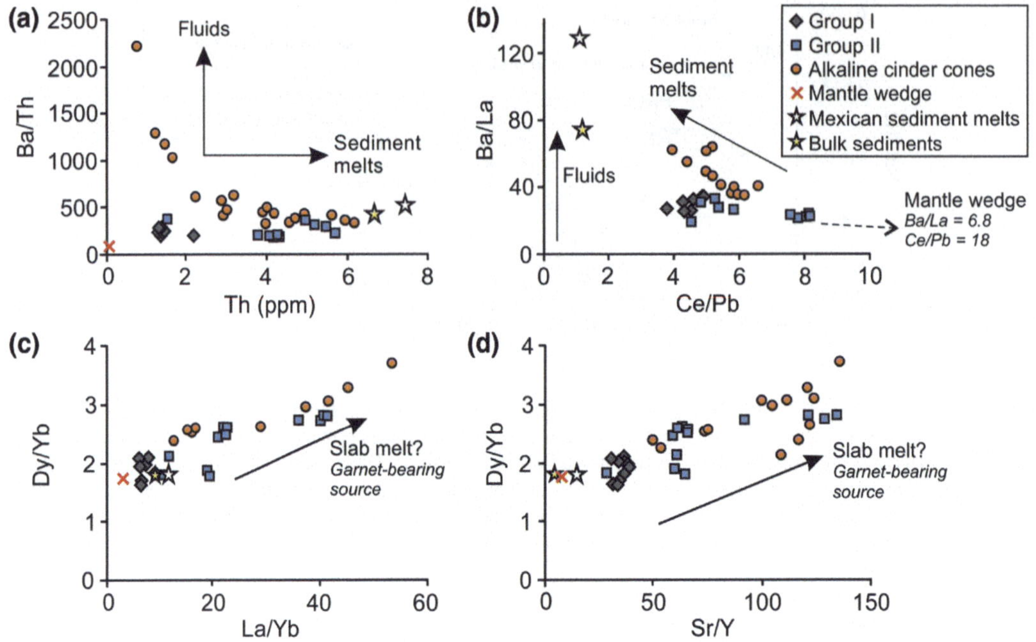

Fig. 12 Incompatible trace element ratio plots for the CVC magmas. The plots reveal the influence of subduction zone fluids, sediments and slab melts on the Group I and Group II calc-alkaline magmas and the alkaline cinder cone magmas (Luhr and Carmichael 1981; Carmichael et al. 2006; Maria and Luhr 2008; Vigouroux et al. 2008; Cai 2009). Only the mafic (<55 wt% SiO_2) samples have been plotted to remove the effects of AFC. The Mexican mantle wedge composition is from Gómez-Tuena et al. (2007). The bulk sediments are from DSDP Site 487 (Gómez-Tuena et al. 2003; LaGatta 2003), and the sediment melt composition are from Gómez-Tuena et al. (2007)

generation through the involvement of enriched (veined) mantle sources in the melting process: i.e. enrichments in the fluid mobile elements, LILE and light REE (Foley 1992; Maria and Luhr 2008; Vigouroux et al. 2008). The calc-alkaline Group I magmas do not show as strong subduction-related component as the alkaline cones; however, they are enriched relative to the estimated unmodified mantle wedge composition of Gómez Tuena et al. (2007). Cai (2009) estimated enrichments by ~5% sediment partial melts.

5.1 Calc-Alkaline—Alkaline Magma Mixing

The Group II eruption deposits are calc-alkaline basalts to basaltic-andesites distinct from the Group I calc-alkaline units, characterized by the presence of phlogopite, and high MgO, K_2O and P_2O_5 and strongly enriched LILE and LREE abundances relative to the Group I deposits (see Fig. 9). The alkaline cinder cones are composed of primitive high-MgO magmas with high TiO_2, FeOt, CaO, K_2O and P_2O_5 over a narrow range of SiO_2, forming a distinct group of CVC eruptive products.

Co-variations between K_2O and P_2O_5 with SiO_2 for the CVC erupted magmas show that the Group II eruption deposits lie compositionally between the calc-alkaline Group I and alkaline cinder cone magmas (Fig. 13). The calc-alkaline and alkaline magmas follow separate differentiation trends, and therefore cannot be linked through fractional crystallization from a common parental melt; however, they could represent a mixing relationship.

This is supported by the Sr and Nd isotopic data, which indicate that there is an involvement

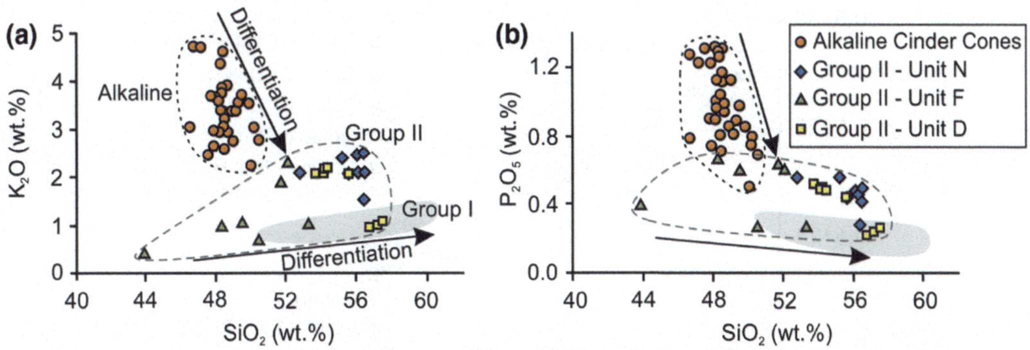

Fig. 13 Whole-rock major element K_2O (**a**) and P_2O_5 (**b**) versus SiO_2 plots showing the relationship between the Group I and Group II eruption deposits and the alkaline cinder cone samples. The Group II samples lie off the fractionation trend of the majority of the CVC calc-alkaline deposits (Group I), with higher K_2O and P_2O_5 concentrations trending towards the alkaline magmas, which may represent mixing between the two magma types

of a slab-derived metasomatic component in Group II magma sources which is absent or extremely dilute in the sources of Group I magmas (Fig. 11). Because of the temporal and spatial proximity of the calc-alkaline CVC central vents (Cántaro, Nevado, Paleofuego and Colima) and the alkaline cinder cones, the potential mixing relationship between these magmas as a possible mechanism to create the Group II eruption deposits should be considered. We have explored this hypothesis by using the two-component mixing model of Langmuir et al. (1978), between end-member samples on a $^{87}Sr/^{86}Sr$ versus $^{143}Nd/^{144}Nd$ variation diagram (Fig. 14). The majority of the Group II samples contain up to 5% alkaline magma component; however, two of the samples from unit F comprise very large (~50%) alkaline component (Fig. 14).

The presence of an alkaline component in the Group II magmas is supported by the whole-rock trace element data. The results of mixing calculations involving the trace elements are shown on Fig. 15. The trace element data reveal that ~20% alkaline component is required to reproduce the measured Group II abundances. However, the trace element abundance patterns for the mixed magmas do not agree perfectly with the Group II eruption deposits. For example, the abundances representing a 50% calc-alkaline—alkaline mixture agrees well with the majority of the elements with the exception of Rb, Ba, Nb, P, Zr and Hf which are all more enriched than in the Group II samples. This may reflect the presence of small amounts of accessory minerals in the alkaline melts, such as sphene, rutile, zircon or apatite which would result in relatively high abundances of Nb, P, Zr and Hf (Tatsumi and Kogiso 1997). Rb and Ba are highly mobile in fluids, therefore intense metasomatism causing enrichments in the alkaline parental melt could account for the elevated Rb and Ba abundances (i.e. Marschall and Schumacher 2012).

The mixing relationship between the two magma types is also supported by the mineralogy of the different eruption deposits. The typical mineral assemblage of the Group I tephra is dominated by plagioclase, orthopyroxene, clinopyroxene and hornblende with olivine only present in the most mafic deposits, while the alkaline (minette) magmas predominantly comprise olivine and clinopyroxene with phlogopite, apatite and sanidine (Table 5).

The Group II deposits are calc-alkaline basalts and basaltic-andesites in composition, yet comprise mineral assemblages dominated by clinopyroxene and olivine with phlogopite, that are more akin to the basanites and minettes of the alkaline cinder cones (Luhr and Carmichael 1981). Plagioclase is absent, or present only in

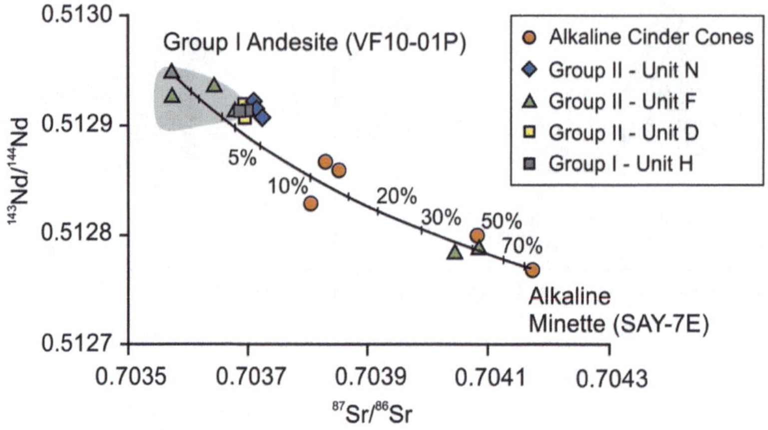

Fig. 14 A simple mixing between the end member magmas on a $^{87}Sr/^{86}Sr$ versus $^{143}Nd/^{144}Nd$ plot following the model of Langmuir et al. (1978). The end-member compositions used were a Group I calc-alkaline andesite (sample VF10-01P) and an alkaline minette (sample SAY-7E from El Carpintero Norte; Cai, 2009). The solid line shows the percentage of alkaline component. The Group II magmas appear to incorporate between 2 and 50% alkaline magma component, which is significant, considering the rarity of CVC-like alkaline magmas (basanites to minettes) in island and continental arc environments

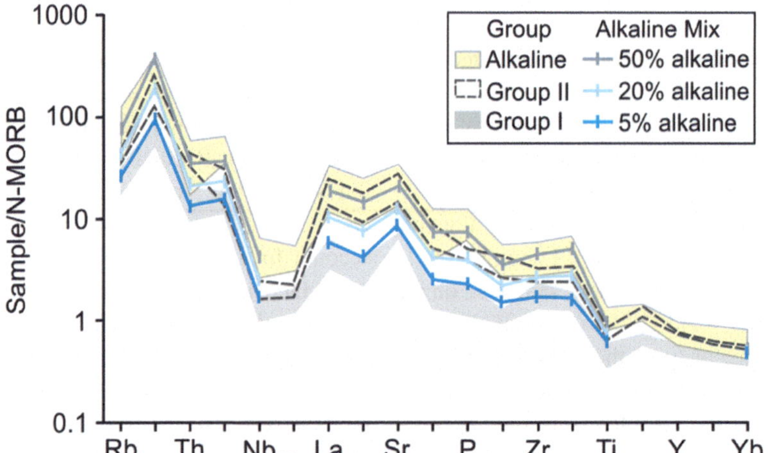

Fig. 15 An N-MORB normalized trace element abundance diagram showing trace element patterns produced by mixing between calc-alkaline (Group I; grey field) and alkaline magmas (pale yellow field). N-MORB normalizing values are from Sun and McDonough (1989). The mixed compositions are calculated using the end-member calc-alkaline andesite (VF10-01P) and alkaline minette (sample SAY-7E from El Carpintero Norte; Cai 2009) following the two-component mixing model of Langmuir et al. (1978). The bright blue line represents compositions consisting of a mixture 95% calc-alkaline and 5% alkaline; the pale blue, 20% alkaline and the grey-blue, 50% alkaline. The Group II magmas (dashed outline) have between 20 and 50% alkaline component

the groundmass, in the alkaline cinder cone magmas (Luhr and Carmichael 1981). Unit F of Group II shows the strongest alkaline geochemical signature, which is consistent with its mineralogy, comprising a mineral assemblage dominated by clinopyroxene, olivine and phlogopite with minor plagioclase and trace hornblende. Units N and D show a weaker alkaline

Table 5 Comparison of the mineral assemblages of Group I, Group II and alkaline cinder cone pumice and scoria samples

Deposit	Mineral assemblage
Group I	plag + opx + cpx + hbd + Fe–Ti oxides + ol
Group II	cpx + ol + hbd + plag + phlog + Fe–Ti oxides + opx
Alkaline (minette)	ol + cpx + phlog + apatite + sanidine + Fe–Ti oxides

A minette is used to represent the alkaline magma
The mineral abbreviations are: *plag* plagioclase; *opx* orthopyroxene; *cpx* clinopyroxene; *hbd* hornblende; *phlog* phlogopite

geochemical signature which, again, is reflected in their mineralogy with lesser amounts of phlogopite and higher plagioclase content.

The geochemical mixing models and mineralogical data reveal there is a petrogenetic relationship between the calc-alkaline and alkaline erupted deposits of the CVC involving physical mixing between the two types of magmas. Evidence from the mineral compositions and zoning patterns of phenocrysts within the Group I and Group II eruption deposits reveal multiple magma mixing events in the magmatic plumbing system of the CVC (Crummy 2013; Crummy et al. 2014). The appearance of the distinct Group II eruption deposits within the calc-alkaline stratigraphy suggests pulses of alkaline melts intercepted the storage region of the volumetrically dominant calc-alkaline magmas (Fig. 16).

The presence of high-K, alkaline magmas (basanites to minettes) at the CVC is unusual in arc settings, although not unique; alkaline magmas have been documented in the Izu-Bonin-Mariana arc (shoshonite volcanic province: see Stern et al. 2003 and references therein), the Sunda-Banda Arc, Indonesia (Wheller et al. 1987) and central Kamchatka (Koloskov et al. 1999), among others. However, the relationship between calc-alkaline and alkaline volcanism in these settings remains poorly understood. The CVC sits in the fore-arc position within the Colima Rift Zone. The extensional setting in the western Trans Mexican Volcanic Belt, which formed the Colima Rift Zone, may have provided pathways for the exotic, low-volume alkaline melts to reach the surface and form monogenetic cinder cones surrounding the CVC (Luhr 1997; Maria and Luhr 2008; Vigouroux et al. 2008).

The alkaline CVC rocks are primitive magmas thought to represent direct melts of veined mantle with little differentiation during ascent (Luhr 1997). The presence of alkaline cinder cones, and the recognition of an alkaline component in magmas erupted from the calc-alkaline CVC stratovolcanoes, allows us to investigate the details of the entire diversity and complexity of the mantle wedge feeding arc volcanism.

6 Concluding Remarks

The CVC eruption deposits are typical calc-alkaline basalts to andesites (Gill 1981) with characteristic subduction-related trace element abundances. During the Holocene, volcanic activity is expressed as both highly explosive (Plinian) and effusive (interplinian) eruptions. Radiocarbon dating of charcoal fragments found within ash horizons separating tephra fallout deposits reveals an explosive history extending as far back as 30,000 yrs B.P. Interplinian activity is predominantly effusive with phases of lava dome growth and collapse leading to lava flows and intermittent small (<12 km a.s.l. plume height; Savov et al. 2008) Vulcanian explosions.

Explosive Plinian eruption deposits can be divided into two groups based on mineralogical and geochemical distinctions. Group I comprise the bulk of eruption deposits and represent magma fractionation from a primitive mantle melt, while Group II units comprise three tephra fallout horizons which are characterized by the presence of phlogopite, and contain high K_2O and P_2O_5 contents, and strongly enriched incompatible trace element abundances.

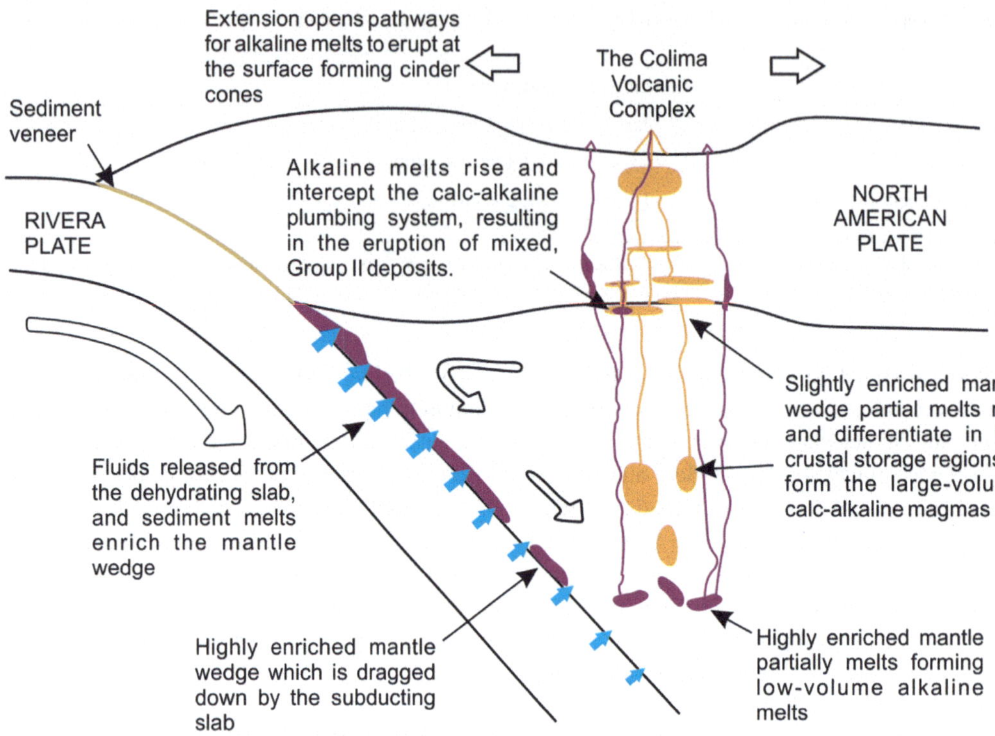

Fig. 16 Cartoon of the petrogenesis of the CVC calc-alkaline and alkaline magmas. Calc-alkaline magmas form from partial melts of the slightly enriched mantle wedge, which evolve through a complex crustal storage system. The alkaline magmas form from highly enriched mantle and rise through the crust forming cinder cones on the Colima Rift floor. The Group II mixed magmas result from the interception of alkaline magmas with the crustal magma chambers, or sills, through which the calc-alkaline magmas differentiate

Whole-rock major and trace element geochemistry, and radiogenic Sr and Nd isotopic compositions reveal an alkaline lamprophyric component in the Group II tephra. These magmas are interpreted to represent mixing between calc-alkaline Group I magmas, and alkaline magmas which formed monogenetic cinder cones on the rift floor to the east and west of the northern CVC. The Group II magmas contain between 5 and 50% alkaline magma component.

The appearance of these mixed units in the tephra record at c.7000, c.12,000 and c.13,000 yrs B.P., interbedded within the Group I calc-alkaline stratigraphy, indicate that pulses of alkaline melts rise and intercept the magmatic plumbing system of the CVC on timescales of a few thousand years.

Acknowledgements We would like to dedicate our work to the living memory of the pioneers of the studies of petrology and geochemistry of the Colima Volcanic Complex—Jim Luhr and Ian Carmichael. This work benefited greatly from the support of staff from the Centro Universitario de Estudios e Investigación en Vulcanología (CUEIV), in particular logistical field support provided by the director, Gabriel Reyes. We thank the curators of the Mineralogy and Petrology Collections of the Department of Mineral Sciences at the National Museum of Natural History (Smithsonian Institution) in Washington DC, USA, for lending us samples and thin sections from Jim Luhr's Colima collection. This work is published with permission of the Executive Director of the British Geological Survey (Natural Environment Research Council).

References

Allan, J.F., Carmichael, I.S.E.: Lamprophyric lavas in the Colima graben, SW Mexico. Contrib. Mineral. Petrol. **88**, 203–216 (1984)

Atlas, Z.D., Dixon, J.E., Sen, G., Finny, M., Lillian, A., Pozzo, M.D.: Melt inclusions from Volcan Popocatepetl and Volcán de Colima, Mexico: melt evolution due to vapor-saturated crystallization during. J. Volcanol. Geoth. Res. **153**, 221–240 (2006)

Bretón González, M., Ramírez, J.J., Navarro-Ochoa, C.: Summary of the historical eruptive activity of Volcán de Colima, México, 1519–2000. J. Volcanol. Geoth. Res. **117**, 21–46 (2002)

Cai, Y.: Tracing upper mantle heterogeneities with radiogenic isotopes at the Mexican Volcanic Belt and the Arctic Gakkel Ridge. Ph.D. thesis, Columbia University (2009)

Carmichael, I., Frey, H., Lange, R., Hall, C.: The Pleistocene cinder cones surrounding Volcán Colima, Mexico re-visited: eruption ages and volumes, oxidation states, and sulfur content. Bull. Volc. **68**, 407–419 (2006)

Centeno-García, E., Ruiz, J., Coney, P.J., Patchett, P.J., Ortegagutierrez, F.: Guerrero terrane of Mexico—its role in the Southern Cordillera from new geochemical data. Geology **21**, 419–422 (1993)

Cortés, A., Garduño, V.H., Macías, J.L., Navarro-Ochoa, C., Komorowski, J.C., Saucedo, R., Gavilanes, J.C., Geologic mapping of the Colima volcanic complex (Mexico) and implications for hazard assessment. In: Groppelli, G., Viereck-Goette, L. (eds.) Stratigraphy and Geology of Volcanic Areas: Geological Society of America Special Paper 464, pp. 249–264 (2010)

Cortés, A., Garduño, V.H., Navarro-Ochoa, C., Komorowski, J.C., Saucedo, R., Macías, J.L., Gavilanes, J. C.: Carta Geológica del Complejo Volcánico de Colima, con Geología del Complejo Volcánico de Colima, vol. 10, Universidad Nacional Autónoma de México, Instituto de Geología, Cartas Geológicas y Mineras (2005)

Crummy, J.M.: Holocene evolution of the Colima Volcanic Complex, Mexico. Ph.D. thesis, University of Leeds, Leeds (2013)

Crummy, J.M., Savov, I.P., Navarro-Ochoa, C., Morgan, D.J., Wilson, M.: High-K Mafic Plinian Eruptions of Volcán de Colima, Mexico. J. Petrol. **55**(11), 2155–2192 (2014)

Defant, M.J., Drummond, M.S.: Derivation of some modern arc magmas by melting of young subducted lithosphere. Nature **347**, 662–665 (1990)

Faure, G., Mensing, T.M.: Isotopes: Principles and Applications. Wiley, Hoboken, NJ (2005)

Foley, S.: Vein-plus-wall-rock melting mechanisms in the lithosphere and the origin of potassic alkaline magmas. Lithos **28**, 435–453 (1992)

Gill, J.B.: Orogenic Andesites and Plate Tectonics: Berlin. Springer, Berlin, Heidelberg (1981)

Global Volcanism Program: Report on Colima (Mexico). In: Venzke, E. (ed.) Bulletin of the Global Volcanism Network, 41:1. Smithsonian Institution (2016)

Gómez-Tuena, A., LaGatta, A., Langmuir, C.H., Goldstein, S.L., Ortega-Gutiérrez, F., Carrasco-Nuñez, G.: Temporal control of subduction magmatism in the Eastern Trans-Mexican volcanic belt: mantle sources, slab contributions and crustal contamination. Geochem. Geophys. Geosyst. **4**, 8912 (2003)

Gómez-Tuena, A., Langmuir, C.H., Goldstein, S.L., Straub, S.M., Ortega-Gutiérrez, F.: Geochemical evidence for slab melting in the trans-Mexican volcanic belt. J. Petrol. **48**, 537–562 (2007)

Hanson, G.N.: Rare earth elements in petrogenetic studies of igneous systems. Ann. Rev. Earth Planet. Sci. **8**, 371–406 (1980)

Hawkesworth, C.J., van Calsteren, P.W.C.: Radiogenic isotopes - some geological applications. In: Henderson, P. (ed.) Rare Earth Element Geochemistry, pp. 375–421. Elsevier, Amsterdam (1984)

Irvine, T.N., Baragar, W.R.A.: A guide to the chemical classification of the common volcanic rocks. Can. J. Earth Sci. **8**, 523–548 (1971)

Kempton, P.D., Fitton, J.G., Hawkesworth, C.J., Ormerod, D.S.: Isotopic and trace-element constraints on the composition and evolution of the lithosphere beneath the southwestern United States. J. Geophy. Res. Solid Earth Planet. **96**, 13713–13735 (1991)

Koloskov, A., Flerov, G., Seliverstov, V., Dorendorf, F., Churikova, T.: Potassic volcanics of Central Kamchatka and the Late Cretaceous-Paleogene Kuril-Kamchatka Alkaline province. Petrologiya **7**, 527–543 (1999)

Komorowski, J.C., Navarro-Ochoa, C., Cortes, A., Saucedo, R., Gavilanes, J.C., Siebe, C., Espindola, J.-M., Rodriguez, S.: The Colima Volcanic Complex, Part I: Quaternary Multiple Debris-Avalanche Deposits. Puerto Vallarta, Mexico (1997)

LaGatta, A.: Arc magma genesis in the Eastern Mexican volcanic belt. Ph.D. thesis, Columbia University, New York (2003)

Lange, R.A., Carmichael, I.S.E.: Hydrous basaltic andesites associated with minette and related lavas in Western Mexico. J. Petrol. **31**, 1225–1259 (1990)

Langmuir, C.H., Vocke Jr., R.D., Hanson, G.N., Hart, S. R.: A general mixing equation with applications to icelandic basalts. Earth Planet. Sci. Lett. **37**, 380–392 (1978)

Le Maitre, R.W., Streckeisen, A., Zanettin, B., Le Bas, M. J., Bonin, B., Bateman, P., Bellieni, G., Dudek, A., Efremova, S., Keller, J., Lameyre, J., Sabine, P.A., Schmid, R., Sorenson, H., Wooley, A.R.: Igneous Rocks A Classification and Glossary of Terms. Cambridge University Press, Cambridge (2002)

Luhr, J.F.: Petrology and geochemistry of stage-I andesites and dacites from the caldera wall of Volcán Colima, Mexico. Geofisica Internacional **32**, 591–603 (1993)

Luhr, J.F.: Extensional tectonics and the diverse primitive volcanic rocks in the Western Mexican Volcanic Belt. Can. Mineral. **35**, 473–500 (1997)

Luhr, J.F.: Petrology and geochemistry of the 1991 and 1998–1999 lava flows from Volcán de Colima, Mexico: implications for the end of the current eruptive cycle. J. Volcanol. Geoth. Res. **117**, 169–194 (2002)

Luhr, J.F., Carmichael, I.S.E.: The Colima volcanic complex, Mexico: part I. Post-caldera andesites from Volcán Colima. Contrib. Mineral. Petrol. **71**, 343–372 (1980)

Luhr, J.F., Carmichael, I.S.E.: The Colima volcanic complex, Mexico: part II. Late-quaternary cinder cones. Contrib. Mineral. Petrol. **76**, 127–147 (1981)

Luhr, J.F., Carmichael, I.S.E.: The Colima volcanic complex, Mexico: part III. Ash and scoria fall deposits from upper slopes of Volcán Colima. Contrib. Mineral. Petrol. **42**, 235–260 (1982)

Luhr, J.F., Carmichael, I.S.E.: Geology of Volcán de Colima. Universidad Nacional Autónoma de México, Instituto de Geología (1990a)

Luhr, J.F., Carmichael, I.S.E.: Petrological monitoring of cyclical eruptive activity at Volcán de Colima, Mexico. J. Volcanol. Geoth. Res. **42**, 235–260 (1990b)

Luhr, J.F., Navarro-Ochoa, C., Connor, C.B., Connor, L.: The 1913 VEI-4 Plinian eruption of Volcan de Colima (Mexico): tephrochronology, petrology and plume modelling. Eos Trans. AGU 87, 52, 1786 (2006)

Luhr, J.F., Navarro-Ochoa, C., Savov, I.P.: Tephrochronology, petrology and geochemistry of late-holocene pyroclastic deposits from Volcán de Colima, Mexico. J. Volcanol. Geoth. Res. **197**, 1–32 (2010)

Maria, A.H., Luhr, J.F.: Lamprophyres, basanites, and basalts of the western Mexican volcanic belt: volatile contents and a vein-wallrock melting relationship. J. Petrol. **49**, 2123–2156 (2008)

Marschall, H.R., Schumacher, J.C.: Arc magmas sourced from Mèlange Diapirs in subduction zones. Nat. Geosci. **5**, 862–867 (2012)

Medina-Martínez, F.: Analysis of the eruptive history of the Volcán de Colima, México (1560–1980). Geofísica Internacional **22**, 157–178 (1983)

Nakamura, N.: Determination of REE, Ba, Fe, Mg, Na and K in carbonaceous and ordinary chondrites. Geochimica Et Cosmochimica Acta **38**, 757–775 (1974)

Navarro-Ochoa, C., Gavilanes, J.C., Cortés, A.: Movement and emplacement of lava flows at Volcán de Colima, México: November 1998–February 1999. J. Volcanol. Geoth. Res. **117**, 155–167 (2002)

Newhall, C.G., Self, S.: The volcanic explosivity index (VEI): an estimate of explosive magnitude for historical volcanism. J. Geophys. Res. **87**, 1231–1238 (1982)

Nimz, G.J., Cameron, K.L., Niemeyer, S.: Formation of mantle lithosphere beneath Northern Mexico: chemical and Sr–Nd–Pb isotopic systematics of peridotite xenoliths from La Olivina. J. Geophys. Res. **100**, 4181–4196 (1995)

Pearce, J.A.: Trace element characteristics of lavas from destructive plate boundaries. In: Thorpe, R.S. (ed.) Andesites: Orogenic Andesites and related Rocks, pp. 525–548. Wiley, Chichester (1982)

Plank, T., Langmuir, C.H.: The chemical composition of subducting sediment and its consequences for the crust and mantle. Chem. Geol. **145**, 325–394 (1998)

Reubi, O., Blundy, J.: Assimilation of plutonic roots, formation of high-K exotic melt inclusions and genesis of andesitic magmas at Volcán De Colima, Mexico. J. Petrol. **49**, 2221–2243 (2008)

Righter, K.: A comparison of basaltic volcanism in the cascades and Western Mexico: compositional diversity in continental arcs. Tectonophysics **318**, 99–117 (2000)

Robin, C., Potrel, A.: Multi-stage magma mixing in the pre-caldera series of Fuego de Colima volcano. Geofisica Internacional **32**, 605–615 (1993)

Robin, C., Mossand, P., Camus, G., Cantagrel, J.M., Gourgaud, A., Vincent, P.M.: Eruptive history of the Colima volcanic complex (Mexico). J. Volcanol. Geoth. Res. **31**, 99–113 (1987)

Robin, C., Camus, G., Gourgaud, A.: Eruptive and magmatic cycles at Fuego de Colima volcano (Mexico). J. Volcanol. Geoth. Res. **45**, 209–225 (1991)

Saucedo, R., Macias, J.L., Gavilanes, J.C., Arce, J.L., Komorowski, J.C., Gardner, J.E., Valdez-Moreno, G.: Eyewitness, stratigraphy, chemistry, and eruptive dynamics of the 1913 Plinian eruption of Volcan de Colima, Mexico. J. Volcanol. Geoth. Res. **191**, 149–166 (2010)

Saunders, A.D., Norry, M.J., Tarney, J.: Fluid influence on the trace element compositions of subduction zone magmas. Philos. Trans. Royal Soc. London. Ser. A Phys. Eng. Sci. **335**, 377–392 (1991)

Savov, I.P., Luhr, J.F., Navarro-Ochoa, C.: Petrology and geochemistry of lava and ash erupted from Volcán Colima, Mexico, during 1998–2005. J. Volcanol. Geoth. Res. **174**, 241–256 (2008)

Schmidt, M.E., Grunder, A.L.: Deep mafic roots to arc volcanoes: mafic recharge and differentiation of basaltic andesite at North Sister Volcano, Oregon Cascades. J. Petrol. **52**, 306–341 (2011)

Schmidt, M.W., Poli, S.: Generation of mobile components during subduction of oceanic crust. In: Rudnick, R. (ed.) The Crust, Volume 3: Treaties on Geochemistry, pp. 567–592. Elsevier-Pergamon, Oxford (2004)

Stern, R.J., Fouch, M.J., Klemperer, S.: An overview of the Izu-Bonin-Mariana subduction factory. In: Eiler, J., Hirschmann, M. (eds.) Inside the Subduction Factory, vol. 138, Geophysical Monograph, American Geohpysical Union (2003)

Straub, S.M., Goldstein, S.L., Class, C., Schmidt, A., Gomez-Tuenca, A.: Slab and mantle controls on the Sr–Nd–Pb–Hf isotope evolution of the post 42 Ma Izu-Bonin volcanic arc. J. Petrol. **51**, 993–1026 (2010)

Sun, S.S., McDonough, W.F.: Chemical and isotopic systematics of ocean basalts: implications for mantle composition and processes. In: Saunders, A.D., Norry, M.H. (eds.) Magmatism in the Ocean Basins, vol. 42, pp. 313–345. Geological Society Special Publication (1989)

Tatsumi, Y., Kogiso, T.: Trace element transport during dehydration processes in the subducted oceanic crust: 2. Origin of chemical and physical characteristics in arc magmatism. Earth Planet. Sci. Lett. **148**, 207–221 (1997)

Tatsumi, Y., Murasaki, M., Nohda, S.: Across-arc variation of lava chemistry in the Izu-Bonin Arc - identification of subduction components. J. Volcanol. Geoth. Res. **49**, 179–190 (1992)

Verma, S.P., Luhr, J.F.: Sr, Nd, and Pb isotopic evidence for the origin and evolution of the Cántaro-Colima volcanic chain, Western Mexican volcanic belt. J. Volcanol. Geoth. Res. **197**, 33–51 (2010)

Vigouroux, N., Wallace, P.J., Kent, A.J.R.: Volatiles in high-K magmas from the Western Trans-Mexican volcanic belt: evidence for fluid fluxing and extreme enrichment of the mantle wedge by subduction processes. J. Petrol. **49**, 1589–1618 (2008)

Wallace, P., Carmichael, I.S.E.: Minette lavas and associated leucitites from the western front of the Mexican volcanic belt: petrology, chemistry, and origin. Contrib. Mineral. Petrol. **103**, 470–492 (1989)

Wheller, G.E., Varne, R., Foden, J.D., Abbott, M.J.: Geochemistry of quaternary volcanism in the Sunda-Banda arc, Indonesia, and three-component genesis of island-arc basaltic magmas. J. Volcanol. Geoth. Res. **32**, 137–160 (1987)

Zindler, A., Hart, S.R.: Chemical geodynamics. Ann. Rev. Earth Planet. Sci. **14**, 493–571 (1986)

Structure of the Colima Volcanic Complex: Origin and Behaviour of Active Fault Systems in the Edifice

Gianluca Norini, Federico Agliardi, Giovanni Crosta, Gianluca Groppelli and Maria Clara Zuluaga

Abstract

The Colima Volcanic Complex (CVC) is one of the most prominent volcanic edifices within the Tran-Mexican Volcanic Belt (TMVB). Its evolution has been characterized by complex interactions among regional tectonics, basement geometry and rheology, and successive volcanic edifices formed in several stages. In the first part of this review chapter, the state-of-the-art in the CVC structural and geological knowledge is summarized and discussed. Such knowledge is based on published structural, geological, geophysical, geodetic and petrological data and models, which allow identification of three main fault systems and definition of the general geometrical properties of the volcano plumbing system. The significance and supporting evidence for the Colima Rift faults, Tamazula Fault and volcano spreading structures, as well as their interactions with the CVC, are presented. In the second part of the paper, numerical models are performed to integrate previous knowledge and test the effects of different structural scenarios and different mechanical assumptions on the onset of gravitational spreading of the volcanic complex. Also, the state of activity of the three fault systems and their control on the magmatism and flank instability are discussed. Finally, the main findings on the CVC structural arrangement are summarized, and the most important requirements for improvement of our understanding of the interaction between regional tectonics, volcanic edifices and basement rheology are suggested. Future investigations are critical, as eruptions and sector failures represent potential high risks for more than 500,000 people.

Keywords

Volcano-basement-tectonics interplay
Gravitational spreading · Active faults
Numerical modelling

G. Norini (✉) · G. Groppelli
Istituto per la Dinamica dei Processi Ambientali, Consiglio Nazionale delle Ricerche, Rome, Italy
e-mail: gianluca.norini@cnr.it

F. Agliardi · G. Crosta
Department of Earth and Environmental Sciences, University of Milano-Bicocca, Milan, Italy

M. C. Zuluaga
Dirección de Recursos Minerales, Servicio Geológico Colombiano, Bogotá, Colombia

1 Introduction

Volcanoes are highly dynamic systems, whose evolution is influenced by many factors, among them the geology of the basement and the

N. Varley et al. (eds.), *Volcán de Colima*, Active Volcanoes of the World,
https://doi.org/10.1007/978-3-642-25911-1_8

regional tectonics. In active volcanic regions, the understanding of the relationship between volcanic edifices and the geological-structural setting of the basement is crucial to project future edifice behaviour. For large volcanic edifices, the regional tectonics, the rheology of the basement, the load of the volcanic pile and the geometry of the plumbing (or magmatic system) system have been recognized as the main factors conditioning their structural architecture (e.g. Voight and Elsworth 1997; Borgia and van Wyk de Vries 2003; Wooller et al. 2004; Norini and Lagmay 2005; Walter et al. 2005; Norini et al. 2008; Norini and Acocella 2011). The interaction among these factors plays a major role in the migration of the magma toward the surface and in the flank instability of large volcanoes or volcanic complexes, with important implications for the potential occurrence of hazardous eruptions and lateral collapses. Thus, the study of a major active volcano is not possible without an adequate knowledge of the structure and the state of stress of the volcanic edifice and its basement. The interpretations of data provided by geomorphological, geological, geophysical, geodetic and geochemical studies and by continuous monitoring of active volcanoes should be integrated to contribute to accurate volcano-tectonic models. Despite this need, a reliable structural model is lacking or incomplete for many of the major active volcanoes worldwide. Such a structural model should be based on the knowledge of the geometry, kinematics and dynamics of the faults and of the fault systems in the volcanic edifice, and of their interactions with the magmatic system of the active volcano. These data, often supported by analogue and/or numerical modelling (e.g. Borgia 1994; van Wyk de Vries and Matela 1998; Apuani et al. 2005; Norini et al. 2010; Norini and Acocella 2011), can contribute to the interpretation of the structural control on magmatic intrusions and volcanic activity, and of gradual and catastrophic edifice deformation.

The Colima Volcanic Complex (CVC) is one of the most active volcanoes in the world (e.g. Macías et al. 2006; Zobin et al. 2015; Capra et al. 2016). The CVC is a prominent volcanic system within the Trans-Mexican Volcanic Belt (TMVB) (Fig. 1). Although the CVC is regarded as a severe hazard for the surrounding populated area, very few works have focused on its structure and on the volcano-basement interplay (e.g. Garduño-Monroy et al. 1998; López-Loera et al. 2010, 2011; Norini et al. 2010). The CVC is a large Pleistocene-Holocene calc-alkaline andesitic volcanic system, whose eruptions and sector

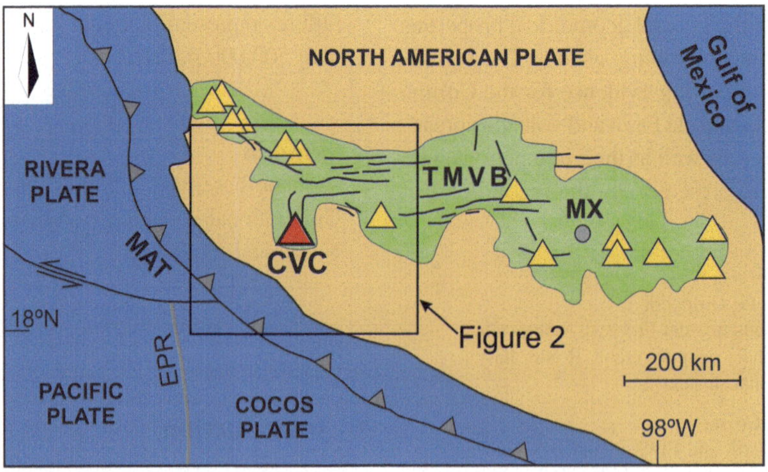

Fig. 1 Location of the Trans-Mexican Volcanic Belt (TMVB, in green) in the geodynamic framework of North and Central America. Triangles in orange show major volcanoes. Solid black lines are major faults. MX: Mexico City. MAT: Middle American trench; CVC: Colima Volcanic Complex; EPR: East Pacific Rise

collapses pose high risk for more than 500,000 people (e.g. Capra et al. 2014). The volcanic complex has been emplaced in the active N-S-trending Colima Rift (e.g. Allan 1986) (Fig. 2). This rift is filled by a sequence of Plio-Quaternary alluvial, colluvial and lacustrine deposits underlying the volcanic complex (Allan 1985, 1986; Allan et al. 1991; Serpa et al. 1992). The understanding of the structural control on the CVC evolution and its potentially hazardous volcanic activity requires an accurate analysis of the interaction among the regional tectonics, the magmatic system, a weak basement and the load of the massive volcanic edifice.

In this review chapter, we present and discuss the previous works on the geodynamic, structural and geological setting of the volcanic complex. New numerical models of volcano deformation under gravitational load are also presented. Finally, we discuss the influence of the structural architecture on the volcano evolution, and the main future needs to improve the knowledge of the volcano-basement-tectonic interplay in the CVC.

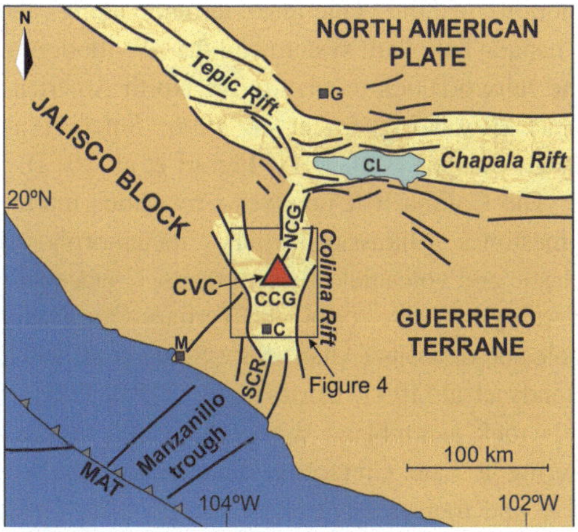

Fig. 2 Sketch map with the location of the Colima-Tepic-Chapala triple junction and the Jalisco Block and the Guerrero Terrane. Areas in light yellow are main tectonic depressions filled by alluvium, colluvium and lacustrine deposits (from Allan et al. 1991; Ferrari et al. 2012). CVC: Colima Volcanic Complex; NCG: Northern Colima Graben; CCG: Central Colima Graben; SCR: Southern Colima Graben; CL: Chapala Lake; G: Guadalajara City; C: Colima City; M: Manzanillo City

2 The Western Trans-Mexican Volcanic Belt and the Colima-Tepic-Chapala Triple Junction

The TMVB is an active continental volcanic arc resulting from the subduction of the Cocos and Rivera oceanic plates beneath the North American continental plate (Fig. 1). A detailed review of the complex evolution and structure of the TMVB has been published by Ferrari et al. (2012). These authors show a subdivision of the volcanic arc in four distinct sectors, with the CVC located in the western sector (e.g. Ferrari et al. 2012). The western sector of the TMVB is separated from the central sector by a change in the subducting plate, from the Rivera plate in the northwest to the Cocos plate in the southeast. These two oceanic plates are separated by the trench-orthogonal Rivera-Cocos Plate Boundary, a tear fault zone subducted under the North America continental crust. Tomographic data show that the Rivera slab dips at 40° in the forearc region and 70° beneath the TMVB, whereas the westernmost Cocos slab dips less, beneath both the forearc and the TMVB (Pardo and Suárez 1993; Yang et al. 2009; Ferrari et al. 2012).

The western TMVB basement comprises the Guerrero Terrane to the east and the continental Jalisco Block to the west (Fig. 2). The Guerrero Terrane (also known as Michoacán block, Rosas-Elguera et al. 1996) is mainly composed of Triassic–Jurassic marine sedimentary sequences and Jurassic-Cretaceous volcanic arcs (e.g. Centeno-García et al. 2011). The Jalisco block is mostly made of Jurassic schists, a Late Cretaceous-Palaeocene batholith, Cretaceous marine sedimentary sequences and Cretaceous-Eocene effusive and explosive volcanic units (Schaaf et al. 1995; Ferrari et al. 2000, 2005, 2012; Frey et al. 2007; Valencia et al. 2009).

The western part of the TMVB is characterized by three main tectonic depressions forming the Colima-Tepic-Chapala triple junction (Demant 1981; Luhr et al. 1985; Allan 1986; Allan et al. 1991; Garduño and Tibaldi 1991; Ferrari et al. 1994, 2012) (Fig. 2). The Chapala

Rift is a structure made of two half grabens with active extensional deformation and a minor left-lateral component of motion (Suter et al. 1992). Some authors suggest that the Chapala Rift was inactive during the Pleistocene–Holocene (Rosas-Elguera et al. 1989; Urrutia-Fucugauchi and Rosas-Elguera 1994; Rosas-Elguera and Urrutia-Fucugauchi 1998; Ferrari et al. 2012). The Colima and Tepic rifts separate the continental Jalisco Block to the west from the North American Plate and the Guerrero Terrane to the north-east and the east, although the exact definition of their geometry and kinematics is still a matter of debate (Luhr et al. 1985; Bourgois et al. 1988; Allan et al. 1991; Garduño and Tibaldi 1991; Serpa et al. 1992; Bandy et al. 1993, 1995, 2005; Ferrari et al. 1994, 2012; Rosas-Elguera et al. 1996; Selvans et al. 2011) (Fig. 2).

The active Tepic Rift, formed in the Mid-Late Miocene, is located along the northern boundary of the Jalisco block (Fig. 2). Two main models have been proposed to explain its geometry and kinematics. In the first model, the Tepic Rift is made of grabens, pull-apart basins and strike-slip faults, and has accommodated right-lateral displacements during Plio-Quaternary times (Nieto-Obregón et al. 1985; Barrier et al. 1990; Bourgois and Michaud 1991; Garduño and Tibaldi 1991; Michaud et al. 1991; Moore et al. 1994). The kinematics of the Tepic Rift would thus be the result of active rifting induced by a relocation of the East Pacific Rise, with the Jalisco block moving toward the NW (Luhr et al. 1985; Allan 1986; Allan et al. 1991; Frey et al. 2007). According to the second and more recent model, the Tepic Rift underwent two phases of shearing in Mid- to Late Miocene time. In the Plio-Pleistocene, the deformation of the Tepic Rift would have been mainly extensional with a NNE–SSW-oriented horizontal minimum principal stress imparted by the Rivera, Cocos and North American plates boundaries (Quintero-Legorreta et al. 1992; Ferrari 1995; Rosas-Elguera et al. 1996; Ferrari and Rosas-Elguera 2000; Ferrari et al. 2012). The second model seems to be supported by recent geodetic GPS data, showing an active SW movement of the Jalisco Block relative to the North American plate (Selvans et al. 2011).

The southern branch of the triple junction at the eastern end of the Jalisco Block is the Colima Rift. The CVC is located inside this rift, so that the structural evolution of the volcanic complex is strictly associated to that of this regional structure and to the geodynamic setting of the area, as described in the following sections (Fig. 2).

3 The Colima Rift

Previous authors have attributed the origin, evolution and geodynamic significance of the Colima Rift to several factors. Among these, an active rifting of the Jalisco block produced by an eastward jump of the East Pacific Rise beneath the North American plate (Luhr et al. 1985; Allan 1986; Allan et al. 1991; Frey et al. 2007), a passive rifting produced by the subduction of the Rivera-Cocos Plate Boundary (Nixon 1982; DeMets and Stein 1990; Ferrari et al. 1994; Bandy et al. 1995), the activity of a hotspot (Moore et al. 1994), and the combined SE motion of the Guerrero terrane along the Chapala-Tula fault system and the SW motion of the Jalisco Block relative to the North American plate (Rosas-Elguera et al. 1996; Suter et al. 2001; Selvans et al. 2011; Ferrari et al. 2012).

The Colima Rift involves Cretaceous marine limestones, Jurassic–Tertiary metamorphosed clastic and volcaniclastic sediments, Cretaceous–Tertiary intrusive rocks, and Tertiary-Quaternary volcanic deposits (Allan 1986; Serpa et al. 1992; Bandy et al. 1995; Cortés et al. 2010). Parts of this rock assemblage were folded and thrusted during a Late Cretaceous–Paleogene compressive and transpressive phase (Serpa et al. 1992; Garduño-Monroy et al. 1998). A subsequent extensional phase generated the rift, whose filling is made of Pliocene–Quaternary lacustrine sediments, alluvium, colluvium and volcanic rocks (Allan 1986; Allan et al. 1991; Serpa et al. 1992).

The Colima Rift consists of three structural depressions, namely the Northern Colima Graben (NCG), the Central Colima Graben (CCG), where the CVC is located, and the Southern Colima Rift (SCR) (Allan 1986; Allan et al. 1991). The NCG is composed of two N-S-aligned and well defined depressions flanked by NNE-SSW-striking faults (Fig. 2). The southern one is bounded to the west by sharp parallel fault scarps that can be traced southward along the aligned volcanic centres of the CVC. The NCG is flanked by a high plateau of Late Miocene-Pleistocene volcanic deposits, unconformably overlying Jurassic-Eocene sedimentary and intrusive rocks. To estimate the thickness of the infilling sediments in the NCG, three-dimensional two-body modelling of gravity data was conducted by Allan (1985). The gravity survey was conducted in the northern portion of the graben and shows that the alluvial, colluvial and lacustrine fill has a total thickness of about 1 km (Allan 1985).

Near the CVC, two topographic depressions from the west and the east merge to form the 60-km-long, 50-km-wide CCG (Allan et al. 1991; Luhr 1997). Well defined N-S- to NNW-trending active normal faults can be detected only along its western boundary, where displaced Cretaceous limestones crop out (Allan et al. 1991; Bandy et al. 1995; Cortés et al. 2005, 2010; Norini et al. 2010). To the east, the boundary of the CCG mainly consists of minor normal faults and a progressive eastward rise into highlands comprising surface exposures of Cretaceous limestone and Cretaceous to Miocene igneous rocks (Bandy et al. 1995; Cortés et al. 2005, 2010; Norini et al. 2010). Inversion of gravity data in the southern portion of the CCG indicate that the Pliocene-Holocene sediment fill of this tectonic depression is about 1–1.5 km thick (Serpa et al. 1992).

The SCR was recognized on the basis of mapped lineaments as a wide tectonic system extending south of the CVC to the pacific coast (Allan 1986; Allan et al. 1991; Luhr 1997). This inferred extensional structure lacks an obvious graben morphology and its existence has been questioned on the basis of geophysical surveys and geologic mapping (Serpa et al. 1992; Rosas-Elguera et al. 1996; Ferrari et al. 2012). Modelling of gravity data in the SCR area imaged two major NE-SW-striking depressions with a total width of 100 km and up to 8 km of sediment fill. To the south, these structures merge with the offshore Manzanillo trough. The NE-SW-trending Manzanillo trough is a 60-km-wide depression with more than 1 km of vertical offset, containing a sedimentary fan and fault-controlled submarine canyons (Bourgois et al. 1988; Michaud et al. 1990; Bourgois and Michaud 1991; Bandy et al. 1995).

The geometry, kinematics and dynamics of the Colima Rift have been studied on the basis of field, seismic and geodetic data, mainly collected in its northern (NCG) and central (CCG) sectors (Fig. 2). Field structural analyses conducted on the NCG and CCG reveal that these grabens are bounded by sub-vertical crustal faults (Allan 1986; Allan et al. 1991; Norini et al. 2010). The amount of vertical displacement of the NCG and CCG is estimated to be at least 2.5 km by adding the topographic relief of the bounding fault scarps (1.5–1.6 km) to the calculated sediment depth (Allan 1986; Serpa et al. 1992). Field data and focal mechanism solutions are consistent with a direction of opening of the NCG and CCG oriented from E-W to NW-SE, with a mainly normal and minor right-lateral displacements of the bounding faults (Barrier et al. 1990; Suárez et al. 1994; Rosas-Elguera et al. 1996; Garduño-Monroy et al. 1998; Norini et al. 2010). In contrast to field and seismic evidence of long-term slightly dextral oblique extension, recent GPS geodetic measurements suggest a possible sinistral oblique extension of the Colima Rift (Selvans et al. 2011). In both cases, the stress regime is mainly extensional, with an approximately E-W orientation of the minimum horizontal stress in the basement of the CVC (Barrier et al. 1990; Suárez et al. 1994; Rosas-Elguera et al. 1996; Norini et al. 2010; Selvans et al. 2011).

The age of the first opening of the Colima Rift has been estimated to be about 3.5 Ma on the basis of the maximum vertical displacement along the bounding faults and the mean vertical movement rate of the NCG (Allan et al. 1991; Rosas-Elguera et al. 1996). Geologic, seismic and geodetic data also show that this extensional tectonic activity has continued into the Pleistocene and Holocene and is still occurring (Suárez et al. 1994; Rosas-Elguera et al. 1996; Norini et al. 2010; Selvans et al. 2011). Thus, the Colima Rift started to form prior to the formation of the CVC and continued its extensional activity throughout the evolution of this volcanic complex up to the present.

4 Structure of the Colima Volcanic Complex

Despite the importance of the volcano-tectonic interplay in determining the potential hazard of the active volcanism and the present and past state of activity of the CVC, very few works have been focused on the volcano structure and the relationships between its evolution and the regional tectonics. In this section we describe the CVC geological setting and the main structures and faults identified on the CVC by means of geological, structural and geophysical data. These structural analyses were based on the remote sensing and field mapping of faults in the basement and in the volcanic edifice (Rosas-Elguera et al. 1996; Garduño-Monroy et al. 1998; Norini et al. 2010), magnetic and aeromagnetic surveys (López-Loera et al. 2010, 2011), and numerical modelling of the volcano gravitational instability (Norini et al. 2010). Other works exploited the available structural data and models to support, interpret and discuss geological maps (Cortés et al. 2005, 2010), reconstructions of past volcanic activity (e.g. Cortés et al. 2009; Roverato et al. 2011), seismic data recorded in the volcano and the surrounding regions (e.g. Núñez-Cornú and Sánchez-Mora 1999; Domínguez et al. 2001; Zobin et al. 2002; Pacheco et al. 2003; Andrews et al. 2011), and geodetic data (Selvans et al. 2011).

4.1 Geology of the Volcanic Complex

Since the 80s, few papers have been devoted to the geology of the CVC and its surroundings areas (e.g. Robin et al. 1987; Luhr and Carmichael 1990; Rodríguez-Elizarraras 1995; Cortés et al. 2005, 2010), producing geological maps at different scales (from 1:500,000 to 1:100,000) and recognizing a long and complex volcanic history. The CVC began its formation about 1.7 Ma (Allan 1986) and consists of several superimposed edifices aligned along a N-S trend. The volcanic complex has a total volume of about 900 km^3, estimated by means of GIS and based on interpolation of the volcano basal surface (Norini et al. 2010). The most recent geological maps (Rodríguez-Elizarraras 1995; Cortés et al. 2010) defined lithostratigraphic units allowing the identification of four stratovolcanoes or volcanic complexes, named from north to south and from the oldest to the youngest active one: (1) El Cántaro, (2) Nevado de Colima, (3) Paleofuego or Ancestral Volcán de Colima, and (4) Volcán de Colima, sometimes referred to as Fuego de Colima or Volcán de Fuego (Figs. 3 and 4).

The following description of the four constructional phases of the CVC is mainly based on the most recent and detailed (1:100,000 scale) geological map of Cortés et al. (2010) covering an area of about 4000 km^2.

El Cántaro volcano is made of lava flows and pyroclastic deposits that are andesitic in composition, with some dacitic domes. Its total volume is 71 km^3 and its age spans from 1.66 ± 0.24 to 0.95 ± 0.17 Ma (Allen 1986). In the final stage, the products of El Cántaro Volcanic Complex are interlayered with those of the subsequent Nevado de Colima Volcanic Complex (Robin et al. 1987; Cortés et al. 2010).

Nevado de Colima is located south of El Cántaro and evolved through different eruptive phases or periods, the details of which vary among authors. Mooser (1961) identified two caldera structures, Robin et al. (1987) four eruptive phases and three caldera structures, Cortés et al. (2010) six eruptive periods. The Nevado de Colima Volcanic Complex consists of a large stratovolcano built during the first eruptive period and was affected by at least four lateral collapses producing debris avalanches, debris flows and pyroclastic deposits. The main morphostructural collapse rim related to this

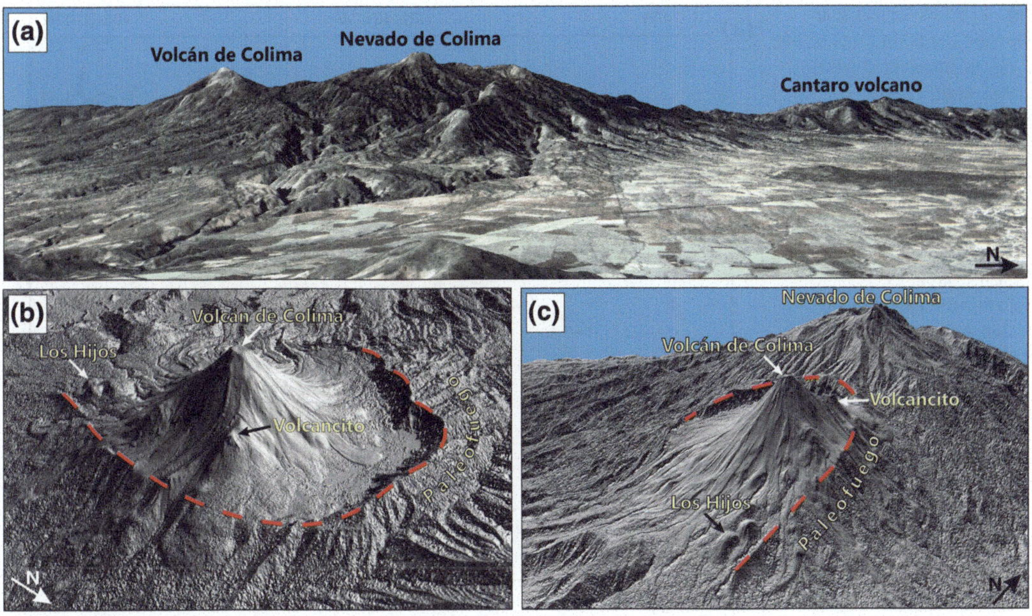

Fig. 3 Perspective views of the CVC. **a** View from the east of a Landsat ETM satellite image over a DEM. **b** View from the NE and **c** view from the SE of a shaded image over a DEM of Nevado de Colima and Volcán de Colima

complex is represented by a 7 km long semicircular caldera rim open to SE and dated between 0.53 and 0.37 Ma (Robin et al. 1987). Estimates for the total volume of the edifice range from 334 km^3 (Cortés et al. 2010) to 450 km^3 (Luhr and Carmichael 1990) and the age spans from 0.53 ± 0.1 Ma (Robin et al. 1987) to about 18 ka (Komorowski et al. 1997; Stoopes and Sheridan 1992).

The Paleofuego volcano or Ancestral Volcán de Colima (hereafter Paleofuego) is located south of the Nevado de Colima volcano and was affected by numerous lateral collapses (six after Komorowski et al. 1997, five after Cortés et al. 2010). The main visible scar is represented by a 5-km-wide semicircular caldera open to the south and related to a large sector collapse associated to a debris avalanche deposit uncertainly dated at 9370 B.P. (Robin et al. 1987), 4300 B.P. (Luhr and Prestegaard 1988), 3600 B.P. (Komorowski et al. 1997) and 2505 B.P. (Cortés et al. 2010, Cortés, this volume). Its total estimated volume is about 80 km^3 (Cortés et al. 2010) and the summit of the volcano is interpreted to have reached 4200 m a.s.l. (Roverato et al. 2011). The estimated age of Paleofuego spans from 50 ka (Robin et al. 1987) to 2500 B.P. (Cortés et al. 2010). Associated with this edifice, two small andesitic lava domes named "Los Hijos del Volcán" grew along its southern flank (Fig. 4).

Volcán de Colima is the youngest, active cone, which was built inside the Paleofuego scarp. This edifice is a stratovolcano, made of pyroclastic flows, fall deposits, and lava flows. Its total volume has been estimated to be between 9 (Cortés et al. 2010) and 10 km^3 (Luhr and Carmichael 1990). During the past 400 years the Volcán de Colima erupted more than 40 times and the last major eruption occurred in 1913 (see chapters by Crummy et al. and Connor et al. this volume).

The Paleofuego and possibly also the Volcán de Colima cones were affected by numerous catastrophic sector collapses, building a huge volcaniclastic fan to the south of the volcanic complex (Fig. 4) (Robin et al. 1987; Luhr and Prestegaard 1988; Stoopes and Sheridan 1992; Komorowski et al. 1997; Cortés et al. 2009, 2010; Cortés et al., this volume; Roverato et al. 2011).

One of the main open and debated problems concerning the CVC geological evolution is the

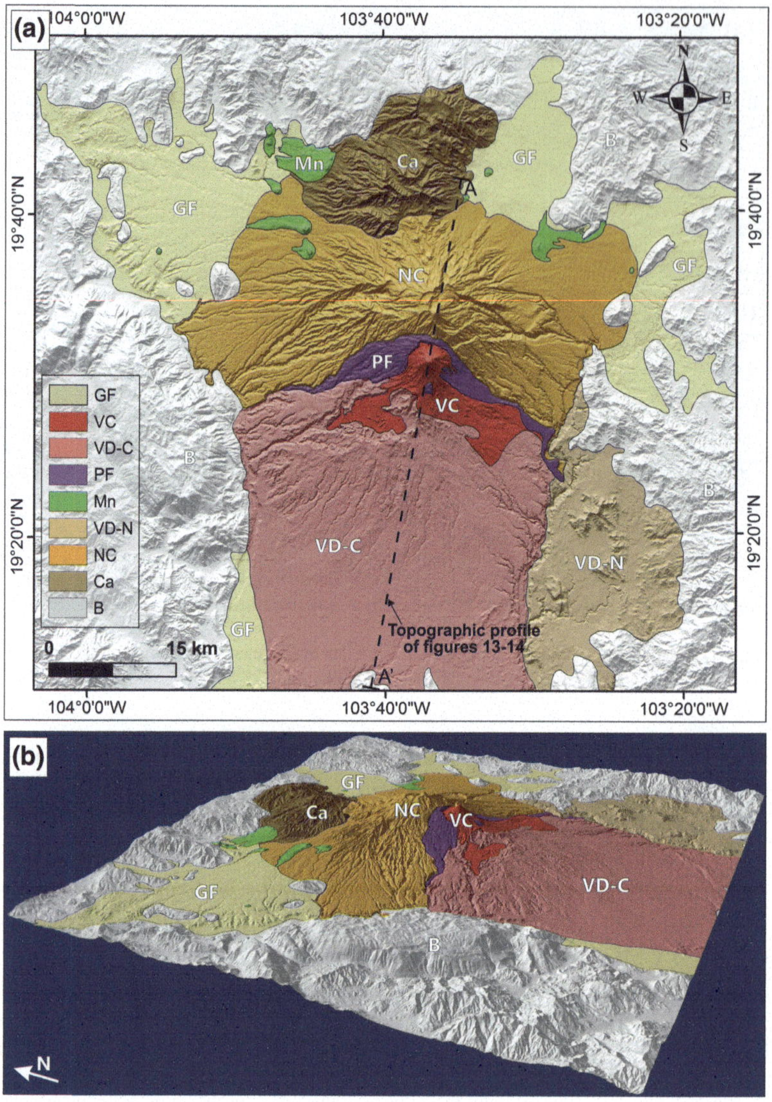

Fig. 4 **a** Geological map showing the main units of the CVC, as interpreted from Cortés et al. (2010). B: basement of the CVC. Ca: El Cántaro. NC: Nevado de Colima. VD-N: debris avalanche deposits from Nevado de Colima. Mn: monogenetic scoria cones in the CVC area. PF: Paleofuego. VD-C: debris avalanche deposits from Paleofuego. VC: Volcán de Colima. GF: Quaternary alluvial, colluvial and lacustrine deposits. The trace of the topographic profile of Figs. 13 and 14 is shown. **b** Perspective view of the geological map of the CVC (**a**)

nature of the boundary between Paleofuego and Volcán de Colima. There is a general agreement that the morphological scar located immediately north to the present active cone represents the final collapse of the Paleofuego volcano, but its estimated age spans from 2500 yr BP (Cortés et al. 2010) to more than 23 ky (Roverato et al. 2011). This difference in age estimates has consequences for the timing of the onset and volcanological evolution of the Volcán de Colima and on the edifice source of the debris avalanches. In fact Cortés et al. (2010) consider that Volcán de Colima has never experienced major collapses, whereas others authors, among them Roverato et al. (2011), assign to Volcán de Colima numerous collapses. Thus, understanding of the Paleofuego-Volcán de Colima boundary has clear implications on the hazard assessment for the potential for future lateral collapses.

4.2 The Rift-Parallel Faults

The Colima Rift, formed by the NCG, CCG and SCR (Fig. 2), is a prominent morphostructural feature, formed by several faults trending roughly N-S. These faults and the CVC were simultaneously active during the Quaternary and remain active today, with strong spatial and temporal interaction (e.g. Cortés et al. 2010; Norini et al. 2010).

The western boundary of the NCG is probably the tectonic structure exerting the largest control on the CVC evolution. The main evidence of this volcano-tectonic interplay is the alignment of the polygenetic volcanic centres of the CVC along the continuation of the main western boundary faults of the NCG. These volcanic centres are (from north to south, and from the oldest to the youngest): Cántaro, Nevado de Colima, Paleofuego and Volcán de Colima. Also, Los Hijos domes are located on the southern flank of the active Volcán de Colima cone, along the same alignment as the main volcanic centres (Figs. 3, 4 and 5).

One of the main sub-vertical faults of the NCG western boundary displaces lava flows and pyroclastic deposits on the northern flank of the Nevado de Colima edifice, with the NW and NE flanks of this volcano located at different elevations (Figs. 5, 6 and 7). The same fault cuts the rim of the summit caldera C3 of Robin et al. (1987) and Cortés et al. (2010), with a maximum fault throw to the east of about 300–400 m (Norini et al. 2010) (Fig. 6). The radiometric age of the faulted volcanic rocks is 0.37–0.14 Ma (Robin et al. 1987; Cortés et al. 2005), resulting in a minimum vertical displacement rate of about 1–3 mm/yr during the Upper Pleistocene and Holocene (Norini et al. 2010). This displacement rate is in agreement with the velocity constraint provided by a recent GPS geodetic study, which allows for up to 5 mm/yr of extension across the NCG (Selvans et al. 2011).

The CVC is mainly located inside the CCG. The western boundary of this graben is formed by a sharp and curved fault scarp, with a maximum height of 1600 m and the fault throw to the east. The eastern boundary of the graben is less defined and composed of an irregular eastward topographic rise and small discontinuous fault scarps, and lineaments (Fig. 5). The recognized boundary faults of this graben are located far from the CVC summit (Fig. 5). Field analysis at the boundaries of this graben shows that recent lacustrine and volcaniclastic deposits of the CVC exhibit N-S-trending normal faulting (Garduño-Monroy et al. 1998; Norini et al. 2010). This field evidence suggests that the whole CVC and the volcano basement are subjected to an active extensional stress regime, with an E-W orientation of the minimum horizontal stress imparted by the tectonic activity of the NCG and CCG (Allan 1986; Allan et al. 1991; Norini et al. 2010).

4.3 The Tamazula Fault

The occurrence of a major NE-SW-striking regional fault system has been suggested in the CVC area by some previous works. This structure was first described by Rodriguez-Elizarraras (1995) and then named Tamazula Fault by Rosas-Elguera et al. (1996) and Garduño-Monroy et al. (1998). The Tamazula Fault was identified through mapping of fault scarps and some parallel lineaments on the CVC flanks and in the surrounding region. This regional fault has been interpreted as an active normal fault system, extending for more than 160 km, bounding the Jalisco Block to the SE and cutting the Colima Rift and the CVC (Rosas-Elguera et al. 1996; Garduño-Monroy et al. 1998; Cortés et al. 2005, 2010). Rosas-Elguera et al. (1996) and Garduño-Monroy et al. (1998) stated that the alignment of some crustal earthquakes in the Manzanillo area confirms the existence and activity of the Tamazula Fault. More recently, various seismic events indicative of E-W and NE-SW faulting in the CVC and Manzanillo areas have been attributed to the reactivation of crustal structures parallel to the Tamazula Fault (e.g. Nuñez-Cornú and Sánchez-Mora 1999; Domínguez et al. 2001; Zobin et al. 2002; Pacheco et al. 2003; Andrews et al. 2011).

Garduño-Monroy et al. (1998) described four stages of tectonic deformation in the CVC area. The first two stages pre-date the emplacement of the CVC and the opening of the Colima Rift. The third stage has been related to E-W-trending tectonic extension along the Colima Rift. The last

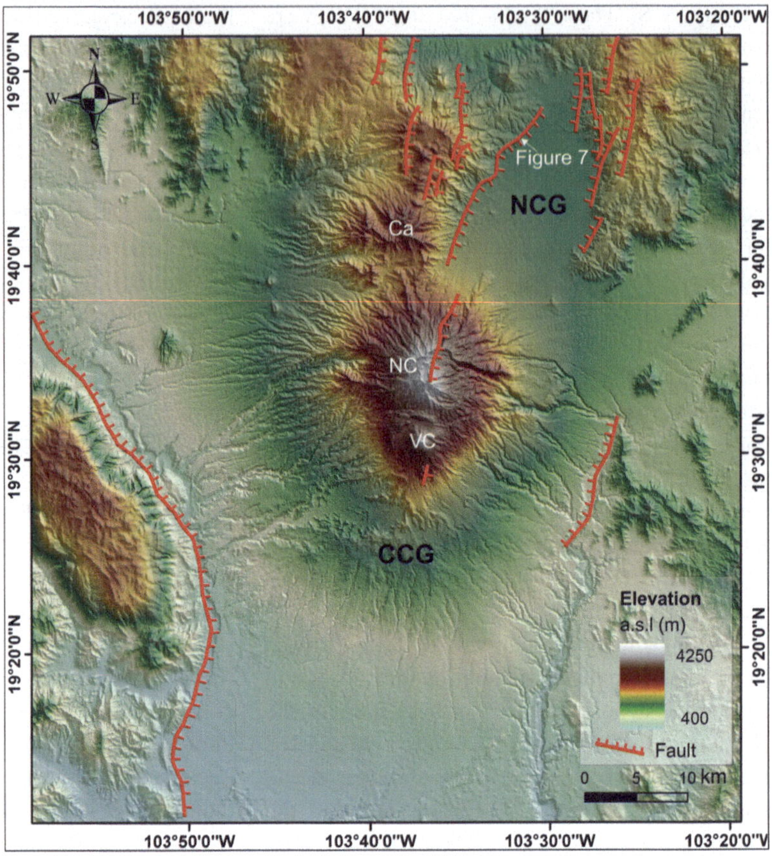

Fig. 5 Morphotectonic map on a shaded relief image showing the trace of the Colima Rift faults (modified from Norini et al. 2010). Ca: Cántaro. NC: Nevado de Colima. VC: Volcán de Colima. NCG: Northern Colima Graben. CCG: Central Colima Graben. Location of Fig. 7 is shown

stage generated a NE-SW-trending depression, named the graben of Alseseca-Atenquique, cutting through the CVC and resulting from the extensional activity of the Tamazula Fault (Fig. 8). This phase is associated with a NW-SE-trending tectonic extension (Garduño--Monroy et al. 1998). Following Garduño--Monroy et al. (1998), the CVC would have been emplaced at the intersection between the Colima Rift and the Tamazula Fault. Garduño--Monroy et al. (1998) also suggest that the NE-SW alignment of Volcán de Colima and the *Volcancito* parasitic cone reflect the occurrence of magma migration paths along the Tamazula Fault (Fig. 3b, c). This fault and the associated stress field were also thought to be responsible of the flank instability of the CVC, with preferred direction of lateral collapse toward the S and SE (Garduño-Monroy et al. 1998) and toward the SW (Cortés et al. 2009).

Some of the fault scarps and linear features attributed to the Tamazula Fault by Garduño--Monroy et al. (1998) and Cortés et al. (2009, 2010) have been alternatively interpreted as the result of active gravitational spreading of the volcanic complex by Norini et al. (2010), as discussed below.

4.4 The E-W Trending Volcano-Tectonic Fault System

An E-W-trending fault system has been described for the first time by Norini et al. (2010). This volcano-tectonic structure is characterized by the most prominent morphostructural features affecting the CVC (Fig. 9). Previously, most fault scarps attributed to this system were associated to the Tamazula Fault (Sect. 4.3)

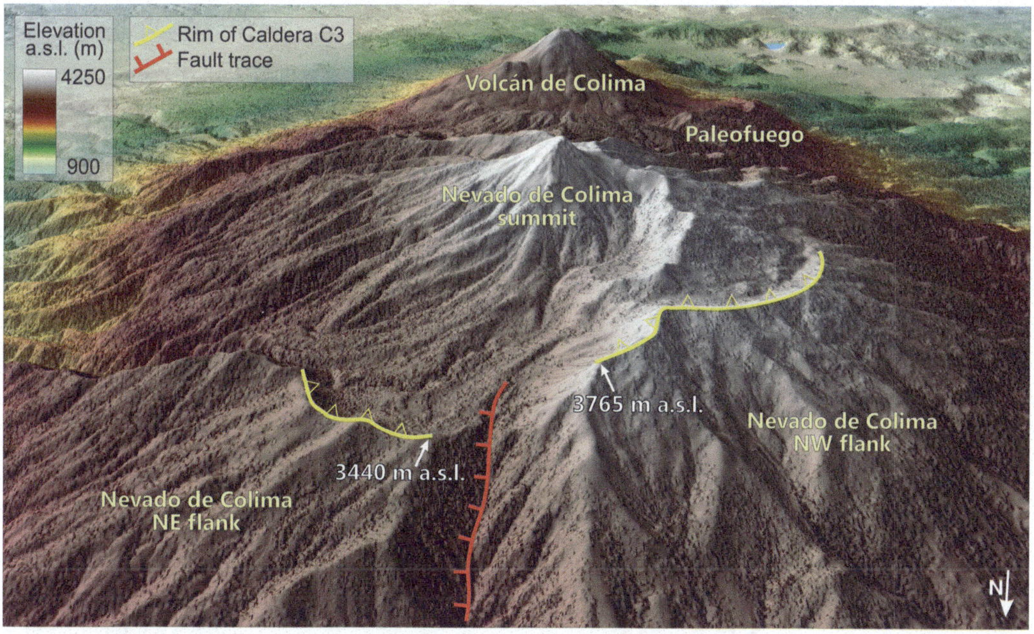

Fig. 6 Perspective view from the NNE of a fault scarp cropping out on the northern flank of Nevado de Colima. The fault displaces the C3 summit caldera rim (Robin et al. 1987; Cortés et al. 2010) and the northern flank of the volcanic edifice. The view was generated from DEM and Landsat ETM satellite image

Fig. 7 Photograph of a structural control site with NNE-trending faults in pyroclastic deposits along the NCG. The sequence has a ^{14}C conventional age of about 20 ka (Norini et al. 2010). Location of the photograph is shown in Fig. 5

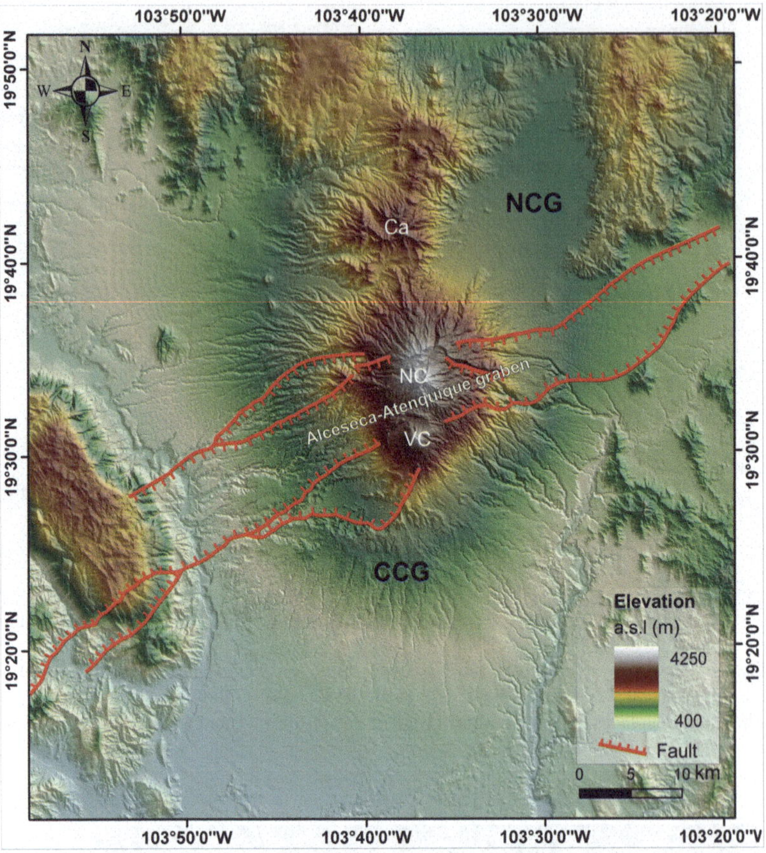

Fig. 8 Morphotectonic map on a shaded relief image showing the trace of the Tamazula Fault (modified from Garduño-Monroy et al. 1998; Cortés et al. 2010). Ca: Cántaro volcano. NC: Nevado de Colima. VC: Volcán de Colima. NCG: Northern Colima Graben. CCG: Central Colima Graben

(Garduño-Monroy et al. 1998). The E-W fault system crosses for 30 km the CVC, vanishing in the area surrounding the volcanic complex. This fault system is made of curved and discontinuous fault scarps across the eastern and western flanks of Nevado de Colima, striking from 70° to 115° azimuth (Fig. 9). The fault scarps present heights up to 500 m that decrease toward the volcano periphery and their overall arrangement shows a graben-like geometry (Fig. 9). Close to the summit of the Nevado de Colima, lava domes and flows of this volcano partially fill the graben and cover its fault scarps (Fig. 9), suggesting a decrease of the displacement rate beginning with the final phase of activity of Nevado de Colima (Norini et al. 2010). The available outcrops of structures along the E-W fault system show fault planes dipping 60° to 80° with pitch angles of 76° to 90°, confirming the normal faulting (Fig. 10) (Norini et al. 2010). Based on age of the faulted deposits and on the height of the fault scarps, Norini et al. (2010) estimate for this fault system a mean rate of vertical displacement of 1–3 mm/yr in the last 500 ka, with a reduction in activity since the final construction stage of Nevado de Colima.

Using a high resolution DEM, GIS software, and morphostructural and field data, Norini et al. (2010) have interpolated the subsurface geometry of the E-W fault scarps obtaining a set of straight fault planes with listric geometry, striking E-W and intersecting each other at shallow depth (Fig. 11). The modelled fault planes confirm the graben-like structure of this fault system (Norini et al. 2010).

The lack of surface continuity of the fault system beyond the volcanic complex indicates that the E-W fault system is a local volcano-tectonic structural feature, characterized by normal kinematics and N-S extension (Fig. 9).

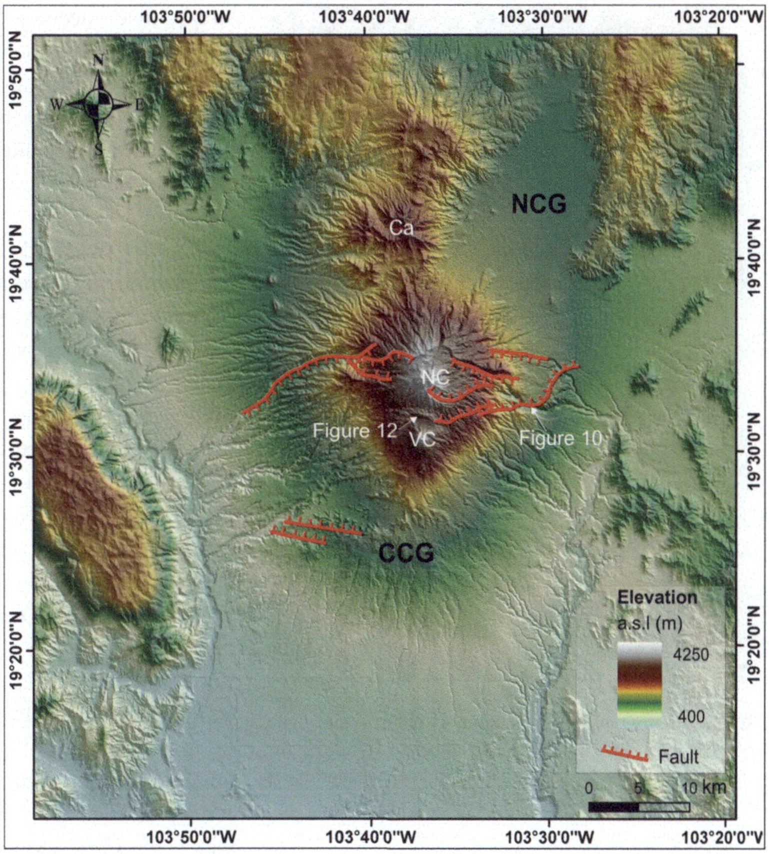

Fig. 9 Morphotectonic map on a shaded relief image showing the trace of the volcanic spreading faults (modified from Norini et al. 2010). Ca: Cántaro. NC: Nevado de Colima. VC: Volcán de Colima. NCG: Northern Colima Graben. CCG: Central Colima Graben. Locations of Figs. 10 and 12 are shown

Norini et al. (2010) suggested that this fault system was generated by the southward gravitational spreading of the volcano, due to the loading of the volcanic edifice (3000 m thick) over a soft lacustrine and alluvial sequence gently dipping to the south. This spreading process and its origin related to the weak basement show a similar structural setting and behaviour of well-known cases described in the literature (e.g. Wooller et al. 2004; Norini and Acocella 2011; Acocella et al. 2012).

4.5 Other Recent Tectonic Structures in the CVC Area

Recent magnetic anomaly surveys of the CVC area revealed some sharp magnetic changes that have been interpreted as linear features of tectonic origin (López-Lorea et al. 2010, 2011). The main four inferred faults have N-S-, E-W-, NE-SW- and NW-SE-trending traces that extend for tens of kilometers across and near the CVC. Three of these faults were identified as magnetic negative anomalies in some magnetic profiles, one fault was recognized by an aeromagnetic survey. The magnetic profiles are relatively far from each other (López-Lorea et al. 2010), so that the trace of the inferred faults is highly interpretative. No data are provided about the deep geometry, the kinematics and the significance of these structures (López-Lorea et al. 2010, 2011). It is worth noting that there is an apparent inconsistency among the faults inferred by magnetic profiles (López-Lorea et al. 2010) and the faults and lineaments identified by an aeromagnetic survey (López-Lorea et al. 2011). In both cases, these main faults are not related to any clear surface expression of faulting. Also, many clearly visible faults in the CVC and the

Fig. 10 Photograph of a structural control site with E-W-trending faults in volcaniclastic deposits along the volcanic spreading faults. Location of the photograph is shown in Fig. 9

surrounding area do not correspond to any magnetic feature in either the magnetic profiles or aeromagnetic survey of López-Lorea et al. (2010, 2011) (e.g. Garduño-Monroy et al. 1998; Norini et al. 2010). This suggests that the nature of the linear magnetic anomalies described by López-Lorea et al. (2010, 2011) is not definitive and could have a different origin than recent faults bisecting the CVC.

4.6 The Feeding System of the CVC

The geometry of the magmatic plumbing system is a fundamental part of the CVC structural architecture. The magmatic system interacts with the tectonic and volcano-tectonic structures in the volcano and its basement and can play an important role in potentially hazardous volcanic events, like lateral eruptions and sector collapses. Thus, the geometry of the CVC magmatic system and its interaction with the basement faults and stress field need to be carefully evaluated. Available data on the CVC magmatic system are based on geophysical and petrologic constraints on the depth and geometry of the magma reservoir and conduits, and on the location, alignment and elongation of exposed volcanic centres and dikes.

For the Volcán de Colima volcano a magma chamber depth of about 6.3 km below the summit was calculated based on the depth of the hornblende stability region for the 1998–1999 erupted lavas (Luhr 2002). The gravimetric model of Medina-Martinez et al. (1996) shows a low-density body between 5 and 10 km depth below the summit of Volcán de Colima. Also, Nuñez-Cornú and Sánchez-Mora (1999) recognized an aseismic zone for the 1991 eruption between 4 and 7 km below the summit, and suggested that this zone may correspond to the depth of the magma chamber below the volcano. Zobin et al. (2002) analyzed the location of earthquake hypocentres for the 1997–2000 activity at Volcán de Colima and identified a zone of possible magma storage at about 5–10 km depth. A viscoelastic model of mass ejection applied to the 1998–1999 eruptive period of Volcán de Colima fits reasonably well with a magma chamber with a volume of 30 km^3, centred at about 5.6 km below the summit crater (Cabrera-Gutiérrez and Espíndola 2010). CVC 2¾D magnetic modelling performed to investigate the deep structure of the CVC revealed that a source of magnetic anomaly is related to a possible N-S-elongated magma chamber, located at a 4.8 km depth beneath the Volcán de Colima summit. Also, this magnetic model suggests that an older magma chamber with a N-S-elongated shape is located at a depth of 5 km below the Nevado de Colima summit (López-Loera et al. 2011). Ambient seismic noise tomography indicates southward migration of magma, correlated with the N-S trend of the stratovolcanoes of the CVC (Escudero and Bandy 2017).

Above the 5–10 km deep magma chamber of Volcán de Colima, the melt would continue its rise through conduits or inter-fingered dikes, as suggested by both the analysis of phenocryst-hosted melt inclusions of the *Volcancito* and

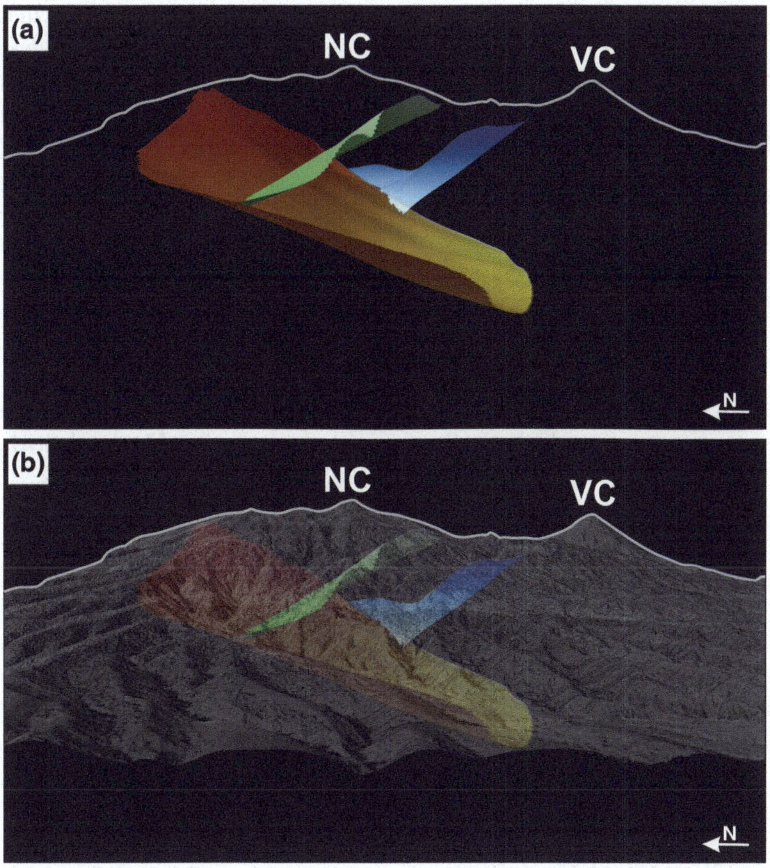

Fig. 11 Perspective views of the E-W-trending spreading fault planes inside the CVC (modified from Norini et al. 2010). The surfaces are depicted with different colours to distinguish the three main fault planes. The topography is represented with a topographic profile (**a**) and a transparent DEM (**b**). NC: Nevado de Colima. VC: Volcán de Colima

1890 lava flow samples, and by ambient seismic noise tomography (Atlas et al. 2006; Escudero and Bandy 2017). Melt inclusions from the 1998–2005 andesite lavas of Volcán de Colima confirmed low-pressure crystallization (130–10 MPa, corresponding to less than 5 km) of the melt feeding the eruptions (Reubi and Blundy 2008). Most volcano-tectonic seismic foci associated with the 1991, 1994, and 1997–1998 eruptions of Volcán de Colima occurred within 5–7 km from the surface, with the epicentres of the 1997–1998 activity located in a slightly N-S-elongated area south of the volcano summit (Núñez-Cornú et al. 1994; Zamora-Camacho et al. 2007).

At the surface, in the inner wall of the Paleofuego collapse scar, three subvertical N-S-trending dikes crop out with thickness of 5–10 m (Fig. 12) (Norini et al. 2010). These dikes lie along the N-S-striking alignment defined by Nevado de Colima, Volcán de Colima, and Los Hijos domes.

The above observations and analyses suggest that the active CVC has one or more, probably N-S-elongated, main magma chambers at a depth of >5 km below the summit, which are drained by a network of N-S-trending dikes.

Moreover, the CVC is not the only Quaternary volcanic feature in the region. The NCG and CCG are punctuated by 11 monogenetic scoria cones of Pleistocene age (Luhr and Carmichael 1981; Carmichael et al. 2006) (Fig. 4). The geometry of the feeding system of these mantle-derived, potassic magmas, was inferred from the azimuths of the maximum elongation axis and alignments of the scoria cones by Norini et al. (2010). This study shows a sharp trend along the N-S strike, indicative of N-S-trending fractures through which these primitive magmas rose to the surface.

Fig. 12 Photograph of one of the three N-S-trending dikes exposed in the Paleofuego sector collapse depression. Location of the photograph is shown in Fig. 9

5 Numerical Modelling of Volcano-Basement Interaction

In the previous sections, the relationships between different fault systems and the evolution of the CVC have been reviewed. While there is a general consensus on the control of Colima Rift-parallel structures on the geometry and evolution of the volcanic complex, two contrasting hypotheses on the origin and influence of the E-W/NE-SW-trending fault system have been proposed in the literature: (1) the fault scarps cutting through the CVC are expressions of the regional Tamazula Fault system (Sect. 4.3; Garduño-Monroy et al. 1998); (2) these structures are originated by the southward gravitational spreading of the volcanic complex (Sect. 4.4; Norini et al. 2010). These two different explanations on the origin of this fault system have a significant impact on the overall interpretation of the structural architecture and evolutionary mechanisms of the CVC. Thus, they were tested by 2D Finite Element stress-strain modelling. The numerical analysis was aimed at investigating the geometrical, stratigraphic and structural conditions promoting the onset and development of this prominent fault system intersecting the volcanic edifice. In particular, modelling was used to test the hypothesis that the fault system intersecting the volcanic complex has been originated by gravitational spreading rather than by regional active tectonics.

5.1 Numerical Modelling Set-up: Method and Assumptions

Modelling was performed using a 2D Finite Element software (Phase2, Rocscience Inc.) designed for rock engineering applications, able to simulate the stress-strain behaviour of soil and rock masses accounting for different constitutive models, boundary and initial conditions, and individual discontinuities (Riahi et al. 2010). Despite some limitations to advanced numerical modelling (e.g. large strain, time-dependent behaviour, thermal effects), this code is able to successfully predict deformation and failure mechanisms in even complex rock masses (Riahi et al. 2010). Continuum-based interface elements (Goodman et al. 1968) can be included to simulate the effects of individual discontinuities (e.g. fault zones, master fractures) or joint networks (Hammah et al. 2008).

We simulated the deformation of the volcanic complex under gravitational load considering two alternative scenarios (Fig. 13) corresponding to two different tectonic settings: (a) regional E-W extension associated to the active Colima Rift (Norini et al. 2010; Model A); (b) N-S regional extension associated to the Tamazula Fault system (Garduño-Monroy et al. 1998; Model B). Both 2D models A and B were based on NNE-SSW trending cross sections through the Nevado de Colima and Volcán de Colima summits (Fig. 4).

Six units forming the CVC system (Fig. 13) were defined by re-interpreting and simplifying the available geological data into the following units (e.g. Cortés et al. 2010): (1) Cretaceous

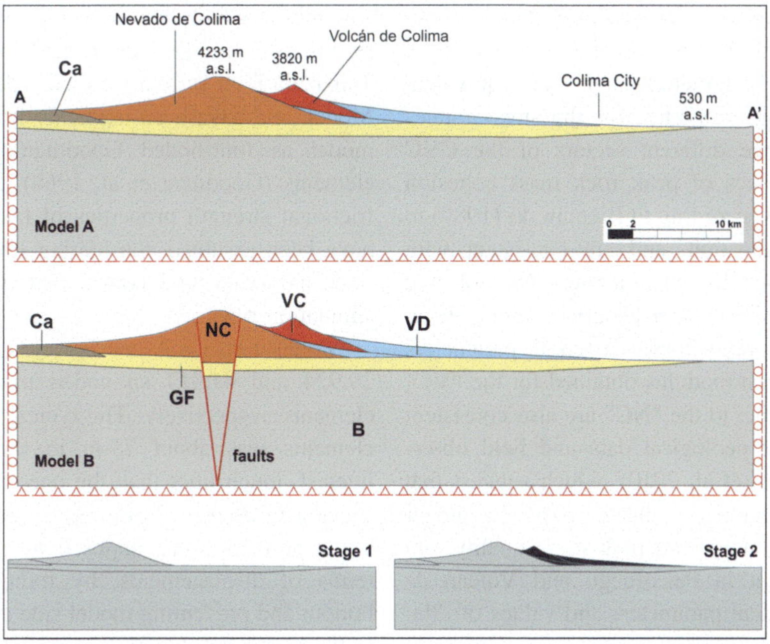

Fig. 13 Finite element models. Model A: volcano deformation under gravitational load with regional E-W extension (after Norini et al. 2010); Model B: volcano deformation under gravitational load accounting for normal faults cutting through the edifice with regional N-S extension (after Garduño-Monroy et al. 1998). Bottom of the page: Stage 1: emplacement of Nevado de Colima over the Colima graben fill deposits; Stage 2: emplacement of the Volcán de Colima edifice and related volcaniclastic sequence on the southern flank of Nevado de Colima. Key to symbols (see text for explanation): B: basement; GF: graben fill; Ca: Cántaro; NC: Nevado de Colima; VC: Volcán de Colima; VD: volcaniclastic deposits

limestones and intrusive rocks forming the bedrock underlying the CVC (unit “B”); (2) Quaternary alluvial, colluvial, and lacustrine deposits filling the NCG and CCG (unit “GF”); (3) volcanic rocks forming the remnants of the old Cántaro volcano (unit “Ca”); (4) andesitic lavas and pyroclastic deposits forming the Nevado de Colima edifice (unit “NC”); (5) andesitic lavas and pyroclastic deposits forming the Paleofuego-Volcán de Colima edifice (unit “VC”); and (6) volcaniclastic (pyroclastic and epiclastic) deposits covering the southern flank of the CVC (unit “VD”). The maximum thickness of the graben fill (about 1 km) was derived from the literature (Allan 1985; Serpa et al. 1992). In the B Model, accounting for the Tamazula Fault, both the bedrock and the graben fill are faulted forming a symmetric graben with geometry and maximum vertical displacement similar to the graben of Alseseca-Atenquique of Garduño-Monroy et al. (1998).

Rock masses have been assumed to exhibit an elastoplastic behaviour according to a Mohr-Coulomb failure criterion with post-peak dilatancy and cohesion loss. Typical material properties for units “B” and “GF” (Fig. 13 and Table 1) were derived from the literature (Hoek and Brown 1997; Marinos and Hoek 2000). For volcanic materials (units “Ca”, “NC,” “VC” and “VD”; Fig. 13 and Table 1), rock mass properties were estimated according to the approach proposed by Del Potro and Hürlimann (2008). Intact rock properties were assumed within the range of literature values (Watters et al. 2000; Hürlimann et al. 2001; Reid et al. 2001; Zimbelman et al. 2004, Moon et al. 2005; Pola et al. 2010, 2012), whereas typical values of GSI (geological strength index) were used to characterize rock mass structure (Hoek et al. 1995). This information allowed us to derive the Hoek-Brown parameters and the equivalent Mohr-Coulomb properties for the stress ranges expected in the different sectors of the CVC (Table 1). Values of peak rock mass cohesion were obtained according to Bieniawski (1989) to avoid overestimation, and are consistent with those published by other authors for volcanic rock masses (Watters et al. 2000; Moon et al. 2005). The lower values of strength parameters and deformation modulus obtained for the "VC" unit with respect to the "NC" are also consistent with available geological data and field observations (Cortés et al. 2010), which suggest that pyroclastic materials, characterized by lower intact rock strength and rock mass quality, are more abundant in Paleofuego and Volcán de Colima. Residual parameters and values of dilation angle were constrained by a series of parametric simulation runs with different assumed post-peak behaviour (e.g. perfectly plastic, strain softening), which however did not affect the results significantly. Normal faults related to the Tamazula fault system were considered in the B Model, and have been explicitly included in the model as four-noded Goodman FE interface elements (Goodman et al. 1968). Stiffness and frictional strength properties of faults (Table 1) have been assumed considering involved materials and scale, and constrained by parametric simulation runs.

Model domains A and B were discretized by 39,924 and 41,917 six-noded triangular finite elements, respectively. The typical size of finite elements was about 75 m in the regions of interest, much finer than the one (about 400 m) previously used by Norini et al. (2010). Boundary conditions were imposed to the model in terms of displacements, by fixing the model bottom and preventing model side displacements in the horizontal direction. A gravitational stress field was then initialized in the model, and different combinations of horizontal pressure

Table 1 Rock mass and joint mechanical properties used in finite-element models

Rock mass parameters	B	GF	Ca	NC	VC	VD
Unit weight, γ (MN/m^3)	0.026	0.018	0.022	0.022	0.017	0.015
Poisson's ratio, ν	0.30	0.35	0.32	0.32	0.30	0.32
Deformation modulus, E_m (MPa)	3.6e^4	1.5e^3	1.0e^4	1.0e^4	1.4e^3	1.7e^3
Failure criterion	Equivalent Mohr-Coulomb (from Hoek-Brown)					
Tensile strength, σ_{tm} (MPa)	1.00	0	0.1	0.1	0.05	0.01
Cohesion, C_m (MPa)	5.00	0.1	0.5	0.5	0.3	0.03
Friction angle, ϕ (°)	36	30	32	32	28	30
Dilation angle, ψ (°)	0	10	0	10	10	0
Residual friction angle, ϕ_r (°)	36	30	32	32	28	30
Residual cohesion, C_{mr} (MPa)	5.00	0	0.01	0.01	0.01	0
Fault parameters (Model B only)						
Normal stiffness (MPa/m)	50,000					
Shear stiffness (MPa/m)	5000					
Slip criterion	Mohr-Coulomb					
Tensile strength (MPa)	0.01					
Cohesion (MPa)	0.05					
Friction angle (°)	30					

Keys to units: *B* basement (cretaceous limestones and intrusive rocks); *GF* graben fill (quaternary alluvial, colluvial and lacustrine deposits); *Ca* El Cántaro; *NC* Nevado de Colima (andesitic lavas and pyroclastic deposits); *VC* Volcán de Colima (andesitic lavas and pyroclastic deposits); *VD* Volcaniclastic deposits (pyroclastic, epiclastic) coefficient K values (ranging between 0.3 and 0.5) were imposed to the two models in order to account for the different postulated regional stress regimes, namely: E-W extension for Model A (i.e. active Colima Rift), and N-S extension for Model B (i.e. active Tamazula Fault). Modelling was performed according to a sequential approach by simulating the stress-strain evolution of the CVC in two stages: Stage (1) emplacement of Nevado de Colima over the Colima Rift fill deposits; Stage (2) emplacement of the Paleofuego-Volcán de Colima edifice and related volcaniclastic sequence on the southern flank of Nevado de Colima (Fig. 13).

5.2 Model Results

Model results are presented in terms of computed values of maximum shear strain, horizontal displacements, and vertical displacements (Fig. 14). In Model A, during the first stage (i.e. emplacement of Nevado de Colima) fairly continuous, slightly asymmetric conjugate shear bands develop under a gravitational stress field in the Nevado de Colima edifice and in the underlying weaker graben fill material (Fig. 14, A1). These shear bands are associated with an asymmetric spreading of the volcano edifice with dominant S-directed horizontal displacements (Fig. 14, A3) and a comparable downward displacement of the volcano summit (Fig. 14, A5). Notwithstanding small absolute values of computed displacements, due to the small-strain solution logic implemented in the code, these results suggest that purely gravitational stress distribution in Nevado de Colima is a convincing candidate to trigger volcano spreading, whose evolution has been constrained by the geometry of the underlying weak basement (i.e. the graben fill).

Explicitly accounting for the postulated occurrence of active faults related to the Tamazula Fault (Model B) results in a more complex pattern of shear strain localization during the first stage (i.e. emplacement of Nevado de Colima), and in enhanced vertical displacements of the volcano summit, without significant differences in the overall kinematic configuration of the volcano spreading (Fig. 14, B1, B3 and B5). The similar overall response of Model A and Model B during the first modelling stage indicates that gravitational spreading occurs in any case. This gravitational spreading can explain the N-S extension of the volcanic complex without the need of invoking active regional tectonics. In this context, tectonic structures related to the Tamazula Fault provide geometrical and mechanical constraints to the geometry of spreading, and on the evolution of morpho-structures observed at the surface.

The emplacement of Volcán de Colima and associated volcaniclastic deposits in the second model stage causes minor differential changes in the strain and displacement patterns of Nevado de Colima with respect to the first stage (Fig. 14, A2, A4 and A6). Model results suggest that the load of the Volcán de Colima cone upon the southern flank of the Nevado de Colima edifice may have exerted a confining effect, resulting in an exhaustion of volcano spreading (Fig. 14, A4 and A6). This provides a dynamic reason for the available geomorphological and neotectonic data, suggesting a decreasing activity of the gravitational spreading in recent times (Sect. 4.4).

On the other hand, the emplacement of Volcán de Colima shifted the gravitational instability of the CVC towards the south: values of maximum shear strain and displacement computed for the Volcán de Colima edifice are one order of magnitude larger than those corresponding to the spreading of Nevado, suggesting a high susceptibility of Volcán de Colima to southward lateral collapse (Fig. 14, A2, A4 and A6). This is consistent with the volcano stratigraphic record, where several southward collapses of both the Volcán de Colima and Paleofuego edifices have been described (Robin et al. 1987; Navarro et al. 1994; Komorowski et al. 1997; Cortés et al. 2009, 2010; Roverato et al. 2011; Cortés et al. this volume).

These numerical models provide evidence that gravitational spreading is almost completely independent of the considered structural setting

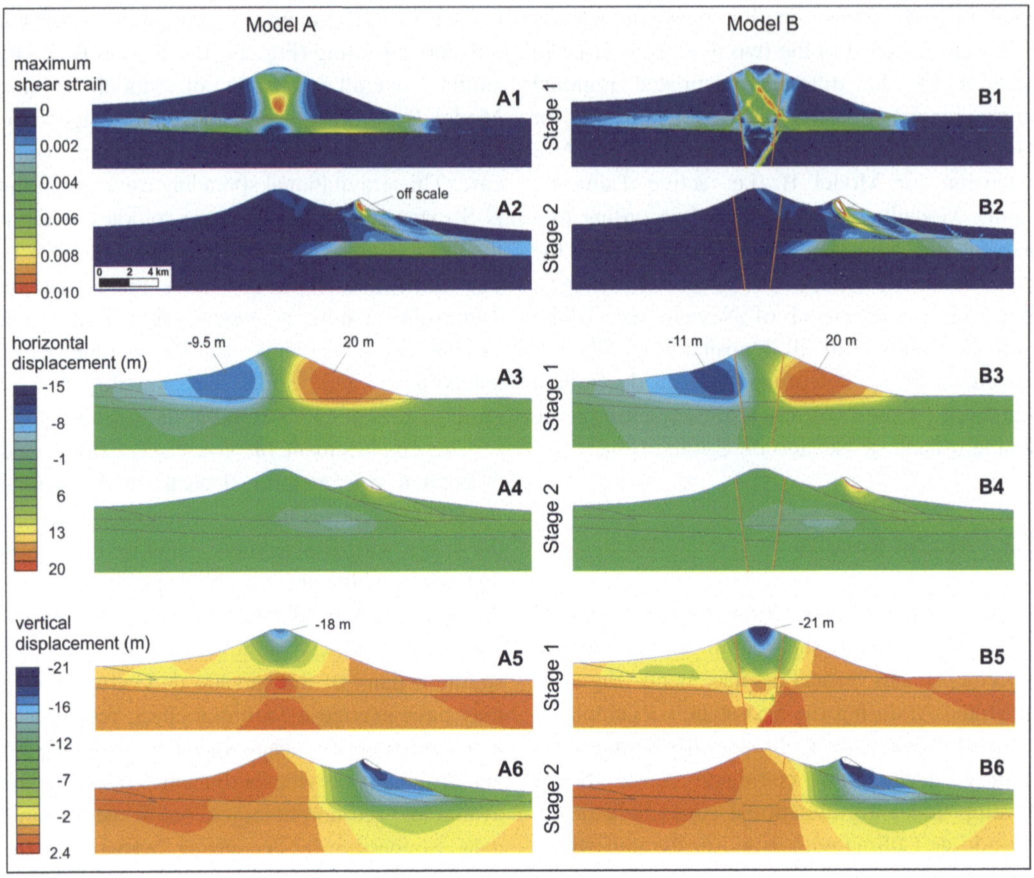

Fig. 14 Finite element modelling results. Maximum shear strain, horizontal displacement and vertical displacement computed for the two considered tectonic settings (Model A and Model B) and for different modelling stages (Stage 1 and Stage 2) are compared. Positive horizontal displacement is to the south (right on the figure) and positive vertical displacement refers to upward movement. Stage 2 results are incremental, i.e. they portray shear strain and displacements accumulated after the emplacement of the Volcán de Colima edifice

and tectonic regime, and is mainly controlled by both the load and morphology of the volcanic pile as well as the geometry and rheology of the graben fill.

6 Discussion

6.1 Active Fault Systems in the Volcanic Complex

Morphostructural, field and geophysical studies of the CVC show that the style of deformation of the volcanic complex and its basement is the result of the coexistence of two nearly-orthogonal active fault systems (e.g. Garduño--Monroy et al. 1998; Norini et al. 2010). These two main faults systems are: (1) the regional NNE-SSW fault system linked to the opening of the Colima Rift, and (2) a series of faults with main NE-SW- or E-W trend, which have been interpreted as a part of the regional Tamazula Fault (Garduño-Monroy et al. 1998) or as the result of the southward gravitational spreading of the volcanic complex (Norini et al. 2010).

The regional Colima Rift generated pervasive faulting of the volcanic complex and its basement. NNE-SSW-trending crustal subvertical faults displaced the whole CVC (Sect. 4.2 and Figs. 5 and 6). The geological, geophysical and geodetic data provides evidence of the long-term and present state of activity of these mainly normal faults, with an E-W-directed extension at 1–3 mm/yr mean vertical displacement rate (Barrier et al. 1990; Suárez et al. 1994; Allan 1986; Allan et al. 1991; Serpa et al. 1992; Bandy et al. 1995; Rosas-Elguera et al. 1996; Norini et al. 2010, Selvans et al. 2011; Ferrari et al. 2012).

The other prominent fault scarps on the CVC flanks have been geometrically modelled by Norini et al. (2010), showing that they represent E-W-striking listric faults forming a shallow and small graben centred beneath Nevado de Colima (Fig. 11). This graben was originated by the southward gravitational spreading of the volcanic complex (Fig. 9; Sect. 4.4). Parts of the same fault scarps have also been attributed to the wider and longer NE-SW-trending Alseseca-Atenquique graben cutting throughout the CVC and resulting from the extensional activity of the Tamazula Fault (Garduño-Monroy et al. 1998) (Fig. 8; Sect. 4.3). These fault scarps sharply decrease in height toward the volcano periphery and lack obvious continuity beyond the volcanic complex (Sect. 4.4) (Norini et al. 2010). Also, the proposed kinematics of the Tamazula Fault would require roughly N-S-trending extension (Garduño-Monroy et al. 1998), with a strong difference in orientation of the horizontal principal stresses compared to the Quaternary and still active tectonic regime of the area (E-W-trending extension associated to the opening of the Colima Rift) (e.g. Barrier et al. 1990; Suárez et al. 1994; Norini et al. 2010; Selvans et al. 2011; Ferrari et al. 2012). Finally, the new numerical modelling presented in this paper (Sect. 5) suggests that the southward volcanic spreading is likely to occur because of the topographic configuration of the volcanic complex, of the weakness of the underlying sedimentary graben fill, and of the geometry of the regional basement. The finite element modelling shows that this process is independent of the tectonic configuration of the region, so that even in the case of pre-existing E-W or NE-SW faults of tectonic origin (e.g. Tamazula Fault), most of the N-S extension is originated by the slow gravitational collapse of the volcanic complex (Fig. 14).

All the above mentioned data are lines of evidence supporting active spreading of the CVC toward the south. In this case, the development of an E-W-striking graben trending perpendicular to the direction of the spreading depends on the southward dipping regional topography and on the buttressing effect of the basement (Fig. 9) (e.g. van Wyk de Vries and Borgia 1996; Borgia et al. 2000; Wooller et al. 2004; Norini et al. 2010; Norini and Acocella 2011).

The occurrence of a NE-SW-trending fault system crossing the basement of the CVC (Tamazula Fault) is not questioned in this review chapter, although the available data point to a shallow N-S extension of the volcanic complex originated by the slow gravitational southward spreading. The mean displacement rate along the spreading faults is in the range of 1–3 mm/yr (Norini et al. 2010). Since 40 ka, this regular spreading rate could have been reduced as a consequence of the stabilizing effect imparted by the lithostatic load of the Paleofuego-Volcán de Colima edifices (Sect. 5) (Norini et al. 2010).

6.2 Growth of the Volcanic Complex

The CVC is made of different volcanic edifices aligned along the NNE-SSW western boundary of the NCG (Sect. 4.1). Available geophysical and petrologic data indicate that these polygenetic volcanoes were fed by one or more magma chambers located at >5 km depth below the summit of the volcanic complex (Núñez-Cornú et al. 1994; Medina-Martinez et al. 1996; Luhr 2002; Zobin et al. 2002; Cabrera-Gutiérrez and Espíndola 2010; López-Loera et al. 2011). The geodetic, geochemical, geophysical and volcano-tectonic data suggest that the magma is drained from this chamber to the surface by N-S-trending dikes (Atlas et al. 2006; Núñez-Cornú et al. 1994; Zamora-Camacho et al. 2007; Norini et al. 2010). The elongation and alignments of the monogenetic cones in the surrounding region also follow the same N-S trend (Norini et al. 2010).

The above data indicate that the geometries of the feeding systems of the Quaternary volcanic activity in the CVC area are controlled by the orientation of the NCG and CCG bounding faults and by the extensional stress regime with an approximately E-W orientation of the minimum horizontal stress (Barrier et al. 1990; Suárez et al. 1994; Allan 1986; Allan et al. 1991; Serpa et al. 1992; Bandy et al. 1995; Rosas-Elguera et al. 1996; Norini et al. 2010; Selvans et al. 2011). On the contrary, the E-W/NE-SW-striking fault system does not show any clear geometric relationship with the feeding system of the polygenetic and monogenetic volcanism. Only the Volcancito cone on the NE flank of the Volcán de Colima cone (Fig. 3b, c) was interpreted as a volcanic feature probably aligned along the trace of the Tamazula Fault (Sect. 4.3) (Garduño--Monroy et al. 1998). It is worth noting that this parasitic cone is located on the higher flank of the main Volcán de Colima cone, where lateral intrusions are controlled by the radial stress field imparted by both the conical topography and the magma pressure, so that a clear relationship with a tectonic or volcano-tectonic structure is more difficult to establish (e.g. Muller and Pollard 1977; van Wyk de Vries and Borgia 1996; Gudmundsson 2011). The consistency of the N-S trend of the Quaternary volcanic features in the CVC area can be explained considering that the shallow and low angle spreading-related E-W-trending faults have much less probability to intercept and guide magma to the surface compared to a sub-vertical crustal fault system associated to the regional far stress field (e.g. the Colima Rift).

6.3 Progressive Versus Catastrophic Instability of the CVC

The stratigraphic data and geological maps of the CVC (e.g. Cortés et al. 2010) show two different instability behaviours of the Nevado de Colima and the Paleofuego-Volcán de Colima edifices:

Nevado de Colima is a massive volcanic edifice made of lava flows and domes, whose evolution has been characterized by caldera-forming destructive events (e.g. Robin et al. 1987; Cortés et al. 2010) and several southward and eastward sector collapses (Robin et al. 1987; Luhr and Prestegaard 1988; Stoopes and Sheridan 1992; Capra and Macias 2002; Cortés et al. 2010). Also, the edifice was shaped by active southward spreading, which probably reduced the inclination and elevation of the volcano flanks, thus diminishing the probability of catastrophic sector and flank failures (Norini et al. 2010). The direction of the few known sector collapses of this volcano could have been controlled by the eastward dipping normal Colima Rift faults (mainly for the eastward sector failures) (e.g. Vidal and Merle 2000; Acocella et al. 2013) and by the Colima Rift topographic confinement (for the southward sector and flank failures) (e.g. Norini et al. 2008).

Paleofuego and Volcán de Colima are steeper cones whose emplacement has been marked by several catastrophic sector collapses directed toward the south (e.g. Komorowski et al. 1997; Roverato et al. 2011; Cortés et al. this volume). The two cones grew rapidly on the southern Nevado de Colima flank (Roverato et al. 2011) (Fig. 3a). The inclination of the southward dipping Nevado de Colima flank is the main topographic constraint generating the marked instability of the Paleofuego-Volcán de Colima cones (Sect. 5). The numerous southward catastrophic sector failures induced by this topographic constraint have been triggered by several concurring factors, such as active spreading and tectonic faults in the edifice interior, magmatic intrusions, rock alteration, and hydrothermal activity (e.g. Roverato et al. 2011; Cortés et al. this volume).

The change in the instability behaviour from the Nevado de Colima to the Paleofuego-Volcán de Colima edifices could have been promoted by the emplacement of the steep Paleofuego and Volcán de Colima cones. As presented in Sect. 5, these two recent edifices and their lithostatic loads may have been responsible for a stabilization of the southern flank of Nevado de Colima, a southward shift of the gravitational instability (from the Nevado de Colima edifice to the Paleofuego-Volcán de Colima edifices) and a change towards more frequent catastrophic events (Norini et al. 2010; Roverato et al. 2011).

7 Final Remarks and Future Needs

The structure of the CVC depicted by the studies reviewed in this paper and by the numerical modelling presented in Sect. 5 can be summarized as follow:

1. Three main fault systems have been described in the CVC and surrounding area: the N-S-trending Colima Rift, the NE-SW-trending Tamazula Fault, and a set of E-W-trending faults generated by the southward spreading of the volcanic complex. The lasts two (Tamazula Fault and spreading faults) are different interpretations of the same set of E-W/NE-SW-trending faults cropping out on the CVC flanks (Garduño-Monroy et al. 1998; Norini et al. 2010);
2. The southward spreading of the volcanic edifice, driven by the load of the volcanic pile above the weak sedimentary fill of the Colima Rift, is controlled by the geometry and rheology of the involved geologic bodies (Norini et al. 2010). The numerical modelling presented in Sect. 5 suggests that this gravitational spreading is a basic active process in the evolution of the CVC;
3. The structural architecture of the volcanic complex was shaped by the interaction between the activity of the regional Colima Rift and the gravitational spreading of the volcano along the E-W-trending fault system. The role of the Tamazula Fault seems to be more controversial and not clearly identifiable on the volcano and its basement (e.g. Ferrari et al. 2012);
4. The two orthogonal Colima Rift and spreading-related fault systems are active, and to some extent independent one from the other, because they originated at different levels in the crust and have different dynamics;
5. The crustal structures of the N-S-trending Colima Rift control the rise of magma from the >5 km deep magma chamber of the CVC, and consequently the eruptions of the volcanic complex. The same tectonic structures and the associated regional stress field control the orientation of the N-S-trending dikes feeding the mantle-derived magmas of the monogenetic cones;
6. The southward catastrophic instability of the Paleofuego-Volcán de Colima cones is mainly promoted by the differential confinement of the Colima Rift and the inclined southern Nevado de Colima flank.

The understanding of the structural control on the CVC evolution and volcanic activity is of paramount importance, not only for the hazard evaluation in the surrounding area, but also because the CVC is an important natural laboratory for the development of general models of volcano-tectonic interaction. The few works published on this important topic show that different structures are interacting in the CVC area, providing an important control on the volcanic activity and gravitational instability of the volcanic edifices.

In our opinion, the next steps to develop knowledge of the CVC should be based on: (1) continuous monitoring of active deformation of the volcano and its surroundings, to track the active volcanic spreading and tectonic processes; (2) geophysical exploration of the volcano interior and its basement to detect the main faults and the feeding system; (3) integration of geophysical, geological and petrological-geochemical data in the light of the known tectonic and volcano-tectonic arrangement of the CVC; (4) analogue and numerical modelling to test the proposed structural models, to assess the possible future structural evolution of the CVC, and to evaluate the volcanic hazard implications.

Acknowledgements We acknowledge Abel Cortés, Lucia Capra, Juan Carlos Gavilanes and Nick Varley for useful discussions on the geological and structural evolution of the CVC. This work was financed by a CONACyT grant to Gianluca Norini (CB-2009-132265). The Ministry of Foreign Affairs of Italy and SRE of Mexico provided travel assistance to Gianluca Groppelli and Gianluca Norini. Marina Manea and Vlad Manea kindly provided the GeoWall 3D Visualization tool at the CGEO-UNAM. We thank Sonia Calvari, Victor Hugo Garduño-Monroy and Charles Connor for their helpful reviews that improved the manuscript.

References

Acocella, V., Neri, M., Norini, G.: An overview of experimental models to understand a complex volcanic instability: application to Mount Etna, Italy. J. Volcanol. Geoth. Res. **251**, 98–111 (2013)

Allan, J.F.: Sediment depth in the NCG from 3-D interpretation of gravity. Geofis. Int. **24**, 21–30 (1985)

Allan, J.F.: Geology of the Northern Colima and Zacoalco grabens, Southwest Mexico: Late Cenozoic rifting in the Mexican Volcanic Belt. Geol. Soc. Am. Bull. **97**, 473–485 (1986)

Allan, J.F., Nelson, S.A., Luhr, J.F., Charmichael, I.S.E., Wopat, M., Wallace, P.J.: Pliocene-Holocene rifting and associated volcanism in Southwest Mexico: an exotic terrane in the making. In: Dauphin, J.P., Simoneit, R.R.T. (eds.) The Gulf and Peninsular Provinces of the Californias, AAPG Mem., vol. 47, pp. 425–445 (1991)

Andrews, V., Stock, J., Ramirez Vázquez, C.A., Reyes-Davila, G.: Double-difference relocation of the Aftershocks of the Tecoman, Colima, Mexico Earthquake of 22 January 2003. Pure. appl. Geophys. **168**, 1331–1338 (2011)

Apuani, T., Corazzato, C., Cancelli, A., Tibaldi, A.: Stability of a collapsing volcano (Stromboli, Italy): Limit equilibrium analysis and numerical modelling. J. Volcanol. Geotherm. Res. **144**, 191–210 (2005)

Atlas, Z.D., Dixon, J.E., Sen, G., Finny, M., Martin-Del Pozzo, A.L.: Melt inclusions from Volcan Popocatepetl and Volcan de Colima, Mexico: melt evolution due to vapor-saturated crystallization during ascent. J. Volcanol. Geoth. Res. **153**, 221–240 (2006)

Bandy, W.L., Mortera-Gutiérrez, C.A., Urrutia-Fucugauchi, J.: Gravity field of the southern Colima graben, Mexico. Geofis. Int. **32**, 561–567 (1993)

Bandy, W.L., Mortera-Gutiérrez, C.A., Urrutia-Fucugauchi, J., Hilde, T.W.C.: The subducted Rivera-Cocos plate boundary: where is it, what is it, and what is its relationship to the Colima Rift? Geophys. Res. Lett. **22**, 3075–3078 (1995)

Bandy, W., Michaud, F., Bourgois, J., Calmus, T., Dyment, J., Mortera-Gutiérrez, C.A., Ortega-Ramírez, J., Pontois, B., Royer, J.Y., Sichlerg, B., Sosson, M., Rebolledo-Vieyra, M., Bigot-Cormier, F., Díaz-Molina, O., Hurtado-Artunduaga, A.D., Pardo-Castro, G., Trouillard-Perrot, C.: Subsidence and strike-slip tectonism of the upper continental slope off Manzanillo. Mexico Tectonophys. **398**, 115–140 (2005)

Barrier, B., Bourgois, J., Michaud, F.: The active Jalisco triple junction rift system. C.R. Acad. Sci. Paris, 310 (II), 1513–1520 (1990)

Bieniawski, Z.T.: Engineering Rock Mass Classifications: A Complete Manual for Engineers and Geologists in Mining, Civil and Petroleum Engineering, p. 251. Wiley, New York (1989)

Borgia, A.: Dynamic basis of volcanic spreading. J. Geophys. Res. **99**, 17791–17804 (1994)

Borgia, A., van Wyk de Vries, B.: The volcano-tectonic evolution of Concepción, Nicaragua. Bull. Volcanol. **65**, 248–266 (2003)

Borgia, A., Delaney, P.T., Denlinger, R.P.: Spreading volcanoes. Annu. Rev. Earth Planet. Sci. **28**(1), 539–570 (2000)

Bourgois, J., Renard, V., Auboin, J., Bandy, W., Barrier, E., Calmus, T., Carfantan, J.C., Guerrero, J., Mammerickx, J., Mercier De Lepinay, B., Michaud, F., Sosson, M.: Active fragmentation of the North American plate: offshore boundary of the Jalisco block off Manzanillo. C.R. Acad. Sci. Paris, 307(II), 1121–1130 (1988)

Bourgois, J., Michaud, F.: Active fragmentation of the North America plate at the Mexican triple junction area off Manzanillo. Geo-Mar. Lett. **11**, 59–65 (1991)

Cabrera-Gutiérrez, R., Espíndola, J.M.: The 1998–1999 eruption of Volcán de Colima, Mexico: an application of Maeda's viscoelastic model. Geofísca Internacional **49**(2), 83–96 (2010)

Capra, L., Macías, J.L.: The cohesive Naranjo debris-flow deposit (10 km3): a dam breakout flow derived from the Pleistocene debris avalanche deposit of Nevado de Colima Volcano (México). J. Volcanol. Geoth. Res **117**, 213–235 (2002)

Capra, L., Gavilanes-Ruiz, J.C., Bonasia, R., Saucedo-Giron, R.: Re-assessing volcanic hazard zonation of Volcán de Colima. México. Nat. Hazards **76**, 41–61 (2014)

Capra, L., Macías, J.L., Cortés, A., Dávila, N., Saucedo, R., Osorio-Ocampo, S., Arce, J.L., Gavilanes-Ruiz, J. C., Corona-Chávez, P., García-Sánchez, L., Sosa-Ceballos, G., Vázquez, R.: Preliminary report on the July 10–11, 2015 eruption at Volcán de Colima: Pyroclastic density currents with exceptional runouts and volume. J. Volcanol. Geoth. Res **310**, 39–49 (2016)

Carmichael, I.S.E., Frey, H.M., Lange, R.A., Hall, C.M.: The pleistocene cinder cones surrounding Volcan Colima, Mexico re-visited: eruption ages and volumes, oxidation states, and sulfur content. Bull. Volcanol. **68**, 407–419 (2006)

Centeno-García, E., Busby, C., Busby, M., Gehrels, G.: Evolution of the Guerrero composite terrane along the Mexican margin, from extensional fringing arc to contractional continental arc. Geol. Soc. Am. Bull. **123**, 1776–1797 (2011)

Cortés, A., Garduño, V.H., Navarro, C., Komorowski, J. C., Saucedo, R., Macías, J.L., Gavilanes, J.C.: Carta Geológica del Complejo Volcánico de Colima, Con Geología del Complejo Volcánico de Colima. Cartas Geológicas Y Mineras, Universidad Nacional Autónoma de México, 10 (2005)

Cortés, A., Garduño, V.H., Macías, J.L., Navarro-Ochoa, C., Komorowski, J.C., Saucedo, R., Gavilanes, J.C.: Geologic mapping of the Colima volcanic complex (Mexico) and implications for hazard assessment. In: Groppelli, G., Viereck-Goette, L. (eds.) Stratigraphy and Geology of volcanic areas, GSA Special Paper, vol. 464, pp. 249–264 (2010)

Cortés, A., Macías, J.L., Capra, L., Garduño-Monroy, V. H.: Sector collapse of the SW flank of Volcán de Colima, México. The 3600 yr BP La Lumbre-Los Ganchos debris avalanche and associated debris flows. J. Volcanol. Geotherm. Res **197**, 52–66 (2009)

Del Potro, R., Hürlimann, M.: Geotechnical classification and characterization of materials for stability analyses of large volcanic slopes. Eng. Geol. **98**(1), 1–17 (2008)

Demant, A.: Interpretación geodinámica del volcanismo del Eje Neovolcánico Transmexicano. Revista Instituto de Geología **5**, 217–222 (1981)

DeMets, C., Stein, S.: Present-day kinematics of the Rivera plate and implications for tectonics in southwestern Mexico. J. Geophys. Res. **95**, 21931–21948 (1990)

Domínguez, T., Zobin, V.M., Reyes-Davila, G.A.: The fracturing in volcanic edifice before an eruption: the June–July 1998 high-frequency earthquake swarm at Volcán de Colima, Mexico. J. Volcanol. Geotherm. Res. **105**, 65–75 (2001)

Escudero, C.R., Bandy, W.L.: Ambient seismic noise tomography of the Colima Volcano Complex. Bull. Volcanol. **79**, 13 (2017)

Ferrari, L.: Miocene shearing along the northern boundary of the Jalisco block and the opening of the southern Gulf of California. Geology **23**, 751–754 (1995)

Ferrari, L., Pasquarè, G., Venegas, S., Romero, F.: Geology of the western Mexican Volcanic Belt and adjacent Sierra Madre Occidental and Jalisco block. In: Delgado-Granados, H., Aguirre-Díaz, G., Stock, J. (eds.), Cenozoic Tectonics and Volcanism of Mexico: Geological Society of America Special Paper, 334, pp. 65–84 (2000)

Ferrari, L., Rosas-Elguera, J., Orozco-Esquivel, T., Carrasco-Nuñez, G., Norato-Cortez, T.: Digital Geologic Cartography of the Trans-Mexican Volcanic Belt and Adjoining Areas. Universidad Nacional Autonoma de México http://digitalgeosciences.unam.mx/maps/TMVB_MXD/viewer.htm (2005)

Ferrari, L., Rosas-Elguera, J.: Late Miocene to Quaternary extension at the northern boundary of the Jalisco block, western Mexico: the Tepic–Zacoalco rift revised. In: Aguirre-Díaz, G., Delgado-Granados, H., Stock, J. (eds.), Cenozoic Tectonics and Volcanism of Mexico: Geological Society of America Special Paper, 334, pp. 42–64 (2000)

Ferrari, L., Garduño, V.H., Pasquarè, G., Tibaldi, A.: Volcanic and tectonic evolution of central Mexico: Oligocene to present. Geofis. Int. **33**, 91–105 (1994)

Ferrari, L., Orozco-Esquivel, T., Manea, V., Manea, M.: The dynamic history of the Trans-Mexican Volcanic Belt and the Mexico subduction zone. Tectonophysics **522–523**, 122–149 (2012)

Frey, H.M., Lange, R.A., Hall, C.M., Delgado-Granados, H., Carmichael, I.S.E.: A Pliocene ignimbrite flare-up along the Tepic-Zacoalco rift: evidence for the initial stages of rifting between the Jalisco Block (Mexico) and North America. Geol. Soc. Am. Bull. **119**, 49–64 (2007)

Garduño, V., Tibaldi, A.: Kinematic evolution of the continental active triple junction of the western Mexican Volcanic Belt. Comptes Rendus de l'Académie des Sciences, Paris **2**, 135–142 (1991)

Garduño-Monroy, V.H., Saucedo-Girón, R., Jiménez, Z., Gavilanes-Ruiz, J.C., Cortés-Cortés, A., Uribe-Cifuentes, R.M.: La Falla Tamazula, límite suroriental del Bloque Jalisco, y sus relaciones con el Complejo Volcánico de Colima, México. Revista Mexicana de Ciencias Geológicas **15**(2), 132–144 (1998)

Goodman, R.E., Taylor, R.L., Brekke, T.L.: A model for mechanics of jointed rock. J. Soil Mech. Found. Div. **94**, 637–659 (1968)

Gudmundsson, A.: Rock Fractures in Geological Processes. Cambridge University Press, Cambridge, 592 pp., ISBN: 978-0-521-86392-6 (2011)

Hammah, R.E., Yacoub, T., Corkum, B., Curran, J.H.: The practical modelling of discontinuous rock masses with Finite Element analysis. In: Proceedings of the 42nd US Rock Mechanics Symposium and 2nd U.S.-Canada Rock Mechanics Symposium, San Francisco, paper 08-180, American Rock Mechanics Association (2008)

Hoek, E., Brown, E.T.: Practical estimates or rock mass strength. Int. J. Rock Mech. Min. Sci. **34**, 1165–1186 (1997)

Hoek, E., Kaiser, P.K., Bawden, W.F.: Support of Underground Excavations in Hard Rock, p. 215. Balkema, Canberra (1995)

Hürlimann, M., Ledesma, A., Martí, J.: Characterisation of a volcanic residual soil and its implications for large landslide phenomena: application to Tenerife, Canary Islands. Eng. Geol. **59**, 115–132 (2001)

Komorowski, J.C., Navarro, C., Cortés, A., Saucedo, R., Gavilanes, J.C., Siebe, C., Espíndola, J.M., Rodríguez-Elizarrarás, S.: The Colima volcanic complex: part I: quaternary multiple debris-avalanche deposits, part II: Historical pyroclastic sequences (1913, 1991, 1994), Special Publication of the Organizing Committee of the IAVCEI (International Association of Volcanology and Chemistry of the Earth's Interior), Puerto Vallarta, México, General Assembly of the IAVCEI, Excursion Guidebook, p. 1–38 (1997)

López-Loera, H., Urrutia-Fucugauchi, J., Alva-Valdivia, J.L.M.: Magnetic characteristics of fracture zones and constraints on the subsurface structure of the Colima Volcanic Complex, western Mexico. Geosphere **6**, 35–46 (2010)

López-Loera, H., Urrutia-Fucugauchi, J., Alva-Valdivia, L.: Estudio aeromagnético del complejo volcánico de Colima, occidente de México – implicaciones tectónicas y estructurales. Revista Mexicana de Ciencias Geológicas **28**(2011), 349–370 (2011)

Luhr, J.: Extensional tectonics and the diverse primitive volcanic rocks in the western Mexican Volcanic Belt. Can. Mineral. **35**, 473–500 (1997)

Luhr, J.F.: Petrology and geochemistry of the 1991 and 1998–1999 lava flows from Volcán de Colima, Mexico: implications for the end of the current eruptive cycle. J. Volcanol. Geotherm. Res. **117**, 169–194 (2002)

Luhr, J.F., Carmichael, I.S.E.: The Colima Volcanic complex, Mexico: part II. Late quaternary cinder cones. Contrib. Mineral. Petrol. **76**, 127–147 (1981)

Luhr, J.F., Carmichael, I.S.E.: Geology of Volcán de Colima, Universidad Nacional Autonma Mexico. Inst. Geol. Bol. **107**, 1–101 (1990)

Luhr, J.F., Prestegaard, K.L.: Caldera formation at Volcán de Colima, Mexico, by large Holocene volcanic debris avalanche. J. Volcanol. Geotherm. Res. **35**, 335–348 (1988)

Luhr, J.F., Nelson, S.A., Allan, J.F., Carmichael, I.S.E.: Active rifting in southwestern Mexico: manifestations of an incipient eastward spreading-ridge jump. Geology **13**, 54–57 (1985)

Macías, J.L., Saucedo, R., Gavilanes, J.C., Varley, N., Velasco, García S., Bursik, M.I., Vargas, Gutiérrez V., Cortés, A.: Flujos piroclásticos asociados a la actividad explosiva del Volcán de Colima y perspectivas futuras. GEOS **25**(3), 340–351 (2006)

Marinos, P., Hoek, E.: GSI: a geologically friendly tool for rock mass strength estimation. In: Proc. GeoEng2000 Conference, Melbourne, 1422–1442 (2000)

Medina-Martínez, F., Espíndola, J.M., De la Fuente, M., Mena, M.: A gravity model of the Colima, México region. Geofis. Int. **35**(4), 409–414 (1996)

Michaud, F., Bourgois, J., Aubouin, J.: Fragmentation continentale en cours dans la zone d'affrontement entre le continent Nord-Américain et la ride d'accrétion océanique de Pacifique est (Mexique). Oceanol. Acta. **10**, 87–96 (1990)

Michaud, F., Quintero-Legorreta, O., Barrier, E., Burgois, J.: La frontiére Nord du Bloc Jalisco (ouest Mexique): localisation et évolution de 13 Ma á l'actuel. Comptes Rendus de l'Académie des Sciences, Paris, II **312**, 1359–1365 (1991)

Moon, V., Bradshaw, J., Smith, R., de Lange, W.: Geotechnical characterization of stratocone crater wall sequences, White Island Volcano, New Zealand. Eng. Geol. **81**, 146–178 (2005)

Moore, G., Marone, C., Carmichael, I.S.E., Renne, P.: Basaltic volcanism and extension near the intersection of the Sierra Madre volcanic province and the Mexican volcanic belt. Geol. Soc. Am. Bull. **106**, 383–394 (1994)

Mooser, F.: Los volcanes de Colima. Univ. Nac. Mexico. Geol. Bol. **61**, 49–71 (1961)

Muller, H.O., Pollard, D.D.: The stress state near Spanish peaks, Colorado Determined From a Dike Pattern. Pageoph **115**, 69–86 (1977)

Navarro, C., Komorowski, J.C., Cortés, A.: Depósitos de avalanchas de escombros en el Complejo Volcánico de Colima: evidencias geológicas y edades C14 para múltiples eventos. Nuevas preguntas a previas interpretaciones, un problema aún no resuelto. Universidad de Colima, Cuarta Reunión Internacional Volcán de Colima: A Decade Volcano Workshop, Colima, pp. 49–50, 24–28 Jan 1994

Nieto-Obregón, J., Delgado-Argote, L., Damon, P.: Geochronologic, petrologic and structural data related to large morphologic features between the Sierra Madre Occidental and the Mexican Volcanic Belt. Geofísica Internacional **24**, 623–663 (1985)

Nixon, G.T.: The relationship between Quaternary volcanism in central Mexico and the seismicity and structure of subducted oceanic lithosphere. Geol. Soc. Am. Bull. **93**, 514–523 (1982)

Norini G., Acocella V.: Analogue modelling of flank instability at Mount Etna: understanding the driving factors. J. Geophys. Res. Solid Earth, **116**, 1–21, B07206 (2011)

Norini G., Capra L., Groppelli G., Agliardi F., Pola A., Cortés A.: Structural architecture of the Colima Volcanic Complex. J. Geophys. Res. Solid Earth, 115, 1–20, B12209 (2010)

Norini, G., Lagmay, A.M.F.: Deformed symmetrical volcanoes. Geology **33**, 605–608 (2005)

Norini, G., Capra, L., Groppelli, G., Lagmay, A.M.F.: Quaternary sector collapses of Nevado de Toluca volcano (Mexico) as consequence of regional tectonics and volcanic evolution. Geosphere **4**, 854–871 (2008)

Núñez-Cornú, F.J., Sanchez-Mora, C.: Stress field estimations for Colima Volcano, Mexico based on Seismic data. Bull. Volc. **60**, 568–580 (1999)

Pacheco, J.F., Bandy, W., Reyes-Dávila, G.A., Núñez-Cornú, F.J., Ramírez-Vázquez, C.A., Barrón, J.R.: The Colima, Mexico earthquake (Mw 5.3) of March 7, 2000: seismic activity along the SCR. Bull. Seismol. Soc. Am. **93**, 1458–1467 (2003)

Pardo, M., Suárez, G.: Steep subduction geometry of the Rivera plate beneath the Jalisco block in western Mexico. Geophys. Res. Lett. **20**, 2391–2394 (1993)

Pola, A., Crosta, G.B., Castellanza, R., Agliardi, F., Fusi, N., Barberini, V., Norini, G.: Relationships between porosity and physical mechanical properties in weathered volcanic rocks. In: Olalla, C., Hernández, L.E., Rodríquez-Losada, J.A., Perucho, A., Gonzales-Gallego, J. (eds.) Rock Mechanics and Geo-engineering in Volcanic Environments, p. 9780415584787. CRC Press/Balkema, ISBN (2010)

Pola, A., Crosta, G., Fusi, N., Barberini, V., Norini, G.: Influence of alteration on physical properties of volcanic rocks. Tectonophysics **566–567**, 67–86 (2012)

Quintero-Legorreta, O., Michaud, F., Bourgois, J., Barrier, E.: Evolución de la frontera septentrional del

bloque Jalisco, México, desde hace 17 Ma. Revista Instituto de Geología **10**, 111–117 (1992)
Reid, M.E., Sisson, T.W., Brien, D.L.: Volcano collapse promoted by hydrothermal alteration and edifice shape, Mount Rainier, Washington. Geology **29**, 779–782 (2001)
Reubi, O., Blundy, J.: Assimilation of plutonic roots, formation of high-K 'exotic' melt inclusions and genesis of andesitic magmas at Volcán De Colima, Mexico. J. Petrol. **49**, 2221–2243 (2008)
Riahi, A., Hammah, R.E. Curran, R.E.: Limits of applicability of the finite element explicit joint model in the analysis of jointed rock problems. In: Proceedings of the 44th U.S. Symposium on Rock Mechanics and the 5th U.S.-Canada Rock Mechanics Symposium, Salt Lake City, Utah, USA, June 2010, paper ARMA 10–336 (2010)
Robin, C., Mossand, P., Camus, G., Cantagrel, J.M., Gourgaud, A., Vincent, P.M.: Eruptive history of the Colima volcanic complex (Mexico). J. Volcanol. Geotherm. Res. **31**, 99–113 (1987)
Rodríguez-Elizarraras, S.R.: Estratigrafía y estructura del Volcán de Colima, México. Rev. Mex. Cienc. Geol. **12**, 22–46 (1995)
Rosas-Elguera, J., Urrutia-Fucugauchi, J.: Tectonic control on the volcano-sedimentary sequence of the Chapala graben, western Mexico. Int. Geol. Rev. **40**, 350–362 (1998)
Rosas-Elguera, J., Urrutia-Fucugauchi, J., Maciel, R.: Geología del extremo oriental del Graben de Chapala; breve discusión sobre su edad: Zonas geotérmicas Ixtlan de Los Hervores-Los Negritos, México. Geotermia-Revista Mexicana de Geoenergía **5**, 3–18 (1989)
Rosas-Elguera, J., Ferrari, L., Garduño-Monroy, V.H., Urrutia-Fucugauchi, J.: Continental boundaries of the Jalisco block and their influence in the Pliocene-Quaternary kinematics of western Mexico. Geology **24**, 921–924 (1996)
Roverato, M., Capra, L., Sulpizio, R., Norini, G.: Stratigraphic reconstruction of two debris avalanche deposits at Colima volcano (Mexico): insights into pre-failure conditions and climate influence. J. Volcanol. Geotherm. Res. **207**, 33–46 (2011)
Schaaf, P., Moran-Zenteno, D., Hernandez-Bernal, M., Solis-Pichardo, G., Tolson, G., Kohler, H.: Paleogene continental margin truncation in southwestern Mexico: geochronological evidence. Tectonics **14**, 1339–1350 (1995)
Selvans, M., Stock, J., DeMets, C., Sanchez, O., Márquez-Azua, B.: Constraints on Jalisco Block motion and tectonics of the Guadalajara triple junction from 1998–2001 Campaign GPS Data. Pure. appl. Geophys. **168**, 1435–1447 (2011)
Serpa, L., Smith, S., Katz, C., Skidmore, C., Sloan, R., Pavlis, T.: A geophysical investigation of the southern Jalisco block in the state of Colima, Mexico. Geofisica Internacional **31**, 247–252 (1992)
Stoopes, G.R., Sheridan, M.F.: Giant debris avalanches from the Colima Volcanic Complex, Mexico: implication for long-runout landslides (>100 km). Geology **20**, 299–302 (1992)
Suárez, G., Garcia-Acosta, V., Gaulon, R.: Active crustal deformation in the Jalisco block, Mexico: evidence for a great historical earthquake in the 16th century. Tectonophysics **234**, 117–127 (1994)
Suter, M., Quintero, O., Johnson, C.: Active faults and state of stress in the central part of the Trans-Mexican Volcanic Belt 1: The Venta de Bravo fault. J. Geophys. Res. **97**, 11983–11994 (1992)
Suter, M., López-Martinez, M., Quintero-Legorreta, O., Carrillo Martinez, M.: Quaternary intra-arc extension in the Central TMVB. Bull. Geol. Soc. Am. **113**, 693–703 (2001)
Urrutia-Fucugauchi, J., Rosas-Elguera, J.: Paleomagnetic study of the eastern sector of Chapala Lake and implications for the tectonics of west-central Mexico. Tectonophysics **239**, 61–71 (1994)
Valencia, V.A., Ducea, M., Talavera-Mendoza, O., Gehrels, G., Ruiz, J., Shoemaker, S.: U-Pb geochronology of granitoids in the north-western boundary of the Xolapa terrane. Revista Mexicana de Ciencias Geológicas **26**(1), 189–200 (2009)
van Wyk de Vries, B., Borgia, A.: The role of basement in volcano deformation. In: McGuire, W.J., et al., Volcano Instability on the Earth and Other Planets. Geol. Soc. Sp. Pub. **110**, 95–110 (1996)
van Wyk de Vries, B., Matela, R.: Styles of volcano-induced deformation: numerical models of substratum flexure, spreading and extrusion. J. Volcanol. Geotherm. Res. **81**, 1–18 (1998)
Vidal, N., Merle, O.: Reactivation of basement faults beneath volcanoes: a new model for flank collapse. J. Volcanol. Geotherm. Res. **99**, 9–26 (2000)
Voight, B., Elsworth, D.: Failure of volcano slopes. Geotechnique **47**, 1–31 (1997)
Walter T.R., Acocella V., Neri M., Amelung F.: Feedback processes between magmatism and E-flank movement at Mt. Etna (Italy) during the 2002–2003 eruption. J. Geophys. Res. **110**, B10205 (2005)
Watters, R.J., Zimbelman, D.R., Bowman, S.D., Crowley, J.K.: Rock mass strength assessment and significance to edifice stability, Mount Rainer and Mount Hood, Cascade Range Volcanoes, pure and applied. Geophysics **157**, 957–976 (2000)
Wooller, L., van Wyk, B., de Vries, J.B., Murray, H. Rymer, Meyer, S.: Volcano spreading controlled by dipping substrata. Geology **32**, 573–576 (2004)
Yang, T., Grand, S.P., Wilson, D., Guzmán-Speziale, M., Gómez-González, J.M., Domínguez-Reyes, T., Ni, J.: Seismic structure beneath the Rivera subduction zone from finite-frequency seismic tomography. J. Geophys. Res. **114**, B01302 (2009)
Zamora-Camacho, A., Espíndola, J.M., Reyes-Davila, G.: The 1997–1998 Activity of Volcán de Colima, Western Mexico: some aspects of the associated seismic activity.

Pure Appl. Geophys. **164**, 39–52 (2007)
Zimbelman, D.R., Watters, R.J., Firth, I.R., Breit, G.N., Carrasco-Nunez, G.: Stratovolcano stability assessment methods and results from Citlaltépetl, Mexico. Bull. Volcanol. **66**, 66–79 (2004)
Zobin, V.M., Luhr, J.F., Taran, Y.A., Bretón, M., Cortés, A., De la Cruz-Reyna, S., Domínguez, T., Galindo, I., Gavilanes, J.C., Muñiz, J.J., Navarro, C., Ramírez, J. J., Reyes, G.A., Ursúa, M., Velasco, J., Alatorre, E., Santiago, H.: Overview of the 1997–2000 activity of Volcán de Colima, Mexico. J. Volcanol. Geotherm. Res. **117**, 1–19 (2002)
Zobin, V.M., Arambula-Mendoza, R., Breton, M., Reyes, G., Placencia, I., Navarro, C., Téllez, A., Campos, A., González, M., León, Z., Martínez, A., Ramírez, C.: Dynamics of the January 2013–June 2014 explosive–effusive episode in the eruption of Volcán de Colima, México: insights from seismic and video monitoring. Bull. Volcanol. **77**, 31–44 (2015)

Late Pleistocene-Holocene Debris Avalanche Deposits from Volcán de Colima, Mexico

A. Cortés, J.-C. Komorowski, J. L. Macías, L. Capra and P. W. Layer

Abstract

Volcán de Colima has experienced numerous partial edifice collapses with associated debris avalanche deposits, widely distributed toward the SW, S and SE sectors. According to new $^{40}Ar/^{39}Ar$ dates, activity began more than 97,000 years ago on the southern flank of Nevado de Colima with the formation of the so-called Paleofuego edifice. Several collapses occurred prior to a catastrophic edifice collapse event 7000 years ago, creating a horseshoe-shaped avalanche crater, 5 km wide, opened towards the south. After this last lateral collapse of Paleofuego, the currently active cone began to grow in the central part of the crater, which, during the Late Holocene, has experienced two lateral collapse events that generated debris avalanches. Based on new fieldwork and stratigraphic correlation of deposits supported by ^{14}C dates, we present a description and approximate distribution of eight debris avalanche deposits generated by Volcán de Colima during the last 30,000 years. These debris avalanche deposits are exposed at 40 km to the S and 25 km in the SW and SE sectors of the volcano, and cover an area of 1200 km^2 with an approximated volume of 86 km^3. Field evidence indicates that at least some sector collapses were accompanied by magmatic activity. The regional tectonic setting that consists of the active N-S regional extension of the Colima graben, as well as E-W, and NE-SW structures, such as the Tamazula Fault, also played an important role in volcano instability. The contribution of a volcanic spreading component was also recently recognized. The emplacement of the most voluminous debris avalanches have obstructed the Armería and Tuxpan-Naranjo rivers producing temporary lakes, where thick lacustrine sediments were accumulated. The recurrence times of these sector collapses vary between 3000 and 6500 years during the Late Pleistocene and 1100–3400 years during the Holocene, with

A. Cortés (✉)
Centro Universitario de Estudios e Investigaciones de Vulcanología, Universidad de Colima, Colima, Mexico
e-mail: abel31@hotmail.com

J.-C. Komorowski
Institut de Physique du Globe de Paris, CNRS, 4 Place Jussieu, B89, Équipe de Volcanologie, 75252 Paris Cedex 05, France

J. L. Macías
Instituto de Geofísica, Unidad Michoacán, Universidad Nacional Autónoma de México, Campus-Morelia, Michoacán 58090 Morelia, Mexico

L. Capra
Centro de Geociencias, UNAM, Campus Juriquilla, 76230 Querétaro, Mexico

P. W. Layer
Geophysical Institute and College of Natural Science and Mathematics, University of Alaska Fairbanks, Fairbanks, AK 99775, USA

N. Varley et al. (eds.), *Volcán de Colima*, Active Volcanoes of the World,
https://doi.org/10.1007/978-3-642-25911-1_4

the youngest one having occurred ~2.5 ka BP. A future sector collapse of Volcán de Colima on the scale of past events could be catastrophic for up to about 350,000 inhabitants (including the city of Colima) that currently are located on top of ancient debris avalanche deposits.

Keywords
Colima Volcanic Complex · Nevado de Colima · Paleofuego · Volcán de Colima Debris avalanche deposits

1 Introduction

After the 1980 eruption of Mount St. Helens, the collapse of a volcanic edifice and the emplacement of the associated debris avalanche have been recognized as common events at numerous volcanoes around the world (Siebert 1984; Ui et al. 1986; Siebert et al. 1987; Vallance et al. 1995; Francis and Wells 1988). Since then, numerous debris avalanche deposits (DADs) have been identified at several volcanoes of the Trans-Mexican Volcanic Belt (TMVB; Fig. 1a) (Capra et al. 2002) that include: Cofre de Perote (Carrasco-Núñez et al. 2006; Popocatépetl (Robin and Boudal 1987; Siebe et al. 1995b), Pico de Orizaba (Carrasco-Núñez et al. 1993; Hoskuldsson and Robin 1993; Carrasco-Núñez and Gomez-Tuena 1997), Jocotitlán (Siebe et al. 1992a), Nevado de Toluca (Macías et al. 1997; Capra and Macías 2000), including the volcanoes of the Colima Volcanic Complex (CVC) (Fig. 1b), (Luhr and Carmichael 1982a, b, 1990; Luhr and Prestegaard 1988; Robin et al. 1987; Stoopes and Sheridan 1992; Rodríguez-Elizarrarás et al. 1995; Komorowski et al. 1994, 1997; Capra and Macías 2002; Cortés et al. 2005, 2010a, b; Roverato et al. 2011).

Debris avalanches are large volumes of rock, ash and soil that move at high speeds, generated by the partial collapse of volcanic (or non-volcanic) edifices (Ui 1983; Siebert 1984; Schuster and Crandell 1984; Siebert et al. 1987). In a few minutes, they can drastically modify the morphology of the surrounding landscape, covering extensive areas and obstructing and filling water courses. Debris avalanche deposits consist of a poorly sorted mixture of brecciated debris. The most distinctive feature is their boulder-rich surface morphology, hummocky topography (mounds) with numerous small hills, longitudinal and transverse ridges and enclosed depressions (Ui et al. 1986; Glicken 1991, 1996) that often form temporary lakes where thick sequences of lacustrine sediments are deposited.

Debris avalanche deposits are characterized by two primary facies: block facies and mixed (matrix) facies (Glicken 1991; Ui et al. 2000). The block facies is composed mainly of debris avalanche blocks, which include large unconsolidated pieces of the volcano transported to its place of deposition with little disruption. The mixed facies consists primarily of debris avalanche matrix and is a mixture of clasts and interclast matrix and may contain clasts of all rock types and of sizes from micrometres to metres. Facies analysis was also recently used to better describe geometrical relationships and relative abundance among their constituents (e.g. Mehl and Schmincke 1999; Roverato et al. 2011).

Debris avalanches are events that potentially represent the most danger for the people living within the zone of a volcano´s influence. Fortunately, they are not commonly occurring and as such are often ignored during risk analysis. In this chapter, we present a summary of the existing information about the debris avalanche deposits (DADs) from Volcán de Colima, mainly those associated with Paleofuego (name sometimes given to previous edifice of Volcán de Colima) and the currently active cone. Also, we present new stratigraphic sections and their relationships along with new $^{40}Ar/^{39}Ar$ and ^{14}C dates that allow us to determine the distribution of some of these debris avalanche deposits and their sources. These new data provide us with a better understanding of the evolution of Volcán de Colima. It should be noted that debris avalanche deposits associated with the evolution of Nevado de Colima that have been discussed elsewhere (Robin et al. 1987; Stoopes and Sheridan 1992; Capra and Macias 2002) are not considered in this chapter.

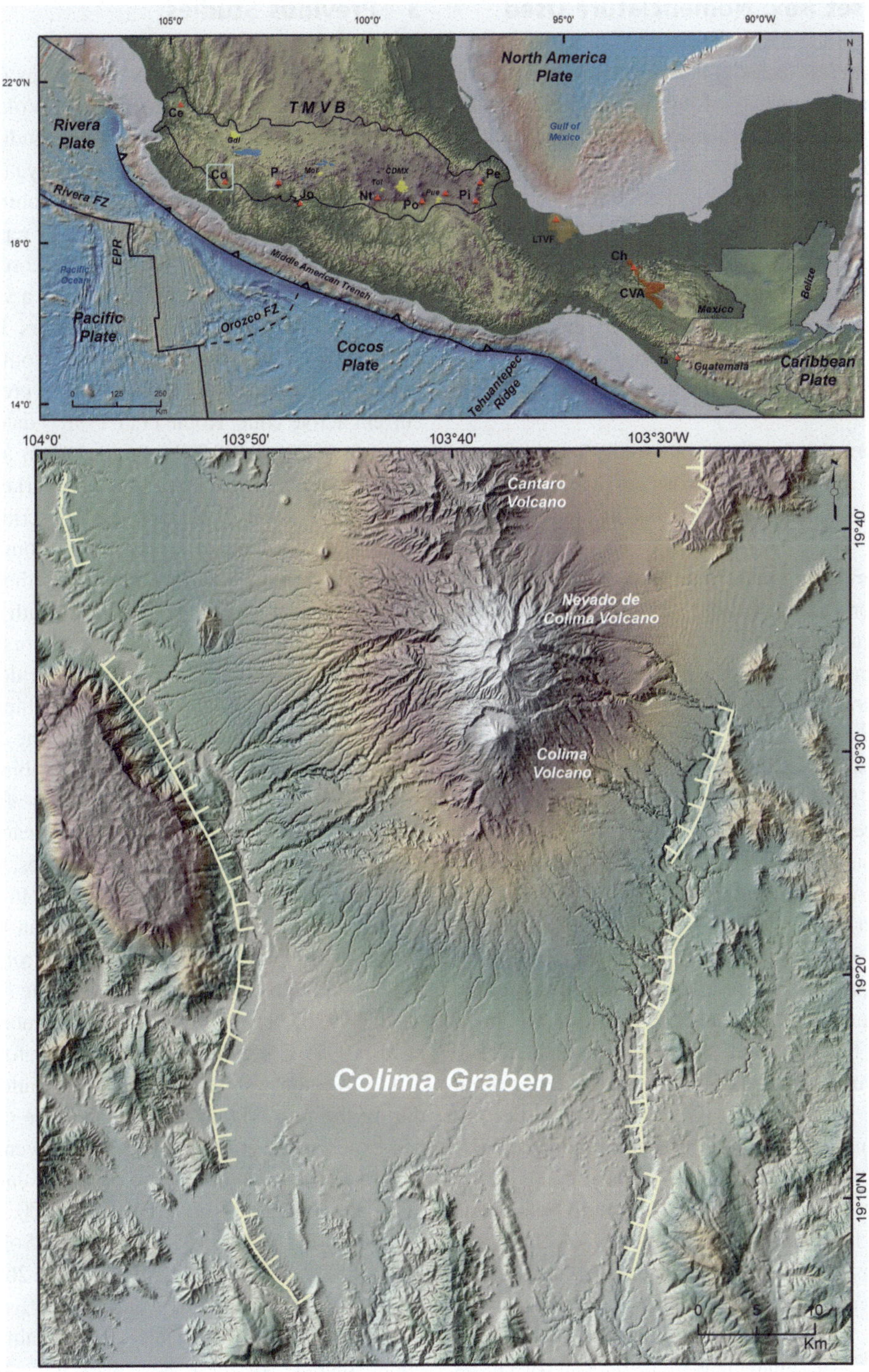

Fig. 1 **a** Location map of the Trans-Mexican Volcanic Belt (TMVB grey area) which shows the location of the Colima Volcanic Complex (CVC). Open triangles represent the stratovolcanoes that have collapsed on several occasions. Abbreviations are: NC, Nevado de Colima; VC, Volcán de Colima; Ta, Tancítaro; Jo, Jocotitlán; To, Nevado de Toluca; Iz, Iztaccíhuatl; Po, Popocatépetl; Ci, Citlaltépetl; Cp, Cofre de Perote; Co, Colima City; RFZ, Rivera Fracture Zone; OFZ, Orozco Fracture Zone; OGFZ, O´Gorman Fracture Zone (after García-Palomo et al. 2004). **b** Digital elevation model that shows the Colima Volcanic Complex within the Colima Graben. The solid yellow lines represent major faults of the Colima graben

2 Inset Box: Nomenclature Used

Block facies: facies of a debris avalanche deposit that mainly consists of debris avalanche blocks, which include large unconsolidated pieces of the volcano transported to its place of deposition with little disruption (Glicken 1991).

BP: Years before present, usually defined as prior to 1950.

Clast: A "clast" is defined as a rock of any size from microns to meters in diameter that would not break if passed through an appropriate sieve or immersed in water: a hard rock (Glicken 1991).

Debris avalanche: Is a rapidly moving incoherent and unsorted mass of rock and soil mobilized by gravity (Schuster and Crandell 1984).

Debris-avalanche block: An unconsolidated (or poorly consolidated) piece of the old mountain transported to its place of deposition (Glicken 1991).

Debris flow: Is a flowing mixture of debris and water having sediment concentration between 70 and 90% by weight (Pierson and Costa 1987).

Lahar: General term for rapidly flowing water-saturated mixture of rock debris and water from a volcano (Smith and Fritz 1989).

Matrix: All components of the deposit comprising material <2 mm in diameter.

Megaclast: All components of the deposit with a dimension from 2 mm to metres. Are often highly fractured and show pervasive shattering, sometimes exhibiting an irregular fracture pattern termed jigsaw cracks (Ui 1983) due to decompression during the sliding stage of movement (Ui et al. 1986; Komorowski et al. 1991).

Mixed facies: Consists primarily of debris avalanche matrix and is a mixture of clasts and interclast matrix and may contain clasts of all rock types and of sizes from micrometres to metres. This facies is prominent in the middle, distal and marginal areas of the deposit (distances are highly variable) (Siebert 1984; Schuster and Crandell 1984; Glicken 1991) and can still show hummocky morphology (Glicken 1991) with megaclasts surrounded by matrix.

3 Previous Studies

The presence of multiple debris avalanche deposits south of the active cone of Volcán de Colima has been a matter of multiple studies and certain controversy during the last 30 years (i.e. Cortés 2002 and references therein). Robin et al. (1987) were the first authors to report a 32 km-long DAD to the south, that according to them covered an area of 500 km^2 with a volume of 10–20 km^3 (Fig. 2a). These authors related the deposit to the 5 km-wide southward horseshoe-shaped scar, within which grew the current active cone. Robin et al. (1987) dated this event at 9370 ± 400 BP, obtained from a pyroclastic flow deposit that directly overlays the DAD. This same DAD was also described by Luhr and Prestegaard (1988) who found the deposit exposed up to 70 km towards the south and covering a surface of 1550 km^2 with a volume of ~10 km^3 (Fig. 2b). These authors dated a charcoal fragment found in fluviatile deposits below the DAD at 4280 ± 110 BP, which provided a revised maximum age for the event. The DAD studied by these authors (Robin et al. 1987; Luhr and Prestegaard 1988) formed superimposed aprons composed of volcanoclastic deposits (several debris avalanches, lahar, fluvio-lacustrine and pyroclastics). In fact, through the years, several authors began to discover new debris avalanche deposits from new fieldwork supplemented by ^{14}C dating (Siebe et al. 1992b; Navarro et al. 1994; Komorowski et al. 1997; Cortés et al. 2005; Roverato et al. 2011). Komorowski et al. (1997) concluded that during the past 45,000 years, Volcán de Colima had experienced at least nine collapse events that generated the same number of debris avalanches, with ages of 2550 +110/−105 BP; 3600 ± 120 BP; 7040 ± 160 BP; 9671 ± 88 BP; 15–16,000 BP; 18,520 ± 260 BP; 21,545 +265/−260 BP; 27,800 +4140/−2720 BP and 45,030 +1470/−125 BP (Table 1, the table also contains calibrated dates). All these new ages indicate that Volcán de Colima experienced a complex evolution with generation and destruction of volcanic edifices that generated debris avalanche deposits as previously thought by Komorowski

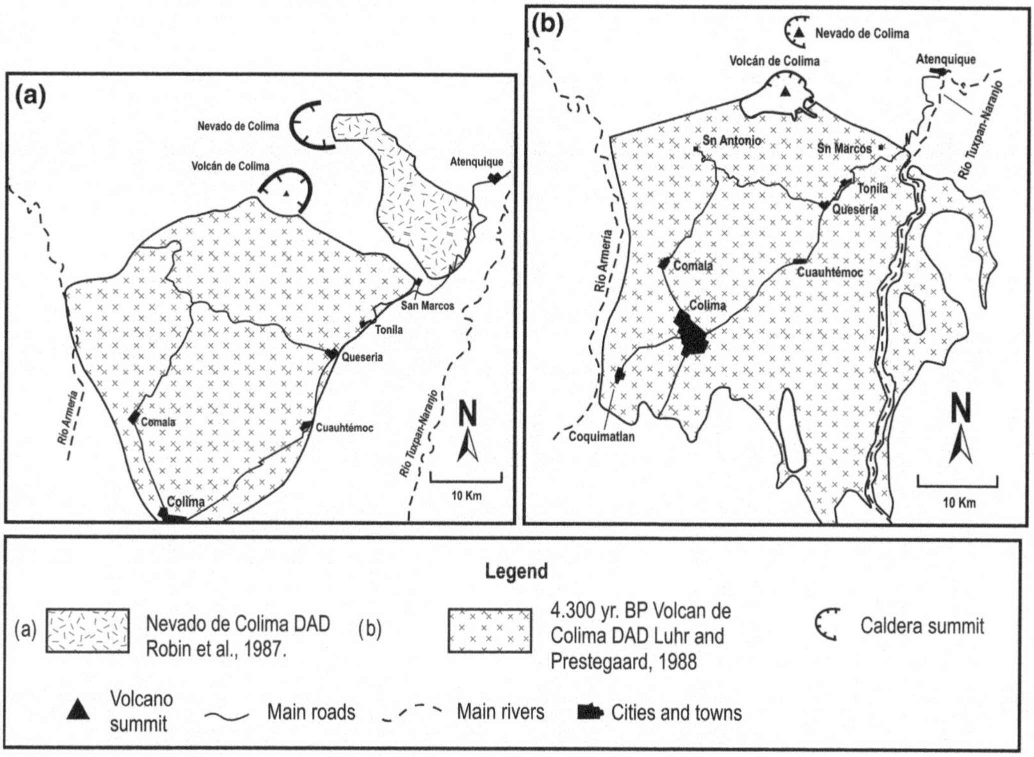

Fig. 2 Distribution of the CVC debris avalanche deposits according to: **a** Robin et al. (1987); **b** Luhr and Prestegaard (1988)

et al. (1997). These authors interpreted that the edifice remnants of the Paleofuego stage that they called "the Playón caldera", had been the result of repeated lateral collapses the youngest of which had occurred 2500 years ago. However, Cortés et al. (2010a) concluded that the DAD of 3600 ± 120 BP, had been generated by the sector collapse of the SW flank of the active cone of Volcán de Colima. This conclusion implied that the present Volcán de Colima edifice is older than 3600 BP New field evidence supports the idea that at least the two younger debris avalanche deposits found (3600 and 2500 BP) are associated with the current edifice of Volcán de Colima.

Based on the analysis of previous works of different authors, additional field evidence, and new radiocarbon and $^{40}Ar/^{39}Ar$ dates, in this chapter we present the distribution of eight DADs generated by Volcán de Colima (Fig. 3). These deposits dated back at least 30,000 years and have ages of 2550 +110/−105 BP; 3600 ± 120 BP; 7040 ± 160 BP; 9671 ± 88 BP; 15–16,000, 18,520 ± 260 BP; 21,545 +265/−260 BP and ≥ 28,000 BP (Table 1).

4 Birth of the Current Volcán de Colima Edifice

Although there were no published K-Ar dates of lavas from Paleofuego, Robin et al. (1987) suggested that this volcano had a maximum age of 50,000 years. Two lava flows from the Paleofuego stage were recently dated with the $^{40}Ar/^{39}Ar$ method at the geochronology laboratory of the University of Alaska at Fairbanks, following the methodology described by Layer et al. (2009) (Table 2). The whole rock analyses of these samples resulted in ages of 97 ± 13 ka (sample M19), and 22 ± 6 ka BP (sample M12). Based on these ages and new stratigraphic

Table 1 Summary of 14C radiocarbon ages obtained from charcoal found within debris avalanche deposits generated by Volcán de Colima during late Pleistocene-Holocene, as well as, from paleosols interbedded with these deposits, reported in previous studies and this work

Edifice	Site	Coordinates		Avalanche (DAD)	Laboratory number	Conventional date (yr BP)	Calibrated date (2σ) yr (AD 1950)	δ¹³CPDR (‰)	Deposit Type	Author
		N	W							
Current edifice of Volcán de Colima	24	19°26'24.03"	103°49'24.41"	El Remate-Armería	13621	1940±90	BC 173-AD 256	-24.5	LD	2
	25	19°22'05.78"	103°48'44.37"	“	13615	2440±40	BC 596-407	-25.8	DAD	1
	“	19°21'21.62"	103°48'34.30"	“	13616	2475±45	BC 772-428	-25.3	DAD	1
	“			“	13617	2505±45	BC 795-482	-23.8	DAD	2
	26			“	A-7094	2550 + 110/-105	BC 895-403	-26.6	DAD	3
	24	19°26'24.03"	103°49'24.41"	La Lumbre-Los Ganchos	A-7721	3600±120	BC 2293-1635	-26.3	DAD	4
	“				A-6956	3925±115	BC 2859-2809	-25.1	DAD	3
Paleofuego (ancient edifice of Volcán de Colima)	22	19°17'12.61"	103°44'30. 99"	Villa de Álvarez-Coquimatlán	A-6745	6990±130	BC 6086-5632	-27.2	DAD	4
	23	19°09'21.42"	103°47'29. 83"		A-7225	7040±160	BC 6223-5641	-27.1	DAD	3
					13623	8005±50	BC 7064-6750	-19.8	P	1
	19	19°23'59.97"	103°48'08. 68"	Mesa Yerbabuena	6139	9370±400	BC 10000-9984	-30.2	PFD	7
	20	19°22'21.02"	103°47'32. 19"		A-7230	9671±88	BC 9281-8806	-27.3	DAD	4
					A-7387	12040 + 155/-150	BC 12409-11522		P	3
	4	19°22'39.67"	103°33'29. 69"	Tonila	A-7647	12,460±70	14,750±160	-23.8	P	6
	16	19°20'43.4"	103°45'34. 66"		A-7641	13,585±135	17,040±140	-22.9	DAD	6
	17	19°20'46.86"	103°45'53. 80"		A-7645	15330 + 965/860	BC 18486-13866	-26.2	DAD	3
						15626 + 285/-275	BC 17509-16518			3
						15675±125	BC 18618-16852			3
	12	19°18'19.01"	103°37'09. 13"	Cuauhtémoc-Aeropuerto	A-7062	17960 + 560/-525	BC 20585-18035	-26.9	DAD	5
	21	19°15'13.49"	103°40'33. 68"		13618	17905±145	BC 20151-19293	-27.0	DAD	1
					A-5304	18520±260	BC 20596-19248	-26.2	DAD	8
	10	19°17'39.23"	103°35'26. 17"	El Tecuán	A-7392	21460 + 400/-380	BC 24640-22760	-25.3	DAD	3
	11	19°28'32.33"	103°42'49. 40"		A-7646	21545 + 265/-260	BC 24388-23324	-26.9	DAD	4
				San Marcos	A-7393	23890 + 1075/−950	28,700±1110	-25.2	P	6
					A-7231	27800 + 4140/-2720	BC 38730-21628	-23.5	DAD	3
						27820 + 2410/-1850	BC 34884-25164		DAD	5
					A-7678	35,220 + 1010/-570	BC 39831-35403	-27.9	DAD	3
					A-7389	39,090 + 760/-700	BC 42317-39983	-26.3	DAD	3
					A-7383	45,030 + 1470/-125	BC 48051-43888	-28.2	DAD	3

1: This work, 2: Cortés et al. (2010), 3: Cortés et al. (2005), 4: Komorowski et al. (1997), 5: Komorowski et al. (1993), 6: Roverato et al. (2011), 7: Robin et al. (1987), 8: Stoopes & Sheridan (1992). Abbreviations are: L=Lahar, PF=Pyroclastic Flow Deposit, P=Paleosol, and Debris Avalanche (DAD). SM = San Marcos, ET = El Tecuán, CA = Cuauhtémoc-Aeropuerto, T = Tonila, MY = Mesa Yerbabuena, VAC = Villa de Álvarez-Coquimatlán, LG = La Lumbre-Los Ganchos, ERA = El Remate-Armería

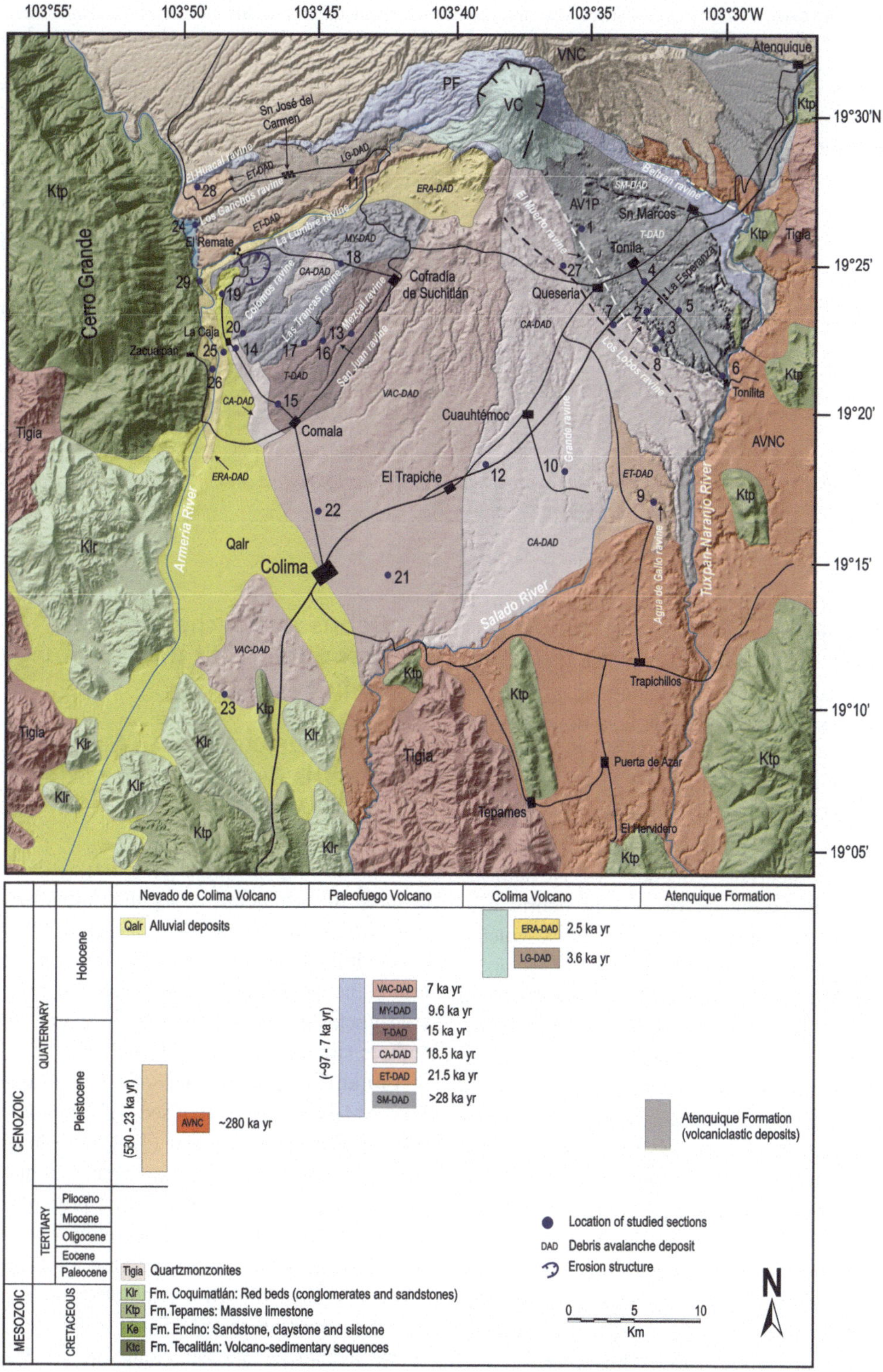

Fig. 3 Simplified geological map that shows the most likely distribution of the debris avalanche deposits from both the Paleofuego and active Volcán de Colima cones after Cortés et al. (2005, 2010b). Abbreviations are: VNC, Volcán Nevado de Colima, PF, Paleofuego, VC, Volcán de Colima. The blue filled dots are the location of the reported sections

Table 2 Summary of the approximated areas and volumes estimated for the different debris avalanche deposits generated by Volcán de Colima since late Pleistocene. The debris avalanche deposits have H/L ratios that vary from 0.08 to 0.21. Abbreviations of Debris Avalanche Deposits (DAD) are: SM = San Marcos, ET= El Tecuán, CA= Cuauhtémoc-Aeropuerto, T = Tonila, MY= Mesa Yerbabuena, VAC= Villa de Álvarez-Coquimatlán, LG = La Lumbre-Los Ganchos, ERA= El Remate-Armería

Volcanic edifice	Avalanche	Age (years B.P.)	Area (km^2)*	Volume (km^3)*	H/L ratio
Current edifice of Volcán de Colima	El Remate-Armería	2505 ± 45	65.5	~ 0.4	~0.15
	La Lumbre-Los Ganchos	3600 ± 120	48	1.7	~0.13
Paleofuego (ancient edifice of Volcán de Colima)	Villa de Álvarez-Coquimatlán	7040 ± 160	400	16	~0.08
	Mesa Yerbabuena	9671 ± 88	65	3	~0.17
	Tónila	15675 ± 125	405	12	~0.14
	Cuauhtémoc-Aeropuerto	18520 ± 260	1188	35	~0.08
	El Tecuán	21545 +265/-260	805	16	~0.09
	San Marcos	>28000	140	1.3	~0.21

The areas and volumes calculated represent approximated values due to the fact that in most cases the deposits are interdigitated making very difficult to obtain exact values

relationships, it can be inferred that Volcán de Colima began its activity more than 97,000 years ago with the construction of the Paleofuego ancestral cone on the southern flank of the Nevado de Colima. Andesitic lava flows reached distances of 21 km toward the SE sector, reaching Tuxpan-Naranjo River and 22 km to the SW reaching Armería River. Close to the source, the lava flows cover debris avalanche deposits and thick volcanoclastic sequences of Nevado de Colima. The existence of several debris avalanche deposits intercalated with thick volcanoclastic sequences in the SW, S and SE sectors of Volcán de Colima, suggests that the Paleofuego cone experienced several lateral collapses associated with explosive eruptions. These volcanoclastic sequences are exposed along the main ravines of Volcán de Colima. Excellent outcrops of these sequences occur along the Armería and Tuxpan-Naranjo rivers forming terraces of several tens of metres in thickness. The Paleofuego stage ended with a lateral collapse of the edifice that occurred 7000 years ago leaving a 5 km wide horseshoe-shaped wall, opened towards the south. After this event, the current edifice began to grow in the central part of the crater.

5 Debris Avalanche Deposits Associated with the Paleofuego Edifice

5.1 The >28,000 BP San Marcos Debris Avalanche Deposit (SM-DAD)

The *SM-DAD* of Paleofuego is exposed 15 km towards the southeast, along the Arena-Rosario and El Muerto ravines (Roverato et al. 2011). It has a limited distribution (white broken line in Fig. 3) because the deposit is covered by thick volcanoclastic sequences. However, these authors suggest that the SM-DAD has an extent of 140 km^2 with a volume ~1.3 km^3 (Table 3). The SM-DAD is composed of angular and sub-angular andesitic clasts (10–25 cm in diameter) with jigsaw fit and jigsaw cracks, supported by an indurated sandy matrix that presents reddish zones of hydrothermal alteration (Roverato et al. 2011). These authors reported a ^{14}C age of 23,890 +1075/−950 BP, obtained from a paleosol that separates the SM-DAD from a younger debris avalanche deposit along the El Muerto

Table 3 Summary of $^{40}Ar/^{39}Ar$ dates obtained from two lava flows of Paleofuego volcano (ancestral cone of Volcán de Colima)

Sample	Coordinates		Integrated Age (ka)	Plateau Age (ka)	Plateau Information			Isochron Age	Isochron Information		
	N	W			# of Fractions	$\%^{39}Ar$ Released	MSWD	(ka)	# of Fractions	$(^{40}Ar/^{36}Ar)_i$	MSWD
M12	19° 31'37.25"	103° 36'21.06"	15 ± 6	**22 ± 6**	4	81	0.19	27 ± 17	5	292 ± 9	0.33
M19	19° 28'46.65"	103° 32'48.74"	91 ± 17	**91 ± 13**	8	92	0.48	90 ± 13	10	298 ± 2	0.69

In bold preferred ages

ravine. This they define as the Tonila debris avalanche deposit (T-DAD), with an estimated volume ~1 km^3. Therefore this age represents a minimum age for the SM-DAD. Similar stratigraphic relationships were observed along the Agua de Gallo ravine (Fig. 3). Here, three debris avalanche deposits separated by paleosols where found. At Sect. 9 (Fig. 3), about 27 km southeast of Volcán de Colima summit, the older DAD is covered by a paleosol dated at 26,210 +1450/−1230 BP, suggesting that it should correlate with the SM-DAD (Fig. 4). In this location the SM-DAD has a thickness of 6 m and is composed of angular and sub-angular andesitic clasts (5–20 cm in diameter). Some clasts have a jigsaw-fit texture and are supported by a grey sandy to silty matrix. The two younger units are separated by a paleosol dated at 22,470 +1590/−1325 BP. The upper unit corresponds to the 21,545 +265/−260 BP, El Tecuan debris avalanche deposit as described below, while the middle unit could not be correlated with other deposits described.

5.2 The 21,500 BP El Tecuan Debris Avalanche Deposit (ET-DAD)

The ET-DAD is exposed in the SE, S and SW sectors of the volcano. This deposit covers an area of about 805 km^2 with an approximate volume of 16 km^3 (Table 2). The ET-DAD is named after El Tecuan village, where the deposit is well exposed. In the southeast sector along the Agua de Gallo ravine (Sect. 9; Fig. 3), the ET-DAD is ~20 m thick, it is composed of angular and sub-angular andesitic clasts 5–20 cm in diameter. Some clasts present a jigsaw-fit texture and are supported by a grey sandy-silty matrix. The ET-DAD rests on top of an 80 cm thick paleosol that was dated at

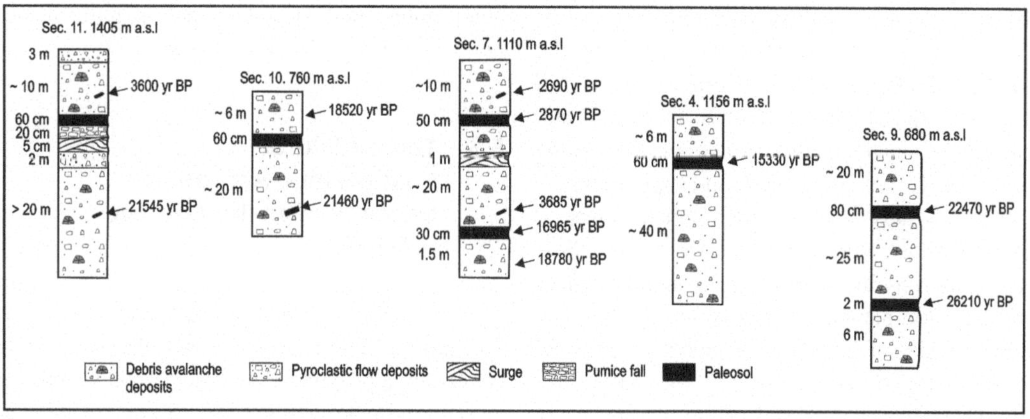

Fig. 4 Correlation of stratigraphic columns of debris avalanche deposits exposed along the different ravines on the SE sector

Fig. 5 The 21,545 BP ET-DAD and 3600 BP LG-DAD separated by a yellow pyroclastic flow deposit (Sect. 11). The 3600 BP LG-DAD is also directly covered by a yellow pyroclastic flow deposit (After Cortés et al. 2010a)

22,470 +1590/−1325 BP, (Sect. 9; Fig. 4). In addition, in the southern sector, along the Grande ravine, 25 km from the Volcán de Colima summit (Sect. 10; Fig. 3), a sample of charcoal found in the ET-DAD was dated at 21,460 +400/−380 BP In this place, the ET-DAD is covered by the younger 18,520 ± 260 BP, DAD (Robin et al. 1987; Stoopes and Sheridan 1992). In contrast, at Sect. 11, 12 km southwest of the Volcán de Colima summit, the ET-DAD is covered by the 3600 ± 120 BP, LG-DAD (Cortés et al. 2010a) (Fig. 5). The ET-DAD is >20 m thick and formed by angular and sub-angular andesitic clasts (10–30 cm in diameter) and supported by a grey sandy-silty matrix with reddish zones of hydrothermal alteration (Fig. 5). Most clasts have a porphyritic texture with phenocrysts of feldspar, hornblende and pyroxene. At this site, Komorowski et al. (1997) dated a charcoal fragment found within the ET-DAD at 21,545 BP, which they considered to be the age of the collapse. The ET-DAD is directly covered by a pyroclastic flow deposit suggesting that this failure was of Mt. St- Helens-type, thus associated with a magmatic eruption (Cortés et al. 2010a). In this sector, the debris avalanche travelled 25 km to Armería River prior to stopping against the Cerro Grande limestone (Cortés et al. 2010a). The deposit dammed the Armería River drainage forming a lake in which thick sequences of lacustrine sediments were deposited outcropping along Los Ganchos and El Huacal ravines, at Sects. 24 and 28 (Fig. 6).

5.3 The 18,500 BP Cuauhtémoc-Aeropuerto Debris Avalanche Deposit (CA-DAD)

This deposit is exposed to the S and SW of the volcano at about 30 km from the source. It covers an approximate area of 1188 km^2 and has an approximate volume of 35 km^3 (Table 2), making it the largest known DAD from Volcán de Colima. The CA-DAD is exposed in matrix

Fig. 6 **a**, **c** Lacustrine sediment deposits with an age of <1940 ± 90 BP, covering the 3600 BP LG-DAD (Sect. 24), **b** the pen points to a layer of evaporitic minerals (gypsum?), **d** lacustrine sediments along the El Huacal ravine, Sect. 28 (modified after Cortés 2002)

facies with abundant reddish zones of hydrothermal alteration (Fig. 7a, b). In some places the CA-DAD presents a coarser texture with some andesitic clasts >2 m in diameter showing jigsaw-fit cracks (Fig. 7a) and scoria clasts with abundant phlogopite phenocrysts. Cortés et al. (2005) reported a ^{14}C age of 18,445 +135/−130 BP, obtained from charcoal found in this deposit exposed at the San Juan ravine (Sect. 13; Fig. 3).

Initially this deposit was related to the evolution of a second stage of Nevado de Colima and the formation of a 4 km wide crater open to the east (Robin et al. 1987). These authors suggested that the avalanche travelled towards the east, then turned towards the southeast about 20 km from the summit, where it graded into lahars between the towns of San Marcos and Atenquique. Stoopes and Sheridan (1992) determined an age of 18,520 ± 260 BP, from a carbonized tree found within a debris avalanche deposit, exposed south of Cuauhtémoc. They concluded that the debris avalanche channelled into the drainage of Naranjo and Salado rivers travelling more than 120 km to the Pacific coast, covering an area of 2200 km^2 with a total volume of 22–33 km^3. Capra and Macias (2002) studied the same deposit and proposed that it had a volume of 7 km^3 and extended towards the southeast only a distance of 45 km from the source and stopped against the topographic barrier (Cerro La Carbonera), obstructing the drainage of Naranjo River, forming a temporary lake. These authors suggested that the deposit exposed between Naranjo River and the Pacific coast (more than 90 km) was a debris flow deposit derived from secondary remobilization of the debris avalanche. However, more recently Cortés et al. (2005, 2010b) and Cortes (2015), with additional detailed field work and new ^{14}C and $^{40}Ar/^{39}Ar$ dates, concluded that the DAD which reached the pacific coast and the DAD dated by Stoopes and Sheridan (1992; 18,500 BP) were generated by two different events and different edifices within the CVC. The debris avalanche that travelled along the Naranjo and Salado rivers as far as the Pacific coast, was generated by the second edifice of Nevado de Colima more than

Fig. 7 **a** Textural characteristics of the 18,500 BP CA-DAD of the Paleofuego cone [Sect. 12, after Cortés et al. (2010b)], **b** road cut near the village of La Caja, southwest sector, showing the texture of the 18,500 BP CA-DAD (Sect. 14; Fig. 3)

184,000 years ago. The DAD dated by Stoopes and Sheridan (1992) was generated by a collapse of Paleofuego and not by Nevado de Colima.

In intermediate areas, 15 km from the source, the CA-DAD is characterized by a pronounced hummocky morphology, whereas in distal zones, 20–40 km from the source presents a more flat morphology. In some places, the CA-DAD is covered by younger debris avalanche deposits discussed below, for example, northwest of Comala, the deposit is overlaid by a 15,625 +285/−275 BP, DAD. On the outskirts of La Caja town, the CA-DAD is covered by the 9671 ± 88 BP, DAD (Fig. 3). In the southeast sector at 10 km from the summit of Volcán de Colima, Sect. 27 (Komorowski et al. 1997; Cortés et al. 2005) a ^{14}C age of 17,960 +215/−210 BP, is reported for a yellow pyroclastic flow deposit with fine ash that is widely distributed in this sector, considered to be associated with the CA-DAD.

Fig. 8 Textural characteristics of the 15–16 ka BP T-DAD (Sect. 15; Fig. 3). **a** Matrix facies of the debris avalanche deposit (~10 m thick) with small sub-angular and sub-rounded andesite clasts and reddish hydrothermal alteration zones. Here the deposit is resting on a 30 cm thick paleosol, **b** small andesite clast with a jigsaw fit texture

5.4 The 15,000 BP Tonila Debris Avalanche Deposit (T-DAD)

The T-DAD was described by Roverato et al. (2011) in the eastern sector of Volcán de Colima (discontinuous black line in the geologic map of Fig. 3) along Los Lobos and Montegrande ravines, with a minimum age of 12,460 ± 70 BP, and 13,585 ± 135 BP, obtained from small charcoal and organic fragments found in the pyroclastic sequence that directly overly the T-DAD. Based only on measured stratigraphic sections, a minimum volume of 1 km^3 for the T-DAD was calculated (Roverato et al. 2011). However, based on ^{14}C ages (Cortes et al. 2005) the same deposit was also recognized in the southwestern sector. Here the T-DAD is up to 10 m thick, monolithologic and composed of sub-angular and

Fig. 9 Textural characteristics of the 9600 BP MY-DAD, exposed at 18 km SW of the source, which was the Paleofuego cone (Sect. 18; Fig. 3). **a** Two units of the same debris avalanche are separated by an irregular contact (dashed line), **b** megaclast with jigsaw cracks

sub-rounded andesite clasts (10–30 cm in diameter) set in an abundant brown-grey silty-sandy matrix (Sect. 15; Figs. 3 and 8a). In some places the deposit presents reddish patches of hydrothermal alteration and clasts with a jigsaw texture (Fig. 8b). At Sect. 15 (Fig. 3) the deposit rests over a 30 cm thick paleosol that separates it from the CA-DAD (Fig. 9a). Cortés et al. (2005) reported two ^{14}C ages for this deposit, obtained from charcoal found imbedded in the DAD, one at Sect. 16 along of the Mezcal ravine of 15,625 +285/−275 BP, and another at Sect. 17, along of Las Trancas ravine of 15,675 ± 125 BP, (Fig. 3). The T-DAD was estimated to be distributed over an area of 402 km^2 with a total volume of 12 km^3 (Table 2).

5.5 *The 9600 BP Mesa Yerbabuena Debris Avalanche Deposit (MY-DAD)*

The MY-DAD extends southwest up to 20 km from Volcán de Colima, covering an approximate area of 65 km^2 with a volume of 3 km^3 (Table 2). In intermediate areas, 10–12 km from the source, the deposit is characterized by a pronounced hummocky morphology of mounds hundreds of metres high. It is massive, matrix-supported, composed of angular to sub-angular andesite clasts (10–30 cm in diameter), set in a grey silty-sandy matrix poorly indurated with reddish zones resulting from hydrothermal alteration (Sect. 18; Fig. 9). At this site, the outcrop exposes two debris avalanche units defined by an irregular contact (dashed line in Fig. 10a). Both units are massive, matrix supported, with angular clasts and megaclasts

Fig. 10 **a** The 7000 BP VAC-DAD overlying the 18,500 BP CA-DAD, towards the S at 30 km from the source (Sect. 21; Fig. 3). The arrow points to a scoria megaclast with abundant phenocrysts of phlogopite. **b** At Sect. 23 (Fig. 3) 43 km from the source, the 7000 BP VAC-DAD, rests on a 30 cm thick paleosol dated at 8005 ± 50 BP Here, the deposit is ~10 m thick and shows some clasts of limestone (10–30 cm in diameter)

with jigsaw fractures (Fig. 9b). Komorowski et al. (1997) reported a ^{14}C age of 9671 ± 88 BP, for charcoal found within the deposit (Sect. 19; Fig. 3). On the outskirts of La Yerbabuena, 7.5 km southwest from the volcano's summit, Robin et al. (1987) reported a similar age of 9370 ± 400 BP, for a DAD. Along the Colomos ravine at Sect. 20 (Fig. 3), the avalanche deposit rests directly on top of a 30 cm thick paleosol dated at 12,040 +155/−150 BP by Cortés et al. (2005), considered as the maximum age for the MY-DAD. The MY-DAD is directly covered by a pyroclastic flow deposit, which could be associated with the failure event.

5.6 *The 7000 BP V. Álvarez-Coquimatlan Debris Avalanche Deposit (VAC-DAD)*

The VAC-DAD is distributed towards the south of Volcán de Colima up to a distance of 43 km from the source, covering an approximate area of 400 km^2 with a volume of 16 km^3 (Table 2). The deposit is massive matrix-supported made of white, grey and reddish angular to sub-angular andesite clasts (10–30 cm in diameter), imbedded in a sandy-silty matrix (Sect. 21 in Figs. 3 and 10a). The DAD rests with a horizontal contact directly over the CA-DAD that is here recognized for its content of scoria clasts with abundant phenocrysts of phlogopite (Fig. 10a). Cortés et al. (2005) reported an age of 7040 ± 160 BP, for a carbonized trunk found within the CA-DAD at Sect. 22, 27 km from the source (Fig. 3). At this locality, the avalanche unconformably overlies thick volcanoclastic sequences formed by sand, gravel and andesitic boulders. The debris avalanche is covered by a yellow pyroclastic flow deposit, suggesting that a magmatic component could have been associated with this collapse event. At Sect. 23, 43 km from its source, the VAC-DAD rests above a 30 cm thick paleosol, which was dated at 8005 ± 50 BP, (Fig. 10b). Here, the VAC-DAD is 10 m thick and consists of sub-angular to sub-rounded andesite clasts and exotic material, such as clasts of limestones (10–30 cm in diameter) set in an indurated sandy-silty matrix. The Colima and Villa de Alvarez cities with a combined population of 260,000 (Instituto Nacional de Estadística Geografía e Informática 2010) are constructed over this DAD. After this lateral collapse, the active cone of Volcán de Colima began its construction in the central part of the crater, with the generation of lava flows that extended 10 km to the SW. At around 4 ka BP, Volcán de Colima had already reached elevations similar to the present cone. Cortés et al. (2010a) have concluded that the current Volcán de Colima cone has experienced at least two lateral collapse events, at 3600 and 2500 BP Both events are associated with current debris avalanche deposits that will discussed below. Unfortunately, no scars or crater remnants of these edifice failures are observable at the cone today, most likely because they have been covered by material of subsequent eruptions.

6 Description and Distribution of the Debris Avalanche Deposits from Current Cone of Volcán de Colima

After the last lateral collapse of the Paleofuego edifice occurred about 7000 years ago, the modern cone of Volcán de Colima began its construction within "the crater", which is breached to the south, focusing deposits in that direction.

6.1 *The 3600 BP Los Ganchos Debris Avalanche Deposit (LG-DAD)*

The LG-DAD is exposed on the southwest flank of Volcán de Colima up to a distance of 25 km (Fig. 3). This DAD has been studied by Cortés (2002) and Cortés et al. (2010a), who referred to it as La Lumbre–Los Ganchos (LG-DAD). The deposit covers a surface area of 48 km^2, with a volume of 1.7 km^3 (Table 2) and an H/L ratio of 0.13 (Cortés 2002). This DAD was previously dated by Navarro et al. (1994) and Komorowski et al. (1997) at 3600 ± 120 BP, with charcoal fragments found within the debris avalanche. In the proximal zones 12 km from the source, the LG-DAD has an irregular topography with large mounds and intermontane lakes, while in the distal zones 20 km from the source, it presents small mounds elongated in the direction of the flow, that form terraces some tens of metres high (Cortés 2002; Cortés et al. 2010a). The deposit is massive, matrix supported, and constituted by black, grey and reddish angular to sub-angular andesite clasts (10–30 cm in diameter), although also presents blocks up to 1.5 m in diameter with jigsaw-fit fractures, embedded in a sandy-silty matrix, again with reddish zones due to hydrothermal alteration. The LG-DAD has both block and mixed facies. The block facies is exposed in proximal areas where the deposit is composed of megaclasts of other pyroclastic flow and lacustrine deposits (up to 30 m long) still preserving the original stratification. In some places, the presence of lacustrine sequences below the LG-DAD suggests the existence of small lakes formed in enclosed depressions of previous debris avalanche deposits. Along La Lumbre ravine, the LG-DAD formed a series of isolated terraces with thicknesses that vary from 20 to 65 m. At Sect. 11 (Fig. 3), the deposit rests either on top of a 60 cm thick paleosol (that in some parts was partially eroded by the avalanche), or directly over a ~0.5 m-thick pyroclastic flow deposit (Fig. 5). This pyroclastic flow deposit lies unconformably over the ET-DAD (Komorowski et al. 1997). The LG-DAD is directly covered by a pyroclastic flow deposit 0.5–1.5 m thick (Fig. 5).

In Sect. 24 (Fig. 3), at the junction between Los Ganchos ravine and Armería River, the LG-DAD is about 60 m thick. It is composed of sub-angular to sub-rounded andesite clasts, with some limestone clasts, all of them with diameters between 20 and 40 cm, immersed in a sandy to silty matrix, which presents abundant mm to cm-sized holes and cavities, made by burrowing bees and reptiles (iguanas). In this zone, Cortés et al. (2005) reported an age of 3925 ± 115 BP, obtained from a carbonized trunk found within the deposit. At this site, the LG-DAD is unconformably covered by a hyperconcentrated flow deposit (HFD) and by a thick (>40 m) lacustrine sequence (Fig. 6a). The HFD contains pottery shards and charcoal fragments dated at 1940 ± 90 BP (Cortés et al. 2010a). According to these authors, the HFD also contains pottery shards likely related to the Capacha people (820–300 years BC) indicating that the region has been inhabited during the last 3000 years. The lower section of the lacustrine sequence presents thin horizontal layers interbedded with evaporitic minerals (gypsum), suggesting the existence of periods of lake droughts (Fig. 6b). About 1 km downstream along the eastern margin of Armería River, LG-DAD is also covered by lacustrine deposits (Fig. 6c).

Field evidence and stratigraphic relationships suggest that during its emplacement, the LG-DA impacted against the eastern flank of the Cerro Grande limestones, which caused the avalanche flow to stop. The deposit was ca. 100 m thick and blocked the channel of Armería River, forming a temporal lake. The absence of lacustrine sediment deposits in the lake zone formed by the DAD, suggests that the lake's permanence was relatively short (Cortés et al. 2010a). The breaking of the dam generated a secondary remobilization of a part of the debris avalanche deposit and the formation of a debris flow that travelled 20 km along Armería River valley (Fig. 3). The debris flow emplaced 70 m thick sequences that today occur as isolated terraces with flat surfaces. The debris flow deposit (DFD) covers a surface of 6.38 km^2 with a volume of 0.191 km^3 (Cortés 2002). According to the clay content in the matrix of the DFD, it can be classified as a non-cohesive DFD (Crandell 1971; Scott 1985).

6.2 *The 2500 BP El Remate-Armería Debris Avalanche Deposit (ERA-DAD)*

The ERA-DAD extends 25 km to the SW of Volcán de Colima resting upon the eastern flank of the Cerro Grande limestones, and 10 km further downstream along the Armería River valley.

Despite its restricted distribution, we estimated that this deposit covers an area of 65.5 km^2 and has a volume of ~ 0.4 km^3 (Table 2).

The ERA-DAD is massive matrix-supported formed by black, grey and reddish angular-subangular andesite clasts (10–30 cm in diameter), some of them with jigsaw fractures, embedded in an indurated sandy-silty matrix, which presents reddish zones by hydrothermal alteration (Fig. 11a). Along La Lumbre gully and Armería River, the deposit is 10–30 m high, whilst along the eastern margin of Armería River, the DAD is covered by thick HFD sequences (over 70 m thick), Sect. 25 (Figs. 3 and 11b), which define a flat terrace. In this region, the ERA-DAD resembles a debris flow deposit in places, although it also features megaclasts with jigsaw fractures and large reddish zones from hydrothermal alteration, typical of DADs. Similar ^{14}C dates of charcoal samples have been reported in different places for this deposit (Komorowski et al. 1997; Cortés et al. 2005, 2010a, b), for instance, along La Lumbre gully and Armería River (Sects. 25 and 26 in Figs. 3 and 11c). This deposit seems to correlate in age with the description made by Siebe et al. (1992b) of a

Fig. 11 **a** Textural characteristics of the 2500 BP ERA-DAD along the western margin of Armería River (Sect. 26; Fig. 3), **b** View of the ERA-DAD along of the eastern margin of Armería River (Sect. 25; Fig. 3) underlying a >70 m thick volcanoclastic sequence, **c** carbonized trunk dated at 2505 ± 45 BP

DAD at Los Lobos bridge on the western margin of Los Lobos gully (Sect. 7; Fig. 3), that yielded an age of 2690 ± 40 BP. Although the main extension of the deposit is to the southwest, according to the age of Siebe et al. (1992b), it is possible that the avalanche may have been distributed over a larger area.

The existence of HFD sequences resting on top of the debris avalanche suggests that a period of intense erosion took place after its emplacement. These deposits could have been generated by secondary remobilization of the 9600 BP MY-DAD forming a semicircular structure located towards the east of the El Remate village (Fig. 3). Along the west margin of Armería River, the HFD sequences form further isolated terraces, upon one of which was established the Zacualpan village.

Five kilometres upstream of this community, at the junction between La Lumbre ravine and Armería River, on the east margin (Sect. 29; Fig. 3), a 150 m thick sequence of HDF overlie the 2500 BP ERA-DAD. According to the terraces′ distribution and their stratigraphic relationship, we consider that their emplacement could have caused the obstruction of Armería River forming an upstream ephemeral lake, in which the <1940 BP thick sequences of lacustrine sediments that overlie the LG-DAD were deposited (Sect. 28; Fig. 6a, c; dashed line in the geologic map, Fig. 3).

7 Discussion

7.1 DAD-Dammed Lake

The emplacement of debris avalanches can drastically change the morphology of the terrain surrounding a volcano, with catastrophic secondary effects (e.g. Mayer et al. 1983; Glicken 1996; Costa and Schuster 1988). The formation of natural dams is a very common phenomenon (Schuster and Crandell 1984; Capra 2007) and their rupture can cause voluminous debris flows (e.g. Pulgarin et al. 2005).

The existence of lacustrine horizons intercalated between debris avalanches deposits in the SW-SE sectors suggests that the Armería and Tuxpan-Naranjo rivers have been repeatedly dammed by debris avalanches, forming temporary lakes where thick sequences of lacustrine sediments were deposited (Komorowski et al. 1997; Capra and Macías 2002; Cortés et al. 2005, 2010a, b). Thick volcanoclastic sequences have been recognized and associated with the rupture of those impoundments (Capra and Macías 2002; Cortés et al. 2010a). This phenomenon is quite common at Volcán de Colima, due to is location within the Colima Graben, along which the two major rivers in the area, which drain into the Pacific (Fig. 1). Stratigraphic evidence points to a short duration of those temporary lakes (Cortés et al. 2010a), probably due to the high discharge rate of the rivers ($\sim$1000 m^3/s at present; SPP 1981), that promotes rapid infilling and subsequent overflowing of the dam. Piping and overtopping are the two main mechanisms of dam failure, after which catastrophic debris flows formed downstream. The last episode at Volcán de Colima was generated by the 2500 BP debris avalanche deposit in the Zacoalpan area, where thick hyperconcentrated debris flow sequences directly overlie the 2500 BP, ERA-DAD along Armería River. A future event of edifice failure at Volcán de Colima will likely be accompanied by this type of secondary event with catastrophic effects in distal areas, far away from the affectation zone of the initial avalanche.

7.2 Triggering Mechanisms

A wide variety of factors contribute to edifice failure (Campbell et al. 1995; McGuire 1996; Siebert et al. 1987). One significant factor is hydrothermal alteration, which can significantly weaken an edifice and increase pore-fluid pressures of rocks contributing to failure (López and Williams 1993; Carrasco-Núñez et al. 1993; van Wyk de Vries and Francis 1997; Reid et al. 2001). Other important factors are gravitational spreading of the edifice (Borgia et al. 1992; van Wyk de Vries and Francis 1997), or the intrusion of a new magma body underneath the volcano (Glicken 1996; Siebert et al. 1987). In the case of

Volcán de Colima, several DADs are associated with juvenile fragments with bread-crust textures (e.g. CA-DAD, VAC-DAD, LG-DAD), imbedded charred logs or emptied casts in the deposits (CA-DAD, VAC-DAD, LG-DAD, ERA-DAD), and underlying pyroclastic flow deposits (e.g. 21,500; 18,500; 7000 and 3600 BP DAD, see Table 2). In fact, paleomagnetic data of some of these DADs suggest that their emplacement temperatures exceed 350 °C (Clement et al. 1993). Therefore, this evidence suggests that most of the edifice failures have been accompanied by magmatic activity. However, it is difficult to define if the magmatic component was the main or only factor inducing the instability of the volcano, or if other elements such as climate or the structural architecture of the volcano promoted its failure.

Volcán de Colima has evolved within in a complex tectonic framework, at the intersection of two main regional structures: the N-S Colima graben and the NE-SW Tamazula fault (TF) (Garduño et al. 1998; Norini et al., this volume). Evidence of volcano-tectonic interplay is widespread (Siebert 1984; Siebe et al. 1992a; van Wyk de Vries and Borgia 1996; van Wyk de Vries and Merle 1998; Garcia-Palomo et al. 2000; Lagmay et al. 2000; Tibaldi et al. 2005; Bellotti et al. 2006; Lagmay and Valdivia 2006; Merle et al. 2006), and apparent at Volcán de Colima (Norini et al. 2010). In the WSW sector, N-S normal faults are affecting fluvial-lacustrine sediments and volcanoclastic sequences associated with the volcanic activity of CVC (Garduño et al. 1998). In addition, field evidence shows that structures with a N-S orientations are affecting both Nevado de Colima and Volcán de Colima, contributing to the evolution and southward migration of magmatic activity of the CVC (Garduño et al. 1998; Loera et al. 2010; Norini et al. 2010). The NE-SW Tamazula fault that intersects both volcanic edifices, displaces thick volcanoclastic sequences and lava flows from both eruptive centres. This fact, suggests that this structure is active, and that it represents an important factor that has contributed to the instability of the volcanic edifices and the generation of debris avalanches (Garduño et al. 1998; Cortés et al. 2010b; Cortés. 2015). More recently, the E-W-trending structures affecting the CVC have been interpreted as promoting a southward spreading of the entire volcanic complex (Norini et al. 2010). That is a common mechanism controlling volcano stability at large stratovolcanoes (e.g. van Wyk de vries and Borgia 1996; Borgia et al. 2000), with this spreading having been identified at Volcán de Colima by means of campaigns of GPS (Wooller et al. 2004).

Finally, the presence of andesite blocks on the surface of the 18,500 BP DAD with evidence of glacial striations, similar to those observed at the Nevado de Colima summit (Luhr et al. 2010), suggest that during late Pleistocene the summit of Volcán de Colima was covered by an ice-snow cap enhancing hydrothermal alteration and promoting volcano instability (Roverato et al. 2011).

7.3 Volume and Frequency of Debris Avalanches at Volcán de Colima

After of the 1980 eruption of Mount St. Helens, USA, giant landslides and debris avalanches have been recognized to play an important role in the evolution of stratovolcanoes throughout the world (Ui 1983; Siebert 1984; Francis and Self 1987; McGuire 1996). Globally, during the past 500 years, major edifice failures have occurred at a rate of about four per century (Siebert et al. 1987). However, there are some volcanoes that have experienced failure events with intervals of a few hundred years. For example, Augustine volcano in Alaska, a low-volume volcano with a high magma effusion rate, has collapsed twelve times over the past 2000 years with an average recurrence interval of about 150–200 years (Begét and Kienle 1992). Similarly, Shiveluch volcano in Kamchatka experienced eight or more failure events during the Holocene, with the emplacement of debris avalanches with volumes between 0.1 and 3 km^3 (Belousov et al. 1999; Ponomareva et al. 1998, 2006). Siebert et al. (2006) documented 40 possible edifice failures at volcanoes in Central America, nine of them occurred during the Holocene, with a regional recurrence of ~1000–2000 years.

Komorowski et al. (1997) suggested that Volcán de Colima collapsed at least nine times during the last 45,000 years yielding a mean recurrence of ~5000 years. Based on new detailed field mapping and supported by additional ^{14}C ages we have defined eight of these debris avalanche deposits that were generated during the last 30,000 years, giving a mean recurrence interval of ~3750 years (Table 2 Considering the two separate stages of the volcano, during the late Pleistocene-early Holocene, the Paleofuego edifice had six failure events with a mean recurrence interval of ~3800 years, whilst the current active cone has experienced two lateral collapse events with a comparable recurrence interval of ~3500 years. Therefore, the recurrence of failure events at Volcán de Colima has been quite consistent, although, the volumes generated were smaller during the late Holocene.

Borselli et al. (2011), based on a stochastic approach, that is to say, each event is taken to be independent, with their occurrence in time defined by a random process, defined a mean failure recurrence interval of 2698 years, with an uncertainty range of 180 years, for the Holocene. In Chap. 11, the authors reassess the data available for defining the reoccurrence rate for large eruptions. Most of their evidence comes from the dating of debris avalanche deposits and assumptions regarding the coincidence of large eruptions. The small reoccurrence interval shows that Volcán de Colima, after Mt. Augustine volcano in Alaska, has produced the second largest number of Holocene debris avalanches in North America.

Volume estimation of DADs at Volcán de Colima always is very difficult. The irregular topography of older DADs favoured that subsequent debris avalanches refilled depressions and low-lying areas, leading to a complicated stratigraphic relationship (Komorowski et al. 1997). This makes it difficult to determine their exact distribution and volume. Where the largest avalanches filled up the main ravines that descend the volcano, their deposits are some hundred metres thick. Whilst in the lateral zones, their thickness is of the order of only metres (Fig. 5). Also, the extent of older DADs is usually obscured by overlying younger deposits. To determine an approximate volume of each debris avalanche (Table 2), a minimum thickness of 30 m was considered, although in some places some deposits are more than 100 m thick.

Zernack et al. (2012) observed that when volcano instability is controlled by the magma supply rate, a linear trend can be observed between volume of DAD and frequency. However, in the case of Volcán de Colima, the cumulative DAD volume over time does not

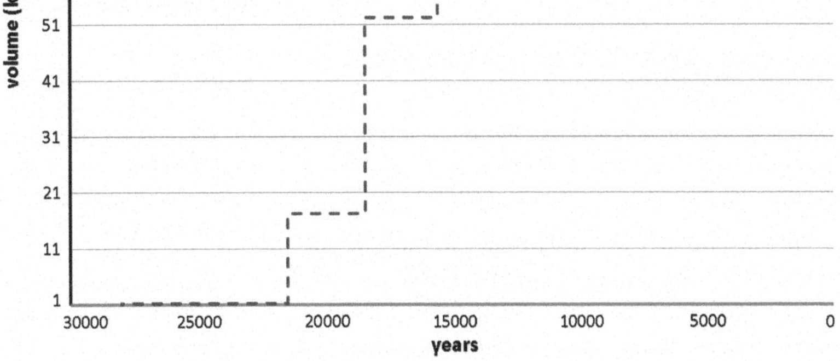

Fig. 12 Linear trend observed between the volume and frequency of the debris avalanches of Volcán de Colima (both the Paleofuego and current cone)

follow a linear trend, instead, there is a large volume deviation between events (Fig. 12). This evidence confirms that even if a magmatic component was recognized in almost all the events, other factors, such as the tectonic architecture of the volcano, obviously played an important role in volcano instability as previously described.

Finally, the cumulative debris avalanche volume can be used as a proxy of the magma supply rate (Zernack et al. 2012). From data here presented, a value of 0.21 km^3/100 years is obtained for the Holocene, in agreement with the average of 0.2 km^3/100 years estimated by Luhr and Prestegaard (1988) for the last ~4000 years.

8 Conclusions

According to recent $^{40}Ar/^{39}Ar$ ages the Paleofuego stage of Volcán de Colima began about 100,000 years ago on the southern flank of Nevado de Colima. According to ^{14}C ages, during the Late Pleistocene-Early Holocene, this edifice experienced six edifice collapses that generated debris avalanches, which vary in length and volume. Some of these avalanches from the SW and SE sectors dammed the Armería and Tuxpan-Naranjo rivers generating temporary lakes where thick sequences of lacustrine sediments were deposited. These catastrophic collapse events present recurrence intervals that vary between 3000 and 6500 years. The Paleofuego edifice terminated with a lateral collapse 7000 years ago, generating a debris avalanche that travelled 47 km toward the south. This event left a 5 km wide crater open in this direction. Afterwards, the current active cone of Volcán de Colima began its construction. This new cone has collapsed twice generating small debris avalanches (3600 and 2500 BP) yielding a recurrence interval of ~3500 years. We consider that the absence of source area remnants of the new cone associated with these events is due to their burial by volcanic material from subsequent eruptions. The existence of abundant large charred tree logs inside of some DADs, as well as the presence of pyroclastic flow deposits directly overlying some avalanche deposits, suggest that some failure events were associated with a magmatic activity.

The presence of andesite blocks on the surface of the CA-DAD, with evidence of glacial striations, suggests the existence of a glacier at the Volcán de Colima summit during the late Pleistocene. This would have favoured hydrothermal alteration and promoted the instability of the volcanic cone before its collapse. Another factor that surely contributed to instability is the active NE-SW Tamazula fault that has played an important role in the generation of debris avalanches of the CVC (Norini et al., this volume).

The secondary remobilization of these debris avalanche deposits has generated voluminous lahar and debris flow deposits, which currently form a series of terraces along of the valley of the Tuxpan-Naranjo and Armería Rivers. Colima, the capital city of the state of Colima, and many smaller cities and towns are built upon primary and secondary deposits of debris avalanches, therefore in the future a massive edifice failure on the scale of previous events could be catastrophic for the inhabitants currently living on these deposits.

Acknowledgements Our gratitude and recognition for the support received from the following Institutions and people in the fulfilment of the current work: University of Colima, for the economic support by means of a grant; Chris Eastoe, for the ^{14}C dating carried out at the University of Tucson, Arizona. Special thanks are given to M. Ursúa and J. Velasco of the Sistema Estatal de Protección Civil of Colima for their logistical support during field work. Thanks to L. Siebert, M. Roverato and B. Bernard for their thorough reviews of the manuscript and insightful comments which helped to make this a better contribution. We thank G. Cisneros for his technical aid to draft some of the figures.

References

Begét, J.E., Kienle, J.: Cyclic formation of debris avalanches at Mount St. Augustine volcano, Alaska. Nature **356**, 701–704 (1992). https://doi.org/10.1038/356701a0

Bellotti, F., Capra, L., Groppelli, G., Norini, G.: Tectonic evolution of the central-eastern sector of Trans Mexican Volcanic Belt and its influence on the eruptive history of the Nevado de Toluca volcano (Mexico). J Volcanol Geotherm Res **158**, 21–36 (2006)

Belousov, A., Belousova, M., Voight, B.: Multiple edifice failures, debris avalanches and associated eruptions in the Holocene history of Shiveluch volcano, Kamchatka, Russia. B. Volcanol. **61**, 324–342 (1999)

Borgia, A., Delaney, P.T., Den Linger, R.P:. Spreading volcanoes. Ann. Rev. Earth Planet. Sci. **28**(1), 539–570 (2000). https://doi.org/10.1146/annurev.earth.28.1.539

Borgia, A.: The dynamic basis of volcanic spreading. J. Geoph. Res. **99**, 17,791–17,804 (1994). https://doi.org/10.1029/94jb00578

Borgia, A., Ferrari, L., Pasquaré, G.: Importance of gravitational spreading in the tectonic and volcanic evolution of Mount Etna. Nature **357**, 231–235 (1992)

Borselli, L., Capra, L., Sarochi, D., De la Cruz-Reyna, S.: Flanck collapse scenarios at Volcán de Colima, Mexico: a relative instability analysis. J. Volcanol. Geotherm. Res. **208**, 51–65 (2011)

Campbell, C.S., Cleary, P.W., Hopkins, M.: Large-scale landslide simulations: global deformation, velocities, and basal friction. J. Geoph. R **100**, 8267–8283 (1995). https://doi.org/10.1029/94JB00937

Campbell, C.S.: Self-lubrication for long run out landslides. J. Geol. **97**, 653–665 (1989)

Capra, L.: Volcanic natural dams: identification, stability and secondary effects. Nat. Hazards **43**, 45–61 (2007)

Capra, L., Macías, J.L: The cohesive Naranjo debris-flow deposit (10 km^3): a dam breakout flow derived from the Pleistocene debris-avalanche deposit of Nevado de Colima Volcano (Mexico). J. Volcanol. Geotherm. Res. **117**, 213–235 (2002)

Capra, L., Macías, J.L., Scott, K., Abrams, M., Garduño, V.H.: Debris avalanches and debris flows transformed from collapses in the Trans-Mexican Volcanic Belt, Mexico— behavior, and implications for hazard assessment. J. Volcanol. Geotherm. Res. **113**, 81–110 (2002)

Capra, L., Macías, J.L.: Pleistocene cohesive debris flows at Nevado de Toluca Volcano, central Mexico. J. Volcanol. Geotherm. Res. **102**, 149–168 (2000)

Carrasco-Núñez, G., Gómez-Tuena, A.: Volcanogenic sedimentation around Citlaltépetl volcano (Pico de Orizaba) and suroundings, Veracruz, Mexico. In: Aguirre-Díaz, G.J., Aranda-Gómez, J.J., Carrasco-Núñez, G., Ferrari, L. (eds.) Magmatism and tectonics in the central and northwestern México—a selection of the 1997 IAVCEI general assembly excursions; México, D.F., UNAM, Instituto de Geol., Excursion, vol. 16, pp. 131–151 (1997)

Carrasco-Núñez, G., Vallance, J.W., Rose, W.I.: A voluminous avalanche-induced lahar from Citlaltépetl volcano, México: implications for hazard assessment. J. Volcanol. Geotherm. Res. **59**, 35–46 (1993)

Carrasco-Núñez, G., Díaz-Castellón, R., Siebert, L., Hubbard, B., Sheridan, M.F., Rodríguez, S.R.: Multiple edifice-collapse events in the Eastern Mexican Volcanic Belt: the role of slopping substrate and implications for hazard assessment. J. Volcanol. Geotherm. Res. **158**, 151–176 (2006)

Clement, B.M., Connor, C.B., Graper, G.: Paleomagnetic estimate of the emplacement temperature of the long-runnout Nevado de Colima volcanic debris avalanche deposit, México. Earth Plan. Res. Lett. **120**, 499–510 (1993)

Cortés, A.: Depósitos de Avalancha y Flujos de escombros originados hace 3 300 años por el Colapso del Sector Suroeste del Volcán de Colima. Ph.D. thesis, UNAM, México (2002)

Cortés, A., Garduño, V.H., Navarro, C., Komorowski, J.-C., Saucedo, R., Macías, J.L., Gavilanes, R.J.C.: Carta Geológica del Complejo Volcánico de Colima con Geología del Complejo Volcánico de Colima. Cartas Geológicas y Mineras 10. Inst. Geol. **10**, 1–37 (2005)

Cortés, A., Macías, J.L., Capra, L., Garduño-Monroy, V. H.: Sector collapse of the SW flank of Volcán de Colima, México: the 3600 yr BP La Lumbre-Los Ganchos debris avalanche and associated debris flows. J. Volcanol. Res. **197**, 52–66 (2010a)

Cortés, A., Garduño, V.H., Macías, J.L., Navarro-Ochoa, C., Komorowski, J.C., Saucedo, R., Gavilanes, J.C.: Geologica mapping of the Colima volcanic complex (Mexico) and implications for hazard assessment. Geol. Soc. Am. Spec. Pap. **464–12**, 1–16 (2010b)

Cortés, A.: Historia eruptiva del volcán Nevado de Colima y su evolución dentro del Complejo Volcánico de Colima: México D.F., Universidad Nacional Autónoma de México. Programa de Posgrado en Ciencias de la Tierra, Tesis de Doctorado, p. 115 (inédita) (2015)

Costa, J.E., Schuster, R.L.: The formation and failure of natural dams. Geol. Soc. Amer. Bull. **100**, 1054–1068 (1988)

Crandell, D.R.: Postglacial lahars from Mount Rainier Volcano, Washington. U. S. Geol. Surv. Prof. Pap. **677**, 75 (1971)

Francis, P.W., Self, S.: Collapsing Volcanoes. Scient Am. **256–6**, 90–97 (1987)

Francis, P.W., Wells: Landsat thematic mapper observations of debris avalanche deposits in the Central Andes. B. Volcanol. **50**, 258–278 (1988)

García-Palomo, A., Macías, J.L., Espíndola, J.M.: Strike-slip faults and K-alkaline volcanism at El Chichón volcano, Southeastern Mexico. J Volcanol Geoth. Res. **136**, 247–268 (2004)

Garcia-Palomo, A., Macías, J.L., Garduño, V.H.: Miocene to recent structural evolution of the Nevado de Toluca volcano región, Central Mexico. Tectonophysics **318**, 281–302 (2000)

Garduño, V.H., Saucedo, R., Jiménez, S., Gavilanes, J.C., Cortés, A., Uribe, R.M.: La Falla Tamazula, limite suroriental del bloque Jalisco, y sus relaciones con el Complejo Volcánico de Colima, México. Revista Mexicana de Ciencias Geológicas **15**, 132–144 (1998)

Glicken, H.: Sedimentary architecture of large volcanic-debris avalanches. In: Fisher, R.V., Smith, G.A. (eds.) Sedimentation in Volcanic Settings, SEPM (Soc Sed Geol) Spec Pub, 45, pp. 99–106 (1991)

Glicken, H.: Rockslide Avalanche of May 18, 1980, Mount St, Helens Volcano, Washington, U.S. Geol Surv, Open File Report 96–677, p. 88 (1996)

Höskuldsson, A., Robin, C.: Late Pleistocene to Holocene eruptive activity of Pico de Orizaba, Eastern Mexico. Bull. Volcanol. **55**, 571–587 (1993)

Instituto Nacional de Estadística Geografía e Informática: Conteo Nacional de población y vivienda (2010). www.inegi.org.mx

Komorowski, J.-C., Glicken, H., Sheridan, M.F.: Secondary electron imagery of microcracks and hackly fracture surfaces in sand-size clasts from the 1980 Mount St. Helens debris avalanche deposit: Implications for particle-particle interactions. Geology **19–3**, 261–264 (1991)

Komorowski, J.-C., Navarro, C., Cortés, A., Siebe, C.: The repetitive collapsing nature of Colima volcanoes (México). In: Problems Related to the Distinction of Multiple Deposits and Interpretation of ^{14}C Ages with Implications for Future Hazards. Universidad de Colima, Cuarta Reunión Internacional Volcán de Colima: A Decade Volcano Workshop, Colima, 24–28 Jan, pp. 12–18 (1994)

Komorowski, J.C., Navarro, C., Cortés, A., Saucedo, R., Gavilanes, J.C., Siebe, C., Espíndola, J.M., Rodríguez-Elizarrarás, S.R.: The Colima Volcanic Complex, Field Guide # 3. IAVCEI, General Assembly, Puerto Vallarta, Mexico (1997)

Lagmay, A.M.F., Valdivia, W.: Regional stress influence on the opening direction of crater amphitheaters in Southeast Asian volcanoes. J. Volcanol. Geotherm. Res. **158**, 139–150 (2006)

Lagmay, A.M.F., de Vries, B.V.W., Kerle, N., Pyle, D. M.: Volcano instability induced by strike-slip faulting. Bull. Volcanol. **62**, 331–346 (2000). https://doi.org/10.1007/s004450000103

Layer, P.W., García-Palomo, A., Jones, D., Macías, J.L., Arce, J.L., Mora, J.C.: El Chichón volcanic complex, Chiapas, México: stages of evolution based on field mapping and 40Ar/39Ar geochronology. Geofis. Int. **48**, 33–54 (2009)

Loera, L.H., Urrutia-Fucugauchi, J., Valdivia, A.L.: Magnetic characteristics of fracture zones and constraints on the subsurface structure of Colima Volcanic Complex. Western Mexico. Geosphere **6**(1), 35–46 (2010). https://doi.org/10.1130/GES00204.1

López, D.L., Williams, S.N.: Catastrophic volcanic collapse: relation to hydrothermal processes. Science **260**, 1794–1796 (1993)

Luhr, J.F., Navarro-Ochoa, C., Savov, I.P.: Tephrochronology, Petrology and geochemistry of Late-Holocene pyroclastic deposits from Volcán de Colima, México. J. Volcanol. Geotherm. Res. **197**, 1–32 (2010)

Luhr, J.F., Carmichael, I.S.E.: Geology of Volcán de Colima Universidad Nacional Autónoma de México. Instituto de Geología. Bol 107. México DF. 1–101 + plate (1990)

Luhr, J.F., Prestegaard, K.L.: Caldera formation at Volcán Colima, México, by large Holocene volcanic debris avalanche. J. Volcanol. Geotherm. Res. **35**, 335–348 (1988)

Luhr, J.F., Carmichael, I.S.E.: The Colima Volcanic Complex, Mexico; part III. Ash and scoria fall deposits from the upper slopes of Volcán Colima. Contrib. Mineral. Petrol. **80**, 262–275 (1982a)

Luhr, J.F., Carmichael, I.S.E.: The Colima Volcanic Complex: I. post-caldera andesites from Volcán de Colima. Contrib. Mineral. Petrol. **71**, 343–372 (1982b)

Macías, J.L., García-Palomo, A., Arce, J.L., Seibe, C., Espíndola, J.M., Komorowski, J.-C., Scott, K.M.: Late Pleistocene-Holocene cataclysmic eruptions at Nevado de Toluca and Jocotitlán Volcanoes, Central Mexico. In: Kowallis, B.J. (ed.) Proterozoic to Recent Stratigraphy, Tectonics, and Volcanology, Utah, Nevada, Southern Idaho and Central Mexico, pp. 493–528. BYU Geol, Stud (1997)

Mayer, W., Sabol, M.A., Schuster, R.L.: Landslide-dammed lakes at Mount St. Helens, Washington. In: Schuster, R.L. (ed.) Landslide dams: processes, risk and mitigation, pp. 21–41. Geotech Specl Publ, New York (1983)

McGuire, W.J.: Volcano instability: a review of contemporary themes. In: McGuire, W.J., Jones, A.P., and Neuberg, J. (eds.) Volcano Instability on the Earth and Other Planets, vol. 110, pp. 1–23. Geol Soc Spec Publ, London (1996)

Mehl, K.W., Schmincke, H.U.: Structure and emplacement of the Pliocene Roque-Nublo debris avalanche deposit, Gran Canaria. Spain. J. Volcanol. Geotherm. Res. **94**, 105–134 (1999)

Merle, O., Barde-Cabusson, S., Maury, R., Legendre, C., Guille, G., Blais, S.: Volcano core collapse triggered by regional faulting. J. Volcanol. Geotherm. Res. **158**, 269–280 (2006)

Navarro, C., Komorowski, J.-C., Cortés, A.: Depósitos de avalanchas de escombros en el Complejo Volcánico de Colima: evidencias geológicas y edades C^{14} para múltiples eventos. In: Nuevas preguntas a previas interpretaciones, un problema aún no resuelto. Universidad de Colima, Cuarta Reunión Internacional Volcán de Colima: A Decade Volcano Workshop, Colima, 24–28 Jan 1994, pp. 49–50 (1994)

Norini, G., Capra, L., Groppelli, G., Agliardi, F., Pola, A., Cortés, A. (2010). Structural architecture of the Colima Volcanic Complex. J. Volcanol. Geotherm. Res. **115**, 1–20 (B12209) (2010). https://doi.org/10.1029/2010jb007649

Pierson, T.C., Costa, J.E.: A rheologic classification of subaerial sediment-water flows. In: Costa, J.E., Wieczorek, G.F. (eds.) Debris Flows/Avalanches: Process, Recognition, and Mitigation (Geol. Soc. Amer. Rev. Eng. Geol.), pp. 1–12 (1987)

Ponomareva, V.V., Pevzner, M.M., Melekestsev, I.V.: Large debris avalanches and associated eruptions in the Holocene eruptive history of Shiveluch volcano, Kamchatka. Russia. Bull. Volcanol. **59**(7), 490–505 (1998)

Ponomareva, V.V., Melekestsev, I.V., Dirksen, O.V.: Sector collapses and large landslides on Late

Pleistocene-Holocene Volcanoes in Kamchatka, Russia. J. Volcanol. Geotherm. Res. **158**, 117–138 (2006)
Pulgarin, B., Macías, J.L., Cepeda, H., Capra, L.: Late Pleistocene deposits associated with a southern flank collapse of the Nevado del Huila Volcanic Complex (Colombia). Acta Vulcanol. **16**(1) (in press) (2005)
Reid, M., Sisson, T., Brien, D.: Volcano collapse promoted by hydrothermal alteration and edifice shape Mount Rainier, Washington. Geology **29**, 779–782 (2001)
Robin, C., Boudal, C.: A gigantic Bezymianny type event at the beginning of modern Volcán Popocatépetl. J. Volcanol. Geotherm. Res. **31**, 115–130 (1987)
Robin, C., Mossand, P., Camus, G., Cantagrel, J.-M., Gourgaud, A., Vincent, P.: Eruptive history of the Colima Volcanic Complex. México. J. Volcanol. Geotherm. Res. **31**, 99–113 (1987)
Rodriguez-Elizarrarás, S.R., Siebe, C., Komorowski, J.C., Espíndola, J.M.: Consideraciones preliminares sobre riesgo en el Volcán de Colima, México. R Mex. Cs. Geol. **12**(1), 47–51 (1995)
Roverato, M., Capra, L., Sulpizio, R., Norini, G.: Stratigraphic reconstruction of two debris avalanche deposits at Colima Volcano (Mexico): Insights into pre-failure conditions and climate influence. J. Volcanol. Geotherm. Res. **207**, 33–46 (2011)
Schuster, R.L., Crandell, D.R.: Catastrophic debris avalanches from volcanoes. Proc. IV Int. Symp. Landslides Proc. **1**, 567–572 (1984)
Scott, K.M.: Lahars and flow transformations at Mount St. Helens Washington, U.S.A. In: International Symposium on Erosion, Debris Flow and Disaster Prevention, pp. 209–214. Tsukuba, Japan (1985)
Siebe, C., Komorowski, J.-C., Sheridan, M.F.: The tectonically induced avalanche deposit at Jocotitlán volcano, Central México: unusual morphology, stratigraphical relations, and possible mode of emplacement. B. Volcanol. **54**, 573–589 (1992a)
Siebe, C., Komorowski, J.-C., Sheridan, M.F.: Morphology and emplacement collapse of an unusual debris avalanche deposit at Jocotitlán Volcano, Central Mexico. Bull. Volcanol. **54**, 573–589 (1992a)
Siebe, C., Abrams, M., Macías, J.L.: Derrumbes gigantes, depósitos de avalanchas de escombros y edad del actual cono del Volcán Popocatépetl. Volcán Popocatépetl. In: Gobernación, C.N.d.D.S.d. (ed.) Estudios realizados durante la crisis de, vol. 1994–1995, pp. 195–220 (1995b)
Siebert, L.: Large volcanic debris avalanche: characteristics of source areas, deposits and associated eruptions. J. Volcanol. Geotherm. Res. **22**, 163–197 (1984)
Siebert, L., Glicken, H., Ui, T.: Volcanic hazards from Bezymianny and Bandai type eruptions. Bull. Volcanol. **49**, 435–459 (1987)
Siebert, L., Alvarado, G.E., Vallance, J.W., de Vries, B.V. W: Large-volume volcanic edifi ce failures in Central America and associated hazards. In: Rose, W.I., Bluth, G.J.S., Carr, M.J., Ewert, J., Patino, L.C., Vallance, J. (eds.) Volcanic Hazards in Central America. Geol. Soc. Am. Spec. Pap. **411**, 1–26 (2006). https://doi.org/10.1130/2006.2412(01)
Smith, G.A., Fritz, W.J.: Volcanic influences on terrestrial sedimentation. Geology **17**, 375–376 (1989)
SPP: Secretaría de Programación y Presupuesto. Carta hidrológica de aguas superficiales, escala 1:250,000. Hoja Colima (E 13–3) (1981)
Stoopes, G.R., Sheridan, M.F.: Giant debris avalanches from the Colima Volcanic Complex, Mexico: implications for long-runout landslides (b100 km) and hazard assessment. Geology **20**, 299–302 (1992)
Tibaldi, A., Lagmay, A.M.F., Ponomareva, V.V.: Effects of basement structural and stratigraphic heritages on volcano behavior and implications for human activities (the UNESCO/IUGS/IGCP project 455). Episodes **28**(3), 158–170 (2005)
Ui, T., Takarada, S., Yoshimoto, M.: Debris avalanches. In: Sigurdsson, H., Houghton, B., McNutt, S.T., Rymer, H., Stix, J. (eds.) Encyclopedia of Volcanoes, pp. 617–626. Academic Press, San Diego (2000)
Ui, T.: Volcanic dry avalanche deposits—identification and comparison with nonvolcanic debris stream deposits. J. Volcanol. Geotherm. Res. V **18**, 135–150 (1983)
Ui, T., Yamamoto, H., Suzuki-Kamata, K.: Characterization of debris avalanche deposits in Japan. J. Volcanol. Geotherm. Res. V **29**, 231–243 (1986)
Vallance, J.W., Siebert, L., Rose W.I., Jr., Girón J.R., Banks, N.G.: Edifice collapse and related hazards in Guatemala. J. Volcano Geoth. Res. **66**, 337–355 (1995)
Van Wyk de Vries, B., Borgia, A.: The role of basement in volcano deformation. In McGuire, W.J., Jones, A. P., Neuberg, J. (eds.) Volcano Instability on the Earth and Other Planets. Geol. Soc. London Spec. Publ. **110**, 95–110 (1996). https://doi.org/10.1144/GSL.SP.1996.110.01.07
Van Wyk de Vries, B., Francis P.: Catastrophic collapse at stratovolcanoes induced by gradual volcano spreading. Nature **387**, 387–390 (1997)
Van Wyk de Vries, B., Merle, O.: Extension induced by volcano loading in regional strike-slip zones. Geology **26**, 983–986 (1998). https://doi.org/10.1130/0091-7613(1998)026%3c0983:eibvli%3e2.3.co;2
Wooller, L., de Vries, B.V.W., Murray, J.B., Rhymer, H., Meyer, S.: Volcano spreading controlled by dipping substrata. Geology **32**, 573–576 (2004). https://doi.org/10.1130/g20472
Zernack, A.V., Cronin, S.J., Bebbington, M.S., Price, R. C., Smith, E.M., Stewart, R.B., Procter, J.N.: Catastrophic stratovolcano collapse: a model based on Mount Taranaki, New Zealand. Geology **40**, 983–986 (2012)

Modelling Tephra Thickness and Particle Size Distribution of the 1913 Eruption of Volcán de Colima, Mexico

C. B. Connor, L. J. Connor, C. Bonadonna, J. Luhr, I. Savov and C. Navarro-Ochoa

Abstract

A crucial problem at most volcanoes is reconstructing past eruptions from the geological record. The rapid erosion of many volcanic terrains results in geologically recent eruptions leaving a relatively sparse record of the event. Here we consider the tephra-stratigraphic record of the 1913 eruption of Volcán de Colima, a recent but greatly eroded tephra fallout deposit. A total of 38 stratigraphic sections of the 1913 deposit have been analysed for thickness, granulometry and geochemistry. The 1913 scoria are hornblende and two-pyroxene andesites with approximately 58 wt% SiO_2, providing a geochemical and petrographic signature that is distinct from earlier (1818) and later tephra fallout deposits. Tephra2, a tephra dispersion computer code based on the advection-diffusion equation, is used to model thickness variation and particle size distribution of the pyroclasts for the 1913 deposit. Based on computer simulations, the observed tephra stratigraphy is best fit with a total eruption mass of $\sim 5.5 \times 10^{10}$ kg. Computer simulations including reports of tephra accumulation from the historical record produces an alternative deposit model with a finer median particle size (~ 1.77 ϕ), a higher eruption column (~ 25 km above mean sea level, amsl), and a greater total eruption mass ($\sim 1.4 \times 10^{11}$ kg). This larger eruption magnitude is supported by modelling the granulometry of the 38 stratigraphic sections. The models suggest a median deposit particle size of at least 2ϕ, a deposit mass of $1–5 \times 10^{11}$ kg (VEI 4), and that significant segregation by particle size as a function of height occurred in the 1913 eruption column. This analysis highlights potential bias in

J. Luhr—Deceased.

C. B. Connor (✉) · L. J. Connor
University of South Florida, 4202 E. Fowler Ave, Tampa, FL 33620, USA
e-mail: cbconnor@usf.edu

C. Bonadonna
Département des Sciences de la Terre, Université de Genève, Rue des Maraîchers 13, CH-1205 Geneva, Switzerland
e-mail: Costanza.Bonadonna@unige.ch

J. Luhr
Department of Mineral Sciences, Smithsonian Institution, MRC 119, PO Box 37012, Washington, DC 20013-7012, USA

I. Savov
School of Earth and Environment, The University of Leeds, Leeds LS2 9JT, UK
e-mail: I.Savov@leeds.ac.uk

C. Navarro-Ochoa
Colima Volcano Observatory and CUICT, Universidad de Colima, Avenida 25 de Julio #965, Col. Villa San Sebastian, COL 28045 Colima, Mexico
e-mail: naoc@ucol.mx

N. Varley et al. (eds.), *Volcán de Colima*, Active Volcanoes of the World,
https://doi.org/10.1007/978-3-642-25911-1_3

eruption magnitude estimates that use only thickness of proximal deposits, and the advantage of modelling the granulometry of the deposit in such circumstances.

Keywords

Tephra · Fallout model · Numerical model Advection-diffusion equation · Tephra stratigraphy · Volatile content · Volcanic eruption · Volcán de Colima · Eruption volume

1 Introduction

Eruption volume and eruption column height are traditionally estimated from interpolated isopach and isopleth maps of tephra fallout deposits (Pyle 1989; Carey and Sparks 1986; Fierstein and Nathenson 1992; Bonadonna and Houghton 2005; Burden et al. 2011; Bonadonna and Costa 2012; Burden et al. 2013). These techniques are based on statistical models, e.g., exponential fitting of the eruption deposit data and simplified physics explaining the erupting column and the dispersing wind field. More recently, numerical simulations of eruptions have gained traction and several models have been developed to forecast the dispersion and deposition of tephra fallout, given specific eruption conditions (Suzuki 1983; Armienti et al. 1988; Searcy et al. 1998; Hurst and Turner 1999; Macedonio et al. 2005; Bonadonna et al. 2005; Jones et al. 2007; Barsotti et al. 2008; Folch et al. 2009). These models can be coupled to inversion algorithms that seek to find a set of best-fit eruption parameters to reproduce tephra thickness variations at sample sites across a deposit (Connor and Connor 2006; Magill et al. 2015; White et al. 2017), thus providing an alternative way of estimating eruption volume and column height from field data.

Reconstruction of the 1913 eruption of Volcán de Colima (Fig. 1) initially relies on the identification and characterisation of its tephra fallout deposit. For more than a decade, two of the authors (J. L. and C. N.-O.) worked within the Colima Volcano Complex to locate and identify remaining deposit outcrops of the 1913 eruption, to measure deposit thicknesses, to collect tephra samples from these outcrops and to analyse the geochemistry and granulometry. Their resulting data set, based on 38 sample locations (Table 1), provides a geological record of this eruption, located mostly on the edifice of Nevado de Colima, immediately north of Volcán de Colima (Fig. 1). All tephra samples have been checked for stratigraphic consistency and trace element characteristics provide unambiguous geochemical identification of the tephra as a product of the 1913 eruption, known as the α layer (Luhr et al. 2010). Using these geochemical observations along with models of magma volatile content and ascent rate, our physical volcanological models can infer fragmentation conditions and potential eruption explosivity.

To improve upon basic statistical techniques for estimating eruption parameters, we use a numerical simulation of tephra dispersion (Tephra2) coupled with an inversion algorithm, and model tephra thickness and particle size variation at individual sites in order to achieve an overall best-fit model of the 1913 eruption in terms of total erupted mass (or tephra volume), eruption column height, total particle size distribution of the pyroclasts, and wind conditions at the time of the eruption. Tephra thickness reports from cities and towns found in various historical accounts are included with the geological data to more accurately estimate realistic atmospheric conditions on the day of this eruption. By using the same modelling technique for each class of particle size in our distribution (i.e., φ size fractions), the uniqueness of the mass and column height estimates are assessed. A comparison of the resulting models, those that include historical data with geological data and models that use only thickness of measured stratigraphic sections, indicates that some bias occurs.

Our results highlight uncertainties associated with the reconstruction of eruption parameters such as volume and column height, and hence eruption magnitude or VEI (volcano explosivity index), from limited outcrop data using any method. We find that by inverting tephra fallout for individual particle size fractions, we can identify

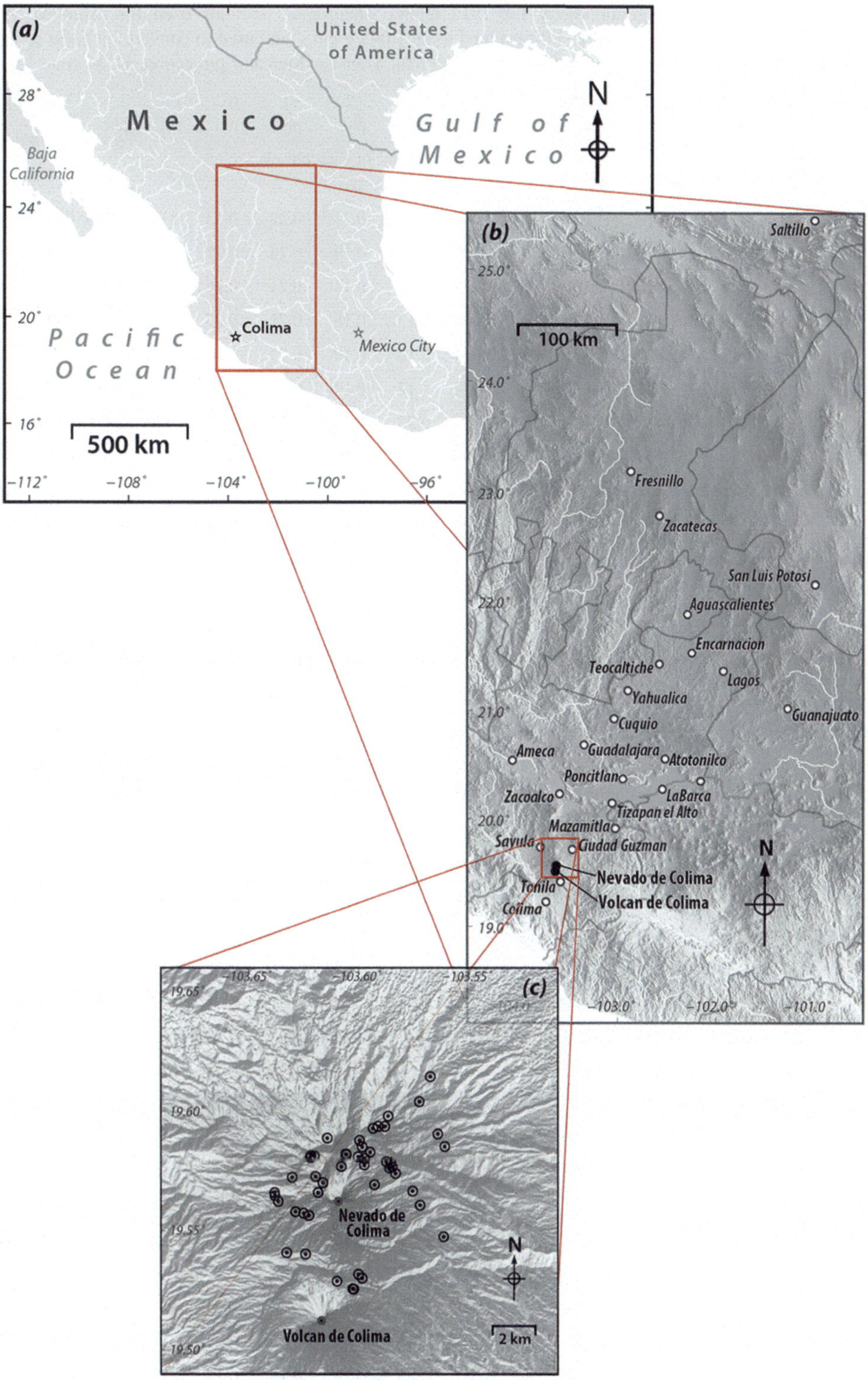

Fig. 1 Location of Volcán de Colima (**a**) Nearby towns with historical (1913) accounts of tephra fallout (**b**) are indicated for reference. The source of the 1913 eruption, Nevado de Colima, where most sampling localities (circles with dots) for the 1913 deposit are located (**c**)

Table 1 Tephra thickness and granulometry for 1913 tephra stratigraphic sections used for inversion. A bulk scoria-fall-deposit density was measured and determined to be 1.043 g cm^{-3} and used to convert thickness to kg m^{-2}. Latitude/Longitude are converted to Easting/Northing, Universal Transver Mercator projection, UTM zone 13

Sample no.	Easting	Northing	Sample measures		−4 phi	−3 phi	−2 phi	−1 phi	0 phi	1 phi	2 phi	3 phi	4 phi
	m	m	kg m^{-2}	cm	(weight percent of each phi class)								
VF95-06a	648,243	2,167,610	417.20	40	0.001	0.11	0.21	0.22	0.23	0.17	0.04	0.002	0.002
VF95-08a	647,017	2,166,184	312.90	30	0.006	0.13	0.19	0.24	0.25	0.16	0.03	0.002	0.002
VF97-20Q	650,610	2,166,811	208.60	20	0.017	0.07	0.14	0.21	0.26	0.21	0.06	0.007	0.006
VF98-01a3	647,775	2,167,142	239.89	23	0.044	0.13	0.18	0.21	0.23	0.15	0.03	0.003	0.002
VF98-02a	646,264	2,165,823	156.45	15	0.014	0.12	0.19	0.23	0.24	0.17	0.03	0.004	0.003
VF98-04a	644,555	2,165,754	104.30	10	0.027	0.12	0.19	0.19	0.20	0.14	0.05	0.020	0.020
VF98-05a	645,142	2,164,486	187.74	18	0.137	0.23	0.18	0.15	0.13	0.08	0.02	0.009	0.006
VF98-07a	646,028	2,165,268	208.60	20	0.008	0.14	0.19	0.23	0.24	0.15	0.02	0.004	0.004
VF98-08a	647,534	2,167,051	187.74	18	0.008	0.07	0.12	0.22	0.29	0.23	0.05	0.005	0.005
VF98-09a	649,727	2,168,287	187.74	18	0.025	0.04	0.10	0.19	0.29	0.26	0.09	0.004	0.003
VF98-10a	650,242	2,169,453	125.16	12	0.011	0.03	0.06	0.17	0.29	0.32	0.11	0.003	0.002
VF99-02B	647,629	2,164,406	260.75	25	0.035	0.15	0.21	0.22	0.22	0.13	0.02	0.002	0.001
VF99-03a	647,129	2,165,343	208.60	20	0.055	0.15	0.23	0.24	0.19	0.10	0.01	0.003	0.003
VF99-04as	645,335	2,166,568	125.16	12	0.017	0.06	0.13	0.21	0.28	0.20	0.06	0.013	0.014
VF99-05a	644,723	2,165,744	83.44	8	0.028	0.13	0.20	0.20	0.18	0.13	0.06	0.017	0.017
VF99-06as	648,089	2,167,144	292.04	28	0.036	0.08	0.15	0.22	0.26	0.20	0.04	0.003	0.002
VF00-04a	646,609	2,159,594	625.80	60	0.192	0.31	0.18	0.11	0.06	0.02	0.01	0.006	0.009
VF00-05a3	646,897	2,160,261	500.64	48	0.101	0.30	0.24	0.16	0.09	0.03	0.01	0.007	0.010
VF00-06a3	647,067	2,160,085	417.20	40	0.067	0.30	0.25	0.17	0.11	0.05	0.01	0.008	0.008
VF01-01a3	647,147	2,165,653	344.19	33	0.031	0.11	0.18	0.22	0.25	0.16	0.02	0.004	0.003
VF01-03a	650,933	2,162,032	135.59	13	0.026	0.13	0.19	0.22	0.24	0.15	0.03	0.005	0.004
VF01-07a5	649,798	2,163,484	365.05	35	0.020	0.17	0.23	0.23	0.21	0.12	0.02	0.003	0.002
VF01-09a3	649,447	2,164,123	584.08	56	0.036	0.18	0.22	0.21	0.19	0.11	0.02	0.002	0.002
VF02-01a	648,336	2,165,198	208.60	20	0.035	0.14	0.20	0.21	0.21	0.12	0.03	0.014	0.015
VF02-02a	648,622	2,164,957	187.74	18	0.021	0.14	0.25	0.23	0.20	0.12	0.02	0.004	0.005
VF02-03a	648,176	2,165,507	354.62	34	0.016	0.06	0.18	0.26	0.29	0.15	0.02	0.004	0.004
VF02-05a	648,462	2,165,255	344.19	33	0.022	0.15	0.22	0.23	0.22	0.13	0.02	0.003	0.003
VF02-06a3	647,397	2,165,932	344.19	33	0.017	0.16	0.22	0.23	0.22	0.12	0.02	0.003	0.003
VF05-02a	644,773	2,164,771	292.04	28	0.027	0.11	0.17	0.21	0.22	0.15	0.05	0.020	0.019
VF05-03a	644,524	2,165,676	73.01	7	0.019	0.08	0.15	0.20	0.23	0.18	0.06	0.018	0.018
VF05-05a3	643,671	2,164,740	62.58	6	0.016	0.10	0.16	0.18	0.19	0.15	0.06	0.026	0.027
VF05-08a3	643,437	2,161,240	125.16	12	0.100	0.22	0.24	0.16	0.10	0.04	0.02	0.008	0.008
VF05-13a	642,838	2,164,013	62.58	6	0.042	0.09	0.17	0.19	0.17	0.11	0.08	0.055	0.042
VF05-14a3	642,860	2,163,859	62.58	6	0.001	0.09	0.13	0.19	0.21	0.14	0.07	0.046	0.056
VF05-16a3	643,009	2,163,605	93.87	9	0.041	0.15	0.19	0.19	0.19	0.12	0.04	0.026	0.028

(continued)

Table 1 (continued)

Sample no.	Easting	Northing	Sample measures		−4 phi	−3 phi	−2 phi	−1 phi	0 phi	1 phi	2 phi	3 phi	4 phi
	m	m	kg m^{-2}	cm	(weight percent of each phi class)								
VF05-17a	643,842	2,163,136	125.16	12	0.098	0.14	0.17	0.20	0.17	0.10	0.04	0.021	0.018
VF05-18a3	644,473	2,162,964	166.88	16	0.031	0.21	0.25	0.22	0.16	0.07	0.02	0.009	0.008
VF0 (6-03/5-09)	644,340	2,161,170	594.51	57	0.113	0.25	0.25	0.19	0.11	0.04	0.01	0.006	0.005

these biases, potentially improving volume estimates from limited outcrop data. Also, by modelling individual particle size fractions, insights into the structure of the eruption column emerge, which begin to suggest eruption mechanisms.

2 The 1913 Eruption

Two major explosive eruptions of Volcán de Colima occurred on 18 February 1818 and 20 January 1913 (Bretón-Gonzalez et al. 2002) (Fig. 2). The 1913 eruption produced the α layer described and modelled in this paper. Both eruptions left a large open steep-walled crater. According to Waitz (1915), the crater following the 1913 eruption was 450 m in diameter and 350 m deep. Both eruptions produced pyroclastic flows on the south flanks of the volcano and eruption columns that led to tephra fallout many hundreds of kilometres to the north. Comparatively, these two eruptions had much in common, which makes it all the more surprising that only the younger eruption has left the α tephra unit with apparently very little of the 1818 tephra unit preserved as a deposit.

Historical accounts of the 1913 eruption indicate that it was short-lived and are detailed in Waitz (1915), Bretón-Gonzalez et al. (2002), and Saucedo et al. (2010). Briefly, on 20 January, at approximately 05:00, the citizens of the town of Ciudad Guzman heard strong noises resembling thunder coming from the direction of Volcán de Colima. Later that morning it became clear that the source was indeed the volcano as the eruption column, and even lightning from within, was visible. Tephra fallout in Ciudad Guzman was strongest until about noon on that day. By about 20:00 the eruption column was no longer visible and by 23:00 the sky was clear of visible ash. During the next several days a few medium (Vulcanian) explosions occurred, but by 28 January these explosions had stopped and tephra was no longer visible in the atmosphere over the volcano.

Saucedo-Girón (1997) and Komorowski et al. (1997) were the first to suggest that unit α was the product of the 1913 eruption, and they went on to calculate the height of the eruption column and volume of magma erupted. The erupted mass of tephra fallout from the 1913 eruption has been most recently estimated to be approximately 2–6 × 10^{11} kg (Saucedo et al. 2011; Bonasia et al. 2011), based on tephra deposit sections (Saucedo-Girón 1997), with a maximum eruption column height of approximately 23 km (Saucedo et al. 2010).

3 Petrology and Geochemistry

The petrology and geochemistry of the Volcán de Colima 1913 eruption products (including ballistically transported pyroclasts, scoria and ash samples) have been covered in significant detail by Luhr and Carmichael (1990), Robin et al. (1991) and more recently by Savov et al. (2008) and Luhr et al. (2010). In addition, geochemical and petrological information for the 1913 tephra are presented in (Crummy et al., this volume) where the 1913 eruption is contrasted with other Holocene tephra deposits from Volcán de

Fig. 2 Field photos of Volcán de Colima stratigraphic sections. The 1913 tephra, unit α, deposited directly on top of the much older (~3000 year BP) unit Z. **a** In other areas, two units are visible above unit Z. **b** These are the 1818 and the 1913 deposits (bottom and top, respectively). A paleosoil horizon lies between these two most recent pumice fall deposits (Luhr et al. 2010)

Colima. Here we briefly describe the major characteristics of the 1913 deposits and provide new data on volatile contents of melt inclusions, aiming to further clarify the nature of the eruption. We concentrate on the petrography and rock textures, mineral and glass chemistry and their volatile contents. These aspects of the 1913 deposits are key for understanding the eruption dynamics and also for contrasting the 1913 tephra with older (Holocene-Pleistocene) deposits found on the slopes of Nevado de Colima (Luhr et al. 2010), which can have a similar appearance in outcrop.

The textures of 1913 tephra are highly vesicular and commonly porphyritic, with phenocrysts (and microphenocrysts) of plagioclase, orthopyroxene, clinopyroxene, hornblende and oxide minerals (usually Ti-magnetite) that may reach 2 mm in size. Importantly, olivine is not present in any of the examined 1913 tephra samples ($n > 100$). The range of mineral abundances (vol %) was estimated after careful point-counting of

Table 2 Mineral abundances in Colima 1913 tephra

Sample		98-12A	98-12B	95-06a	95-09a	97-20Q	98-01a3	98-02a	99-02B
(vol %)									
Pl	ph	9.4	15.7	10.6	7.1	5.8	5.0	8.6	7.5
Pl	mp	2.7	2.9	4.0	5.1	3.8	5.3	5.1	5.2
Pl	tot	12.1	18.6	14.6	12.2	9.6	10.3	13.7	12.7
Opx	ph	4.4	3.9	0.7	1.0	0.9	1.3	1.5	2.2
Opx	mp	1.0	1.3	1.2	1.7	1.9	1.0	2.4	1.6
Opx	tot	5.4	5.2	1.9	2.7	2.8	2.3	3.9	3.8
Cpx	ph	5.7	2.4	1.1	1.4	1.7	1.1	1.2	0.8
Cpx	mp	0.2	0.2	1.1	1.6	1.0	1.1	0.8	2.1
Cpx	tot	5.9	2.6	2.2	3.0	2.7	2.2	2.0	2.9
Ox	ph	0.1							
Ox	mp	0.9	1.0	1.1	0.9	1.5	0.7	0.4	0.8
Ox	tot	0.9	1.0	1.1	0.9	1.5	0.7	0.5	0.8
Hb	ph	0.2	0.1	1.8	1.4	2.1	1.7	1.0	2.5
Hb	mp	0.3	0.2	0.2	0.4	0.8	0.9	1.7	0.6
Hb	tot	0.5	0.3	2.0	1.8	2.9	2.6	2.7	3.1
Ol	ph	0.0	0.0	0.0	0.0	0.0	0.0	0.0	0.0
Ol	mp	0.0	0.0	0.0	0.0	0.0	0.0	0.0	0.0
Ol	tot	0.0	0.0	0.0	0.0	0.0	0.0	0.0	0.0
Grndmass	75.1	72.2	78.1	79.4	80.4	81.8	77.8	76.6	80.5
Sum Phen	19.7	22.1	14.2	10.9	10.5	9.1	12.4	13.0	9.6

(continued)

Table 2 (continued)

Sample		98-12A	98-12B	95-06a	95-09a	97-20Q	98-01a3	98-02a	99-02B
Sum crystals	24.8	27.7	21.8	20.6	19.5	18.1	22.8	23.3	19.5
Total	99.9	99.9	99.9	100.0	99.9	99.9	100.6	99.9	100.0

Sample	99-03a	99-06as	99-06ai	00-03a	00-05a00	00-05a1	00-05a3	00-06a1	00-06a3	01-01a3
(vol%)										
Pl	4.9	10.1	8.7	11.5	11.3	17.1	7.6	14.6	9.0	13.1
Pl	5.4	4.1	5.8	6.5	2.3	6.8	4.2	2.5	5.6	6.5
Pl	10.3	14.2	14.5	18.0	13.6	23.9	11.8	17.1	14.6	19.6
Opx	1.7	1.0	1.7	2.5	2.3	1.6	0.8	2.5	1.3	2.1
Opx	1.8	1.2	2.9	2.1	2.6	2.2	1.1	2.7	1.4	1.6
Opx	3.5	2.2	4.6	4.6	4.9	3.8	1.9	5.2	2.7	3.7
Cpx	1.6	0.7	0.7	2.3	1.6	3.7	0.7	3.1	1.6	1.8
Cpx	0.7	0.8	0.6	1.2	0.9	1.0	0.9	0.9	1.0	1.1
Cpx	2.3	1.5	1.3	3.5	2.5	4.7	1.6	4.0	2.6	2.9
Ox										
Ox	1.4	0.5	0.3	0.3	0.4	0.4	0.7	1.2	0.6	1.3
Ox	1.4	0.5	0.3	0.3	0.4	0.4	0.7	1.2	0.6	1.3
Hb	1.4	1.8	0.9	2.1	2.9	1.7				
Hb	0.6	0.4	0.1	0.2	0.3	0.5	0.2			
Hb	2.0	2.2	0.0	0.0	0.1	0.0	1.1	0.0	2.4	3.4
Ol	0.0	0.0	0.0	0.0	0.0	0.0	0.0	0.0	0.0	0.0
Ol	0.0	0.0	0.0	0.0	0.0	0.0	0.0	0.0	0.0	0.0
Ol	0.0	0.0	0.0	0.0	0.0	0.0	0.0	0.0	0.0	0.0
Grndmass	79.4	79.3	73.7	78.5	67.2	82.9	72.5	77.1	69.1	76.5
Sum Phen	13.6	11.1	16.3	15.2	22.4	10.0	20.2	14.0	19.9	14.9
Sum crystals	20.6	20.7	26.4	21.5	32.8	17.1	27.5	22.9	30.9	23.6
Total	100.0	100.0	100.1	100.0	100.0	100.0	100.0	100.0	100.0	100.1

19 thin sections, with at least 1000 mineral grains counted from each (phenocrysts and microphenocrysts) (Table 2). The mineral abundances (in vol%) for the phenocrysts and microphenocrysts, respectively, are as follows: plagioclase (9.6 and 4.8), orthopyroxene (1.6–1.9), clinopyroxene (1.5 and 1.1), hornblende (1.8–0.5) and oxide mineral (0.6–0.8). The total phenocryst content of the 1913 tephra is 14% and the sum of all crystals is 23%. The total phenocryst content of the 1913 dense lithics that are present in pyroclastic density current deposits and in the fall deposits of unit α (representing the lithics of the conduit or the former lava dome in the summit before the explosive eruption), is 18.7% and the sum of all crystals is 26.2%. The relative abundance of plagioclase crystals against the sum of all other crystals is ~0.6 and the ratio, plagioclase/(sum of mafic minerals) is ~1.8, essentially the same ratio in the 1913 lithics as in the 1913 tephra samples. Overall, when compared with the post-1913 eruption products (Luhr 2002; Savov et al. 2008), the mineral modes characteristic for the 1913 samples reveal less abundant plagioclase, much more abundant hornblende and oxides, and comparable orthopyroxene and clinopyroxene content.

Using the major element composition of the volcanic glass and assuming it is in equilibrium with the measured major element composition of pyroxenes and plagioclase crystal rims, one can reconstruct the physical conditions of the eruption. These include important parameters such as the melt temperatures, pressures (source depth) and the volatile contents of the magmas as they were ejected from the conduit and quenched. The 18 orthopyroxene rims analyzed from 1913 tephra samples show MgO = 26.3–27.0(±0.33) wt%; FeO = 16.4–16.9(±0.48) wt%; and CaO = 1.3–1.34(±0.1) wt%.; with Mg # ~75. The 19 analyzed clinopyroxene rims from the 1913 Colima samples have MgO = 15.3–15.4(±0.4) wt%; FeO = 8.5–8.7(±0.5) wt%; and CaO = 20.7–20.9(±0.3) wt%.; with Mg # ~76. The plagioclase rims analyzed in 1913 tephra samples ($n = 24$) have CaO contents of 12.1 wt% and Na_2O of 4.6 wt%, making their rims of anorthite [An] content of 59.5 mol%. The volcanic glass analyses of the 1913 tephra reveal andesite composition and contains the following average major element compositions (all in wt%): SiO = 62.93(±0.5), MgO = 1.78(±0.08); FeO = 4.63 (±0.25); CaO = 4.11(±0.28); Na_2O = 5.07 (±0.32) and K_2O = 1.96(±0.09) (Table 3). These tephra glass analyses are in very good agreement with those from 1913 lithic samples ($n = 24$).

All of the above, the description of major element compositions and the use of various thermometers and published calibrations, reveal the following temperatures for the 1913 melt: 984 ± 1 °C (plagioclase-glass; Putirka 2005); 982 ± 12 °C (2 pyroxenes; Wells 1977); 995 ± 7 °C and 1006 ± 13 °C (orthopyroxene-glass and clinopyroxene-glass; Putirka 2008).

Using the Putirka (2005) plagioclase-glass hygrometer we estimate that the minimum H_2O content of the 1913 melt was in the order of

Table 3 Groundmass glass composition of Colima 1913 scoria

Sample	SiO_2	TiO_2	Al_2O_3	FeO	MnO	MgO	CaO	Na_2O	K_2O	P_2O_5	Total
VF9506a2gm-3	63.56	0.78	16.49	4.91	0.17	2.28	3.82	4.93	2.06	0.25	99.25
VF9506a2gm-7	63.30	0.75	17.28	4.34	0.10	1.30	4.46	5.53	1.50	0.23	98.79
VF9506a2gm-8	62.18	0.68	17.85	4.42	0.11	1.43	4.43	5.34	1.65	0.20	98.29
VF9506a1gm-1	62.70	0.80	16.25	4.76	0.11	1.94	3.83	4.52	1.89	0.23	97.03
VF9506a1gm-2	63.68	0.78	16.52	4.78	0.11	1.77	3.90	4.89	2.04	0.23	98.70
VF9506a1gm-4	62.93	0.80	16.72	4.32	0.12	1.93	4.10	4.95	1.93	0.23	98.03
VF9506a1gm-5	62.44	0.78	17.18	4.60	0.16	1.82	4.39	5.34	1.90	0.25	98.86
VF9506a1gm-9	62.64	0.78	17.03	4.91	0.05	1.77	3.95	5.03	2.03	0.22	98.41

Table 4 Major element composition of orthopyroxene-hosted glass inclusions

Inclusion	VF9506aA	VF9506aB1	VF9506aB2	VF9506aC	VF9506aD	VF9506aE	VF9506aF	VF9506aG	VF9506aH
Mean	n = 3	n = 3	n = 3	n = 3	n = 3	n = 3	n = 3	n = 3	n = 3
SiO_2	62.10	62.41	61.31	62.33	61.32	59.66	60.51	59.23	58.84
TiO_2	0.40	0.64	0.55	0.37	0.53	0.70	0.83	0.66	0.85
Al_2O_3	15.60	15.58	15.69	15.64	15.99	16.25	16.24	16.73	16.46
FeO_t	4.23	4.24	4.43	3.99	4.58	4.24	4.09	4.60	4.42
MnO	0.08	0.10	0.08	0.09	0.16	0.09	0.11	0.14	0.08
MgO	1.43	1.18	1.19	1.11	1.42	1.56	1.35	1.61	1.54
CaO	3.25	3.29	3.93	2.98	3.52	4.06	3.47	4.33	3.79
Na_2O	3.95	3.92	4.26	4.42	4.26	4.35	4.42	4.47	4.35
K_2O	1.77	1.96	1.66	1.93	1.76	1.75	2.04	1.48	1.96
P_2O_5	0.21	0.20	0.23	0.18	0.25	0.30	0.20	0.24	0.32
SO_3	0.187	0.187	0.215	0.151	0.219	0.271	0.233	0.279	0.272
Cl	0.185	0.201	0.208	0.201	0.182	0.171	0.202	0.166	0.202
Total	93.38	93.91	93.74	93.40	94.19	93.39	93.68	93.92	93.09
s.d.									
SiO_2	0.23	0.10	0.14	0.16	0.23	0.18	0.17	0.25	0.19
TiO_2	0.00	0.02	0.01	0.02	0.01	0.03	0.01	0.01	0.02
Al_2O_3	0.06	0.19	0.18	0.11	0.08	0.10	0.05	0.03	0.06
FeO_t	0.15	0.10	0.17	0.10	0.02	0.10	0.21	0.18	0.08
MnO	0.02	0.05	0.05	0.02	0.04	0.06	0.01	0.02	0.04
MgO	0.08	0.06	0.09	0.04	0.05	0.12	0.08	0.02	0.06
CaO	0.10	0.08	0.03	0.10	0.08	0.06	0.06	0.04	0.04
Na_2O	0.25	0.09	0.19	0.21	0.22	0.05	0.25	0.09	0.06
K_2O	0.02	0.03	0.04	0.02	0.05	0.03	0.02	0.07	0.05
P_2O_5	0.00	0.01	0.03	0.02	0.01	0.04	0.02	0.03	0.01
SO_3	0.001	0.006	0.004	0.009	0.006	0.009	0.005	0.003	0.004
Cl	0.002	0.006	0.003	0.003	0.002	0.001	0.004	0.001	0.004

3.9 ± 0.1 wt%. Use of the Housh and Luhr (1991) hygrometer (based on the plagioclase Na–Ca–K concentrations) yields 4.3 ± 0.1 wt%. When normalized, the orthopyroxene-hosted melt inclusions have dacitic major element composition. They have elevated K_2O contents (up to 2 wt%) and relatively low MgO (1.1–1.6 wt%) (Table 4). Direct FTIR measurements of these orthopyroxene- hosted melt inclusion glasses reveals H_2O contents reaching 4.4 wt% in the 1913 magmas, also in excellent agreement with the insights from the hygrometry based on the Putirka (2005) and Housh and Luhr (1991) methods. Finally, high quality ion probe (SIMS) data for the same melt inclusions also reveals H_2O concentrations as high as 6.05 wt% (Luhr et al. 2006).

The bulk rock composition of the 1913 tephra is presented in Luhr et al. (2010) and references therein, where it is discussed as unit *alpha*. Briefly, the major element chemistry reveals that the magmas erupted in 1913 are all low silica andesites ($SiO_2 = 57.7 - 59.9$ wt%) with $K_2O = 1.2 \pm 0.1$ wt%, classifying them as belonging to the medium K–series. Plotted on a total alkali silica (TAS) diagram (Le Bas et al. 1986), all 1913 eruption products fall within the andesite field. The MgO and FeO contents of the 1913 samples are elevated and, importantly, often exceed 4.0 and 4.8 wt%, respectively, abundances that are increasingly more common in the post-1999 Volcán de Colima ash and lava samples (Savov et al. 2008). Overall, the bulk rock major element chemistry agrees very well with insights from the petrography of the tephra and the mineral modes. In particular, the high MgO and overall low SiO_2 are in agreement with the observed lower plagioclase content and the elevated abundance of hornblende in the 1913 samples.

When it comes to trace elements, the 1913 deposits contain light rare earth element (LREE)-enriched chondrite-normalized patterns that are very similar to all other products of Volcán de Colima. However, it is important to note that the 1913 samples have slightly elevated Yb (always >1.58 ppm) and lower to much lower Ba (~430 ppm) and Rb (~16 ppm) contents, which Savov et al. (2008) suggested is a very reliable tool for discrimination between the 1818 and 1913 pyroclastic fallout deposits. Otherwise they are very difficult to discriminate both in the field and in thin section. The high field strength element (HFSE) systematics of the 1913 tephra include distinct Nb and Ta negative anomalies that are a characteristic feature of typical continental arc-derived andesites (Kelemen et al. 2003; Savov et al. 2008). For further discussions on the geochemistry of the Colima arc andesites, both modern and ancient, please refer to (Reubi et al. and Crummy et al., this volume).

Cumulatively, high H_2O contents of the 1913 eruption products are completely consistent with the high abundance and lack of reaction rims around hornblende crystals in this magma (Fig. 3), because hornblende in equilibrium with mafic melt necessitates H_2O contents in access of 4.3 wt% (Carmichael 2002). Assuming that the Colima melt inclusions are usually trapped at pressures 10–150 MPa (see Ruebi and Blundy, this monograph), we can safely assume that the melt both below the conduit of Volcán de Colima and at the point of eruption was H_2O -rich (i.e., this melt experienced very little degassing from source en route to the surface). The petrology and geochemistry of the 1913 deposits therefore supports the ideas outlined in Savov et al. (2008), that the explosive nature of the 1913 eruption was due to the very fast ascent of volatile-rich andesitic melts originating deep in the mantle wedge (Luhr and Carmichael 1990).

4 The Tephra2 Algorithm

A basic conclusion from geochemical analyses and derived magma properties is that the 1913 eruption was explosive, with explosivity driven by the exsolution of volatiles during rapid magma ascent. The Tephra2 model (Bonadonna et al. 2005; Connor and Connor 2006; Volentik et al. 2010) was used to investigate the dispersion and sedimentation of the 1913 deposit and hence, its explosivity, from a physical volcanological perspective. This model calculates total mass per unit area, M (kg m^{-2}), of tephra accumulated at a

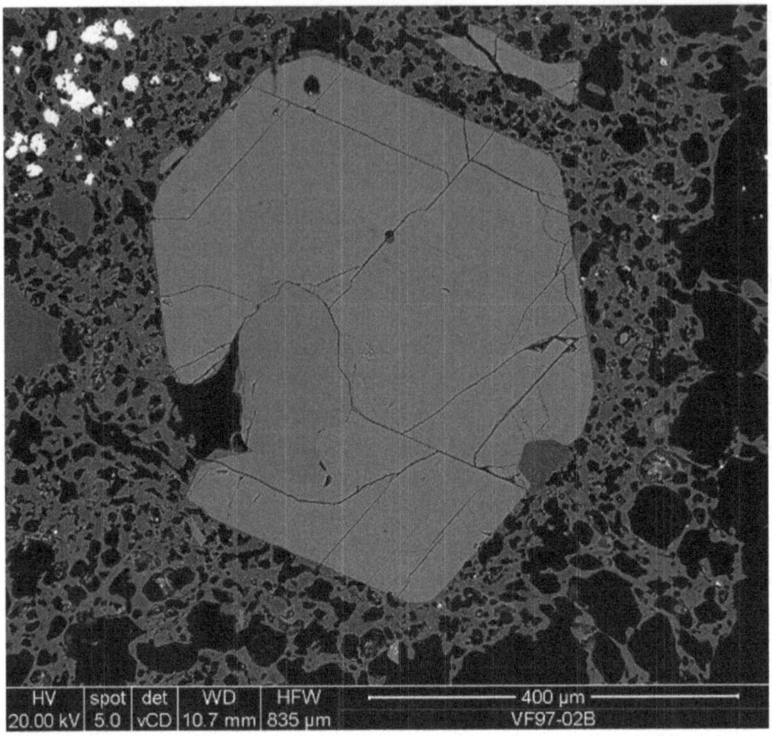

Fig. 3 BSE image of Volcán de Colima tephra sample VF97–02B. This sample is representative for the 1913 tephra. The image shows a large hornblende crystal surrounded by very highly vesiculated volcanic glass (see Savov et al. 2008). Note that the hornblende crystal lacks reaction rim with the surrounding glass, indicating that the crystal was in equilibrium with volatile-rich melt (glass). Analysis of the glass volatile content indicates H_2O content of at least 4.5 wt%. Direct SIMS measurement of the volatile contents of melt inclusions trapped in orthopyroxene crystals from the 1913 tephra reveal H_2O contents of up to 6 wt%. Overall the high volatile contents of the 1913 glasses are consistent with the explosive nature of this andesitic eruption

given location on the ground at map coordinates (x, y) due to an explosive eruption.

The numerical simulation of tephra accumulation is based on the advection-diffusion equation (Suzuki 1983; Armienti et al. 1988; Connor et al. 2001), which is expressed in Tephra2 by a simplified mass-conservation equation for which an analytical solution can be determined (Suzuki 1983; Lim et al. 2008). This analytical solution is written as

$$M(x,y) = \sum_{i=0}^{H_t} \sum_{j=\varphi_{min}}^{\varphi_{max}} M^o_{i,j} f_{i,j}(x,y) \qquad (1)$$

to illustrate the relationship between particle size distribution, column geometry, and mass distribution of tephra within the mapped deposit. The right hand side of Eq. (1) consists of two parts. $M^o_{i,j}$ describes the distribution of tephra mass within the eruption column as a function of height and particle size. The eruption column is discretized into i atmospheric layers. Initially, tephra is distributed among these layers within the eruption column between the vent and the maximum column height, H_t. For example, a uniform random distribution might be used to reflect the structure of a well-mixed strong plume (i.e., a volcanic plume that does not bend in the wind, but is rather a strong enough jet to deflect the wind; Sparks 1986), or a beta distribution might be used to model particle segregation in the plume as a function of particle size and settling velocity. Thus, $M^o_{i,j}$ is a one-dimensional source term for tephra dispersion in the atmosphere, only varying as a function of height. Different assumptions about the particle size distribution in the eruption column, total mass, and maximum height of the eruption column will affect the distribution of mass in the deposit. Numerical methods are used to investigate the ranges of possible values of these eruption column parameters for a given deposit.

The variable $f_{i,j}(x, y)$ in Eq. (1) describes the atmospheric diffusion of tephra particles of size j that are released from column layer i and then advected by the wind away from the erupting column. Tephra particles of varying sizes have varying settling velocities in the atmosphere, which is ultimately reflected in their total fall time; fall time is measured from a particle's release point in the eruption column to deposition on the ground surface. For example, large particles with faster settling velocities will be less dispersed in the deposit than small particles with slower settling velocities, if both are released from the same height. Similarly, particles of a given settling velocity will be more dispersed when released from higher in the eruption column than when released lower in the column. The distribution of mass in the deposit is strongly affected by the total fall time of particles, which in turn, depends on their settling velocities. Modelling particles within a limited range of settling velocities (particles of similar size and density) allows values of $f_{i,j}(x, y)$, and hence $M^o_{i,j}$, to be better constrained for a given mass fraction $M(x, y)$ mapped in a specific deposit.

The settling velocity of a particle is a function of multiple parameters including particle density, particle volume (here characterized as an equivalent sphere diameter based on sieve measurements), particle shape, atmospheric properties, gravitational acceleration, and local Reynold's number (Kunii and Levenspiel 1969; Bonadonna et al. 1998; Bonadonna and Phillips 2003). Tephra2 employs a formulation of settling velocity proposed by Bonadonna et al. (1998) that changes based on whether a particle settles via a laminar, intermediate or turbulent regime at different heights in the atmosphere,

$$v_j = \begin{cases} \dfrac{\rho_j g d_j^2}{18\mu} & \text{if laminar, Re } < 6, \\ d_j \left[\dfrac{4g^2\rho_j^2}{225\mu\rho_a}\right]^{1/3} & 6 \leq \text{Re } < 500, \\ \left[\dfrac{3.1\rho_j g d_j}{\rho_a}\right]^{1/2} & \text{if turbulent,Re} \geq 500, \end{cases} \tag{2}$$

where ρ_j is the density and d_j the diameter of particles of size class j, $\mu = 0.000018325$ Pa s is the air viscosity, $g = 9.81$ m s^{-2} is gravity, ρ_a is the air density of layer k in the atmosphere, with

$$\rho_a = \rho_{std} e^{-h_k/8200} \tag{3}$$

where $\rho_{std} = 1.293$ kg m^{-3} at sea-level, and h_k is the height above mean sea level (AMSL) of the centre of atmospheric layer k. Note the difference in notation, i represents layers in the eruption column and k represents layers in the atmosphere, because these layers need not represent the same elevation or have the same total number of divisions.

Tephra particle density also varies as a function of particle size given that coarse particles contain numerous vesicles and fine particles are more likely to be made up of individual crystals. Tephra2 uses a linear model to approximate this change in particle density as a function of particle size (Bonadonna and Phillips 2003),

$$\rho_j = \begin{cases} \rho_{min} & \text{if coarse}, \varphi_j < \varphi_C, \\ \rho_{max} - \dfrac{(\rho_{max} - \rho_{min})(\varphi_j - \varphi_F)}{(\varphi_C - \varphi_F)} & \varphi_C \leq \varphi_j < \varphi_F, \\ \rho_{max} & \text{if fine}, \varphi_j \geq \varphi_F, \end{cases} \tag{4}$$

where ρ_{min} and ρ_{max} are the mean densities of the coarsest and finest fractions of the deposit, respectively. The coarse particle threshold is defined as φ_C, the fine particle threshold is defined as φ_F, with particle density varying linearly between the two. Tephra2 assumes fixed densities for fine and coarse particles, so the only factor affecting settling velocity is, d_j, the particle diameter.

The total fall time of a particle is thus estimated by summing particle fall times through multiple atmospheric layers, starting from a particle's initial release height from the eruption column, H_i, to its final resting place on the ground,

$$t_{i,j} = \sum_{k=0}^{H_i} \frac{z_k}{v_j} \tag{5}$$

where z_k is the thickness of layer k. The settling velocity is assumed to be constant within a given atmospheric layer, k, so the summation must be done in sufficiently small steps to avoid large settling velocity errors. Also, it is assumed that a particle instantaneously reaches its settling velocity within layer k. This assumption is reasonable for low levels in the atmosphere, but not necessarily true for high levels in the atmosphere (e.g., >40 km), where low atmospheric density creates very high settling velocities.

The total diffusion of particles of size j that are released from height H_i with a total fall time, $t_{i,j}$, is estimated using a linear, Fickian, model for coarse particles and a nonlinear model for fine particles, with the switch between the two models determined by a parameter, τ, known in the Tephra2 code as the fall time threshold (FTT),

$$\sigma_{i,j}^2 = \begin{cases} 4K(t_{i,j} + t_i') & \text{if } \mathrm{t_{i,j}} < \tau, \\ \frac{8C}{5}(t_{i,j} + t_i')^{5/2} & \text{if } \mathrm{t_{i,j}} \geq \tau, \end{cases} \tag{6}$$

where K is the Fickian diffusion coefficient, C is a turbulent eddy constant, t_i' is the increased horizontal diffusion time accounting for the broadening of the eruptive column with height. For relatively coarse particles, $t_{i,j} < \tau$, the diffusion is assumed to be linear with respect to total fall time. It has been shown (Suzuki 1983; Bonadonna et al. 2005) that for small particles found in tephra fallout deposits, $t_{i,j} \geq \tau$, diffusion is nonlinear with respect to total particle fall time. Analysis of deposits suggests that a value for τ of approximately 1 h is often appropriate (Bonadonna et al. 2005). In this formulation diffusion is assumed to be isotropic in the horizontal (X, Y) direction; vertical diffusion is neglected.

The variable t_i' reflects a change in width of the eruption column with height, a feature of eruption columns which effectively broadens the area impacted by the resulting tephra fallout deposit. Suzuki (1983) used,

$$t_i' = \left[0.2h_i^2\right]^{2/5} \tag{7}$$

and this formula is used here, acknowledging that more complex formulations associated with laterally spreading plumes are possible (Bursik et al. 1992; Bonadonna and Phillips 2003; Mannen 2014).

Given these important, albeit complex, details, it is possible to estimate diffusion and advection away from the eruptive column,

$$f_{i,j}(x,y) = \frac{1}{2\pi\sigma_{i,j}^2} exp\left[-\frac{(x - \bar{x}_{i,j}^2) + (y - \bar{y}_{i,j}^2)}{2\sigma_{i,j}^2}\right] \tag{8}$$

where

$$\bar{x}_{i,j} = x_0 + \sum_{k=0}^{H_i} \frac{w_{x,k} z_k}{v_j, k} \tag{9}$$

and

$$\bar{y}_{i,j} = y_0 + \sum_{k=0}^{H_i} \frac{w_{y,k} z_k}{v_j, k} \tag{10}$$

Particles originate from a point-release source in the eruption column (x_0, y_0, H_i) and are advected by the wind (velocity components $w_{x,k}$ and $w_{y,k}$). Thus, the wind direction and speed may vary as a function of height, but for simplicity do not vary as a function of horizontal distance.

Equation (8) shows that $\sigma_{i,j}^2$ represents the variance of a Gaussian distribution, and Eqs. (5) and (6) indicate that this dispersion term is directly controlled by a diffusion model that depends on particle settling velocity, and hence upon particle size.

Tephra2 assumes that the total particle size distribution (for the entire erupted mass of tephra) follows a lognormal distribution (i.e., is normally distributed in units of φ) (Inman 1952; Fisher and Schmincke 1984). Therefore the fraction of particles in size class j is

$$f_j(\varphi) = \frac{1}{\sqrt{2\pi\bar{\sigma}_\varphi}} exp\left[-\frac{\varphi - \bar{\mu}_\varphi}{2\bar{\sigma}^2_\varphi}\right] \quad (11)$$

where $\bar{\mu}_\varphi$ and $\bar{\sigma}^2_\varphi$ are the estimated mean and variance of the particle size distribution, respectively. A simple plume model (Sparks et al. 1992; Woods 1995; Ernst et al. 1996; Sparks et al. 1997; Bonadonna and Phillips 2003),

$$f_i(z) = U[H_{min}, H_t] \quad (12)$$

assumes that all particles are uniformly distributed within the eruptive column between H_{min} and H_t, regardless of particle size; for strong plumes, H_{min} and H_t represent the base and top of the umbrella region, respectively. In this case, tephra fallout associated with the development or waning stages of the plume are not accounted for, nor is sedimentation from the margins of the plume during steady eruption (Ernst et al. 1996) or ballistic transport (Fagents and Wilson 1993). Alternatively, a probability density function may be used to describe the distribution of tephra released from the erupting column as a function of height above the vent. Tephra2 employs a plume model using the beta distribution,

$$f_i(z) = \frac{1}{B(\alpha,\beta)} z^{\alpha-1}(1-z)^{\beta-1} \quad (13)$$

where

$$B(\alpha,\beta) = \frac{\Gamma(\alpha+\beta)}{\Gamma(\alpha)\Gamma(\beta)}, \quad \alpha > 0, \beta > 0 \quad (14)$$

and $\Gamma()$ is the gamma function. If $\alpha = \beta = 1$, then the beta distribution is the same as a uniform random distribution (Eq. 12). For cases where $1 \leq \alpha \leq 2$, when $\beta < 1$, then particle release heights are skewed toward the top of the erupting column; when $\beta \geq 1$, particle release heights are skewed toward the lower part of the column. Thus the equation,

$$M^o_{i,j} = M^{\varphi_{max}}_{\varphi_{min}} f_j(\varphi) f_i(z) \quad (15)$$

describes the total mass of tephra, $M^{\varphi_{max}}_{\varphi_{min}}$, between size classes φ_{min} and φ_{max}.

Given these model assumptions, everything is in place to calculate tephra mass loading, $M(x,y)$, using the model parameters just described. Table 5 summarises the necessary model inputs for 2 classes of simulations of the 1913 Colima deposit, models using collected stratigraphic samples and models using collected stratigraphic samples plus historical reports. Output of Tephra2, $M(x,y)$, is in units of mass loading per unit area (kg m^{-2}). This output is appropriate for hazard assessments involving mass loading effects such as roof collapse, crop destruction, failure of power, water, and/or mechanical systems, etc. Outputs can be compared with field measurements of deposit thickness by dividing $M(x,y)$ by a bulk deposit density.

Tephra2 can be used in a forward modelling approach where model inputs (Table 5) are selectively changed so that model outputs match the thicknesses of a tephra unit observed in stratigraphic sections over a region. This approach is laborious, however, and the problem is ill-posed and ill-conditioned, meaning that many different combinations of model inputs result in outputs that match field observations. For example, acceptable matches to observed deposit thicknesses might be found by adjusting model wind speed, eruption column height, or both. In these circumstances, understanding the range of input parameters that might account for observed deposit thicknesses is difficult to quantify. This situation is common in geophysics and can be better addressed using nonlinear inversion methods.

The nonlinear inversion method implemented with Tephra2 is based on the downhill simplex algorithm (Nedler and Meade 1965; Press et al. 1992; Connor and Connor 2006; Magill et al. 2015). This algorithm uses a simplex of $N+1$ vertices or parameter sets, where N is the number

Table 5 Estimating isomass, $M(x, y)$, for a simulated explosive eruption of Volcán de Colima using Tephra2. Inversion outputs include best-fit eruption parameters using geological data only (Table 1) and using geological data plus data from historical accounts of tephra accumulation

Model	Variable	Model parameters and/or ranges		
Parameter (units)	Name	Input	Geological	Geological + historical
Volcano				
Easting (m)	x_0	645,110	–	–
Northing (m)	y_0	2,158,088	–	–
Elevation (AMSL)	z_0	3085	–	–
Sample file		–	38 samples	59 samples
Outputs				
Total eruption mass (kg)	$M_{\varphi_{min}}^{\varphi_{max}}$	$10^8 - 10^{13}$	5.5×10^{10}	1.4×10^{11}
Eruption column				
Integration steps	i	200	–	–
Outputs				
Column level (km ASL)	H_t	*fixed*	24	24
β Distribution	α	1.0 – 2.0	1.02	1.0
β Distribution	β	0.01 – 3.0	1.56	0.96
Particle size				
Integration steps	j	100	–	–
Maximum (φ)	φ_{max}	−7.0	–	–
Minimum (φ)	φ_{min}	7.0	–	–
Density-fines (kg m^{-3})	ρ_{max}	2700	–	–
Density-coarse (kg m^{-3})	ρ_{min}	1024	–	–
Threshold-fines (φ)	φ_F	7.0	–	–
Threshold-coarse (φ)	φ_C	−1.0	–	–
Outputs				
Median (φ)	$\bar{\mu}_\varphi$	−2.5 – 2.5	−1.8	1.8
Sigma (φ)	$\bar{\sigma}_\varphi$	0.5 – 2.5	2.4	2.4
FTT (s)	τ	$10 - 1 \times 10^6$	6458	5525
Atmosphere model				
Eddy constant (m^{-2} s^{-1})	C	0.04	–	–
Wind velocity	$w_{x,k}, w_{y,k}$			
Wind levels (AMSL)	k	12	–	–
(*Outputs*)				
Wind speed (m s^{-2})		0 – 24.6	*fixed*[a]	*fixed*[a]
Wind direction (° from N)		10.1 – 156.0	*fixed*[a]	*fixed*[a]
DC (m^2 s^{-1})	K	$10 - 1 \times 10^5$	4016	4501

[a]Atmospheric model is the REANALYSIS Wind Model, 1-Jan-2013, mid-morning, found by running the inversion with a set of 123 possible wind profiles and selecting the best-fit profile. Subsequent inversions use this wind model (Figs. 8 and 10; wind profile shown in Fig. 11)

of parameters to be estimated (Table 5), to iteratively search for an optimal set of model input values that minimizes the error between model output and measured observations. To evaluate goodness of fit, a cost function is minimized. The cost function is an equation that directly compares the calculated and observed mass per unit area for each sample location. Here we use the root mean squared error (RMSE) as the goodness of fit metric,

$$RMSE = \min_{t \to 0} \sqrt{\frac{\sum_{p=1}^{P} (M_{p,obs} - M_{p,calc})^2}{P}} \tag{16}$$

where $M_{p,calc}$ are the calculated values and $M_{p,obs}$ are the measured values at each sample location p, for a total of P sample points where observations were made (e.g., Table 1), and t is the tolerance of the downhill simplex algorithm. The RMSE gives a greater weight to larger observations. An alternative cost function described by (Mannen 2014),

$$E = \min_{t \to 0} \sum_{p=1}^{P} \left[\log \left(\frac{M_{p,calc}}{M_{p,obs}} \right) \right]^2 \tag{17}$$

can be chosen when equal weighting of observations is desired. The quality of the resulting solutions is compared using a normalized root mean squared error

$$NRMSE = \frac{\sqrt{\sum_{p=1}^{P} (M_{p,obs} - M_{p,calc})^2 / P}}{\max(M_{p,obs}) - \min(M_{p,obs})}. \tag{18}$$

In practice, some eruption parameters are more uniquely constrained than others by inversion (Connor and Connor 2006; Volentik et al. 2010; White et al. 2017). Mass accumulated at a point is linearly dependent on the total mass, $M_{\varphi_{min}}^{\varphi_{max}}$ (Eq. 15); therefore total eruption mass is often well constrained through inversion (Scollo et al. 2008). Maximum eruption column height, particle size distribution, and wind field are more interdependent (Scollo et al. 2008) and model output is non-linearly dependent on these parameters; therefore unique minima in the cost function do not necessarily occur when modelling these parameters. This non-uniqueness of model parameters increases uncertainty in the model. Using inversion is one way to bound these uncertainties.

Wind velocity and wind direction are a major source of model uncertainty. For example, low eruption columns in high velocity wind fields might produce similar deposits to high columns in lower velocity wind fields. The wind field during the 1913 Colima eruption is basically unknown, although seasonal wind profiles can be inferred. During inversion, it is possible to extract one possible wind profile or to run the inversion using a fixed, pre-selected wind profile. Because of the non-unique nature of determining the most appropriate wind field, we chose to use a fixed, pre-selected wind profile for all of the inversions to reduce the atmospheric uncertainty.

Model uncertainty is further reduced by inverting each sieved phi (φ) size separately (Volentik et al. 2010); this methodology reduces the degrees of freedom in the inversion procedure. Inversion using a single sieve size reduces the uncertainty related to large variations in settling velocities from particles of different sizes. This approach adds additional information while constraining eruption dynamics (e.g., the height of the erupting column and the duration of eruption) because model results from multiple inversions can be compared, each representing a fraction of the total parameter space. For example, it may be possible to determine if fine particles are released from a greater height in the eruption column than coarse particles.

Of course, a single binned phi (φ) class does not correspond to a unique settling velocity. Particles within a given phi (φ) class have a range of sizes, densities, and shapes (Alfano et al. 2011a), but these correspond to a much narrower range of settling velocities than represented by the entire range of particle sizes in the deposit. In practice, a single mean ($\bar{\mu}_\varphi$) and standard deviation ($\bar{\sigma}_\varphi$) is assumed for each binned phi (φ) range. This is done by choosing appropriate values for $\bar{\mu}_{\varphi}$ and $\bar{\sigma}_{\varphi}$, and fixing these parameters as constants in the inversion. For example, to model the mass loading of −1 phi ($\varphi = -1$) particles in a deposit (particles trapped in a -1φ size sieve having passed through a -2φ size sieve), one can set $\bar{\mu}_{\varphi} = -1.5$ and $\bar{\sigma}_{\varphi} = 0.5$. This yields a range of model settling velocities that is assumed to reflect actual settling velocity variation for this size fraction.

In the following section, the 1913 tephra-stratigraphic data set is used to model eruption parameters for the 1913 eruption of Volcán de Colima. First, physical parameters of the eruption, i.e. erupted volume and plume height, were derived based on deposit thinning (Pyle 1989; Bonadonna and Houghton 2005; Bonadonna and Costa 2012) and on the distribution of the largest scorias and lithics around the volcano (Carey and Sparks 1986), respectively. Second, the eruption style was classified based on the scheme of Pyle (1989). Third, the deposit was modelled using Tephra2 with inversion based on the total deposit thickness. Fourth, observed granulometry data was compared with inversion results obtained from individual phi (φ) size modelling as outlined above. Overall, this approach provides an improved uncertainty estimate for the eruption parameters that might have produced our 1913 tephra-stratigraphic data set.

5 Statistical Models of Deposit Volume and Column Height

Total volume estimates of the 1913 tephra deposit and the associated maximum eruption column height were made by interpolating the mapped deposits. First, an isopach map of unit α was contoured based on data collected by Luhr et al. (2010) (Fig. 4). Deposit volume was calculated based on the integration of various thinning fits plotted on semi-log plots of thickness versus square root of isopach areas. Here, we use

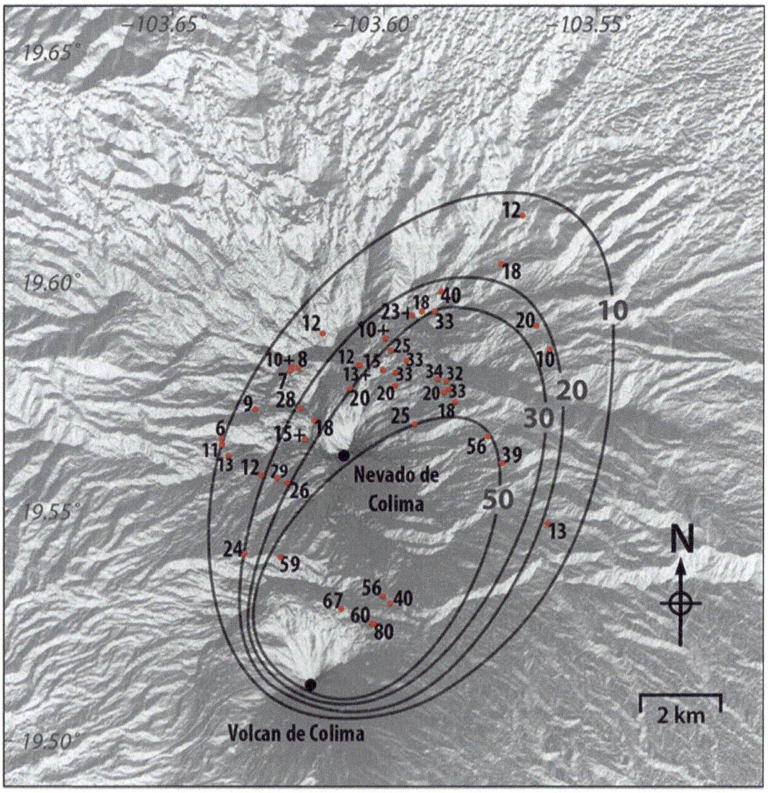

Fig. 4 Isopach map of the 1913 Colima deposit, layer α (red dots), contoured using deposit thickness (cm) reported in Table 1

the integration of one exponential segment (Pyle 1989) and the Weibull fit (Bonadonna and Costa 2012). For the exponential model, the estimated volume is 5.1 ± 1.1 × $10^7 m^3$, with 95% confidence. This volume is considered to be a minimum value for two reasons: (i) the space discontinuity of deposit thinning suggests that the deposit might have been subjected to erosion, especially on the steep slopes of Nevado de Colima and (ii) the poor deposit exposure in distal areas only allows for one exponential segment to be identified. As a result, the deposit could be characterized only to the 10 cm isopach line located approximately 13 km from vent. The Weibull fit gives a similar volume estimate, 6.2 ± 2.4 × $10^7 m^3$, with 95% confidence. Again, this volume is likely a minimum because of the erosion of the distal part of the deposit. The larger volume found with the Weibull method may reflect a more realistic model of the proximal part of the deposit, but again this is poorly constrained by the stratigraphic data.

Second, isopleth maps of both the largest scorias and the largest lithics of unit α were also compiled (average of the 5 largest clasts; Figs. 5 and 6) and an associated plume height of 14 and 15–17 km, respectively, above the sampling height (i.e., about 18 km and 19–21 km amsl) and maximum wind at the tropopause during the eruption of 30 and 20 m s^{-1}, respectively, were inferred based on the application of the method of Carey and Sparks (1986). Diameters of 0.8 and 1.6 cm and a density of 2700 kg m^{-3} were used for the calculation with the largest lithics. Given that the density of scoria clasts is ~1000 kg m^{-3} the method of Carey and Sparks (1986) is applied using the 2 and 4 cm isopleth lines (Fig. 5). A plume height of 13 and 15 km above sampling height (17–19 km amsl) was determined applying the Weibull-based empirical equation of Bonadonna and Costa (2013) to the lithic and scoria isopleth maps, respectively.

Third, using the method described by Pyle (1989), thickness and clast half-distance are

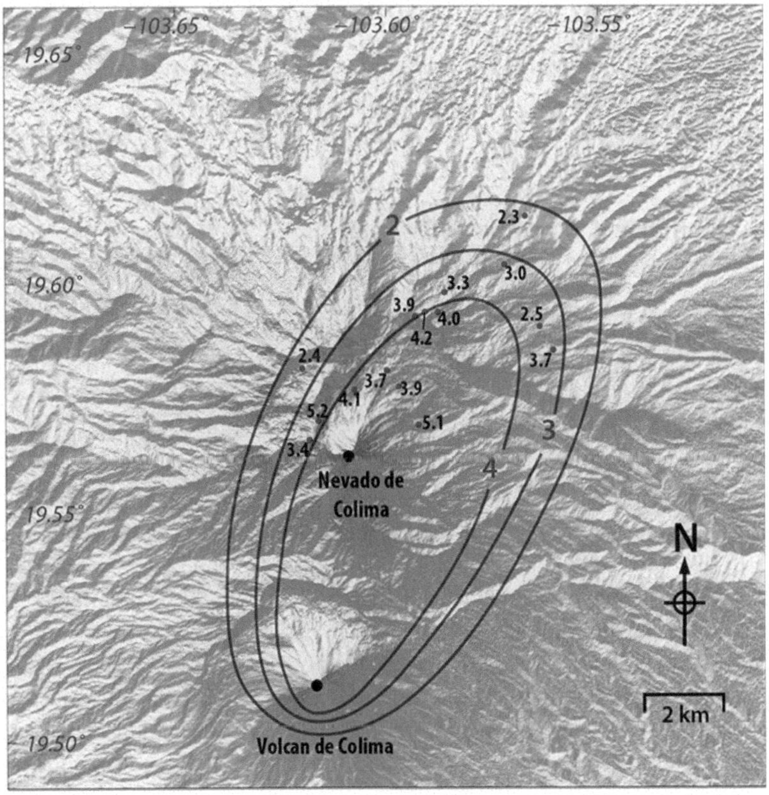

Fig. 5 Isopleth map of the 1913 Colima deposit, layer α, contoured using largest scoria clasts (red dots), values are largest clast dimension in cm

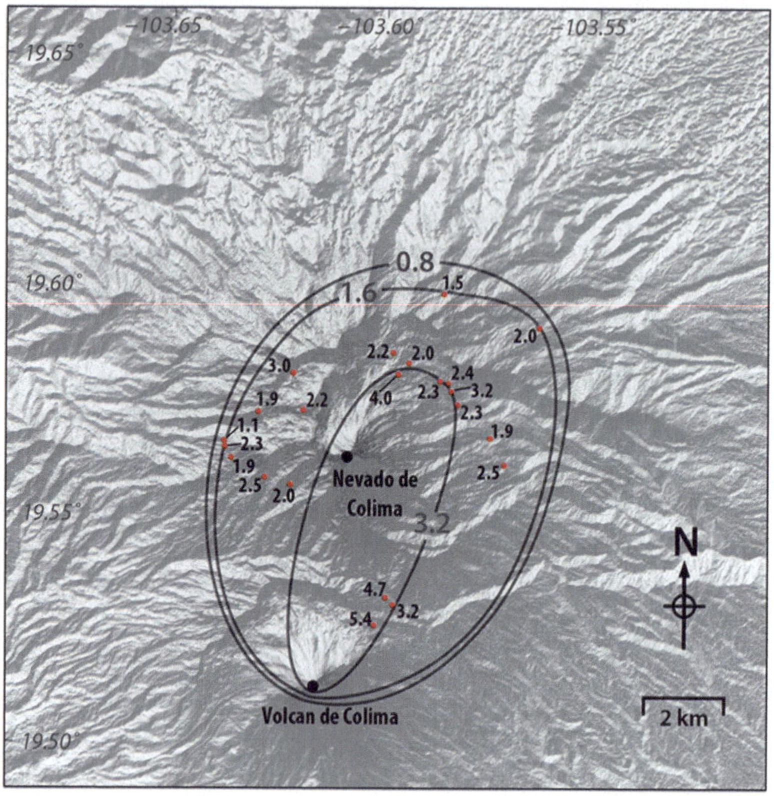

Fig. 6 Isopleth map of the 1913 Colima deposit, layer α, contoured using largest lithic clasts (red dots), values are largest clast dimension in cm

estimated to be 1.0 and 1.6 km, respectively, allowing us to classify the eruption that produced unit α as sub-Plinian.

6 Numerical Models of Deposit Volume and Column Height

Inversion was initially performed using the $P = 38$ sample points of unit α, identified in Table 1, and the range of input parameters specified in Table 5. For these inversions, goodness-of-fit was 0.16 (NRMSE) while total eruption mass varied between 2.7×10^{10} and 5.6×10^{10} kg, median grainsize varied between −2.68 and −0.81 ϕ, and cumulative probability plots (99% tephra release height) show that most tephra was released at heights less than from between 14 and 26 km. In other words, a wide range of model parameters could fit the deposit with equivalent goodness-of-fit. This great uncertainty is not unexpected, given the rather limited spatial distribution of outcrops relative to the expected area of the deposit. The best-fit models find a relatively coarse deposit, also consistent with the proximity of stratigraphic section locations to the vent.

Historical accounts of tephra thickness are available for the 1913 eruption, largely derived from newspaper reports. We explored the impact of incorporating historical data on the eruption model by combining reported tephra thicknesses with the measured geological data. These reported localities are much more widely dispersed than the preserved geological sections. Trace amounts of tephra fallout were recorded at some localities located hundreds of kilometres from the erupting volcano (Fig. 1). However, no locality could be identified, where the distal facies of the deposit is preserved in the geological record. The eruption model changes significantly (Table 6) when the historical estimates are added to the measured data values. Again, a wide range of possible eruption conditions emerge during

Table 6 Comparison of estimated eruption parameters for inversion by particle size fraction, with inversions on the total deposit thickness (Geol + historical and geological). Note that median release height increases in the eruption column for finer particle size fractions. For all size fractions, most particles are released at <21 km

Data set	Total mass (kg)	α	β	$H_{50\%}$ (km)	$H_{95\%}$ (km)	$H_{99\%}$ (km)	*NMRSE*
Geol + Historical	1.4×10^{11}	1.04	1.46	11.7	21.4	23.1	0.14
Geological	5.5×10^{10}	1.0	0.96	14.2	23.1	23.9	0.16
$-4\ \varphi$	2.2×10^{9}	1.0	3.0	8.0	16.6	19.7	0.09
$-3\ \varphi$	5.9×10^{9}	1.0	2.69	8.4	17.4	20.4	0.09
$-2\ \varphi$	6.3×10^{9}	1.23	2.90	9.1	17.5	20.3	0.15
$-1\ \varphi$	6.4×10^{9}	1.66	4.05	9.1	16.3	18.9	0.18
$0\ \varphi$	9.3×10^{9}	2.0	4.25	9.8	16.6	19.1	0.20
$1\ \varphi$	9.0×10^{9}	1.46	3.0	9.8	17.9	20.5	0.20
$2\ \varphi$	1.1×10^{10}	1.97	3.0	11.4	18.9	21.1	0.16

inversion due to model indeterminacy. Including historical data, eruption masses range from 3.5×10^{10} to 1.5×10^{11} kg, including a wide range of potential column heights and potential median particle sizes.

As described previously, one source of model uncertainty is the wind velocity profile. Tephra2 inversion can search for a best-fit wind profile as a function of column height. This approach, however, can lead to solutions that allow rapid changes in wind direction or speed with height, which is not necessarily realistic, especially when these complex wind profiles are inferred by modelling a limited sample point distribution. Alternatively, the wind profile can be fixed using 12 atmospheric levels, and a particular wind speed and wind direction specified for each atmospheric level. We adopted this approach. Selection of the most appropriate wind profile was performed during initial inversions using 123 unique wind fields based on one month (January 2013) of NOAA REANALYSIS wind data (Kalnay et al. 1996). The inversion was performed and a best-fit result obtained for each of the 123 profiles. The REANALYSIS wind profile giving the best-fit inversion solution determined among all of these results became the one wind profile used throughout the analysis. In practice, we identified several wind profiles that yielded equally good model fits. We selected one of these and used this wind profile in subsequent inversions to gain insights into the uncertainty of the remaining model parameters. We note that the model fits using the REANALYSIS wind data are virtually the same as those obtained using wind velocities estimated from the inversion only, but the REANALYSIS profiles are deemed to be more realistic.

Example best-fit inversion results with the fixed wind profile are reported in Table 5. We emphasize that for each data set a range of solutions are of essentially equal best-fit. These results are used to plot isomass maps and goodness-of-fit plots. Inversion using the $P = 38$ stratigraphic sample points, yields a model fit with a $NRMSE = 0.16$ (Fig. 7) and the isomass map shown in Fig. 8. Inversion using the combined geological and the historical data, yields a model fit with a $NRMSE = 0.14$ (Fig. 9) and the isomass map shown in Fig. 10.

This difference in goodness-of-fit between the two data sets, exemplified by these two solutions, reflects uncertainty in characteristics of the eruption column and particle size distribution used in these models. Recall that these maps are produced using the same wind profile, so model differences must be related to eruption parameters. Incorporating the historical data allows for using a greater range of eruption parameters. For example, excellent fits can be obtained using a range of erupted masses, with finer median particle size estimates corresponding to greater

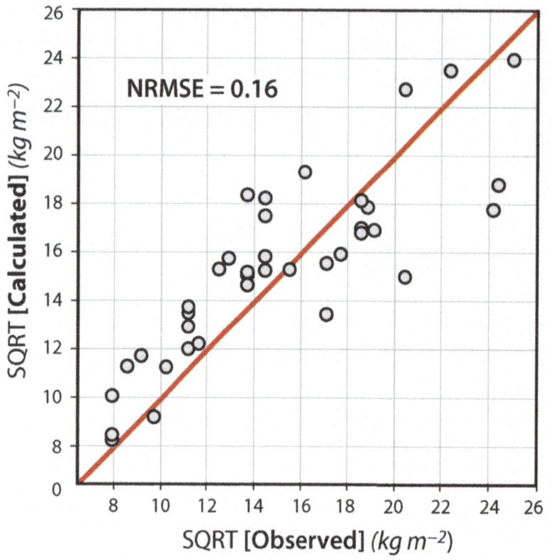

Fig. 7 Equiline plot showing agreement between observed mass loading and calculated mass loading (*square root*) of tephra for the 1913 eruption based on thickness measurements of geologic sections of the deposit (Table 1). See Table 5 for estimated model parameters. Normalized Root Mean Squared Error (NRMSE) is 0.16

erupted masses. Using the historical data allows for a higher eruption column with particle release concentrated in the upper part of the plume. These inversions typically yield maximum column height of $H_t > 22$ km AMSL. Column shape parameters, α and β, are consistent with particle release from a narrow umbrella region at the top of the plume (Table 5). The implications of this comparatively high release height from the eruption column and fine particle size distribution are clear in the isomass map (Fig. 10). Much of the modelled deposit for the geological + historical data is distal to the geological sample localities, also reflecting high uncertainty.

To help differentiate between end-member models of the eruption, one that is relatively coarse grained and of shorter column height, the other relatively fine grained and higher column height, we modelled the mass accumulation at the 38 stratigraphic sampling points by individual particle size. Again, in this procedure the

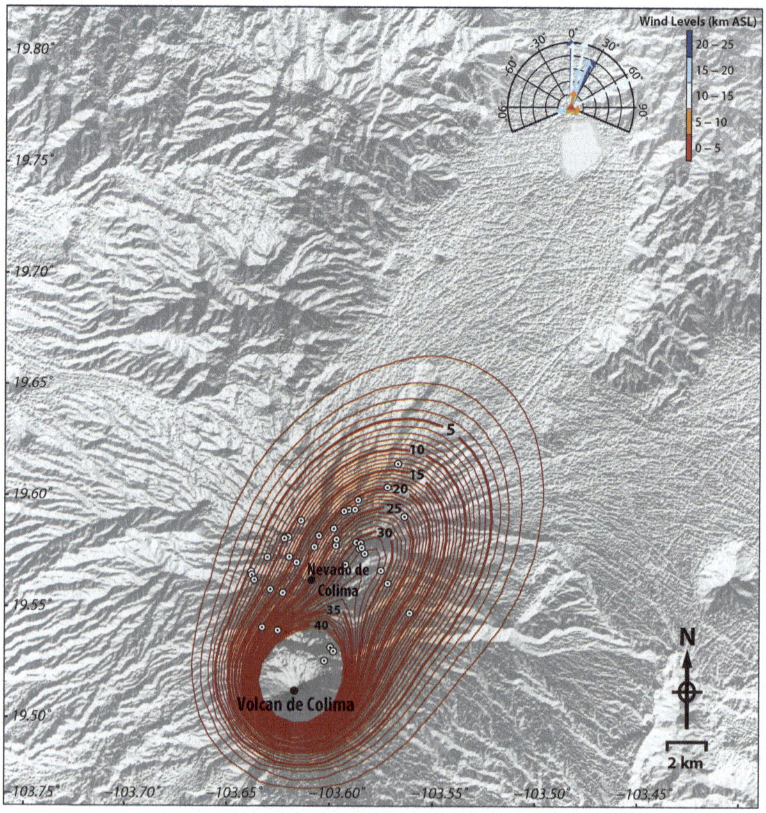

Fig. 8 Isopach map of the 1913 eruption estimated using data from the geologic record (Table 1) only. Contours represent tephra accumulation in cm based on best fit parameters from inversion results. Modelled values in kg m^{-2} are converted to cm using a measured bulk scoria-fall-deposit density of 1.043 g cm^{-3}. Wind speeds and directions for this model were fixed and based on inversion with NOAA REANALYSIS wind data using the combined geological/historical data set

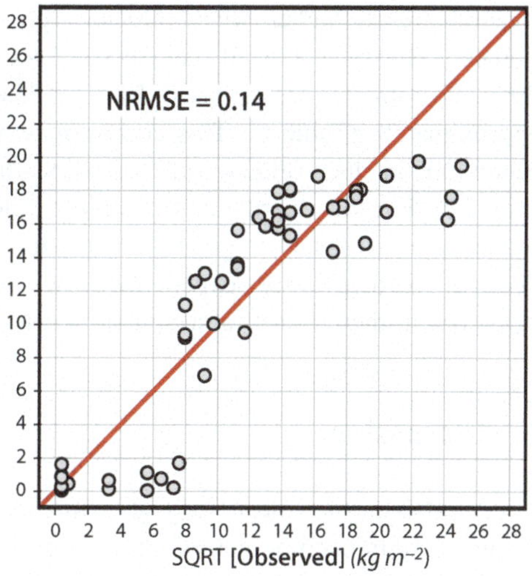

Fig. 9 Equiline plot showing agreement between observed mass loading from geological sections (Table 1) and data derived from historical accounts (Table 7) and calculated mass loading of tephra for the 1913 eruption (see Table 5 for estimated model parameters)

wind profile remained constant (i.e. the fixed wind field originally selected from January 2013 NOAA REANALYSIS data). The total eruption column height (H_t) was fixed at 24 km AMSL, the column shape parameters were given a fixed range, α was allowed to vary between 1 and 2, and β was allowed to vary between 0.01 and 3.0. These value ranges allowed for the possibility of a variety of particle release patterns, from major release at the top of the erupting column to larger releases nearer the vent. Total mass of erupted tephra of a given size fraction (φ class) was allowed to vary between a potential minimum volume of 1×10^8 to a potential maximum volume of 9×10^{11}, the diffusion coefficient and FFT were allowed to vary between 10 and 100,000.

Particle size fractions $\phi = -4, -3, -2, -1, 0, 1$ and 2 were each modelled independently. The quality of model fit, indicated by

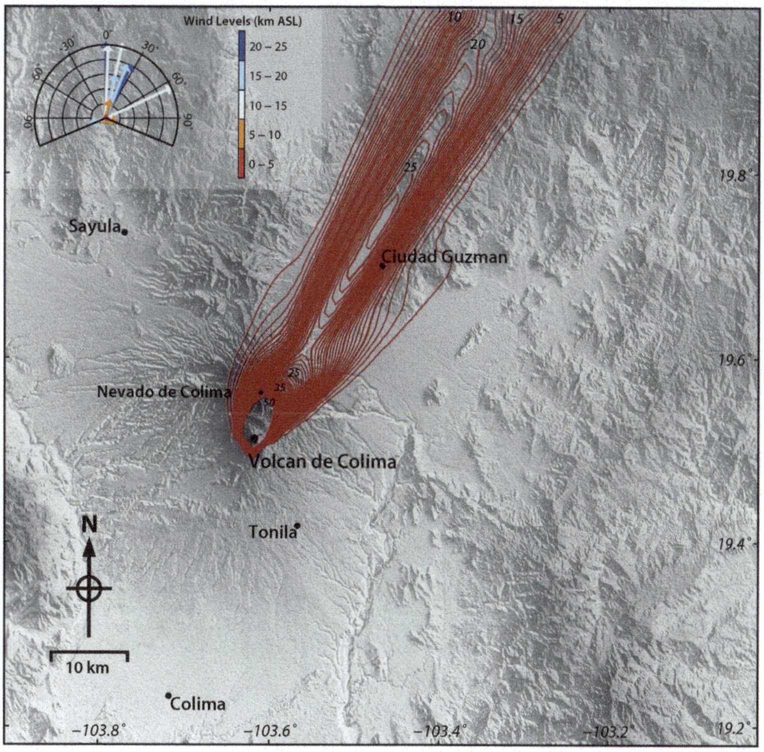

Fig. 10 Isopach map of the 1913 eruption estimated using data from the geologic record (Table 1) and data derived from historical accounts (Table 7). Contours represent tephra accumulation in cm based on best fit parameters from inversion results. Modelled values in kg m^{-2} are converted to cm using a measured bulk scoria-fall-deposit density of 1.043 g cm^{-3}. The wind speed and direction were determined from inversion, selecting the best-fit NOAA REANALYSIS wind profile among 123 samples, all from January 2013

NRMSE, deteriorates significantly for finer particle sizes outside of this range, so they were not included. This result makes sense because all measured stratigraphic sections are located relatively close to the volcano and, therefore, show little variation in the fine particle size fractions.

Inversion results are summarized in Table 6. The modelled mass by particle size fraction increases from −4 to 2ϕ, suggesting that the median particle size for the eruption may be closer to 2ϕ, significantly finer than the median found based on the granulometry of the 38 sample stratigraphic sections.

Models by particle size fraction also indicate there is some separation by particle size vertically within the eruption plume. For relatively coarse particles ($\phi = -4$–-2), modal release height is <6 km; $\phi = -1$ –1 particle size classes have modal release heights of 7–9 km; and $\varphi =$ 2 modal release height is approximately 11 km. The modal release height for each modelled particle size fraction is located below the level of maximum wind velocity, but finer particle size fractions reach higher wind velocities proportionately (Fig. 11). For all particle size models, 99% of particles are released from <21 km, approximately (Table 6). This corresponds to a minimum in the velocity of the wind field (Fig. 11). For this distribution of sample points, models are insensitive to greater column heights because particles will experience low wind velocities at these higher atmospheric levels.

7 Discussion

A central issue in volcanology is estimation of the uncertainty in eruption magnitude, commonly characterized using the volcano explosivity index, from deposits preserved and identified in the geological record (e.g., Klawonn et al. 2014; Bonadonna et al. 2015; White et al. 2017). At Volcán de Colima, the 1913 eruption products are found over a very limited area proximal to the vent. This situation is typical for many eruptions (e.g., Longchamp et al. 2011; Burden et al.

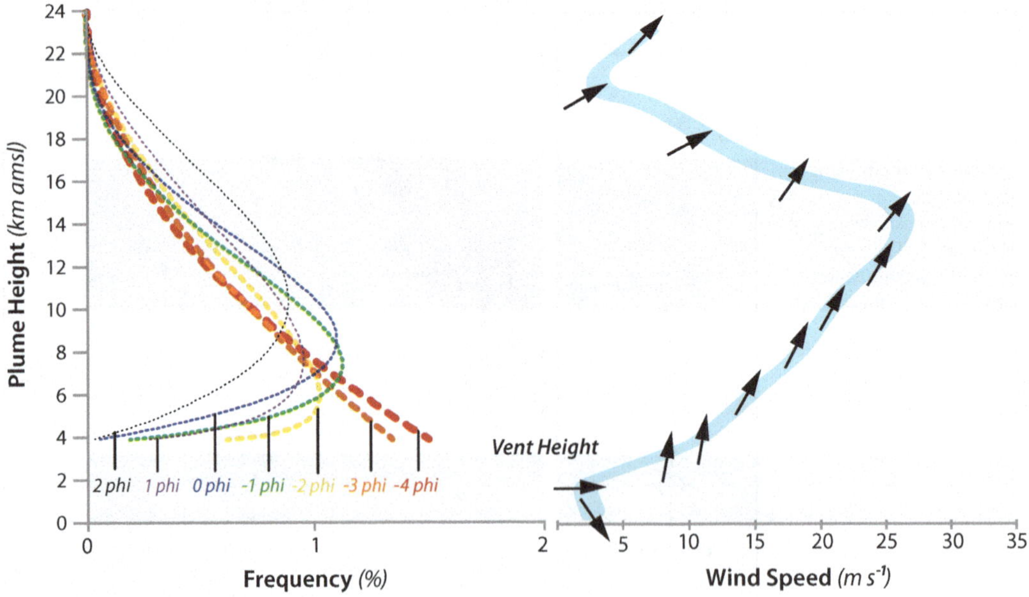

Fig. 11 Left panel: modelled concentration of particle release height from the eruption column as a function of particle size for best-fit models shown in Table 6. Note the tendency for finer particles to be released higher in the column. Right panel: model wind velocities shown as blue curve (best-fitting profile of NOAA REANALYSIS data, see Fig. 10 for map of total deposit thickness produced by this wind profile); arrows indicate wind direction as a function of height, predominantly blowing to the NE

2011), which therefore makes it a pertinent case study. Although the α layer is only 100 years old, we have simply not been able to identify distal facies.

We have used a variety of methods to reconstruct eruption parameters from the 38 geologic sections we have located and studied around the volcano (Table 1) and have *fingerprinted* this event using trace element geochemical analyses. Maximum clast methods (Carey and Sparks 1986; Burden et al. 2011; Bonadonna et al. 2013) and integration and extrapolation of the interpolated isopach map (Pyle 1989; Bonadonna and Costa 2012), yield a minimum column height of 18 km amsl (i.e. 15 km above vent height) and volume of 5–7 × 10^7 m^3. These results suggest, confirmed by the classification scheme of Pyle (1989) that the eruption was sub-Plinian. Results of the numerical inversion of the total deposit, and inversion by individual particle sizes rely on completely different model assumptions and techniques. Nevertheless, for the coarse fraction of the deposit these methods yield similar column heights, with 99% of particles are released from <21 km, and a similar range of volumes.

Results of modelling the 38 stratigraphic sections are at odds with the modelling of Saucedo et al. (2010) and Bonasia et al. (2011), who found higher columns (approximately 23 km) and much larger eruption volumes (approximately 2–6 × 10^{11} kg, or approximately 2–6 × 10^8 m^3). This difference in volume and column height estimates arises partly from differences in section measurements on Nevado de Colima, with the measured sections reported in Saucedo et al. (2010) generally being thicker in any given area. In addition, Saucedo et al. (2010) identified the α layer in stratigraphic sections at distances of >40 km from the volcano, which we were unable to identify. Such sampling uncertainty, illustrated by discrepancies between the two data sets, is common and gives rise to uncertainty in tephra volume estimates (Engwell et al. 2013). Despite the inconsistency in measured section thickness, we can also obtain greater deposit volumes and column height estimates when we incorporate historical reports into our inversions and use relatively fine median particle size. Specifically, best-fit models at the upper part of the range include volumes of 1 × 10^8 m^3 and column height to approximately 25 km. That is, if historical data (Table 7) are included in the analysis, and a fine particle size distribution is assumed, the volume and column height estimates can increase substantially, and the 1913 eruption should be classified as Plinian. Based on our measurements of the thicknesses of the preserved deposit (Table 1), we appear to have large uncertainty in the magnitude of the eruption.

There are, however, clues in the granulometry of measured stratigraphic sections that support the finer total particle size distribution and larger deposit volume. First, the modelled mass by particle size fraction increases from -4ϕ to 2φ, suggesting that the median particle size for the eruption may be closer to 2φ. Second, the total mass from these models, found by summing total mass from −4 to 2φ (Table 6), is 5.2 × 10^{10} kg. This suggests that the total deposit mass might be double this value, assuming a roughly normal distribution of particle size. Thus, if we had no historical record at all, the key to understanding the deposit particle size distribution and mass is modelling of the granulometry. Based on these arguments we favour a deposit volume of 1–5 × 10^8 m^3, meaning the 1913 eruption should be classified as VEI 4. For comparison, this estimated volume is smaller than the 1982 Plinian El Chichón eruption (VEI 5, De la Cruz-Reyna and Martin Del Pozzo 2009; Bonadonna and Costa 2013) or the 2008 Chaitén (Chile) eruption (Alfano et al. 2011b), and larger than the 2010 Merapi (Indonesia) eruption (classified as VEI 3–4; Solikhin et al. 2015).

Uncertainty in this estimate is high, because we do not know the total particle size distribution from the measured sections and uncertainty in the modelled volume increases from coarse particles to fine particles because of the section locations are close to the vent.

Features of the geochemistry of the eruption appear to favour the finer total particle size distribution. The relatively high silica content of the groundmass (Table 3) and comparatively high magma viscosity favour a higher production rate

Table 7 Locations and thickness of tephra reported in historical accounts of the 1913 eruption. Data from Waitz (1935) and Saucedo-Girón (1997). Thickness reported as 0.01 cm where "trace" and similar terms used to describe the deposit

Location	Lat. (N)	Long. (W)	Thickness (cm)
Cd. Guzman	19.700	103.467	8
La Piedad to Yurecuaro	20.350	102.150	0.01
Tonila	19.418	103.567	0.01
Zacoalco	20.227	103.577	0.01
Zacatecas	22.782	102.577	0.01
Aguascalientes	21.881	102.288	1
Sayula	19.736	103.769	1
Teocaltiche (Jal)	21.436	102.577	3
Cuquio (Jal)	20.927	103.029	5
La Barca (Jal)	20.273	102.548	4
Lagos (Jal)	21.363	101.923	0.01
Ameca (Jal)	20.545	104.048	0.01
Fresnillo (Zac)	23.182	102.865	0.01
Yahualica (Jal)	21.182	102.894	0.01
Guadalajara	20.683	103.333	0.05
Guanajuato	21.018	101.260	0.01
Saltillo	25.436	100.990	0.01
San Luis Potosi	22.153	100.980	0.01
Mazamitla (Jal)	19.909	103.019	0.01
Tizapan el Alto (Jal)	20.145	103.048	0.01
Poncitlan (Jal)	20.373	102.942	3

of fine ash (Rose and Durant 2009). The high volatile content of the magma (up to 6.05 wt% H_2O; Luhr et al. 2006), the high abundance of small bubbles in individual scoria, and evidence of rapid ascent rate from reaction rim studies all point to deep and efficient magma fragmentation. This deep fragmentation may also provide increased opportunity for the generation of fine particles due to high energy particle-particle impacts in the conduit above the level of fragmentation (Dufek et al. 2012).

Interestingly, the high volatile contents of melt inclusions from the Colima 1913 tephra (up to 6.05 wt% H_2O; Luhr et al. 2006) mimic the elevated volatile contents of melt inclusions from the Pleistocene La Erita alkaline scoria cone (>5 wt% H_2O; Maria and Luhr 2008), which is associated with Volcán de Colima. However, La Erita scoria cone erupted hot but highly degassed ($H_2O \sim 0.2$ wt%) basanite magma (Carmichael et al. 2006), erupted as mostly low viscosity lava flows. It is suggested here that the combined study of the mafic Volcán de Colima magmas (like the 1913 deposit) in combination with the associated alkaline cones on the flanks of this volcano may provide fruitful insights about the impact of magma ascent rates and degassing on mafic eruption style.

Finally, the question arises: why do the geological sections appear to yield biased results toward a smaller eruption column and smaller total volume, especially since maximum clast studies and inversion of coarse particle sizes generally yield consistently lower columns? One possibility is that particle segregation by settling velocity in the upper part of the plume was relatively efficient. That is, rather than a well-mixed plume ascending to a total column height, upward velocity dropped sufficiently for the coarse particles to fall out without reaching to top of the plume (Fig. 11). Such segregation, and departure from a bulk plume behaviour, may be evidence of the role of post-fragmentation fining of particles through collisions in the conduit (Dufek et al. 2012).

Alternatively, it is possible that our understanding of the time varying nature of the 1913 eruption is simply insufficient. Given the imprecise nature of historical reports, an early sub-Plinian eruptive phase may have occurred, producing the α layer. Following this, fine ash fallout that occurred in Ciudad Guzman and elsewhere that was unrelated to the α layer fallout event, but rather was due to the commutation of particles in pyroclastic flows, and the generation of a co-pyroclastic flow plume following the sub-Plinian eruption. Modelling by individual particle size fraction allows us to discount this hypothesis, since we would not expect to see the trends in volume and column structure by particle size (Table 6).

Overall, modelling the α layer deposits, in light of the historical record, indicates there are two ways to approach modelling such a layer which can yield substantially different results. In the first, inversion proceeds with the eruption parameters (including particle size distribution) estimated through best-fit. For the 1913 Colima eruption, is appears to bias the result toward lower volume, likely due to the limited spatial distribution of reliable outcrops. Alternatively, individual particle size fractions are modelled. For 1913 Colima, this latter approach better agrees with historical accounts, albeit with great uncertainty in the estimated volume. A better understanding of magma fragmentation and its relationship to total particle size distribution might also greatly improve our understanding of the range of eruption magnitudes at many volcanoes where outcrops are limited.

8 Conclusions

Statistical and numerical models of the thickness and maximum clast size distribution of the α layer yield consistent volumes ($\sim$4–6×10^7 m^3) and column heights ($\sim$18 km AMSL) compatible with a subPlinian eruption. Much greater volumes and eruption column heights for the 1913 Colima eruption are permissible by including the historical reports in the inversion, or by fixing the total particle size distribution to $\mu_{\varphi} = 2.0$. Modelling by particle size fraction of the α layer supports a fine-grained, large volume eruption, with estimated range 1–5×10^8 m^3 tephra (VEI 4). This suggests that if the deposit had not been widely and largely eroded, maybe also the empirical methods and the classification scheme of Pyle (1989) could result in a Plinian eruption. Segregation of mass by particle size in the eruption column is suggested by modelling particle size fractions separately, and can account for the lower volumes found by modelling α layer thickness, and lower column heights estimated using maximum clast methods.

Geochemical analyses indicate magma volatile content of approximately 4–6 wt%, rapid ascent velocities, limited degassing during ascent, and a high concentration of small bubbles in scoria fragments, all consistent with the occurrence of relatively deep fragmentation and efficient production of fine ash.

These results indicate that for deposits found in the geological record, especially those of limited spatial distribution of outcrops, including inversion by particle size fraction can improve model resolution and help reduce potential bias in estimation of eruption parameters.

References

Alfano, F., Bonadonna, C., Delmelle, P., Costantini, L.: Insights into tephra settling velocity from morphological observations. J. Volcanol. Geoth. Res. **208**, 86–98 (2011a)

Alfano, F., Bonadonna, C., Volentik, A.C., Connor, C.B., Watt, S.F., Pyle, D.M., Connor, L.J.: Tephra stratigraphy and eruptive volume of the May, 2008, Chaitén eruption. Chile. Bull. Volcanol. **73**(5), 613–630 (2011b)

Armienti, P., Macedonio, G., Pareschi, M.T.: A numerical-model for simulation of tephra transport and deposition—applications to May 18, 1980, Mount-St-Helens eruption. J. Geophys. Res. **93**, 6463–6476 (1988)

Barsotti, S., Neri, A., Scire, J. S.: The VOL-CALPUFF model for atmospheric ash dispersal: 1. Approach and physical formulation. J. Geophys. Res. **113**(B03209) (2008). https://doi.org/10.1029/2006jb004623

Bonadonna, C., Cioni, A., Pistolesi, M., Connor, C., Scollo, S., Pioli, L., Rosi, M.: Determination of the largest clast sizes of tephra deposits for the characterization of explosive eruptions: a study of the IAVCEI commission on tephra hazard modelling. Bull. Volcanol. **75** (2013). https://doi.org/10.1007/s00445-012-0680-3

Bonadonna, C., Connor, C., Houghton, B., Connor, L.J., Byrne, M., Laing, A., Hincks, T.: Probabilistic modelling of tephra-fall dispersal: hazard assessment of a multiphase rhyolitic eruption at Tarawera, New Zealand. J. Geophys. Res. **110**, B03203 (2005). https://doi.org/10.1029/2003JB002896

Bonadonna, C., Costa, A.: Estimating the volume of tephra deposits: a new simple strategy. Geology **40**, 415–418 (2012)

Bonadonna, C., Costa, A.: Plume height, volume and classification of volcanic eruptions based on the Weibull function. Bull. Volcanol. **75**(8) (2013)

Bonadonna, C., Biass, S., Costa, A.: Physical characterization of explosive volcanic eruptions based on tephra deposits: propagation of uncertainties and sensitivity analysis. J. Volcanol. Geoth. Res. (2015)

Bonadonna, C., Ernst, G.G.J., Sparks, R.S.J.: Thickness variations and volume estimates of tephra fall deposits: the importance of particle Reynolds number. J. Volcanol. Geoth. Res. **81**, 173–187 (1998)

Bonadonna, C., Houghton, B.F.: Total grain-size distribution and volume of tephra-fall deposits. Bull. Volc. **67**, 441–456 (2005)

Bonadonna, C., Phillips, J.C.: Sedimentation from strong volcanic plumes. J. Geophys. Res. **108** (2003). https://doi.org/10.1029/2002jb002034

Bonasia, R., Capra, L.A., Costa, A., Macedonio, G., Saucedo, R.: Tephra fallout hazard assessment for a plinian eruption scenario at Volcán de Colima (Mexico). J. Volcanol. Geoth. Res. **203**(1), 12–22 (2011)

Bretón-Gonzalez, M., Ramirez, J.J., Navarro, C.: Summary of the historical eruptive activity of Volcán de Colima, Mexico 1519-2000. J. Volcanol. Geoth. Res. **117**, 21–46 (2002)

Burden, R.E., Phillips, J.C., Hincks, T.K.: Estimating volcanic plume heights from depositional clast size. J. Geophys. Res. **116**, B11206 (2011). https://doi.org/10.1029/2011JB008548

Burden, R.E., Chen, L., Phillips, J.C.: A statistical method for determining the volume of volcanic fall deposits. Bul. Volcanol. **75**(6) (2013)

Bursik, M.I., Carey, S.N., Sparks, R.S.J.: A gravity current model for the May 18 1980 Mount-St-Helens plume. Geophys. Res. Lett. **19**, 1663–1666 (1992)

Carey, S.N., Sparks, R.S.J.: Quantitative models of the fallout and dispersal of tephra from volcanic eruption columns. Bull. Volc. **48**, 109–125 (1986)

Carmichael, I.S.E., Frey, H.M., Lange, R.A., Hall, C.M.: The Pleistocene cinder cones surrounding Volcan Colima, Mexico re-visited: eruption ages and volumes, oxidation states, and sulfur content. Bull. Volc. **68**, 407–419 (2006)

Carmichael, I.S.E.: The andesite aqueduct: perspectives on the evolution of intermediate magmatism in west-central (105–99°W) Mexico. Contribuitons Miner. Petrol. **143**, 641–663 (2002)

Connor, C.B., Hill, B., Winfrey, B., Franklin, N., LaFemina, P.: Estimation of volcanic hazards from tephra fallout. Nat. Hazards. Rev. **2**, 33–42 (2001)

Connor, L.J., Connor, C.B.: Inversion is the key to dispersion: understanding eruption dynamics by inverting tephra fallout. In: Mader, H., Coles, S.C., Connor, C.B., Connor, L.J. (eds.) Statistics in Volcanology, pp. 231–242. Geological Society, London (2006)

De la Cruz-Reyna, D., Martin Del Pozzo, A.L.: The 1982 eruption of El Chichón volcano, Mexico: eyewitness of the disaster. Geofísica Int. **48**(1), 21–31 (2009)

Dufek, J., Manga, M., Patel, A.: Granular disruption during explosive volcanic eruptions. Nat. Geosci. (2012)

Engwell, S.L., Sparks, R.S.J., Aspinall, W.P.: Quantifying uncertainties in the measurement of tephra fall thickness. J. Appl. Volcanol. **2**(1), 1–12 (2013)

Ernst, G.G., Bursik, M.I., Carey, S.N., Sparks, R.S.J.: Sedimentation from turbulent jets and plumes. J. Geophys. Res. **101**, 5575–5589 (1996)

Fagents, S.A., Wilson, L.: Explosive volcanic eruptions: VII. The ranges of pyroclasts ejected in transient volcanic explosions. Geophys. J. Int. **113**, 359–370 (1993)

Fierstein, J., Nathenson, M.: Another look at the calculation of fallout tephra volumes. Bull. Volc. **54**, 156–167 (1992)

Fisher, R.V., Schmincke, H.-U.: Pyroclastic Rocks. Springer 1984

Folch, A., Costa, A., Macedonio, G.: FALL3D: a computational model for transport and deposition of volcanic ash. Comput. Geosci. **35**, 1334–1342 (2009)

Housh, T.B., Luhr, J.F.: Plagioclase-melt equilibria in hydrous systems. Am. Miner. **76**, 477–492 (1991)

Hurst, A.W., Turner, R.: Performance of the program ASHFALL for forecasting ashfall during the 1995 and 1996 eruptions of Ruapehu volcano, New Zealand. J. Geol. Geophys **42**, 615–622 (1999)

Inman, D.L.: Measures for describing the size distribution of sediments. J. Sediment. Petrol. **22**, 125–145 (1952)

Jones, A., Thomson, D., Hort, M., Devenish, B.: The U. K. Met Office's next-generation atmospheric dispersion model, NAME III. In: Borrego, C., Norman, A.-L. (eds.) Air Pollution Modelling and Its Application XVII, pages 580–589. Springer (2007)

Kalnay, E., Kanamitsu, M., Kistler, R., Collins, W., Deaven, D., Gandin, L., et al.: The NCEP/NCAR 40-year reanalysis project. Bull. Am. Meteor. Soc. **77** (3), 437–471 (1996)

Kelemen, P.B., Hanghoj, K., Greene, A.R.: One view of the geochemistry of subduction-related magmatic arcs, with emphasis on primitive andesite and lower crust. In: Holland, H.D., Turekian, K.K. (eds.) Treaties in Geochemistry, pp. 593–659. Elsevier-Pergamon, Oxford (2003)

Klawonn, M., Houghton, B.F., Swanson, D.A., Fagents, S.A., Wessel, P., Wolfe, C.J.: From field data to volumes: constraining uncertainties in pyroclastic eruption parameters. Bull. Volcanol. **76**(7) (2014)

Komorowski, J.C., Navarro, C., Cortes, A., Saucedo, R., Gavilanes, J.C., Siebe, C., Espindola, J.M., Rodriguez, S.: The Colima volcanic complex: part I: quaternary multiple debris-avalanche deposits, part II: historical pyroclastic sequences (1913, 1991, 1994). Fieldtrip Guidebook Excursion No. 3, IAVCEI General Assembly, Puerto Vallarta, Mexico (1997)

Kunii, D., Levenspiel, O.: Fluidization Engineering. Wiley, New York (1969)

Le Bas, M., Le Maitre, R.W., Woolley, A.R.: The construction of the total Alkali-Silica chemical classification of the volcanic rocks. Mineral. Petrol. **46**(1), 1–22 (1986)

Lim, L.L., Sweatman, W.L., McKibben, R., Connor, C. B.: Tephra fallout models: the effect of different source shapes on isomass maps. Math. Geosci. **40**, 147–157 (2008)

Longchamp, C., Bonadonna, C., Bachmann, O., Skopelitis, A.: Characterization of tephra deposits with limited exposure: the example of the two largest explosive eruptions at Nisyros volcano (Greece). Bull. Volc. **73** (9), 1337–1352 (2011)

Luhr, J.F., Carmichael, I.S.E.: Petrological monitoring of cyclical eruptive activity at Volcán Colima, México. J. Volcanol. Geoth. Res. **42**, 235–260 (1990)

Luhr, J.F., Navaro, C., Savov, I.P.: Tephrochronology, petrology and geochemistry of Late-Holocene pyroclastic deposits from Volcán Colima, México. J. Volcanol. Geoth. Res. **197**, 1–32 (2010)

Luhr, J.F., Navarro, C., Connor, C.B., Connor, L.J.: The 1913 VEI–4 Plinian eruption of Volcán Colima, México: Tephrochronology, petrology, and plume modelling. Eos, Transactions of the American Geophysical Union, vol. 52, pp. V43B–1786. San Francisco, CA (2006)

Luhr, J.F.: Petrology and geochemistry of the 1991 and 1998-1999 lava flows from Volcán Colima, México. J. Volcanol. Geoth. Res. **117**, 169–194 (2002)

Macedonio, G., Costa, A., Longo, A.: A computer model for volcanic ash fallout and assessment of subsequent hazard. Comput. Geosci. **31**, 837–845 (2005)

Magill, C., Mannen, K., Connor, L., Bonadonna, C., Connor, C.: Simulating a multi-phase tephra fall event: inversion modelling for the 1707 Hoei eruption of Mount Fuji. Japan. Bull. Volcanol. **77**(9), 81 (2015)

Mannen, K.: Particle segregation of an eruption plume as revealed by a comprehensive analysis of tephra dispersal: theory and application. J. Volcanol. Geoth. Res. **284**, 61–78 (2014)

Maria, A., Luhr, J.F.: Lamprophyres, basanites, and basalts of the western Mexican Volcanic Belt: Volatile contents and a vein—wallrock melting relationship. J. Petrol. **49**, 2123–2156 (2008)

Nedler, J.A., Meade, R.: A simplex method for function minimization. Comput. J. **7**, 308–313 (1965)

Press, W.H., Flannery, B.P., Teukolsky, S.A., Vetterling, W.T.: Numerical Recipes in C. Cambridge University Press (1992)

Putirka, K.D.: Igneous thermometers and barometers based on plagioclase + liquid equilibria; tests of some existing models and new calibrations. Am. Miner. **98**, 336–346 (2005)

Putirka, K.D.: Thermometers and barometers for volcanic systems. In: Putirka, K., Tepley, F.J. (eds.) Minerals, Inclusions and Volcanic Processes. Rev. Mineral. Geochem. **69**, 61–120 (2008)

Pyle, D.M.: The thickness, volume and grainsize of tephra fall deposits. Bull. Volc. **51**, 1–15 (1989)

Robin, C., Camus, G., Gourgaud, A.: Eruptive and magmatic cycles at Fuego de Colima volcano (Mexico). J. Volcanol. Geoth. Res. **45**, 209–225 (1991)

Rose, W.I., Durant, A.J.: Fine ash content of explosive eruptions. J. Volcanol. Geoth. Res. **186**(1), 32–39 (2009)

Saucedo-Girón, R.: Reconstrucción de la última erupción explosiva del Volcán de Colima en 1913. Master's Thesis, Instituto de Geofsica, Universidad Nacional Autónoma de México, vol. 85 (1997)

Saucedo, R., Macas, J.L., Gavilanes, J.C., Arce, J.L., Komorowski, J.C., Gardner, J.E., Valdez-Moreno, G.: corrigendum to Eyewitness, stratigraphy, chemistry, and eruptive dynamics of the 1913 plinian eruption of Volcan de Colima, Mexico. [J. Volcanol. Geoth. Res. 191 149–166 (2010)] J. Volcanol. Geoth. Res. **207**, 67 (2011)

Saucedo, R., Macias, J.L., Gavilanes, J.C., Arce, J.L., Komorowski, J.-C., Gardner, J., Valdez-Moreno, G.: Eyewitness, stratigraphy, chemistry, and eruptive dynamics of the 1913 plinian eruption of Volcán de Colima, Mexico. J. Volcanol. Geoth. Res. **191**, 149–166 (2010)

Savov, I.P., Luhr, J.F., Navarro-Ochoa, C.: Petrology and geochemistry of lava and ash erupted from Volcán Colima, Mexico, during 1998–2005. J. Volcanol. Geoth. Res. **174**, 241–256 (2008)

Scollo, S., Tarantola, S., Bonadonna, C., Coltelli, M., Saltelli, A.: Sensitivity analysis and uncertainty estimation for tephra dispersal models. J. Geophys. Res. **113**, 1–17 (2008)

Searcy, C., Dean, K., Stringer, W.: PUFF: a high-resolution volcanic ash tracking model. J. Volcanol. Geoth. Res. **80**, 1–16 (1998)

Solikhin, A., Thouret, J.C., Liew, S.C., Gupta, A., Sayudi, D.S., Oehler, J.F., Kassouk, Z.: High-spatial-resolution imagery helps map deposits of the large (VEI 4) 2010 Merapi Volcano eruption and their impact. Bull. Volc. **77**(3), 20 (2015)

Sparks, R.S.J., Bursik, M.I., Ablay, G.J., Thomas, R.M. E., Carey, S.: Sedimentation of tephra by volcanic plumes: 2. Controls on thickness and grain-size variations of tephra fall deposits. Bull. Volc. **54**, 685–695 (1992)

Sparks, R.S.J., Bursik, M.I., Carey, S.N., Gilbert, J.S., Glaze, L.S., Sigurdsson, H., Woods, A.W.: Volcanic Plumes. Wiley, Chichester, UK (1997)

Sparks, R.S.J.: The dimensions and dynamics of volcanic eruption columns. Bull. Volc. **48**, 3–15 (1986)

Suzuki, T.: A theoretical model for dispersion of tephra. In: Shimozuru, D., Yokoyama, I. (eds.) Arc Volcanism, Physics and Tectonics, pages 95–113. Terra Scientific Publishing Company (TERRAPUB), Tokyo (1983)

Volentik, A.C.M., Bonadonna, C., Connor, C.B., Connor, L.J., Rosi, M.: Modelling tephra dispersal in absence of wind: insights from the climactic phase of the 2450 BP Plinian eruption of Pululagua volcano (Ecuador). J. Volcanol. Geoth. Res. **193**, 117–136 (2010)

Waitz, P.: El estado actual de los volcanes de México y la ultima erupción del Volcán de Colima (1913). Revista Volcanológica, pp. 259–268 (1915)

Waitz, P.: Datos historicos y bibliograficos acerca del Volc an de Colima. Mem. Soc. Cient. Antonio Alzate (México) **53**, 349–384 (1935)

Wells, P.R.A.: Pyroxene thermometry in simple and complex systems. Contrib. Miner. Petrol. **62**, 129–139 (1977)

White, J.T., Connor, C.B., Connor, L., Hasenaka, T.: Efficient inversion and uncertainty quantification of a tephra fallout model. J. Geophys. Res.: Solid Earth **122**(1), 281–294 (2017)

Woods, A.W.: The dynamics of explosive volcanic eruptions. Rev. Geophys. **33**, 495–530 (1995)

Pyroclastic Density Currents at Volcán de Colima

R. Saucedo, J. L. Macías, J. C. Gavilanes-Ruiz, M. I. Bursik and V. Vargas-Gutiérrez

Abstract

In the last ~500 years, Volcán de Colima has generated numerous small-volume pyroclastic density currents (PDCs) that have been associated with dome emplacement, either by a partial collapse or by their explosive destruction. Large PDCs were generated by eruption column collapse in 1690, 1818, and 1913. The 1913 eruption generated pyroclastic density currents with maximum runouts of 15 km. The small PDCs that occurred in 1991, 1994, 1998–1999, 2003, 2004, 2005, 2013–2015 have been extensively videoed and photographed, and their deposits have been widely studied. In this chapter, events that have generated PDCs during the past 200 years are summarized. Special attention is given to the generation of small pyroclastic density currents between 1991 and 2005. We provide new, relevant information regarding the development of eruptive episodes, stratigraphy, textural characteristics, and granulometric-component variations of the deposits, as well as inundation areas, mobility (H/L), volumes, and runouts. New observations allow us to understand the effects of topography, fragmentation processes, and better define some of the volumes, which will improve the understanding of the mechanisms of transport and deposition of these types of pyroclastic density currents.

Keywords

Volcán de Colima · Small volume block-and-ash flow · Pyroclastic density current

R. Saucedo (✉)
Instituto de Geología, Universidad Autónoma de San Luis Potosí, Dr. M. Nava No 5, 78240 San Luis Potosí, Mexico
e-mail: rgiron@uaslp.mx

J. L. Macías
Instituto de Geofísica Unidad Morelia, Universidad Nacional Autónoma de México, Antigua Carretera a Pátzcuaro 870, 58089 Morelia, Michoacán, Mexico

J. C. Gavilanes-Ruiz
Facultad de Ciencias, Universidad de Colima, Bernal Díaz Del Castillo No 340, 2845 Colima, Mexico

M. I. Bursik
Department of Geology, University of Buffalo, SUNY, Buffalo, NY 14260, USA

V. Vargas-Gutiérrez
Posgrado en Ciencias de la Tierra Centro de Geociencias, Universidad Nacional Autónoma de México, Blvd. Juriquilla No. 3001, 76230 Querétaro, Mexico

1 Introduction

The generation of pyroclastic density currents (PDCs) is a frequent and dangerous volcanic phenomenon, particularly at andesitic volcanoes

N. Varley et al. (eds.), *Volcán de Colima*, Active Volcanoes of the World,
https://doi.org/10.1007/978-3-642-25911-1_11

associated with subduction zones (Cas and Wright 1987; Druitt 1998). Volcán de Colima is characterized by the semi-continuous generation of small-volume (<0.3 km^3) PDCs (Saucedo et al. 2004). Since 1576, this volcano has had about 50 major eruptions, of which ~28 formed small-volume PDCs that originated by the collapse of external parts of the dome (Merapi type; Sato et al. 1992) or by the explosive destruction of domes and unsteady collapse of the column (Soufrière type; Sato et al. 1992). During the same period, Volcán de Colima generated at least three Plinian eruptions, in 1690, 1818 and 1913 (Luhr and Carmichael 1990; Saucedo et al. 2005, 2010) that generated PDCs with maximum runouts of 15 km from the summit, produced by the collapse of an eruptive column (Saucedo et al. 2005, 2010).

The generation of PDCs at Volcán de Colima has been the subject of numerous works describing this volcanic phenomenon from different perspectives. Some of these works emphasized the external anatomy and general features of the PDC deposits (Rodríguez-Elizarrarás et al. 1991; Saucedo et al. 2002; Macías et al. 2005; Capra et al. 2016), while others enabled the articulation of conceptual models of transport and emplacement mechanisms, as well as the role of the slope angle in the mechanism of PDC deposition. These were based on detailed stratigraphic and sedimentological analyses of the deposits, laboratory experiments and modelling (Saucedo et al. 2002, 2004; Sarocchi et al. 2011). In the past two decades, and particularly since 1998, an increase in the frequency of PDCs associated with the destruction of domes has been of particular interest for volcanologists and risk evaluation. Deposits have been studied in association with seismic data, photographs and videos, as well as various published studies on different aspects of the events (Rodríguez-Elizarrarás et al. 1991; Lermo et al. 1993; Nuñez--Cornu et al. 1994; Saucedo et al. 2002, 2004; Sarocchi et al. 2011; Capra et al. 2016). Some of this data have been used to develop numerical models aimed at understanding the mobility of block-and-ash flows (Bursik et al. 2005). In addition, computer codes like Flow3D and Titan2D have been used to generate hazard maps to define the areas that could be affected by the emplacement of PDCs (Saucedo et al. 2005; Rupp et al. 2006; Sulpizio et al. 2010; Varley, this volume).

Pyroclastic density currents (PDCs) are fast-moving, inhomogeneous mixtures of hot volcanic particles and gas that move across the ground under the influence of gravity (Druitt 1998; Branney and Kokellar 2002). PDCs may form by the gravitational collapse of lava domes, by the fallback or continuous fountaining of vertical eruption columns, or by lateral blasts. The generation of PDCs at Volcán de Colima has produced four different types of deposits: (1) valley-confined block-and-ash flow deposits (emplaced from *block-rich PDCs*), (2) unconfined dilute ash cloud surge deposits (emplaced from *detached PDCs*), (3) pumice-and-ash flow deposits (emplaced from *pumice-rich PDCs*), and (4) ash fall deposits (emplaced from *buoyant co-ignimbrite ash clouds*). Here, we use the term flow unit to describe individual beds with distinct vertical grading bounded by erosive or gradational contacts that are interpreted as deposited from individual PDC or individual pulses in a continuous but unsteady flow (Druitt 1998; Charbonnier and Gertisser 2011). In this chapter, a summary is presented of the eruptive activity of Volcán de Colima between 1818 and 1990, in which PDCs were generated. Special attention is given to the stratigraphic and granulometric variations of the deposits generated by the 1991 to 2005 eruptive episodes, as well as inundation areas, mobility (H/L), volume and runout, which have varied over time.

2 Historic PDCs (1818–1990)

The earliest eruption for which there is clear evidence of the generation of PDCs at Volcán de Colima is the 1818 Plinian eruption (Bárcena 1887). The historical accounts mention that on 15 February, at around 20:00, strong detonations were heard at the volcano that increased in intensity until ~22:00 (Sartorious 1871; Arreola 1915; Waitz 1920). The explosions were heard up to a distance of 235 km (Bárcena 1887), and were accompanied by the formation of PDCs that reach between 13 and 15 km from the summit on the southwestern flank of the volcano (Bárcena 1887). During this eruption eyewitnesses named this ravine as "La Lumbre," which means fire in Spanish, due to the incandescence generated by the PDCs. The eruption developed an eruptive column that dispersed ash in central Mexico. In fact, at Ciudad Guzmán, located 25 km northeast of the volcano inhabitants were forced to sweep ash from house roofs (Bárcena 1887; Arreola 1915).

Dollfus and Montserrat (1867) described the morphology of the volcano summit as a crater occupied by a blocky lava dome. A similar dome was completely destroyed during the 15 February 1818 eruption. After the explosion, the summit acquired a funnel-shape crater with depths varying between 50 and 230 m. In plain view, the crater had an oval shape with diameters of 500 and 450 m, to the northeast and northwest, respectively.

Between 1818 and 1913, there were many observations of PDCs originating from the explosive destruction of domes (1885, 1890, 1891, 1903, 1908, and 1909). At the beginning of the 20th century, Volcán de Colima showed an increase of activity that reached its climax on 17 January 1913, with a Plinian eruption (Ortiz 1944; Saucedo et al. 2010). The eruption occurred in three main phases, all of which generated PDCs. Phase I started with explosions that destroyed the external parts of the dome with the generation of small-volume PDCs of limited runout (4 km) toward the ravines located south-southwest of the volcano (Fig. 1a). Explosions increased in frequency and violence destroying the dome and generating PDCs by small collapse column (phase II) with runouts of 9–12 km. The activity peaked on the afternoon of 20 January with the establishment of a Plinian eruption column (Waitz 1915; phase III) that lasted ca. 8 h. The Plinian phase generated a ~23 km high plume dispersed to the northeast blanketing important cities in central Mexico. The fall deposit covered an area of ~191,000 km^2 and had a D.R.E. volume of 0.26 km^3 (Saucedo et al. 2010). The column collapse produced pumice-and-ash flows with runouts up to 15 km (Saucedo et al. 2010; Fig. 1b). These flows filled canyons with up to 35 m of mixed pumice, dense juvenile, and accessory material (0.24 km^3). The intensity of the eruption declined through the night of 20 January and the eruption stopped on 24 January. Eight human casualties were indirectly related to the eruption, as were the destruction of several farms and the death of cattle (Saucedo 1997; Saucedo et al. 2010). This event represents a multi-stage eruption that generated PDCs. Based on column height and the volume of the ejected magma, the 1913 eruption has a VEI (Volcano Explosivity Index) of 4. After the 1913 eruption, the summit area of Volcán de Colima had a huge crater that was 450 m wide and 350 m deep (Waitz 1935). In 1932, 29 years after the eruption, Waitz (1935) observed that the upper surface of a new lava dome was around 150 m from the crater rim. In 1958, a new central dome already plugged the crater, and grew as a blocky lava flow at a rate of 20 cm/day (Mooser and Reyes 1961). Between 1961 and 1962, that dome began to spill over the northern flank of the volcano with the formation of a blocky lava flow that reached "El Playón" a small plain between Volcán de Colima and the Paleofuego collapse crater wall (Cortés et al. 2010; Cortés et al. this volume).

The next most important event after the 1913 eruption occurred in 1975–1976 with the extrusion of several blocky lava flows. The collapse of the external parts of the lava flow fronts developed PDCs (Thorpe et al. 1977). These lava flows moved down the southeast and northeast flanks of the volcano to an altitude of 2000 masl (to the SE) and reached 4.5 km from the summit. Six years later, in 1982, a new 1-km long blocky lava flow descended the south flank of the volcano. This flow generated PDCs that covered that flank to an undetermined distance (Luhr and Carmichael 1990).

3 The 1991–2005 PDCs

The 1991 eruption marked the beginning of a particularly active stage in the generation of small-volume PDCs of Volcán de Colima.

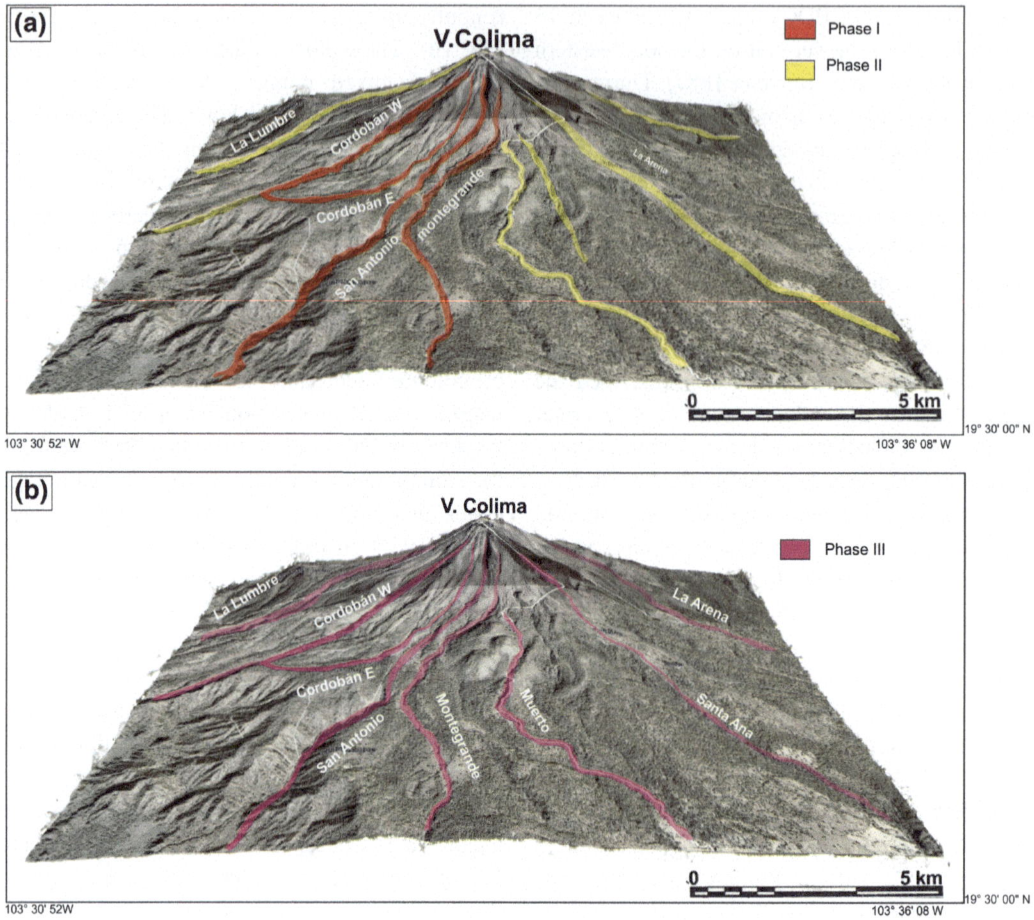

Fig. 1 View from the south of Volcán de Colima showing the maximum extent of the pyroclastic density currents generated during the 1913 Plinian eruption. **a** Phase I and phase II, and **b** phase III

Between 1991 and 2005 seven sets of PDCs were generated, with runouts between 3 and 6.1 km for a total volume of ca. 0.01 km^3 (Table 1), which we describe below.

4 Dome Collapse PDCs

The last four effusive episodes of Volcán de Colima that generated PDCs occurred in 1991, 1998, 2004, 2013–2015. These events involved fast-growing domes that in a few weeks overfilled the crater generating dome collapse PDCs.

On 1 March 1991, a new scoriaceous dark-gray andesitic (59.7% SiO_2) lava dome was exposed at the old dome surface (Fig. 2a). On 16 April 1991 (16:00 local time) after a series of shallow earthquakes, the southwestern portion of the dome (formed from 1958 to 1962), collapsed intermittently during a five-hour period producing a series of small-volume PDCs (Rodríguez-Elizarrarás et al. 199), which moved as granular mass flow (Iverson and Vallance 2001; Saucedo et al. 2004; Bursik et al. 2005). According to Rodríguez-Elizarrarás et al. (1991) the 16 April dome collapse PDCs immediately segregated into a dense basal avalanche and an unconfined dilute ash cloud surge (Fig. 2b, c). Transport of the basal part of the PDC was controlled by topography and followed the Cordobán ravines (west, central and east) reaching a maximum runout of 4 km with a Heim

Table 1 Volume-thickness, runout and coefficient of friction (H/L) of the PDC deposits of Volcán de Colima

Eruption	Ravine	H (km)	Length (km)	H/L	Thickness (m)	Area (km^2)	Volume (km^3)	Volume (m^3)	Eruption type
1913									
Phase I	Playón		2.00		1.56	1.28	0.002	2,000,000	Partial dome callapse
	Cordobán E	1.60	4.00	0.39	1.08	0.12	0.00013	130,000	
	Cordobán W	1.63	3.60	0.43	0.93	0.11	0.0001	100,000	
	Montegrande				2.00	0.11	0.000216	216,000	
					1.86	1.62	0.003	2,446,000	
Phase II	La Lumbre	2.47	9.00	0.27	6.00	0.27	0.00162	1,620,000	Column collapse
	El Zarco	2.48	9.50	0.26	6.00	0.285	0.00171	1,710,000	
	Cordobán	1.60	4.40	0.36	9.00	0.132	0.001188	1,188,000	
	San Antonio	2.35	9.00	0.26	4.00	0.27	0.00108	1,080,000	
	Montegrande	2.25	7.20	0.31	4.00	0.216	0.000864	864,000	
	El Muerto	2.15	7.90	0.27	3.00	0.237	0.000711	711,000	
	La Tuna	2.07	6.80	0.30	3.00	0.204	0.000612	612,000	
	La Arena	2.25	7.90	0.28	5.00	0.237	0.001185	1,185,000	
	Beltrán	2.35	8.65	0.27	7.00	0.2595	0.001817	1,816,500	
	El Playón		2.60		4.00	2.5	0.01	10,000,000	
					9.52	2.1	0.02	20,786,500	
Phase III	La Lumbre	2.85	13.50	0.21	12.00	4.05	0.02835	28,350,000	Plinian
	El Zarco	2.70	12.50	0.22	10.00	3.75	0.0225	22,500,000	
	Cordobán	2.75	13.20	0.21	10.00	3.96	0.02376	23,760,000	
	San Antonio	2.39	10.20	0.23	7.00	3.06	0.01224	12,240,000	
	Montegrande	2.60	15.70	0.17	9.00	4.71	0.02826	28,260,000	
	Muerto	2.67	15.10	0.18	6.50	4.53	0.029445	29,445,000	
	Fresnal	2.55	9.70	0.26	2.00	2.91	0.00582	5,820,000	
	Tuna	2.65	14.70	0.18	4.00	4.41	0.01764	17,640,000	
	Cafesito	2.30	9.00	0.26	5.00	2.7	0.0135	13,500,000	
	Arena	2.77	15.00	0.18	9.00	4.5	0.02475	24,750,000	
	Beltrán	2.85	15.90	0.18	7.00	4.77	0.02385	23,850,000	
	Playón		2.50		5.00	1.8	0.009	9,000,000	
					4.42	45.2	0.2	239,115,000	
1991	Cordobán W	1.76	4.00	0.44	5.00	0.2	0.001	1,000,000	Partial dome collapse
1994	Cordobán E	1.56	3.75	0.42	6.25	0.08	0.0005	500,000	Column collapse
1998	Cordobán W	1.90	4.50	0.42	5.71	0.14	0.0008	800,000	Partial dome collase
	Cordobán E	1.42	3.00	0.47	0.56	0.09	0.00005	5000	
					3.91	0.23	0.0009	805,000	

(continued)

Table 1 (continued)

1999	San Antonio-Montegrande	1.48	3.30	0.45	3.48	0.23	0.0008	790,000	Column collapse
	La Lumbre	1.44	3	0.48	4.00	0.09	0.00036	360,000	Column collapse
					3.13	0.32	0.001	1,150,000	
2003	San Antonio-Montegrande	1.46	3.5	0.42	2.73	0.11	0.0003	300,000	Column collapse
2004	La Lumbre	2.26	6.11	0.35	4.17	0.24	0.001	1,380,000	Partial dome collapse
2005	Montegrande	1.76	5.1	0.35	4.90	0.204	0.001	1,000,000	Column collapse
	La Arena	1.82	5.9	0.30	3.80	0.236	0.000897	862,000	
					4.55	0.44	0.002	1,862,000	
2015[a]	Montegrande			0.20				4,500,000	Partial dome collapse

Abbreviations are: *H* Drop in altitude, *L* Maximum runout, *H/L* Heim coefficient
[a]Capra et al. (2016)

Coefficient (ΔH/L) = 0.44 (Figs. 3a and 4; Table 1). The use of the ratio ΔH/L to describe the mobility of this type of gravity driven mass flow is widespread (Heim 1932; Sheridan et al. 1979; Hayashi and Self 1992; Sheridan and Macías 1995; Charbonnier and Gertisser 2011), where ΔH is the difference in height between the source and the distal limit and L is the maximum runout distance. The deposits of the 1991 collapse had a total volume of $\sim 1.0 \times 10^6$ m^3, including the unconfined dilute ash cloud surge deposits (Rodríguez-Elizarrarás et al. 1991; Saucedo et al. 2004). The 1991 eruption ended with the extrusion of a lava flow that advanced 2.5 km down the southwestern flank of the volcano.

During the first stages of generation of the 16 April 1991 PDCs, the currents moved down the steep, upper slopes of the volcano (above a prominent break in slope) launching large blocks ahead of the moving flow front. PDC fronts were often obscured by a dilute cloud of fine ash (Fig. 2c). When observed, the flow front was thin (of the order of a metre), and appeared as an ash-rich cloud with a snout at ground level, with poorly developed turbulence. Below the break slope (in distal areas), the PDCs were fully turbulent with an ash-rich flow front that overhung the ground surface inside ravines, but was not confined within them. Thick, buoyant ash clouds often evolved behind the PDCs flow heads, although these were not always observed (Fig. 2c; Saucedo et al. 2004). Ashfall from these clouds was reported as far as an airport located $\sim$30 km from the volcano.

On 20 November 1998, a new dome had formed in the southwestern part of the Volcán de Colima summit (Fig. 5a). It was 30 $\times$ 50 m in diameter and 15 m high with a volume of 3.8×10^5 m^3 (Navarro et al. 2002). On 21 November the dome had developed into a lava flow that was 150 m long. Three days later, it was 370 m long, advancing toward the west, central, and eastern branches of El Cordobán ravines. On 22 November blocks collapsing from the flow margins produced the first dome collapse PDCs (Fig. 5b) with runouts of 3–4.5 km towards the west, central and eastern Cordobán and La Lumbre ravines (ΔH/L = 0.42–0.47, Fig. 3) producing deposits with a total volume of 8.1×10^5 m^3 (Table 1). The PDCs consisted of two parts, a basal part composed of blocks and

Fig. 2 **a** Aerial photograph of the summit dome at Volcán de Colima that grew between February and April 1991. **b** Aerial picture that captured the moment when a dome collapsed (Merapi-type) and immediately segregated into a dense basal avalanche and an upper diluted turbulent ash cloud. **c** Detail of the frontal part of the BAF obscured by a diluted cloud of fine ash taken on 16 April 1999. (Photographs B and C by Ernesto Gómez Hoffman)

ash, which was valley confined and an unconfined dilute ash-rich cloud of greater mobility, composed of ash, later identified outside of the ravines. On 1 December 1998, the lava flow had divided into three advancing fronts (2.5, 2.8, and 3.1 km long), for a total estimated volume of 39×10^6 m^3 (Navarro et al. 2002; Fig. 5c). On 8 December 1998, several BAFs deposits were observed at El Cordobán east ravine (Saucedo et al. 2002). On 9 February 1999, seismic activity increased, ending with a strong explosion at the summit dome in 10 February. This explosion marked a change of activity from effusive to explosive, which finished with the 17 July 1999 explosion that generated column collapse PDCs that are described below (Saucedo et al. 2002).

On 30 September 2004, an overflight detected the presence of a summit dome. Later, this dome spilled over on the north and west-northwest borders of the summit, forming two lava flows, which on 31 October were 2.5 km long (Fig. 6a, b) (Macías et al. 2005). On 6 October part of the dome collapsed, generating PDCs that were channelled through La Lumbre ravine, reaching a distance of 6.1 km, the longest since 1913 (Macías et al. 2005; Sulpizio et al. 2010).

5 Deposits from Gravitational Dome Collapses

The 1991 small-volume PDCs segregated into valley-confined block-and-ash flows and unconfined, dilute ash cloud surges (Fig. 2b, c). The stratigraphic sequence consisted of a series of deposits emplaced by at least three different pulses, leaving deposits 0.5–2 m thick, generated by dome collapses towards the Cordobán ravines (Figs. 3a and 7a). These pulses were likely generated by the retrogressive collapse of the dome similar to that reported during the 2010 eruption of Merapi volcano (Komorowski et al. 2013). The first two pulses had runouts of 1.5 and 2.2 km ($\Delta H/L = 0.72$ and 0.6), but the last pulse was the longest of the 1991 PDCs with a runout of 4.0 km (Table 1; Rodríguez-Elizarrarás et al. 1991; Saucedo et al. 2004). The top of each flow unit was characterized by a concentration of aligned andesite blocks in the direction of the flow (Fig. 7b). All deposits were composed of the same components rich in grey old andesites and juvenile clasts (Table 2). The later bock-and-ash flow (BAFs) deposits were enriched in red andesite clasts of the older dome and crater, with respect to early BAFs, suggesting a disruption of larger parts of the summit during

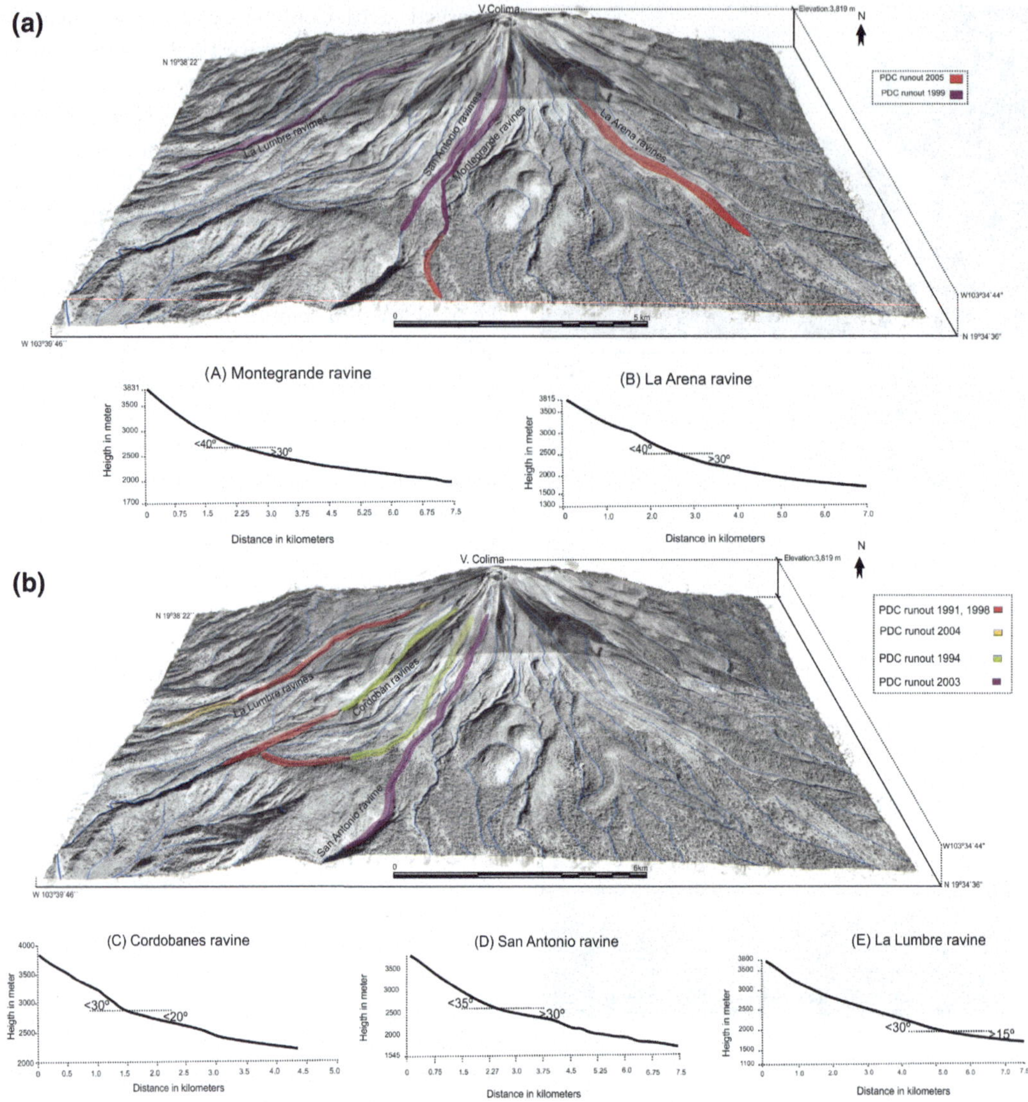

Fig. 3 Runout of the different PDC deposits generated between 1991 and 2005 and topographic profiles of the main ravines. **a** Runouts of the 2005 and 1999, deposits, **b** 1998 and 2004, 1994 and 2003

this event, likely coupled with incorporation of substratum clasts during the flow transport (Table 2). The block-and-ash deposits produced in 1991 are generally polymodal or bimodal (Table 3), finer and better sorted as the distance from source increases, although in general the deposits are poorly sorted (3.5–2.8φ; $\varphi = -\mathrm{Log}_2 d$ where d is the diameter of the particle in millimetres) with reverse grading (Fig. 7b). They are non-welded, being an unconsolidated mixture of sub-rounded blocks suspended in a medium-to-coarse ash sand-sized matrix (Table 3). Abrasion marks on trees have been identified on the flow path margins, suggesting that the deposits were compacted in relation to the height of the original moving debris. Vegetation was totally stripped, sand-blasted, and bent in the direction of flow, but was not burnt. The valley-confined block-and-ash flow deposits were characterized by superimposed, digitate

Table 2 Componentry analyses of the PDC deposits of Colima Volcano

Eruption	Deposit type	Sample (section #)	Main components (wt %)							
			Juvenile lithics 1913	Crystals	Accessories	Altered	Juvenile Pumice (1913)	Glass	Juvenile Scoria	Accesory Pumice
1913	BAF	F1	1.00	0.00	92.00	2.00	0.00	?	5.00	0.00
	Surge	S1	0.00	0.00	95.00	4.00	0.00	?	1.00	0.00
	BAF	F2	19.00	0.00	73.00	7.00	1.00	?	0.00	0.00
	BAF	F3	0.00	0.00	65.00	34.00	0.00	?	1.00	0.00
	Surge	S2	0.00	1.00	82.00	13.00	4.00	?	0.00	0.00
	BAF	F4	0.00	0.00	77.00	3.00	17.00	?	3.00	0.00
	Ash fall	C1	4.00	1.00	46.00	7.00	41.00	?	0.00	0.00
	Surge	S3	0.00	3.00	81.00	6.00	9.00	?	0.00	0.00
	BAF	F5	0.00	0.00	35.00	4.33	11.66	?	49.00	0.00
1991	BAF	F1	0.73	0.21	66.11	14.14	0.00	14.47	0.00	0.00
	Surge	S1	1.47	0.00	27.67	23.47	0.00	45.14	0.00	0.00
	Surge	S2	0.97	0.20	14.63	27.66	0.00	55.36	0.00	0.37
	BAF	F1	0.68	0.00	71.97	14.86	0.00	9.58	0.00	0.00
	BAF	F2	1.02	0.09	55.12	26.50	0.00	13.96	0.00	1.14
	BAF	1	0.36	0.13	66.60	10.69	0.00	18.30	0.00	0.00
	BAF	F3	1.02	0.09	55.13	26.56	0.00	13.96	0.00	1.14
	BAF	F7	2.25	0.10	56.79	12.78	0.00	15.74	0.00	0.04
2003	Surge	S	42.85	32.63	12.67	20.81	0.00	?	0.00	0.68
	BAF	F1	42.85	32.63	12.67	20.81	0.00	?	0.00	0.68
	BAF	F2	28.02	30.92	24.00	25.42	0.00	?	0.00	1.20
	Surge	S1	39.77	35.20	6.85	17.37	0.00	?	0.00	1.40
	Surge	S2	39.77	35.20	6.85	18.37	0.00	?	0.00	1.40
	BAF	F3	47.05	32.93	7.81	15.38	0.00	?	0.00	1.27
	BAF	F4	47.05	32.93	7.81	15.38	0.00	?	0.00	1.27
2004	BAF	FA	35.55	19.60	6.81	17.28	0.00	19.77	0.00	1.00
	BAF	FB	41.39	13.41	5.13	18.05	0.00	20.70	0.00	1.32
	BAF	F2A	32.63	13.79	9.04	19.95	0.00	18.64	0.00	5.96
	BAF	F3A	37.69	12.67	4.80	19.91	0.00	21.62	0.00	3.30
	BAF	F5A	35.34	16.90	5.81	15.59	0.00	24.37	0.00	1.99
	BAF	F5B	37.08	16.34	13.85	20.47	0.00	10.57	0.00	1.70
	BAF	F5C	29.34	18.27	13.74	24.03	0.00	12.96	0.00	1.66
	BAF	F6A	35.42	16.47	9.40	16.15	0.00	21.57	0.00	0.99
	Surge	F7A	38.54	20.93	5.98	15.28	0.00	19.27	0.00	0.00
	BAF	F7B	41.18	10.15	11.69	18.13	0.00	17.09	0.00	1.76
	BAF	F7C	35.09	17.74	11.78	19.51	0.00	14.00	0.00	1.76
	BAF	F7D	35.16	17.39	10.49	22.34	0.00	13.18	0.00	1.44
	BAF	F7E	30.53	17.14	10.96	24.12	0.00	17.25	0.00	0.00
	BAF	F8A	41.66	12.37	8.45	17.37	0.00	18.66	0.00	1.49

(continued)

Table 2 (continued)

2005	Surge	S42(B)	77.54	?	21.70	0.00	0.00	?	0.00	0.75
	Surge	S42(B) LOW	77.54	?	21.70	0.00	0.00	?	0.00	0.75
	Surge	F6-45	82.18	?	15.48	0.00	0.00	?	0.00	2.35
	BAF	F8-08	84.75	?	12.43	0.00	0.00	?	0.00	2.83
	Surge	F804(B1)	93.30	?	6.53	0.00	0.00	?	0.00	0.18
	Surge	F84(B2)	92.94	?	7.06	0.00	0.00	?	0.00	0.00
	BAF	F729(A)	62.98	?	33.95	0.00	0.00	?	0.00	3.08
	BAF	F8-05 (A)	88.20	?	8.73	0.00	0.00	?	0.00	3.08
	BAF	F8-05 (B1)	85.48	?	11.45	0.00	0.00	?	0.00	3.08
	Surge	F702(A)	86.80	?	10.84	0.00	0.00	?	0.00	2.36
	BAF	F703(A)	81.20	?	15.26	0.00	0.00	?	0.00	3.54
	Surge	S703(B)	86.52	?	12.74	0.00	0.00	?	0.00	0.74
	BAF	F705(A)	77.48	?	22.28	0.00	0.00	?	0.00	0.24
	BAF	F705(B)	88.66	?	10.52	0.00	0.00	?	0.00	0.82
	BAF	F720(A1)	69.10	?	29.70	0.00	0.00	?	0.00	1.20
	BAF	F720(A2)	84.64	?	11.94	0.00	0.00	?	0.00	3.42
	BAF	F77(A)	81.87	?	17.57	0.00	0.00	?	0.00	0.56
	Surge	S708(A)	72.84	?	4.48	0.00	0.00	?	0.00	22.68
	Surge	S78(B)	86.00	?	13.10	0.00	0.00	?	0.00	0.90

Abbreviatons are: BAF block-and-ash flow deposit, ? Not determined
2005 data, taken from Vargas-Gutiérrez (2009)

lobes with steep fronts and fines-poor marginal levees (Fig. 7a). The characteristics of these deposits suggest that each of them was emplaced by the passage of a single flow pulse, due to the retrogressive collapse of the dome. These, accreted to form metre-thick layered units inside ravines, and where fluidization and the major break in slope could have been important factors in the mobility and sedimentation (Saucedo et al. 2004; Schwarzkopf et al. 2005; Bursik et al. 2005; Lube et al. 2007; Sulpizio et al. 2007, 2010; Komorowski et al. 2013). The detached upper dilute turbulent ash cloud surge that accompanied the development of the PDC left a partly stratified horizon tan-pinkish in colour, which was composed of medium to fine ash size particles. Its thickness varied from 0.3 to 5.5 cm. Occasionally, it showed inverse grading and a lower erosive contact. In some sites, it was composed of three thin horizons (1, 3, and 1.5 cm thick from bottom) and contained burned vegetation (agave leaves) aligned in the direction of the flow.

On 22 November 1998, a collapse of the lava flow front generated a PDC that descended the western, central and eastern ravines (Fig. 3b). The 1998 PDCs rapidly segregated, as they did in 1991 (Saucedo et al. 2002; Fig. 8a). During the 1998 events, the accompanying, detached dilute ash cloud surge had a higher mobility, being emplaced outside of the ravines along with an ashfall deposit (Fig. 8b). The PDCs were valley confined in the Cordobán East ravine, which contained five superimposed lobes, corresponding to the same number of flow units with a total thickness of 8 m, each with nearly the same texture (Fig. 8c). All deposits observed had similar componentry, in which grey, juvenile clasts of andesitic composition predominated over altered clasts of reddish tone of the same composition. Each depositional frontal lobe was 1.5–2 m thick, matrix-supported with blocks up to 1.7 m in diameter (Table 3), and having 0.7-m thick levees with blocks up to 60 cm in diameter with rare matrix (Fig. 8c). The small-volume block-and-ash deposits were confined in El

Table 3 Granulometric characteristics and statistical parameters of the PDC deposits of Colima Volcano

Eruption	Type	Sample	Runout (km)	Mode (s)	Median Md (φ)	Sorting (σ)	Skewness (φ)	Kurtosis (φ)	Gravel wt. %	Sand wt. %	Silt and Clay wt. %	F1	F2
1913[a]	BAF	F1	3.2	−3, 2, 4	−1.14	3.05	−0.040	0.750	49.76	50.24	?	59.62	17.68
	Surge	S1	3.2	4	2.24	1.66	−0.441	1.248	8.08	91.92	?	94.57	64.47
	BAF	F2	3.2	−2, 2	−0.31	2.65	−0.120	0.754	39.08	60.92	?	70.55	21.49
	BAF	F3	4	−4, −3, −2	−2.37	3.13	0.375	0.630	64.78	35.22	?	41.28	12.35
	Surge	S2	2	2	0.77	2.51	−0.425	1.184	21.55	78.45	?	84.43	34.02
	BAF	F4	12	−4, −3, 2	−0.52	2.80	−0.077	0.789	43.51	56.49	?	66.51	20.84
	Ash fall	C1	13	−2, −1	−0.50	1.39	−0.046	0.929	53.86	46.14	?	67.28	2.58
	Surge	S3	9.4	3	2.20	1.23	−0.155	0.922	1.25	98.75	?	99.66	58.21
	BAF	F5	15	−4, −3, 0, 3	−2.20	2.98	0.185	0.639	61.90	38.10	?	44.58	8.02
1991	BAF	F1	1.8	−4, 0	−3.20	2.78	0.404	0.777	55.12	41.67	3.20	35.54	6.58
	BAF	F2	2	−4, 0	−2.75	2.80	0.307	0.763	52.61	44.32	3.08	37.10	6.35
	Surge	S1	1.8	2, 4, 5	4.45	0.42	0.001	0.727	?	73.68	26.32	100.00	34.21
	Surge	S2	2	2, 4, 5	5.25	0.72	0.287	0.672	?	73.45	26.55	100.00	34.88
	Surge	S3	2.4	2, 4, 5	5.25	0.82	−0.273	0.823	?	48.09	51.92	100.00	69.72
	Surge	S4	3.1	1, 4	5.30	0.53	0.294	0.558	35.93	38.97	25.10	58.91	28.44
	Surge	S5	3.1	1, 4	5.20	2.63	?	?	12.89	75.51	11.62	?	?
	BAF	1	1.8	−4, 0	−0.90	2.95	0.033	0.837	34.19	59.71	5.24	55.38	11.73
	BAF	2	2	−4, 0, 2	−1.30	3.18	0.095	0.747	38.39	54.57	7.05	51.89	12.96
	BAF	3	2.4	−4, 0, 2	1.20	3.05	−0.574	0.677	42.99	53.96	3.06	71.32	49.45
	BAF	1	1.8	−4, 0	−1.95	3.03	0.168	0.733	45.55	51.83	5.63	45.22	9.86
	BAF	2	2	−4, -2, 2	−2.45	2.82	0.319	0.753	53.18	44.48	2.36	41.94	15.10
	BAF	3	2.4	−3, 0, 4	−0.60	2.73	−0.048	0.883	31.34	63.01	5.66	57.66	11.65
	BAF	4	2.6	−4, 2	−0.25	0.25	−0.103	1.055	21.33	70.29	8.38	71.45	17.25
	BAF	5	3.3	−4, −3, 1, 2	0.40	3.10	?	?	36.79	55.20	8.01	?	?
	BAF	6	3.8	−5, 1	−0.95	3.53	?	?	48.30	47.30	4.38	?	?
	BAF	7	4	−5, 1	−0.65	3.35	−0.045	0.661	41.60	51.60	6.70	56.07	20.63
	Ash fall	Fa1	1.8	0, 1	4.90	1.60	?	?	0.00	45.24	54.76	?	?

(continued)

Table 3 (continued)

Eruption	Type	Sample	Runout (km)	Mode (s)	Median Md (φ)	Sorting (σ)	Skewness (φ)	Kurtosis (φ)	Gravel wt. %	Sand wt. %	Silt and Clay wt. %	F1	F2
1998[a]	BAF	1	3	−5 y 1	−1.29	3.29	−0.160	0.599	47.06	52.94	?	52.94	8.79
	BAF	2	3	-4 y 2	−0.32	3.00	−0.279	0.972	31.77	68.23	?	68.23	11.53
	BAF	4	3	−5 y 1	−0.68	3.01	−0.241	0.824	38.10	61.90	?	61.90	8.58
	BAF	5	3	−4, 0 y 1	−0.70	2.71	−0.086	0.779	42.53	57.47	?	57.47	7.86
	Surge	6	3	−4 y 1	−0.68	2.74	−0.114	0.769	41.20	58.80	?	58.80	8.17
	Surge	11A	3	4	2.58	1.21	−0.380	1.114	0.84	99.16	?	99.16	43.92
	Surge	11B	3	4	2.26	1.51	−0.513	0.796	0.00	100.00	?	100.00	46.59
	Surge	11C	3	4	2.56	1.08	−0.276	0.966	0.00	100.00	?	100.00	41.06
1999[a]	BAF	12	3.3	−4, 0	−1.83	3.04	0.081	0.666	57.65	42.35	?	42.35	5.00
	Surge	15A	3.3	4	2.32	1.21	−0.206	0.835	0.22	99.78	?	99.78	34.51
	Surge	15B	3.3	4	2.30	1.28	−0.251	0.954	2.11	97.89	?	97.89	32.63
	Surge	13A	5	3	2.35	1.16	0.155	0.688	1.04	98.96	?	40.97	4.65
	BAF	13B	5	−4, −2, 1	−1.81	2.93	−0.190	0.904	59.03	40.97	?	98.96	31.99
2003	BAF	1a	1.65	−6, 2	−4.00	3.32	0.60	?	66.68	29.71	3.51	28.13	1.41
	BAF	1b	1.65	−6, −3, 2	−3.95	3.30	?	0.51	72.64	25.62	1.60	22.69	1.60
	BAF	1c	1.65	−6, 2 y 4	−3.28	3.85	0.73	?	72.10	26.21	1.41	34.44	3.50
	Surge	2a	2.65	−7, −4, 0	−4.28	3.33	0.08	0.82	76.95	21.58	1.31	87.99	23.30
	BAF	2b	2.65	−8, −5, 0	−5.93	2.95	0.47	?	85.81	13.42	0.69	17.58	0.70
	BAF	2c	2.65	2, 4	1.65	3.16	0.44	?	14.58	61.99	23.34	26.61	1.30
	BAF	3a	2.95	−7, −4, 2	−3.07	3.88	0.11	?	56.39	40.43	2.54	40.51	2.54
	BAF	3b	2.95	−7, −4, 2	−3.83	3.83	−0.27	?	66.48	30.02	1.85	28.57	1.85
	BAF	4a	3.05	−8, −6, 2	−4.55	3.88	0.53	?	73.53	24.42	1.67	24.20	1.70
	BAF	4b	3.05	−7, 2	−6.35	2.67	0.64	?	87.27	11.86	0.72	13.36	0.70
	BAF	5a	3.15	−7, −4, 2	−4.47	3.57	0.32	?	58.76	39.70	1.24	40.63	1.25
	BAF	5b	3.15	−6, −4, 2	−3.52	3.49	?	?	67.29	31.05	1.56	25.51	1.60
	BAF	5c	3.15	−6, −3, 2	−3.07	3.51	0.64	?	75.46	22.92	1.53	28.25	1.53
	Surge	6a	3.22	2, 4, 6	3.00	2.68	0.05	1.09	4.22	63.35	32.37	97.51	32.37

(continued)

Table 3 (continued)

Eruption	Type	Sample	Runout (km)	Mode (s)	Median Md (φ)	Sorting (σ)	Skewness (φ)	Kurtosis (φ)	Gravel wt. %	Sand wt. %	Silt and Clay wt. %	F1	F2
	Surge	6b	3.22	2, 4	2.52	1.95	−0.06	0.94	0.45	81.37	18.10	100.00	18.10
	BAF	7a	3.3	−9, −8, −5, 1	−5.98	4.17	0.71	?	75.98	20.37	1.19	18.22	1.20
	BAF	7b	3.3	−9, −7, −4, 1	−4.60	4.33	0.23	?	69.93	26.00	1.86	26.22	1.86
	BAF	8a	3.4	−6, −4, 2	−3.80	3.59	0.37	?	78.82	20.39	0.70	20.19	0.70
	BAF	8b	3.4	−6, −4, 2	−4.13	3.83	0.23	?	69.12	29.26	1.50	30.85	1.50
	BAF	8c	3.4	−6, −1, 2	−4.53	3.10	0.68	?	70.17	28.34	1.38	29.00	1.40
2004	BAF	1A	6.12	1, 3,−7	−2.006	4.175	−2.006	4.175	52.24	45.48	2.28	47.76	10.91
	BAF	1B	6.12	−4, −6, 1	−2.270	3.393	−2.270	3.393	61.68	36.43	1.89	38.32	8.45
	BAF	2A	5.94	−5, 1, 3	−1.385	3.603	−1.385	3.603	50.82	45.79	3.39	49.18	11.87
	BAF	3A	5.87	−6, 1, 3	−1.426	3.949	−1.426	3.949	49.87	45.89	4.24	49.96	13.17
	BAF	4A	5.59	2, −6, 4	−1.816	3.772	−1.816	3.772	51.17	45.54	3.29	48.57	11.10
	BAF	4B	5.59	−7, 2, 4	−1.956	3.730	−1.956	3.730	52.64	44.05	3.31	47.36	10.95
	BAF	4C	5.59	−6, 2, 4	−2.038	3.563	−2.038	3.563	55.69	41.08	3.23	44.31	9.33
	BAF	5A	5.35	3, 1, −6	−2.377	3.960	−2.377	3.960	56.32	41.18	2.50	43.68	8.65
	Surge	6A	4.85	2, 4	3.612	2.733	3.612	2.733	5.56	50.16	44.29	94.44	66.67
	BAF	6B	4.85	−3, 4, 2	0.337	2.681	0.337	2.681	29.63	63.17	7.20	70.37	17.51
	BAF	6C	4.85	−7, 2, 4	−2.544	3.768	−2.544	3.768	58.83	38.29	2.88	41.17	9.11
	BAF	6D	4.85	−10, −7, 2	−3.154	4.327	−3.154	4.327	61.14	36.31	2.55	38.86	8.59
	BAF	6E	4.85	−7, 2, 4	−3.926	3.858	−3.926	3.858	69.23	29.17	1.60	30.77	7.73
	BAF	7A	4.47	−8, −5, 1	−2.061	3.974	−2.061	3.974	53.22	44.81	1.97	46.78	9.24
2005	Surge	PF07-42(B)	2.4	1, 3	−2.74	3.86	0.146	0.709	64.18	33.87	1.95	36.50	6.30
	Surge	PF07-42(B) LOW	2.4	1, 3	−1.18	3.09	−0.073	0.741	49.22	48.72	2.06	49.00	6.00
	Surge	PF-06-45(3)	2.44	1, 3	−4.09	3.87	0.330	0.721	73.09	25.41	1.50	26.80	3.50
	BAF	PF-08-08(B1)	2.59	6, 3	−2.73	3.53	0.309	0.728	65.36	33.00	1.64	33.60	5.60
	Surge	PF08-04(B1)	3.42	6, 3	−2.46	3.38	0.256	0.719	63.64	34.35	2.01	36.00	5.00
	Surge	PF08-04(B2)	3.42	6, 3	−2.06	3.48	0.233	0.744	62.38	34.66	2.96	38.00	7.00

(continued)

Table 3 (continued)

Eruption	Type	Sample	Runout (km)	Mode (s)	Median Md (φ)	Sorting (σ)	Skewness (φ)	Kurtosis (φ)	Gravel wt. %	Sand wt. %	Silt and Clay wt. %	F1	F2
	BAF	PF07-29(A)	3.92	6, 3	−0.67	3.40	−0.147	0.773	49.00	47.02	3.98	51.00	10.60
	BAF	PF-08-05 (A)	4.09	1, 3	−3.05	3.65	0.393	0.700	65.66	31.32	3.02	34.00	7.00
	BAF	PF-08-05 (B1)	4.67	1, 3	−2.15	3.57	−0.114	0.647	56.00	42.01	1.99	44.00	5.00
	Surge	PF07-02(A)	3.14	5, 3, 0, 1	−1.77	4.36	−0.198	0.845	52.08	43.77	4.14	46.00	11.00
	BAF	PF07/03 (A)	3.47	5, 3, 0, 1	−3.08	3.93	0.073	0.847	64.68	29.87	5.45	35.50	8.50
	Surge	PF07/03 (B)	3.47	5, 3, 0, 1	−2.12	3.35	0.137	0.663	59.41	40.59	0.00	41.00	4.00
	BAF	PF07/05 (A)	5.12	2, −3, −4	−1.72	3.56	−0.085	0.704	52.48	44.57	2.96	48.00	8.00
	BAF	PF07/05 (B)	5.12	2, −3, −4	−1.74	3.50	0.134	0.768	63.64	32.35	4.02	36.00	6.00
	BAF	PF07/20 (A)	5.16	2, −3, −4	−3.72	4.37	0.053	0.624	63.68	34.83	1.49	36.50	5.50
	BAF	PF07/20 (B)	5.16	2, −3, −4	−2.56	4.16	−0.054	0.702	56.16	40.42	3.43	44.50	9.50
	BAF	PF07/07 (A)	5.6	2, 3	−1.23	3.09	−0.119	0.815	45.63	49.97	4.39	55.40	11.20
	Surge	PF07/08 (A)	5.77	−4, 4	−0.94	2.83	0.086	0.842	49.00	45.02	5.98	51.60	10.84
	Surge	PF07/08 (B)	5.77	−4, 0	−0.84	3.27	−0.146	0.748	45.00	52.01	2.99	55.80	11.00

[a] Data of Md (φ) and sorting of the 1913, 1998–1999 updated with GRADISTAT
2005 data taken from Vargas-Gutiérrez (2009)
Abbreviations are: *BAF* block-and-ash flow deposit, ? not determined, F1 = %wt < 1 mm, F2 = %wt > 1/16 mm

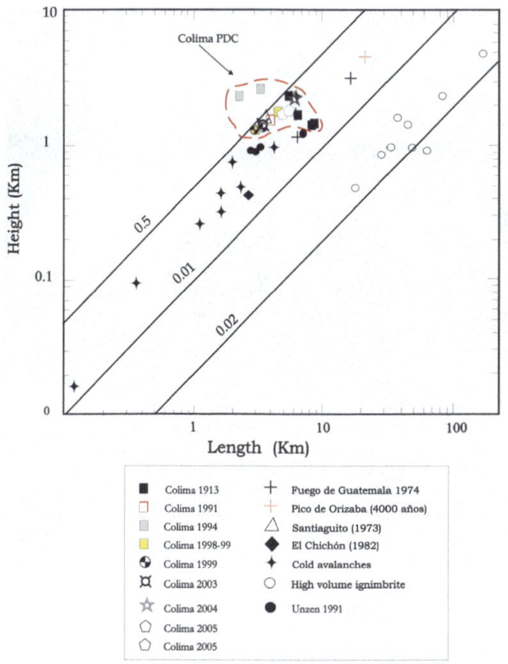

Fig. 4 Plot of the H/L [Height (km)/Length (km)] coefficient of pyroclastic density currents. Data for non-volcanic avalanches and high-volume ignimbrites were taken from Druitt (1998), other data belong to Fuego de Guatemala, Santiaguito, Unzen and Pico de Orizaba were taken from Davis et al. (1978), Hayashi and Self (1992), Rose (1977), Fuji and Nakada (1999), Yamamoto et al. (1993) and Siebe et al. (1993). Data of Volcán de Colima was summarized from previous studies (Saucedo et al. 2002, 2004, 2010; Muñoz-Martinez 2007; Vargas-Gutiérrez 2009)

Fig. 5 a View of a dome already extruded on 20 November 1998, in the southwestern part of the Volcán de Colima summit. **b** Blocks collapsing from the lava flow margins produced small-volume Merapi-type BAFs. **c** The lava flow had divided into three advancing fronts by 01 December 1998, and January 1999

Cordobán East and West ravines with runouts of 3 and 4.5 km ($\Delta H/L = 0.42$), and a total volume of 0.8×10^5 m^3 (Fig. 3b; Table 1). Most deposits did not present important granulometric variations, being bimodal, poorly sorted with a median (Md) of -1.2 to -0.3φ (Table 3). The unconfined dilute ash cloud surge deposits were divided into three layers according to textural characteristics (Fig. 8b) from base to top consisting of a grey, massive, medium to coarse ash layer (4 cm thick), a beige dune-bedded ash layer (1 cm thick), and a tan, massive, well-sorted fine ash layer 0.5 cm thick (Saucedo et al. 2002).

Eruptive activity of 6 October 2004 produced a series of valley-confined block-and-ash flows that were channelled along the bottom of La Lumbre ravine, reaching a 6.1 km runout ($\Delta H/L = 0.33$; Figs. 3b and 4) with a total volume of 1.4×10^6 m^3 (Table 1). The detached PDC dilute ash cloud surge that developed over the blocky basal avalanche (Fig. 9a, b) had a high mobility. At around 3.5 km from the crater,

Fig. 6 **a** Dome observed in the crater of Volcán de Colima in 2004 (photo September 2004). **b** PDC generated towards Montegrande ravine during the March 2005 activity of Volcán de Colima (Photos by CUEIV, Universidad de Colima and by Sergio "Tapiro" Velasco)

the dilute ash cloud reached the break in slope (from 20° to 10°), where it spread laterally to a distance of ~100 m out of the ravine (Fig. 9b), bending the vegetation in the flow direction. Here, the cloud was up to 50 m in height and trunks 15–20 cm in diameter were broken (Fig. 9c). Beyond the break in slope and inside the ravines, the walls were completely eroded and polished, presenting erosive marks ~5 m above the top of the deposits of the basal avalanche (Fig. 10a). At ~4 km from the vent, some BAF units were separated by dilute ash cloud surge deposits of 2.5 cm in thickness. The trunks of trees located in the margins of the ravines presented evidence of impacts and lost foliage up to 1.5 and 2.7 m above the surface of the deposits (Fig. 10b). Mainly at distal sites, the deposits have ~40-cm thick leveés (Fig. 10c). The stratigraphy consisted of between three and four matrix-supported units (each one emplaced by a BAF) varying in thickness from 1.4 to 3 m, and having volumes of between 3.1 and 3.7×10^5 m^3. Flow velocity ranged from 70 to 83 m/s (Sulpizio et al. 2010). The units are separated by concentrations of blocks (0.45–1.6 m in diameter) and occasionally by an unconfined, dilute ash cloud surge deposit (Fig. 9a). The distal deposit (5–6 km) presents two flow units (2–4 m thick) composed of angular to subangular blocks set in an ash matrix. The valley-confined deposit is monolithologic and matrix-supported (54–74 vol.%) (Tables 2 and 3). The grain size distribution of the deposits is polymodal and poorly sorted with median diameters of (Md = −0.9 to 0.1 phi). The unconfined dilute ash cloud surge deposit is light-grey in colour, 0.7–10 cm thick, and consists of clay and silt size ash, massive and

Fig. 7 a Aerial view of the front and levees of the 1991 BAF deposits. b Detail of individual units of the 1991 BAFs in proximal facies (Western Cordobán ravine, photo January 1998). The units are massive and chaotic with a rare ash matrix

laminar, bimodal, with regular sorting and Md = 3.5 phi (Table 3).

Fig. 8 a View of one of the BAFs generated in December 1998 that flowed towards the south of Volcán de Colima (Western Cordobán ravine, photo December 1998), the flow had a tan-colour front and a light-grey body. b The dilute ash clouds emplaced overbank deposits that presented sand dune structures. c The resulting BAF deposits had at least two fronts and levees (box)

6 PDCs from Dome Explosion and Column Collapse

The generation of PDCs from dome explosions and column collapse at Volcán de Colima has occurred approximately 25 times since 1576 (De la Cruz 1993; Saucedo et al. 2005); and four times during the past two decades. This type of PDC has occurred during an eruptive episode that begins with the growth of a summit dome, followed by the emplacement of dome collapse PDCs. These events ended with summit explosions that destroyed the dome with the simultaneous development of 3–10 km high eruptive columns that collapsed to generate PDCs

Fig. 9 View of the different facies of the block-and-ash flow deposits (Montegrande-San Antonio ravines, photo February 2000). **a** A basal massive matrix-supported avalanche with dispersed blocks, **b** Fine massive to stratified ash-rich layers with sand dune structures, and **c** Broken tree trunks by the passage of the dilute ash cloud surge, the trees are either broken or bend in the direction of the flow

Fig. 10 **a** The La Lumbre ravine wall was completely eroded and polished by the passage of the flow. Erosive marks appeared ~5 m above the surface of the basal avalanche deposit. **b** Tree trunks had impact marks and lost foliage up to heights of 1.5 and 2.7 m above the deposit surface. **c** Levees with ~40 cm thicknesses were formed above the 2004 BAF deposits

(Macías et al. 2005; Saucedo et al. 2002, 2005; Sarocchi et al. 2011). During the last two decades this type of eruption has occurred in 1994, 1999, 2003, 2005, and 2013.

The 21 July 1994 explosion was preceded by a seismic swarm that started on 17 July. On 21 July at 8:00 (local time) a sudden increase in seismicity and small rock falls were registered, ending at 20:00 with an explosion that completely destroyed the dome formed during the 1991 eruption. The explosion left a 135 m wide and 40 m deep crater (Fig. 11). The granulometric and components analyses of the deposits indicate that the activity of 21 July 1994 was a phreatic explosion, which generated small-volume blocks-and-ash that were confined into the central and east El Cordobán ravines, reaching distances of 2–3.75 km, respectively (Fig. 3c; Saucedo et al. 1994), with a total estimated volume of $\sim 5 \times 10^5$ m^3 ($\Delta H/L = 0.42$ Fig. 4; Table 1).

During the first months of 1999, an increase in seismic activity was noticed (Saucedo et al. 2002) followed by a shift in volcanic activity from effusive to explosive. On 9 February 1999 explosions occurred at the summit dome. The noise produced by these explosions was heard at the cities of Colima and Ciudad Guzmán located 32 and 26 km from the volcano, respectively. The explosions generated an ash-laden column that rose 3–4 km above the crater. Ballistic projectiles thrown by the explosion reached distances of 3.5–4 km to the north and northeast. Their impacts formed craters up to 2 m in diameter. Some eyewitness reports suggested that this event emplaced PDCs that reached the southwest and south-oriented San Antonio and Montegrande ravines. On 10 May another explosion occurred and was again heard at the city of Colima, followed by the formation of a 6.5-km high ash column above the crater (Saucedo et al. 2002). Ballistic projectiles were launched up to 4.5 km from the summit, causing forest fires. An eyewitness located at ca. 8 km from the summit reported BAFs heading towards La Lumbre and El Cordobán ravines. Seismic activity and discrete explosions continued until 26 May followed by a repose period of a few days. On 3 June 1999, an inspection flight revealed that part of the 1998 lava dome was destroyed and a new crater about 180–200 m in diameter, and $\sim$35 m deep had formed (GVP 1999; Varley, this volume). On 17 July 1999, after 13 h of a seismic swarm, at 12:41 local time, the inhabitants of Colima heard the volcano rumbling: a Vulcanian-type column formed, reaching more than 10 km above the volcano (Fig. 12a). This column deposited ash as far as 30 km from the volcano. At a distance of 13 km, the ash layer was 3–5 mm thick. Small-volume block-rich PDCs were produced by this explosion and emplaced along the San Antonio and Montegrande ravines up to 3.5 km from the summit (Fig. 3b), although some reports indicated that vegetation within La Lumbre ravine was burnt up to distances of 5.5 km (Saucedo et al. 2002). The explosion left a 230-m wide and 70–80 m deep crater (Fig. 12b), enlarging the previous 10 May crater. After 19 July 1999, monitoring indicated that volcanic activity returned to the levels recorded prior to the eruption.

Fig. 11 Crater formed after the explosion of 21 July 1994

Fig. 12 **a** A 10-km high eruptive column generated during the 17 July 1999 explosion. **b** Crater produced by this explosion that estimated 230 m in diameter and 70–80 m in depth (GVP 1999, 24:8)

The eruptive activity between May 2001 and July 2003 began with the formation of eight short lava flows ($\sim 8.3 \times 10^6$ m^3) followed by a period of explosions (see Varley, this volume); the explosion that occurred on 2 May developed a ~0.5-km-high column (Muñoz-Martinez 2007). This eruptive phase ended with a series of three explosions in July and August. The first, on 17 July developed a Vulcanian eruptive column higher than 3 km. The column was sustained for a few moments prior to collapse, producing small-volume block-rich PDCs that were emplaced within the San Antonio ravine up to a distance of 3.5 km (Fig. 3c). The deposits had a total volume of 3×10^5 m^3 and a ΔH/L = 0.42 (Fig. 4; Table 1). The explosion destroyed the 2001–2003 dome (Fig. 13), launching ballistics up to 4 km from the vent, and initiating a series of fires on the southwest flank of the volcano (Muñoz-Martinez 2007).

An explosive phase of Volcán de Colima began in February 2005 and finished with some smaller explosions in January and February, 2006 (Vargas-Gutiérrez 2009). This phase consisted of a series of Vulcanian explosions that developed 2.5–5 km high columns that launched ballistics beyond to 2.5 km from the crater, producing small-volume block-rich PDCs that reached distances of 1.5–5.4 km in the southwest and southeast ravines of the volcano (Vargas-Gutiérrez 2009; Table 1). In May and June, the explosive violence increased significantly; the most violent explosion occurred on 5 June with shock waves that broke glass windows of towns located 9 and 14 km away from the volcano. Ash fall (wind speeds 30–35 km/h) reached the City of Colima from a 5 km high eruptive column. In addition, the explosion produced small-volume block-rich PDCs which were channelled in the Montegrande ravine (Fig. 3a), and reached up to 5.1 km from the source (ΔH/L = 0.35; Fig. 4; Table 1). On 5 July a new explosion produced BAFs with 4.8–5.9 km run-outs (ΔH/L = 0.30; Fig. 4) that were confined to the La Arena ravine on the southeast flank of the volcano (Fig. 3a). This explosion was the last large event reported in 2005 (Fig. 14).

7 Deposits from Dome Explosion and Column Collapse

The 21 July 1994 BAF deposit had a maximum extension of 3.7 km from the source and was channelled in the El Cordobán ravines (Fig. 3b). The deposit was 4 m thick (average) and consisted of blocks with maximum diameters of 2–4 m set in a medium-coarse ash matrix with reverse grading (Saucedo et al. 1994). The marginal and frontal parts of the 1994 deposit had

Fig. 13 Dome captured in January 2002 in filling the summit area of Volcán de Colima

levees and lobes 0.50 m and 2 m-thick on average, respectively. Small degassing pipes generated craters, of 0.30–0.40 m in diameter, observed on the surface of the deposit, and 5 days after the explosion (26 July) the deposit temperature ranged from 116 °C to 282 °C at a depth of 0.30 m (Saucedo et al. 1994). The detached PDC ash cloud or dilute phase generally formed a thin layer of ash a few millimetres thick that spread laterally ~50 m outside the channels, blanketing the vegetation with a 1–1.5 cm thick ash layer, including the bark of trees up to a height of >5 m. The vegetation affected by the passage of the ash cloud was bent in the flow direction and dried without being burned.

The deposits generated by the explosion of 17 July 1999 were emplaced by a small-volume, block-rich PDC that reached 3.5 km from the source along La Lumbre ravine (3.6×10^5 m^3), Montegrande and San Antonio (7.9×10^5 m^3) ravines, located in the southern and southwestern sector of the volcano (Fig. 3a; Table 1). The latter two ravines were completely filled by deposits, so that it was impossible to differentiate them at altitudes between 2400 and 2500 masl. The recognition of fronts or lobes was also impossible. In some places erosion partially cut the deposit, usually exposing a single flow unit (Saucedo et al. 2002), although, subsequent work reported at least two flow units (Sarocchi et al. 2011).

The valley-confined block-and-ash flow deposits in the San Antonio-Montegrande ravines had blocks of angular form with average diameters of 0.70 m embedded in a medium to coarse ash matrix (Saucedo et al. 2002; Table 3). Flow units had thicknesses of 1.2–5 m (Fig. 15 a), reverse grading, monolithologic componentry, and were composed of at least two pulses separated by a concentration of blocks at the top (Sarocchi et al. 2011; Tables 1 and 3). In general, deposits were polymodal, poorly sorted and showed an increase of ash particles with distance (Table 3). The lower unit had a consistent increase in both blocks and fines with distance. In general, samples showed an increase in the fines content toward the deposit bottom with values of Md = 1.4–0.2 phi (Table 3; Sarocchi et al. 2011). The unconfined dilute ash cloud surge deposits formed two units, a basal layer of 2 cm thick ash, massive with an erosive contact at the bottom, and a layer of ash ~9 cm thick with massive and cross-bedded stratification that also had a lower erosive contact (Fig. 13b). These deposits in the proximal facies displayed a unimodal grain-size distribution, asymmetric towards the fine fraction, with a peak at 4 phi (Table 3). In the distal facies, the deposits have similar modes and medians (Table 3).

The valley-confined BAF deposits of the 17 July 2003 explosion reached a distance of 3.5 km

Fig. 14 Explosion and development of BAFs during the 2005 eruptive crisis of Volcán de Colima (Photo March 2005 by Sergio "Tapiro" Velasco)

in the San Antonio ravine (Fig. 3b). This was a single flow unit of angular blocks (0.25–1 m in diameter), supported by an ash matrix (Table 3). The deposit had thicknesses varying from 3 to 1.5 m in the proximal to the distal facies, respectively. It showed reverse grading and had an estimated volume of 3.0×10^5 m^3 ($\Delta H/L = 0.42$; Fig. 4), with a polymodal grain size distribution, median grain size decreasing slightly with distance, and in general poor sorting (Table 3). The deposit decreases in matrix content with distance from 28 to 25.5%, and increases in fine ash from 23 to 32% (Table 3).

The unconfined dilute ash cloud surge deposits covered foliage up to 10 m in height, and the vegetation was inclined in the flow direction. These deposits consisted of fine ash, 1–5 cm thick, showing laminar and cross stratification, erosional contacts and slight consolidation. These deposits changed from bimodal to polymodal and the median decreased with distance (Table 3).

The valley-confined 2005 block-and-ash flow deposits were emplaced in the Montegrande and La Arena ravines located on the south and southeast flanks of the volcano (Fig. 3a). In the Montegrande ravine BAFs had maximum runouts of 5.1 km following the 5 June explosion (Table 1). These deposits consisted of a series of overlapping flow units (thicknesses of 0.5–7 m). The deposits were composed of reversely graded angular to subangular andesitic blocks of 0.15–3.5 m in diameter sustained by medium to coarse ash matrix (Table 3). The granulometric distribution of the deposit was bimodal and poorly sorted, with Md = −3.7 to 0.67 phi (Vargas-Gutiérrez 2009; Table 3).

The valley-confined block-and-ash flow deposits in La Arena ravine reached a maximal runout of 5.4 km during the 9 June explosion (Fig. 3a). The deposits present several reversely graded flow units of 0.7–3.5 m thickness, composed of angular and subangular andesitic blocks (0.05–1 m in diameter), embedded in a medium to coarse ash matrix (Table 3). A few days after the 5 July explosion, some blocks in the deposits retained temperatures >300 °C. The deposits are bimodal, poorly sorted and monolithologic (Tables 2 and 3). In both ravines (Montegrande and La Arena), the segregation of blocks was common, as were the development of levees and a deposit front, rich in blocks with maximum diameters of 2–4 m. Unconfined dilute ash cloud surge deposits were occasionally observed. The deposits were 6 cm thick, and unimodal, with Md = −0.03 to 1 phi (Vargas-Gutiérrez 2009; Table 3).

Fig. 15 Deposits emplaced by the 17 July 1999 block and ash flow along the Montegrande-San Antonio ravines. **a** Basal avalanche and **b** Ash massive layer with a lower erosive contact and ingested roots, scale is 20 cm long (photo of the San Antonio-Montegrande ravines)

8 Discussion

Davies et al. (1978) described pyroclastic flows as "glowing avalanches" produced during the 1974 eruption of Volcán de Fuego, Guatemala. They concluded that the flows moved in two parts: (a) the main mass of debris flowing in a laminar fashion as a high concentration dispersion similar to normal debris flows, and (b) a superjacent ash cloud. Since blocks 5 m in diameter were transported up to 7 km from the source, they suggested that flows moved as a high yield-stress fluid. Nairn and Self (1978) considered that hot avalanches produced by the 1975 eruption of Ngauruhoe volcano, New Zealand, moved as a density modified grain flow, since these flows had a fluid phase denser than the ambient air between particles (Bagnold 1954; Lowe 1976). These authors further advanced the idea that convolute vortices produced on top of the basal avalanche were caused by a low density turbulent mixing layer. Beyond a break in slope, the basal avalanche reduced its speed, causing the turbulent layer to separate from the avalanche and to rise as a cloud depositing fall material. Since these pioneering studies, several authors have considered the basal avalanches of these pyroclastic flows to move as density modified grain flows, in which grains are separated by intergranular dispersive pressure (Denlinger 1987; Mellors et al. 1988; Boudon et al. 1993; Fujii and Nakada 1999; Iverson and Vallance 2001) where there is some degree of fluidization due to the gas liberated from particle break up.

Fluidization can play an important role in the mobility of this type of pyroclastic density current (Bursik et al. 2005). Fujii and Nakada (1999) considered that dome collapse (sometimes referred to as Merapi-type) pyroclastic flows produced on 15 September 1991, at Mt. Unzen, Japan, were formed by a basal turbulent current of high concentration (main body), an intermediate fluidization zone, and an overriding convecting dilute cloud. Entrained cold air, gas and ash particles released from the basal zone produced a fluidization zone capable of moving ahead or to the sides of the main body as an independent gravity current. According to these authors, the fluidization zone can move faster that the main body on steep slopes, enhancing detachment of the former when a flow encounters gentle slopes and/or topographic barriers. Fisher and Heiken (1982) were the first authors to propose that a dilute density current decoupled from the basal avalanche, destroyed the town of Saint Pierre during the 1902 eruption of Mt. Pelée, Martinique Island. Based on field observations, experimental work and theoretical calculations, Takahashi and Tsujimoto (2000) concluded that the 3 June 1991 Merapi-type pyroclastic flow of Mt. Unzen, moved in the proximal zone (0–1.6 km) as an inertial granular flow, where the effect of gas released from particle collision was minimal, and deposition occurred over a short distance on steep slopes. The remaining flow, mainly composed of mm-sized particles, was supported by upward flow of gas released from the same particles. The authors proposed that the flow consisted of a lower, insufficiently fluidized layer and an upper fluidized layer, where particles were sustained in suspension by the upward gas flow.

9 Transport and Deposition Processes

Recently, the frequent eruptive phenomena of Volcán de Colima (De La Cruz 1993; Saucedo et al. 2005) included the generation of small-volume, block-rich PDCs produced from collapse of external parts of fresh lava domes, active lava flow fronts, or partial collapse of the lower parts of an eruptive column. The deposits generated by the different types of block-and-ash flows (Sato et al. 1992), are similar in structure, composition and internal features (Fisher and Heiken 1982; Davies et al. 1978), independent of the mechanism that generated them. However, certain recent studies disagree with this statement (Charbonnier and Gertisser 2011; Komorowski et al. 2010; Charbonnier et al. 2013; Stinton et al. 2014; Ogburn 2008). The PDCs at Volcán de Colima as well as other volcanoes immediately separate into a basal dense flow and an upper dilute turbulent ash cloud (Fisher and Heiken 1982; Mellors et al. 1988; Fuji and Nakada 1999; Saucedo et al. 2002, 2004; Schwarzkopf et al. 2005; Lube et al. 2007). The basal flows, also called basal avalanches, move as inertial granular mass flows on steep slopes, where the interaction between particles, pore pressure and confined or unconfined channel morphology play an important role in their propagation (Iverson and Vallance 2001; Lube et al. 2007). Given sufficient volume, an initial steep slope, sufficient fragmentation to produce fine material in the matrix, and suitable confined flow conditions, PDCs can continue to move on gentler slopes, where there is evidence for extensive clast break-up, and fluidization either during or following propagation (Saucedo et al. 2004; Bursik et al. 2005; Lube et al. 2007; Lube et al. 2011; Andrews and Manga 2011; Cronin et al. 2013; Komorowski et al. 2013; Kokelaar et al. 2014). In the proximal zone, most of the PDC travels as an inertial granular flow or quasi-steady granular flows (Schwarzkopf et al. 2005; Saucedo et al. 2004; Sulpizio et al. 2007; Lube et al. 2007; Sarocchi et al. 2011). A PDC in this zone may travel on a slope that is greater than the angle of repose, and so there can only be minor deposition (Saucedo et al. 2004; Lube et al. 2007). These flows, once moving on slopes below the angle of repose, consist of a basal layer of depositing particles and an upper flowing layer. These layers are separated by a sharp internal interface, which propagates with time towards the free surface. Once it arrives, this signals the end of flow (Takahashi and Tsujimoto 2000; Saucedo et al. 2004; Bursik et al. 2005; Schwarzkopf et al. 2005; Sulpizio et al. 2007: Lube et al. 2007). The interface concept is the same as the flow boundary layer used to explain progressive aggradation in ignimbrites (Branney and Kokelaar 1992, 2002). Between the internal interface and the free surface, PDCs move like granular mass flows and develop levees, steep fronts, and reversely graded boulders, where kinetic sieving, squeezing-out and differential setting speeds through the flow play important roles in the segregation of coarse clasts during transport (Lube et al. 2007; Roux 2003; Sulpizio et al. 2007; Sarocchi et al. 2011). These features were observed at valley-ponded deposits after the 1991, 1998, 1999, and 2004 eruptions along the Cordobans, La Lumbre, and Montegrande ravines (Saucedo et al. 2002, 2004).

When PDCs decelerate and eventually stop, the superjacent turbulent ash cloud detaches, and is able to move independently as a newly formed, unconfined dilute PDC (Fujii and Nakada 1999; Takahashi and Tsujimoto 2000; Woods et al. 2002; Carn et al. 2004; Saucedo et al. 2004; Bursik et al. 2005: Sulpizio et al. 2010). These PDCs are able to surmount topographic barriers, emplacing an unconfined dilute ash cloud surge deposit, with fine-grained, massive/laminated, and dune-bedded layers (Woods et al. 2002; Saucedo et al. 2004; Dellino et al. 2004; Charbonnier and Gertisser 2011; Komorowski et al. 2013). When the detached ash cloud surge stops or traverses a topographic obstacle, a rising cloud of ash forms, depositing a terminal ash fall (Saucedo et al. 2002, 2004; Sulpizio 2007, Sulpizio et al. 2010). Locally, unconfined dilute ash cloud surge deposits may occur overlying the valley-confined block-and-ash flow deposits or lying just outboard of the channel (Saucedo et al. 2002, 2004; Bursik et al. 2005). This process has been observed during the 1902 eruption of Mt. Pelée (Fisher and Heiken 1982) the 1991 events of Mt. Unzen (Fujii and Nakada 1999) as well as the eruptions of Volcán de Colima in 1991, 1998–1999 and 2004 (Saucedo et al. 2002, 2004).

The development of the different facies derived from the emplacement of BAF deposits, associated with the destruction of domes at Volcán de Colima, takes place when PDCs propagate downslope, and particle break up is sufficiently extensive to generate a significant matrix fraction. This matrix increment is associated to the transition from clast support (inertial granular flow in proximal facies) to matrix support (pseudo viscous flow in distal facies) within the valley-ponded BAFs (Saucedo et al. 2002, 2004; Bursik et al. 2005; Lube et al. 2007; Sarocchi et al. 2011). Pyroclast break up is more common as a result of the following processes: as dome destruction becomes progressively more explosive, resulting from the movement of a fragmentation wave within the dome (retrogressive dome collapse), an increase in decompressing gas volume, with progressively greater pyroclast fragility, or with a greater volume (hence weight) of flowing debris (Fink and Kieffer 1993; Hayashi and Self 1992; Woods et al. 2002; Komorowski et al. 2013). The predominant fraction of matrix supported BAF deposits results from a greater degree of particle break up, where both the kinetic sieving and kinetic squeezing could be very important in the segregation of coarse particles during transport (Roux 2003; Schwarzkopf et al. 2005; Sulpizio et al. 2007; Lube et al. 2007; Sarocchi et al. 2011). Analog deposits have been described at numerous volcanoes, which show normal and inverse grading, as well as segregation structures, and having a lateral extent frequently defined by levees and fronts lobes, where sorting is always poor (Saucedo et al. 2002; Schwarzkopf et al. 2005; Lube et al. 2007; Figs. 7a, 8c, and 10c).

The superjacent detached density currents that emplace unconfined dilute ash cloud surge deposits are laterally widespread and occur in both proximal and distal facies (Saucedo et al. 2002, 2004; Schwarzkopf et al. 2005; Sulpizio et al. 2007, 2010). These deposits have only been observed in sections up to a few centimetres thick, as those described at Unzen and others volcanoes, including Volcán de Colima in the eruptions of 1991, 1998–1999, 2004 and 2005 (Nakada and Fujii 1993; Saucedo et al. 2002, 2004; Schwarzkopf et al. 2005; Vargas-Gutiérrez 2009), where they can have an erosional lower contact with underlying soil units (Figs. 9b and 15b). Singe and plant blow-down features are often found at Volcán de Colima, which extend a short distance (in the order of 100 m) beyond the basal flow unit margins. Wherever the detached PDC stops or surmounts topographic barriers, it develops ash cloud deposits that are rarely more than 1 cm in thickness. It is possible to find thin coatings of probably correlative ash cloud fall to distances of several kilometres from the flow margin proper.

During the preparation of this chapter, Volcán de Colima erupted on 10–11 July 2015. The volcano generated block-rich PDCs with a maximum runout of 10.5 km (Capra et al. 2016; Varley, this volume). These PDCs were similar to those generated during the second eruptive phase of the 1913 Plinian eruption (Saucedo et al. 2010; Table 1). This activity that began in 2007 and ended in 2015 has continued with increasingly longer runout pyroclastic density currents of the volcano recorded since 1991. Such a scenario undoubtedly obliges scientists and local authorities to re-evaluate eruptive scenarios associated with the gravitational collapse and explosion of lava domes in terms of their destructive potential and the hazards they can pose to the population.

Acknowledgements We are grateful to Edgar Rojas who always shared his time with us during laboratory work and preparation of figures for this paper. This research was supported by grants from CONACYT-Ciencia Básica (101548 to R. Saucedo), Dirección General de Asuntos del Personal Académico, UNAM (DGAPA-IN116911) to J.L. Macías, and NASA grant NNX12AQ10G to M. Bursik. We are grateful for the constructive comments provided by P. Cole, and S. Charbonnier that improved the ideas stated in this contribution. We also thank the editorial handling of J.C. Komorowski, N, Varley, and Chuck Connor.

References

Andrews, B.J., Manga, M.: Effects of topography on pyroclastic density current runout and formation of coignimbrites. Geology. **39**(2), 1099–1102 (2011)

Arreola, J.M.: Catálogo de las erupciones antiguas del Volcán de Colima. Mem. Soc. Cient. Antonio Alzate T. **32**, 443–481 (1915)

Bagnold, R.A.: Experiments on a gravity-free dispersion of large solid spheres in a Newtonian fluid under shear. Proc. Roy. Soc. Lond. **A225**, 49–63 (1954)

Bárcena, M.: Informe sobre el estado actual del Volcán de Colima. Anales del Ministerio de Fomento, México, pp. 328–365 (1887)

Boudon, G., Camus, G.A., Lajoie, J.: The 1984 nuée-ardente deposits of Merapi volcano, Central Java, Indonesia: stratigraphy, textural characteristics, and transport mechanisms. J. Volcanol. Geotherm. Res. **55**, 327–342 (1993)

Branney, M.J., Kokelaar, P.: A reappraisal of ignimbrite emplacement: progressive aggradation and changes from particulate to non-particulate flow during emplacement of high-grade ignimbrite. Bull. Volcanol. **54**(6), 504–520 (1992)

Branney, M., Kokelaar, P.: Pyroclastic density currents and the sedimentation of ignimbrites. Geol. Soc. Mem. **27**, 144 (2002)

Bursik, M.I., Patra, A., Pitman, E.B., Nichita, C., Macias, J.L., Saucedo, R., Girina, O.: Advances in studies of dense volcanic granular flows. Rep. Prog. Phys. **68**, 271–301 (2005)

Capra, L., Macías, J.L., Cortés, A., Davila, N., Saucedo, R., Osorio-Ocampo, S., Arce, J.L., Gavilanes-Ruíz, J. C., Corona-Chávez, P., García-Sánchez, L., Sosa-Ceballos, G., Vázquez, R.: Preliminary report on the July 10-11, 2015 eruption at Volcán de Colima: pyroclastic density currents with exceptional runouts and volumes. J. Volcanol. Geotherm. Res. **310**, 39–49 (2016)

Carn, S.A., Watt, R.B., Thompson, G., Norton, G.E.: Anatomy of a lava dome collapse: the 20 March 2000 event at Soufriere Hill volcano, Montserrat. J. Volcanol. Geotherm. Res. **131**, 241–264 (2004)

Cas, R.A.F., Wright J.V.: Volcanic successions: modern and ancient. Allen & Unwin, London, p. 528 (1987)

Charbonnier, S.J., Gertisser, R.: Deposit architecture and dynamics of the 2006 block-and-ash flows of Merapi Volcano, Java, Indonesia. Sedimentology. **58**(6), 1573–1612 (2011)

Charbonnier S.J., Germa, A., Connor C.B., Gertisser, R., Preece, K., Komorowski, J.C., Lavigne, F., Dixon, T., Connor, L.: Evaluation of the impact of the 2010 pyroclastic density currents at Merapi volcano from high-resolution satellite imagery, field investigations and numerical simulations. J. Volcanol. Geoth. Res. **261**, 295–315 (2013)

Cortés, A., Garduño, V.H., Macias, J.L., Navarro-Ochoa, C., Komorowski, J.C., Saucedo, R., Gavilanes, J.C.: Geologic mapping of the Colima volcanic complex (Mexico) and implications for hazard assessment. In: Groppelli, G., Viereck-Goette, L. (eds.). Stratigraphy and Geology of Volcanic Areas: Geological Society of America Special Paper, vol. 464, p. 116 (2010)

Cronin, S.J., Lube, G., Dayudi, D.S., Sumarti, S., Subrandriyo, Surono, S.: Insights into the October–November 2010 Merapi eruption (Central Java, Indonesia) from the stratigraphy, volume and characteristics of its pyroclastic deposits. J. Volcanol. Geoth. Res. **261**, 244–259 (2013)

Davies, D.K., Quearry, M.W., Bonis, S.B.: Glowing avalanches from the 1974 eruption of volcano Fuego, Guatemala. Geol. Soc. Amer. Bull. **89**, 369–384 (1978)

De la Cruz, S.: Random patterns of occurrence of explosive eruptions at Colima Volcano, México. J. Volcanol. Geotherm. Res. **55**, 51–68 (1993)

Dellino, P., Isaia, R., Veneruso, M.: Turbulent boundary layer shear flow as an approximation of base surges at Campi Flegrei (Southern Italy). J. Volcanol. Geotherm. Res. **133**, 211–228 (2004)

Denlinger, R.P.: A model for generation of ash clouds by pyroclastic flows, with application to the 1980 eruptions at Mount St. Helens. J. Geophys. Res. **92**, 10284–10298 (1987)

Dollfus, A., Montserrat, E.: Arcive de La commission Scientifique du Mexique. Du Ministére de L'instruction Publique. Tomo II, 1–43 (1867)

Druitt, T.H.: Pyroclastic density currents. In: Gilbert, J.S., Sparks, R.S.J. (eds.). Geological Society, London, Special Publications, vol. 145, pp. 145–182 (1998)

Fink, J.H., W. Kieffer, S.W.: Estimate of pyroclastic flow velocities resulting from explosive decompression of lava domes. Nature. **363**(6430), 612–615 (1993)

Fisher, R.V., Heiken, G.: Mt. Pelée, Martinique: May 8 and 20, 1902, pyroclastic flows and surges. J. Volcanol. Geotherm. Res. **13**, 339–371 (1982)

Fujii, T., Nakada, S.: The 15 September 1991 pyroclastic flows at Unzen volcano (Japan): a flow model for associated ash-cloud surge. J. Volcanol. Geotherm. Res. **89**(1–4), 159–172 (1999)

Global Volcanic Program: Report on Colima (Mexico). In: Wunderman, R. (ed.) Bulletin of the Global Volcanism Network, vol. 24, p. 8. Smithsonian Institution (1999)

Hayashi, J.N., Self, S.: A comparison of pyroclastic flow and debris avalanche mobility. J. Geophys. Res. **97** (B6), 9063–9071 (1992)

Heim, A.: Der Bergsturz und Menschenleben. Fretz und Wasmuth Verlag, Zürich, p. 218 (1932)

Iverson, R.M., Vallance, J.W.: New view of granular mass flows. Geology **29**(2), 115–118 (2001)

Komorowski, J.C., Legendre, Y., Christopher, T., Bernstein, M., Stewart, R., Joseph, E., Fournier N, Chardot, L., Finizola, A., Wadge, G., Syers, R., Williams, C., Bass, V.: Insights into processes and deposits of hazardous vulcanian explosions at Soufriere Hills Volcano during 2008 and 2009 (Montserrat, West Indies), Geophys. Res. Lett. **37** (L00E19), 1–6 (2010)

Komorowski, J.C., Jenkins, S., Baxter, P.J., Picquout, A., Lavigne, F., Charbonnier, S.,Gertisser, R., Cholik, N., Budi-Santoso, A., Surono.: Paroxysmal dome explosion during the Merapi 2010 eruption: processes and facies relationships of associated high-energy pyroclastic density currents. J. Volcanol. Geoth. Res. **261**, 260–294 (2013)

Kokelaar, B.P., Graham, R.L., Gray, J.M.N.T., Vallance, J.W.: Fine-grained linings of leved channels facilitate ronout of granular flows. Earth. Planet. Sci. Lett. **385**, 172–180 (2014)

Lermo, J., Cuenca, J., Monfret, T., Hernández, F., Nava, E.: Algunas características espectrales de la sismicidad asociada a la actividad del Volcán de Colima. Geof. Int. **32**(4), 683–697 (1993)

Lowe, D.R.: Grain flow and grain flow deposits. J. Sediment. Petrol. **46**, 188–199 (1976)

Lube, G., Cronin, S.J., Platz, T., Freundt, A., Procter, J. N., Herderson, C., Sheridan, M.F.: Flow and deposition of pyroclastic granular flows: a type example from the 1975 Ngauruhoe eruption, New Zeland. J. Volcanol. Geotherm. Res. **161**, 165–186 (2007)

Lube, G., Cronin, S.J., Thouret, J.-C., Surono.: Kinematic characteristics of pyroclastic density currents at Merapi and controls on their avulsion from natural and engineered channels. Geol. Soc. Am. Bull. **123**(5–6), 1127–1140 (2011)

Luhr, J.F., Carmichael, I.S.E.: Geology of Volcán de Colima. Instituto de Geología UNAM Publicación especial. Bol. No 107, pp. 1–101 (1990)

Macías, J.L., Saucedo, R., Gavilanes, J.C., Varley, N., Velasco, S., Bursik, M., Vargas, V., Cortés, A.: Flujos piroclásticos asociados a la actividad explosiva del Volcán de Colima y perspectivas futuras. GEOS **25** (3), 340–351 (2005)

Martinez, E.: Estudio del mecanismo eruptivo de emplazamiento del flujo piroclástico de bloques y ceniza del 17 de julio del 2003 en el volcán de Colima. Bacherlor Thesis, Universidad Autónoma de San Luis Potosí, 88 p (2007)

Mellors, R.A., Waitt, R.B., Swanson, D.A.: Generation of pyroclastic flows and surges by hot-rock avalanches from the dome of Mount St. Helens volcano, USA. Bull. Volcanol. **50**, 14–25 (1988)

Mooser, F., Reyes, A.: Los volcanes de Colima. Universidad Nacional Autónoma de México. Inst. Geol. Bol. **61**, 49–71 (1961)

Nairn, I.A., Self, S.: Explosive eruptions and pyroclastic avalanches from Ngauruhoe in February 1975. J. Volcanol. Geotherm. Res. **3**, 39–60 (1978)

Nakada, S., Fujii, T.: Preliminary report on the activity at Unzen Volcano (Japan), November 1990–1991: dacite lava domes and pyroclastic flows. J. Volcanol. Geoth. Res. **54**, 319–333 (1993)

Navarro, C., Gavilanes, J.C., Cortés, A.: Movement and emplacement of lava flows at Volcán de Colima, México: Nov. 1998–Feb. 1999. J. Volcanol. Geotherm. Res. **117**, 155–167 (2002)

Nuñez-Cornu, F., Nava, A., De la Cruz, S., Jiménez, Z., Valencia, C., García, R.: Seismic activity related to the 1991 eruption of Colima Volcano, Mexico. Bull. Volcanol. **56**, 228–237 (1994)

Ogburn, S.E.: Potential hazards at Soufrière Hills Volcano, Montserrat: Northwards-directed dome-collapses and major explosive eruptions. SUNY at Buffalo, M.S. thesis, 178 p (2008)

Ortiz, G.: La zona volcánica "Colima" del estado de Jalisco. Edición del Instituto de Geografía de la universidad de Guadalajara. Monografía, 1–44 pp (1944)

Rodríguez-Elizarrarás, S.R., Siebe, C., Komorowski, J.C., Espíndola, J.M., Saucedo, R.: Field observation of pristine block-and-ash-flow deposits emplaced April 16-17, 1991 at Volcán de Colima, México. J. Volcanol. Geotherm. Res. **48**, 399–412 (1991)

Rose, W.I.: Nuée ardente eruption the foot of a dacite lava flow Santiaguito, Guatemala. Bull. Volcanol. **40**, 23–38 (1977)

Roux, J.P.: Can dispersive pressure cause inverse grading in grain flows? J. Sediment. Res. **73**(2), 333–334 (2003)

Rupp, B., Bursik, M., Namikawa, A., Patra, A., Saucedo, R., Macías, J.L.: Computational modeling of the 1991 block and ash flow at Colima Volcano, México. Geol. Soc. Am. Spec. Pap. **402**, 223–238 (2006)

Sarocchi, D., Sulpizio, R., Macias, J.L., Saucedo, R.: The 17 July 1999 block-and-ash flow (BAF) at Colima Volcano: new insights on volcanic granular flows from textural analysis. J. Volcanol. Geotherm. Res. **204**, 40–56 (2011)

Sartorious, C.: Eruption of the Volcano of Colima in June 1869. American Journal of Science and Article 3rd Serie II, 381. In: Smithsonian Report 1869: 423 (1871)

Sato, H., Fujii, T., Nakada, S.: Crumbling of dacite dome lava and generation of pyroclastic flows at Unzen volcano. Nature **360**, 664–666 (1992)

Saucedo, R., Gavilanes-Ruiz, J.C., Cuevas, A, Cortes, A., Navarro-Ochoa, J.C.: Síntesis de la actividad del Volcán de fuego de Colima de Julio de 1994 a marzo de 1995. Universidad de Colima, Observatorio Vulcanológico, pp. 2–25 (1994)

Saucedo, R.: Reconstrucción de la Erupción de 1913, del Volcán de Colima, México. Master thesis, Universidad Nacional Autónoma de México, Distrito Federal, 197 p (1997)

Saucedo, R., Macías, J.L., Bursik, M.I., Gavilanes, J.C., Mora, J.C., Cortés, A.: Emplacement of pyroclastic flow during the 1998–1999 eruption of Volcán de Colima, México. J. Volcanol. Geotherm. Res. **117**, 129–153 (2002)

Saucedo, R., Macías, J.L., Bursik, M.: Pyroclastic flow deposits of the 1991 eruption of Volcán de Colima, México. Bull. Volcanol. **66**, 291–306 (2004)

Saucedo, R., Macías, J.L., Sheridan, M.F., Bursik, M.I., Komorowski, J.C.: Modeling of pyroclastic flow of Colima Volcano, Mexico: implications for hazard assessment. J. Volcanol. Geotherm. Res. **139**, 103–115 (2005)

Saucedo, R., Macías, J.L., Gavilanes-Ruiz, J.C., Arce, J. L., Valdez-Moreno, G.: Eyewitness stratigraphy chemistry and eruptive dynamics of the 1913 Plinian eruption of Volcán de Colima, México. J. Volcanol. Geotherm. Res. **191**, 149–166 (2010)

Schwarzkopf, L.M., Schmincke, H.U., Cronin, S.B.: A conceptual model for block-and-ash flow avalanche transport and position, based on deposit architecture of 1998 and 1994 Merapi flow. J. Volcanol. Geotherm. Res. **139**, 117–134 (2005)

Sheridan, M.: Emplacement of pyroclastic flow. Geol. Soc. Spec. Pub. **180**, 125–136 (1979)

Sheridan, M.F., Macías, J.L.: Estimation of risk probability for gravity-driven pyroclastic flows at Volcan Colima, Mexico. J. Volcanol. Geoth. Res. **66**(1–4), 251–256 (1995)

Siebe, C., Abrams, M., Sheridan, M.F.: Major Holocene block-and-ash fan at the western slope of ice-capped Pico de Orizaba volcano, Mexico: implications for future hazards. J. Volcanol. Geoth. Res. **59**(1–2), 1–33 (1993)

Stinton, A.J., Cole, P.D., Odbert, H.M., Christopher, T., Avard, G., Bernstein, M.: Dome growth and valley fill during Phase 5 (8 October 2009–11 February 2010) at the Soufrière Hills Volcano, Montserrat. In: Wadge, G., Robertson, R.E.A., Voight, B. (eds.) The Eruption of Soufrière Hills Volcano, Montserrat from 2000 to 2010. Geological Society, London, Memoirs, vol. 39, pp. 113–131 (2014)

Sulpizio, R.: High variability of sedimentology and physical properties of pyroclastic density currents during complex Subplinian eruptions: the example of the AD 472 (Pollena) eruption of Somma-Vesuvius, Italy. Sedimentology (2006), 1–29 (2007)

Sulpizio, R., Mele, D., Dellino, P., La Volpe, L.: Deposits and physical properties of pyroclastic density currents during complex Subplinian eruptions: the AD 472 (Pollena) eruption of Somma-Vesuvius, Italy. Sedimentology **54**, 607–635 (2007)

Sulpizio, R., Capra, L., Sarocchi, D., Saucedo, R., Gavilanes-Ruiz, J.C., Varley, N.R.: Predicting the block-and-ash flow inundation areas at Volcán de Colima (Colima, Mexico) based on the present day (February 2010) status. J. Volcanol. Geotherm. Res. **193**, 49–66 (2010)

Takahashi, T., Tsujimoto, H.: A mechanical model for Merapi-type pyroclastic flow. J. Volcanol. Geotherm. Res. **98**, 91–115 (2000)

Thorpe, R.S., Gibson, I.L., Vizcaíno, J.S.: Andesitic pyroclastic flows from Colima Volcano. Nature **265**, 724–725 (1977)

Vargas-Gutiérrez, V.: Estratigrafía de los flujos de bloques y ceniza asociadosa la actividad 2005 en el volcán de Colima. Master thesis, Universidad Nacional Autónoma de México, Distrito Federal, pp. 1–230 (2009)

Waitz, P.: Der gegenwarting Stand der Mexika nischen Vulkans von Colima. Zeit crift fur Volcanologie **1**(4), 247–274 (1915)

Waitz, P.: Nubes ardientes observadas en las erupciones del Jorullo (1759) del Ceboruco (1870) y del Volcán de Colima (1913). Mem. Soc. Cien. Antonio Alzate T. **37**, 267–295 (1920)

Waitz, P.: Datos históricos y bibliográficos acerca del Volcán de Colima. Mem. Soc. Antonio Alzate T. **53**, 349–383 (1935)

Woods, A.W., Sparks, R.S.J., Ritchie, L.J., Batey, J., Gladstone, C., Bursik, M.I.: The explosive decompression of a pressurized volcanic dome: the 26 December 1997 collapse and explosion of Soufriere Hills Volcano, Montserrat. Geol. Soc. Lond. **21**, 457–465 (2002)

Yamamoto, T., Takarada, S., Suto, S.: Pyroclastic flows from the 1991 eruption of Unzen volcano, Japan. Bull. Volcanol. **55** (3), 166–175 (1993)

Origin, Behaviour and Hazard of Rain-Triggered Lahars at Volcán de Colima

L. Capra, J. C. Gaviláncs-Ruíz, N. Varley and L. Borselli

Abstract

Lahars represent a common phenomenon at Volcán de Colima, directly associated with an eruptive event, or occurring up to several years after. Most common events are rain-triggered lahars, which are generated under different meteorological and eruptive conditions in the main ravines on the flanks of the volcano. Triggering mechanisms and flow behaviour are highly influenced by rainfall intensity and contemporaneous pyroclastic activity, which can control both lahar volume and frequency. Relationships have been established between the textural features of lahar deposits and the origin and behaviour of past flows, using visual, rainfall and seismic monitoring data. Physical characteristics of the slope surfaces are important factors in determining the rate and amount of water runoff, which controls the initial bulking of the lahars. Since the last Plinian eruption in 1913, rain-triggered lahars represent the volcanic hazard that has caused the most damage to the surrounding areas. Notable damage to bridges, electricity-distribution pylons, and houses has occurred during the last fifteen years. We present a lahar hazard-zonation map to provide updated data useful for decision-making regarding possible future damage to inhabited areas around the volcano.

1 Introduction

Lahars (sediment-water, gravity-driven flows) are common phenomena on active or quiescent volcanoes, with their frequency depending upon the availability of loose material on the slopes and the amount of precipitation or volume of water from a glacial outburst or the breaching of a crater lake. Commonly they are syn-eruptive, formed coevally with volcanic activity, post-eruptive, occurring after the termination of the eruptive phase, but where the landscape remains in disequilibrium regarding the sediment balance as a consequence of the eruption, or they can be inter-eruptive, without direct influence of eruptive activity (Manville et al. 2009). During the 20th century, the most catastrophic lahars

L. Capra (✉)
Centro de Geociencias, Campus UNAM-Juriquilla, 76230 Querétaro, Mexico
e-mail: lcapra@geociencias.unam.mx

J. C. Gaviláncs-Ruíz · N. Varley
Facultad de Ciencias, Universidad de Colima, Colima, Mexico

L. Borselli
Instituto de Geología, Universidad Autónoma de San Luis Potosí, San Luis Potosí, Mexico

N. Varley et al. (eds.), *Volcán de Colima*, Active Volcanoes of the World,
https://doi.org/10.1007/978-3-642-25911-1_2

were associated with magmatic activity, such as during the 1985 Nevado de Ruiz eruption (Colombia) and the 1991 Pinatubo eruption (Philippines). In the former case, a small explosive eruption partially eroded and melted ice from several glaciers, initiating a sequence of lahars that converged within the main drainage of the Langunillas River, devastating the town of Armero and killing more than 20,000 people (Naranjo et al. 1986; Williams 1987). For more than 10 years after the large Plinian Mt. Pinatubo eruption, lahars have continued to occur during every monsoon season, descending the volcanic slopes, remobilizing pyroclastic material, and inundating an area of hundreds of square kilometres (Rodolfo and Arguden 1991; Torres et al. 2004).

In other cases, large lahars may not be associated with eruptive activity. For example, the 1998 Hurricane Mitch triggered the collapse of a portion of the upper flank of inactive Casita volcano (Nicaragua), moving initially as a landslide and then suddenly transforming into a lahar that devastated several towns and killing approximately 2000 people (Van Wyk Vries et al. 2000; Scott et al. 2005). Similarly, on 6 June 1994, after a period of heavy rainfall, a tectonic earthquake at the base of Nevado del Huila volcano triggered $\sim$3000 shallow landslides, which coalesced into a massive lahar flowing down the Páez River, killing almost 1000 people, most of them in the towns of Irlanda and Tóez, which were buried by lahar deposits (Scott et al. 2001).

The above examples represent extraordinary events, large in magnitude but infrequent. Smaller but more frequent post-eruptive lahars, generally triggered by heavy rains, are observed every rainy season at a number of active volcanoes that have recently had eruptions, such as Merapi (Lavigne et al. 2000; Lavigne and Thouret 2002; De Belizal et al. 2013) and Semeru (Lavigne and Suwa 2004; Doyle et al. 2010) both on Java (Indonesia), and Soufriere Hills on Montserrat (West Indies; Carn et al. 2004). Volcán de Colima has a history of both larger eruption-related lahars and smaller rainfall-triggered lahars following eruptive activity. The purpose of the present chapter is to present data regarding the triggering mechanism and characteristics of post-eruptive rain-triggered lahars at this particular volcano. The related hazards are also discussed.

2 Volcano Hydrography and Hydrogeological Characteristics

The edifice of Volcán de Colima is 1640 m in relief with respect to the surrounding area, and the flanks of the volcano vary in slope from 40° to 10°. Main drainages on Volcán de Colima are on its E, S, and W flanks (Fig. 1). Only few of these drainages host permanent water discharge; these include Lower El Zarco, Cordobán (Santa Cruz) and La Lumbre ravines (the latter has water springs at 8 km from the volcano summit) (Fig. 2a). The other ravines are not spring-fed and contain ephemeral streams, discharging water-sediment flows only during rainfall events.

The rainy season extends from June to October, with some isolated precipitation events in January or February (for example, one initiated a lahar in 1992, Table 1) producing an average rainfall accumulation of 1000 mm/yr (Fig. 2a). At the beginning of the rainy season, rainfall is characterized by a low duration (less than one hour) and high peak intensity events (>50 mm/hr.), whilst in the late season, they are associated with tropical cyclones with moderate rainfall intensity (<50 mm/hr.) but with longer duration (several hours). Based on rainfall monitoring on the volcanic edifice, the rainfall totals received are variable across the slopes of the volcano, with an apparently higher amount on the southern sector (Fig. 2a).

Eight main watersheds can be identified (Fig. 2a, Capra et al. 2010), which feed the most active ravines where lahars originate: La Lumbre, El Zarco, Cordobán, San Antonio, Montegrande, El Muerto, and La Arena. These have variable surface areas, from 2 to 10 km^2, and each can be divided into smaller watersheds that during rainfall events contribute to the overall volume of water, with inflow at different distances along the

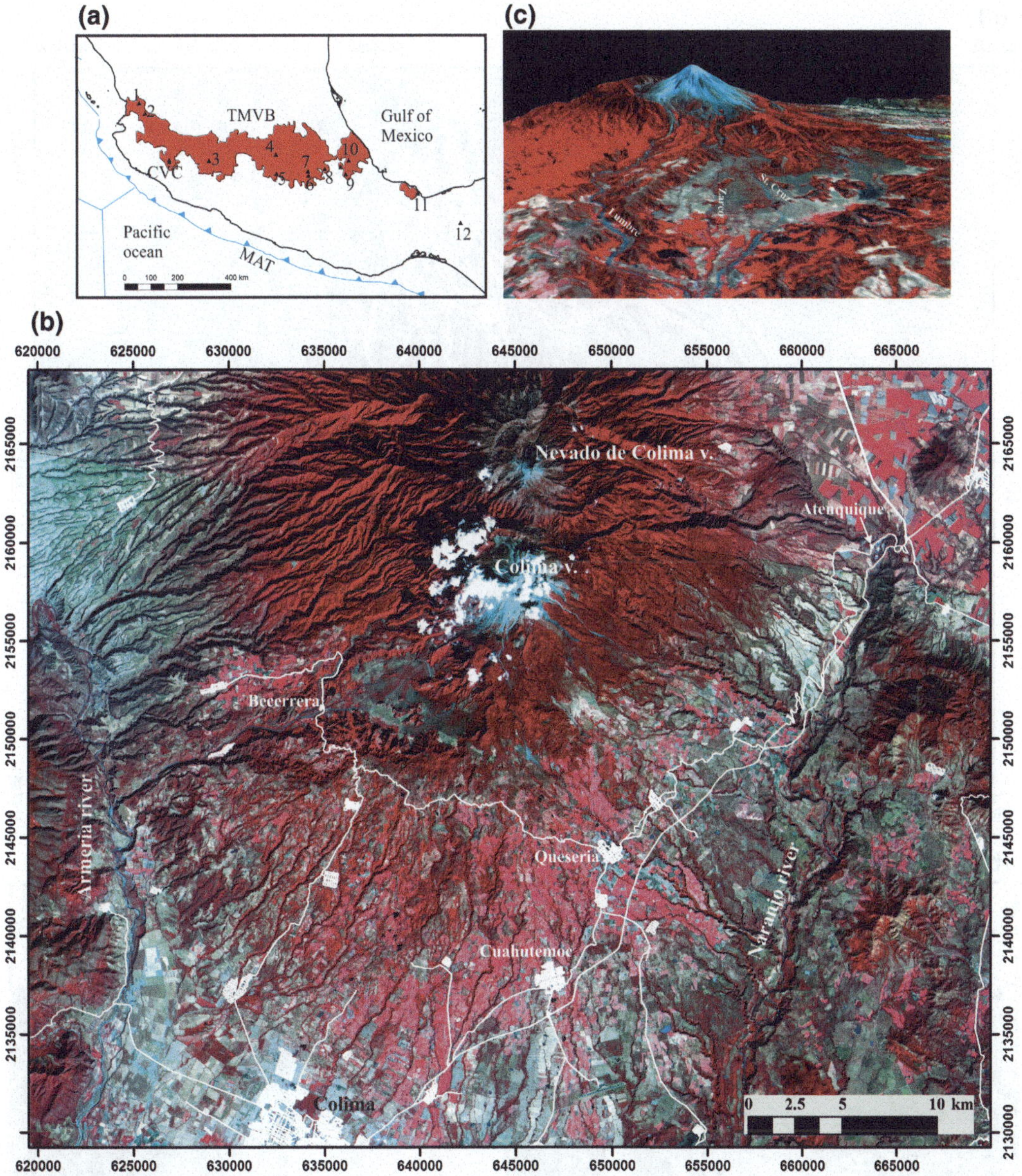

Fig. 1 **a** Simplified map of the Trans-Mexican Volcanic Belt with the location of the Colima Volcanic Complex (CVC) and other active or important volcanoes: 1: San Juan; 2: Ceboruco; 3: Tancitaro; 4: Jocotitlán; 5: Nevado de Toluca; 6: Popocatépetl; 7: Ixtaccíhuatl; 8: Malinche; 9: Pico de Orizaba; 10: Cofre de Perote; 11: San Martín; 12: Chichón. **b** Oblique, false-colour Aster image (RGB composition of 123 N VNIR ASTER image) showing the Colima Volcanic Complex. **c** 2-D view of the ASTER image of the SW sector of the volcano showing main ravines and where the steep south slope of the edifice can be appreciated

ravines (Fig. 2b). The material making up the river beds consists of loose material, mostly gravel and sand in size, but punctuated by meter-sized blocks. This material is highly permeable with infiltration rates of 1–3 mm/hr. but, with the presence of large boulders, which rapidly concentrate water around them, thus contributing to water runoff (Capra et al. 2010). The main cone is densely vegetated, including pine trees and is mantled by organic soils, with

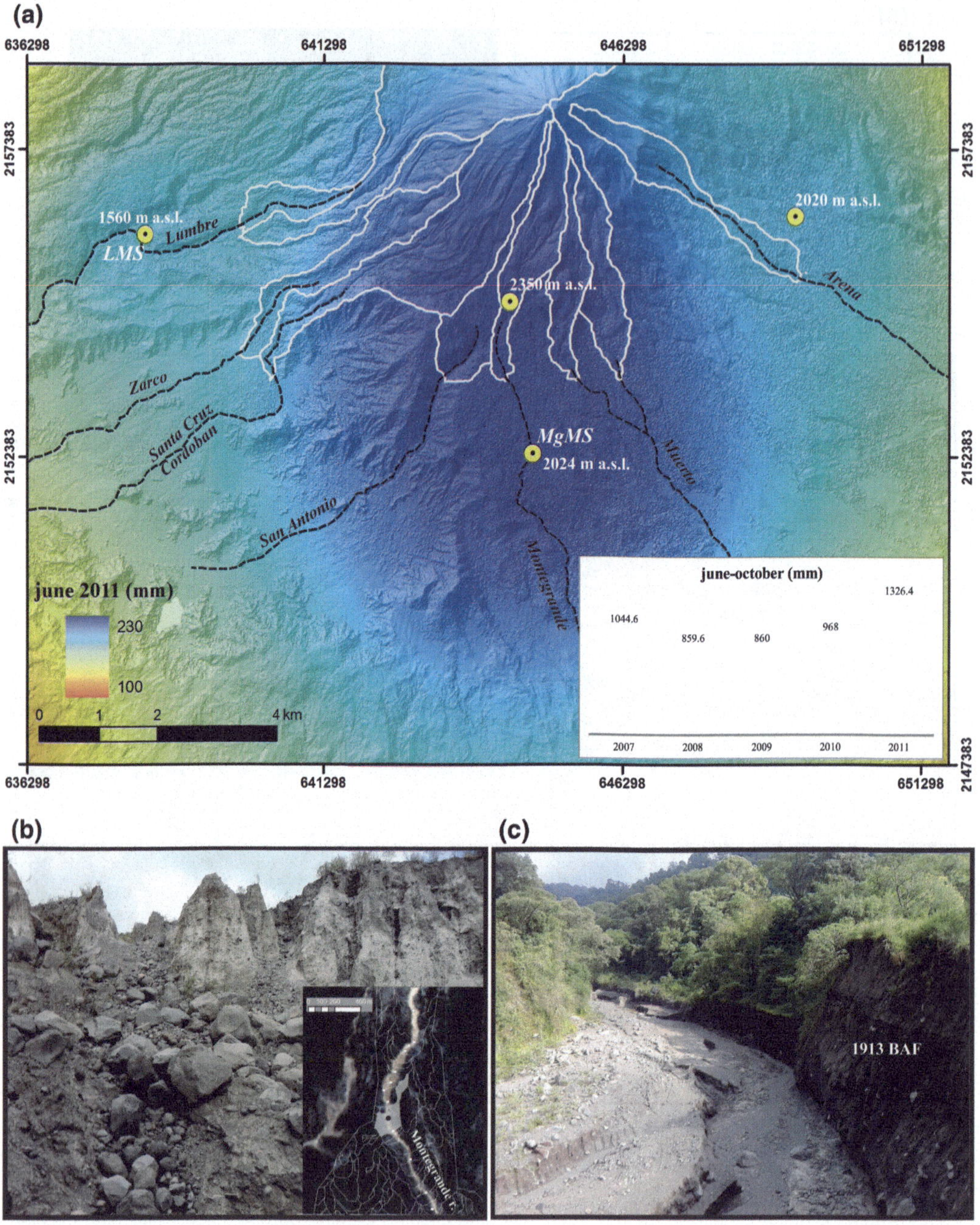

Fig. 2 Hydrological characteristics of Volcán de Colima. **a** Watersheds (white lines) on the volcano feed the main ravines, shown on a shaded relief map. Rainfall distribution is represented for June 2011 (obtained by interpolating the total rainfall accumulation measured at each rain gauge), showing a major concentration on the southern portion of the cone. Yellow dots represent the location of the rain gauge stations and their altitudes. LMS: Lumbre Monitoring Station; MgMS: Montegrande Monitoring Station The inset represents the rainfall accumulated during the past five years. **b** Photo showing the erosional features that are typically observed within the lateral terraces of the main ravines, associated with water inflow from sub-basins (inset). The black dot on the grey sub-basin in the inset refers to site illustrated in the photo. **c** Panoramic view of the La Lumbre ravine showing terraces with a different degree of vegetation

Table 1 Rain-triggered lahars events during the last 100 years

Year	Contemporaneous eruptive activity	Rainfall amount	Ravine affected	Volumes (m^3)	Comments	Ref.
1913–1916	Plinian (1913)	n.a.	All main ravines on the southern flank	$5x10^6$–$1.5x10^6$	Lahars occurred following the Plinian eruption and for several years more. First lahars reached up to the Armeria and Tuxpan river, more than 20 km from the cone	Saucedo (1997), Gavilanes-Ruiz (2004), Saucedo et al. (2010)
14–16/10/1955	–	140 mm in three days	Atenquique	$3.2x10^6$	Several small flow surges joined and flowed into the Atenquique ravine to form a major debris flow that reached the Atenquique village causing 23 death	Saucedo et al. (2008)
1976	1975–'76 Vulcanian and Merapi-type pyroclastic flow	1400 mm/yr; 428 mm/July	Beltran, Arena, Cafecito, La Lumbre	$3x10^6$	–	Gavilanes-Ruiz (2004)
1982	1981 Merapi-type pyroclastic flow	1400mm/yr	El Muerto, La Tuna	$0.2x10^6$	–	Gavilanes-Ruiz (2004)
1988	1987 Phreatic explosion and Merapi-type pyroclastic flow	1600mm/yr	Beltrán, El Cafecito, La Arena	$1.5x10^6$	–	Gavilanes-Ruiz (2004)
25/06/1991	1991 Merapi-type pyroclastic flow	1517/yr; 314mm/June; 70 mm/22 to 24-june	Cordobán	$0.5x10^6$	Originated form a $1x10^6$ m^3 BAF emplaced on 16–17 April. The lahar killed several cows and extended up to 9.9 km from the volcano summit inundating small ranches	BGVN (1992), Rodriguez-Elizarrarás et al. (1991)
1992 (January)	–	1500 mm/yr; 400 mm/January, 108 mm/24 January	Montegrande	$0.83x10^6$	The lahar resulted in different depositional units, up to 1.5 m thick, travelled within 4 km from the town of Queseria	BGVN (1993)
1993	–	1300mm/yr	La Lumbre, Cordobán	$0.16x10^6$	–	Gavilanes-Ruiz (2004)

(continued)

Table 1 (continued)

Year	Contemporaneous eruptive activity	Rainfall amount	Ravine affected	Volumes (m^3)	Comments	Ref.
1994 Three events, one on Oct 7	21/07/1994 Phreatic explosion and Merapi-type pyroclastic flow	500 mm/yr 145 mm/oct	San Antonio, Santa-Cruz, Cordobán, La Arena	$0.5x10^6$	Damages to fences. Water mark of 2 m high on trees	Saucedo et al. (2005), Komorowski et al. (1997)
1997/1998	–	1000/1200 mm/yr	Santa Cruz Cordobán	$0.2x10^6$	Damages to El Jabalí ranch	This work
18/07/1999	17/07/1999 Soufriere-type pyroclastic flow	-	Montegrande	$1x10^6$	Flow destroyed an electric pylon	Cortes, Navarro, personal com., Davila et al. (2007)
01/07/2000	–	955 mm/yr	El Zarco, El Remate	$1.5x10^6$	The lahar buried a house in the La Becerrera village	Cortes and Navarro, personal com
200314 events	–	877/year	Montegrande and San Antonio	$0.1–2.7x10^6$	The largest events reached distances up to 12 km and destroyed an electric pylon in the Montegrande ravine	This work
2004–2006	2004–2005 Merapi and Soufriere-type pyroclastic flow	800–1200 mm/yr	Montegrande, La Lumbre, La Arena	$0.2–0.5x10^6$	Up to 13 km	Gavilanes-Ruiz (2004), Davila et al. (2007)
2007	–	1100 mm/yr	Montegrande, La Lumbre, La Arena	–	Up to 15 km	Capra et al. (2010)
2008–2011	See Table 2					

Note The rainfall amount refers to the available value obtained by the ERIC database. Where indicate, the total rainfall is reported for the entire year. The volume refers to the largest lahar reported the upper 10 cm often consisting of pine needles, generating a hydrophobic behaviour mechanism that enhances water runoff and lahar generation even during low rainfall events (Capra et al. 2010). Along the ravine, this type of vegetation is present on top of the older terraces (20–30 m high) whilst younger terraces (from lahar deposits emplaced during the last ~10 years) are vegetated with low subtropical vegetation (Fig. 2c; Alanís 2011).

3 Lahar Events at Volcán de Colima

Lahars represent a very common phenomenon at Volcán de Colima (Fig. 1b, c). Various types of lahars and triggering mechanisms have been documented: (1) sector collapses that create debris avalanches, which in turn can transform to lahars or form natural impoundments of rivers, whose subsequent failures can form lahars (volumes $10^9 - 10^{10}$ m^3 and runout up to 100 km) (e.g., Capra and Macías 2002; Cortés et al. 2010); (2) post-eruptive (rain-triggered) lahars that follow major eruptions, whose magnitude (volumes from 10^5 to 10^6 m^3) and frequency depend on the degree of landscape modification by the eruption (Plinian eruptions with unconfined, radially distributed pumice/scoria pyroclastic flow deposits or dome collapse producing well confined block-and-ash flow; e.g. Saucedo et al. 2010; Capra et al. 2010); (3) intra-eruptive rain-triggered lahars from extreme hydrometeorological events that mobilize landslides in older deposits (volumes $10^6 - 10^7$ m^3 and runout up to 22 km; e.g., Saucedo et al. 2008).

This chapter focuses on rain-triggered, post-eruptive lahars at Volcán de Colima, associated with dome collapse, dome emplacement having represented the most common eruptive scenario since the 1991 volcanic crisis. A summary of post and intra-eruptive lahars at Volcán de Colima is presented in Tables 1 and 2, starting with the 1913 Plinian eruption, after which several lahars occurred starting a few months following the eruption and continued up to 1916. Based on the record of past events, lahars are more frequently post-eruptive, occurring after dome destruction episodes that have produced pyroclastic flow deposits, either non-explosive gravitational collapse (Merapi-type events) or accompanied by an explosive component (Soufriere-type events). Vulcanian explosions (small to moderate-sized volcanic outbursts producing dense unsustained eruptive columns with both accidental and juvenile fragments) are also frequently followed by lahars. Detailed rainfall data are available only for lahars that have occurred since 2007, when rain gauge stations were installed on the volcano, recording data at 1 min. intervals (Table 2 and Fig. 2a). In contrast, for pre-2007 lahars, the ERIC II database (Extractor Rápido de Información Meteorológica, Instituto Mexicano de Tecnología del Agua 2008) was used to report the daily or monthly rainfall accumulation as available for each reported event. Since 2007, the systematic detection of lahars has been carried out using seismic data from the RESCO network, the seismological network managed by Universidad de Colima, based on signal characteristics unique to lahars (Zobin et al. 2009; Arámbula et al., this volume). More recently, two monitoring stations were installed in the Montegrande and La Lumbre ravines, during 2011 and 2013, respectively. Both stations include a camcorder, a geophone and a rain gauge (Fig. 1b) (Vazquez et al. 2016b). This has made it possible to correlate lahar occurrences with rainfall amount (Capra et al. 2010) and to define their magnitude, duration and a quantitative discrimination of flow behaviour based on their seismic characteristics (Vazquez et al. 2014, 2016b). Based on detailed data obtained since 2007, more frequent episodes occur during the rainy season (from June to October) and they are most frequent during the first years following a major eruptive phase, when the ravines are filled with fresh unconsolidated material, then gradually decrease in frequency during the following years (Dávila et al. 2007). More catastrophic events that caused material damage to infrastructure were recorded in 2000, when a house in the Becerrera village located 12 km from the crater was buried by a lahar which descended the Zarco ravine (Fig. 3a) or during the

Table 2 Characteristics of selected rainfall and lahar events during 2008–2011

	Rainfall					Lahar		Ravine
date	start	R3 (mm)	R1 (mm)	R (mm)	I (mm/h)	lahar start	duration	
01-07-2011	20:25	40	24	12.6	7	22:14	00:43	Montegrande
01-07-2011	21:00	15	0	2.6	26	22:14	00:36	Lumbre
02-07-2011	01:00	16	9	21.06	13	01:51	01:44	Lumbre
02-07-2011	00:55	54	37	34.2	21	01:36	01:21	Montegrande
06-07-2011	14:46	0	0	21	20	15:09	00:51	Lumbre
06-07-2011	17:45	21	21	35.8	32	18:08	01:38	Lumbre
06-07-2011	18:00	0	0.4	25.4	22	18:08	00:56	Montegrande
10-07-2011	06:56	27	21	21.8	21	07:20	01:03	Lumbre
13-07-2011	13:15	47	2.4	26.6	23	13:15	01:14	Lumbre
13-07-2011	12:55	62	0.6	31	17.4	13:16	01:16	Montegrande
26-07-2011	15:11	33.4	25	38	16	15:16	01:43	Montegrande
06-09-2010	15:17	33.8	13.4	37.8	28	16:08	01:36	Montegrande
15-09-2010	17:18	15.6	8.6	30.8	20	18:48	00:46	Montegrande
26-09-2010	13:54	9	0	6	6	13:42	01:02	Montegrande
21-07-2009	14:39	2.8	0	5.2	14	14:56	00:30	Montegrande
13-08-2009	18:32	0	0	6.2	17	18:12	01:23	Arena
26-08-2009	17:39	1.6	1.6	7	9	17:18	01:52	Arena
05-09-2009	12:05	25.6	0	39.4	44	12:10	01:09	Montegrande
09-09-2009	17:10	31	13.4	31.4	22	17:50	00:41	Montegrande
10-09-2009	12:14	57	36.8	14.8	18	13:12	00:24	Montegrande
15-09-2009	18:14	0	1	4.6	3	17:21	00:29	Montegrande
25-09-2009	18:11	1.2	0	6.6	18	18:09	00:30	Montegrande
27-09-2009	15:22	28.6	0	21.6	15	16:25	01:21	Montegrande
30-06-2008	16:40	8.4	8.4	40	40	15:05	00:44	Montegrande
17-07-2008	23:08	37	21	19	38	23:59	00:51	Montegrande
20-07-2008	17:02	2.8	2.8	6.8	24	17:26	00:38	Montegrande
25-07-2008	14:34	1	1	4.4	6	14:38	00:51	Montegrande
26-07-2008	17:50	14	13.8	23	23	18:18	01:30	Montegrande
20-08-2008	16:07	26	17	15	16	17:05	01:04	Montegrande
23-08-2008	17:16	28	0	23	16	17:40	00:28	Montegrande
29-08-2008	12:05	10	0	45	18	12:40	00:58	Montegrande
10-09-2008	15:32	89	38	24	18	16:44	01:32	Montegrande
20-09-2008	14:06	45	13	12	6	14:12	01:36	Montegrande

R3: rain accumulated 72 hrs before the event; R1: rain accumulated 24 hrs before the event; R: rainfall during the event; I: rainfall intensity

2015 Patricia hurricane, when a bridge and road were completely destroyed leaving the same village cut-off (Fig. 2b). An electricity pylon located on the alluvial fan below the Montegrande ravine was destroyed on two occasions, in 1999 and 2007 (Fig. 3b, c).

Fig. 3 Photos showing: **a** house at Becerrera village buried by a lahar in 2000; **b** damage to an electricity pylon within the Montegrande ravine, which was rebuilt at the same location (**c**)

4 Lahar Frequency and Magnitude Versus Rainfall Characteristics

Lahars at Volcán de Colima are triggered under variable meteorological conditions. Figure 4a shows the occurrences of events based on the accumulated rainfall and the average intensity (Table 2). As noted in earlier studies (Capra et al. 2010), during the early rainfall season, with an absence or very low amount of antecedent rainfall (<2 mm in the 24 h. prior to the lahar—"dry" antecedent conditions), lahars can be triggered by low rainfall accumulation (<10 mm) at variable values of rainfall intensities (Fig. 4a). But if antecedent conditions are "wet" (> ~ 10 mm rain in the preceding 24 h), lahars are triggered by >10 mm of rain and with intensities that increase as the total rainfall amount increases (Fig. 4a). The high frequency of lahars at the beginning of the rainy season has been ascribed to the hydrophobic characteristics of soils under dry conditions that favour rapid water runoff (Capra et al. 2010) but also to the large amount of dry and loose sediment that is available at the beginning of the rainy season. This material accumulates during the dry season when instabilities are common on steep slopes. It should be noted that the rainfall is recorded only at up to four localities (Fig. 2a) and therefore does not necessarily represent all the meteorological variability on the volcano slopes. The rainfall intensity is here calculated as the average intensity during the rainfall episode (Fig. 4a), which is much higher if computed in a smaller time interval. In fact, by looking in detail at some of these rainfall events (Fig. 4b) it seems that lahars are triggered during the peak intensity (maximum rainfall intensity calculated at 5 min intervals), usually less than 10–20 min after the beginning of the rainfall, and they last from a minimum of 30–60 min during short rainfall events (dry antecedent conditions) to up to 2–3 h during long duration rainfall episodes (wet antecedent conditions) as recorded at the monitoring stations (Table 2; Vazquez et al. 2014, 2016b). Lahars are characterized by single or multi-pulses, consisting of a block-rich front with detected seismic frequency ranging between 10–30 Hz, followed by the main flow body fluctuating between a debris flow (>50% in sediment volume) and a hyperconcentrated flow (20–50% sediment volume) with higher frequencies (30–50 Hz; Vazquez et al. 2016b). The sedimentation rate varies between 5 and 7 cm/min. (Vázquez et al. 2014). Based on camcorder images and as observed during field work, lahars left flooded cross-section areas of 20–25 m^2 on average with a maximum depth of 3 m (Vázquez et al. 2016a). Considering a mean flow velocity of 4 m/s, an average peak discharge of 60 m^3/s is estimated (Capra et al. 2010; Vazquez et al. 2014, 2016b). Average flow volumes were estimated to be in the range 100–250 $\times$ 10^3 m^3 (Dávila et al. 2007; Capra et al. 2010).

5 Textural Characteristics

Deposits emplaced by post-eruptive rain-triggered lahars have variable textures. In general, the only evidence left by lahars in the upper sections of ravines with river bed slope around 11° are lateral levees or remnants of old terraces, demonstrating the dominance of erosional flow behaviour. The most voluminous deposits typically can be observed in the medial-distal portion of the ravines, where slopes decrease to around 4°–5° and the widening of gullies enhances deceleration of the flow and hence deposition. Figure 5 illustrates characteristics of the 2 July 2011 events in La Lumbre ravine at a distance of 6 km from the crater, close to the monitoring station (Fig. 2a). Pictures were taken the day after the event, so the depositional features are well preserved. The associated rainfall event started at 1:30 local time and reached a maximum intensity of 13 mm/hr. (Table 2). The lahar was detected in the seismogram at 1:51 local time and lasted for approximately 1.45 h (Fig. 4b). The high-water mark was approximately 2–2.5 m, measured within the middle portion of the channel (Fig. 5b), where its maximum width is 4 m. The well-preserved levee was 1.2 m thick (Fig. 5c), with clasts up to 40 cm in diameter. The flat fill terrace in the central portion of the channel was up to 70 cm

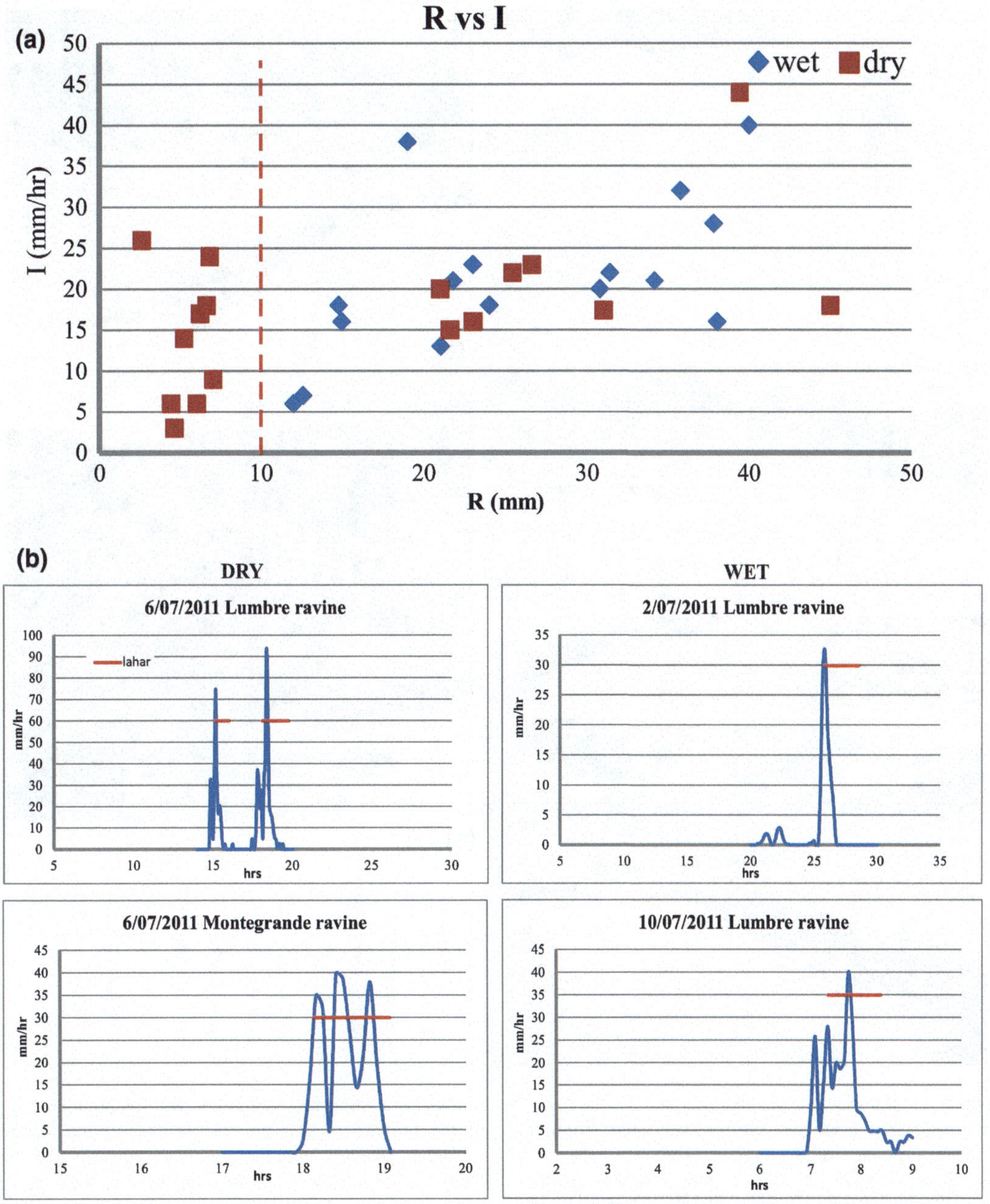

Fig. 4 Characteristics of the rainfall events that triggered lahars. Data correspond to the rainfall measured at 1 min intervals at the nearest gauge station, and rainfall intensity represents the average value calculated for each episode. **a** Diagram showing the relation between the rainfall accumulated during each event and the average intensity. Note as most of the "dry" episodes (with no or <2 mm of rainfall during the previous 24 h.) are triggered by less than 10 mm of rainfall and variable intensities, **b** examples of individual lahars, comparing the rainfall characteristics with the lahar duration

thick, and was characterized by a massive deposit with a bimodal grain-size distribution, with peaks at −4 and 3 phi (Fig. 5d, e). These features represent the main pulse of the flow, and as it was diminishing, it left minor layers that were of decreasing clast size, until finally it was covered

Fig. 5 Photos showing the characteristics of the deposit that originated from the 2 July 2011 lahar within La Lumbre ravine. **a** Panoramic view of the site; **b** high-water mark on an older terrace; **c** lateral levee; **d** section of the main portion of the deposit that shows a bimodal granulometric distribution (**e**); **f** panoramic view of the site, where it is possible to observe the feature related to the flow diminishing (black arrows) and the "milky"-silty horizon from the diluted tail of the flow

by a "milky"-silty horizon from the diluted tail of the flow (Fig. 5f). These textural features are very common in all the ravines. Depending on the magnitude of the episode, massive or normally-graded units can be observed, but studying older terraces it is very difficult to recognize a single event and correlate it with deposits downstream (Fig. 6a). Typically, fill terraces consist of massive debris flows, alternating with intercalated fluvial horizons. But, the vertical juxtaposition of these contrasting deposits does not necessarily indicate that they represent different pulses from the same event. Instead they could represent a sequence of lahars

Fig. 6 Lahar deposits along the Montegrande ravine. **a** Section along the Montegrande ravine where different depositional terraces can be observed; **b** pine needles separating two different lahar deposits; **c** oblique representation of the Montegrande ravine where it is possible to appreciate the development of the ravine up to the fan where the material transported by the lahar accumulates. The white star indicates the location of the electricity pylons that are periodically damaged by lahars

separated by periods of normal streamflow and exposed through erosion by a younger event. In some cases, a layer of pine needles below a lahar deposit clearly indicates a time gap between events (Fig. 6b), but in general it is difficult to interpret if the sequence represents a major event or discrete single episodes.

The Montegrande ravine is the best place to observe the frontal deposition of lahars. In fact, major ravines such as La Lumbre and La Arena gradually widen and are joined by other drainage systems, so lahars that emplace deposits within these ravines are considerably diluted within the distal zone. In contrast, the Montegrande ravine, at 15 km from the crater, opens abruptly onto a plain (Fig. 6c), where the flow fronts rapidly stop, depositing large blocks that were concentrated at the flow fronts. Such blocks have caused damage to electricity pylons (Fig. 3b).

The post-eruptive lahar deposits associated with the 1913 Plinian eruption, can be easily distinguished, because they directly overlay the pumice-flow deposits (Fig. 7a, b), and they comprise the same components as the primary pyroclastic-flow deposits, devoid of rounded andesitic rocks eroded from the river bed. This lack of bed-material contamination indicates that lahars formed by eroding the surface of the recently emplaced pyroclastic-flow deposits that filled major ravines. They flowed as far as 20 km from the volcano (Table 1, Saucedo et al. 2010).

Fig. 7 **a** and **b** Photos showing the deposit that originated from lahars during the 1913 Plinian eruption

6 Hazard Assessment

A hazard assessment for post-eruptive rain triggered lahars has already been performed by Dávila et al. (2007), based on LaharZ code (Iverson et al. 1998), considering a range of volumes from 0.5 × 10^5 to 5 × 10^5 m^3, values that represent the most common scenarios observed since 2007, corresponding to the class size 4 based on the Jakob's classification (Jakob 2005). But those volumes are considerably smaller than catastrophic post-eruptive lahars associated with Plinian eruptions that have reached volumes of between 1 and 5 × 10^6 m^3 (corresponding to classes size 5–6, Jakob 2005) as observed just after the 1913 Plinian eruption (Saucedo et al. 2008; Table 1). Large volume (>1 km^3) debris flows related to volcano collapses, are not here considered, since they represent "secondary" effects and result in devastation of the entire area (Cortés et al., this volume).

We present here a new hazard map that include two possible post-eruptive lahar inundation scenarios considering the maximum observed volume for typical rain-triggered lahars (5 × 10^5 m^3), which are more frequent after dome collapse episodes; and secondly larger post-eruptive lahars resulting from Plinian eruptions (5 × 10^6 m^3) (Fig. 8). Smaller lahars that are generated every rainy season can damage bridges, electrical power lines, and small ranches. For this smaller lahar scenario large settlements are not threatened with inundation, although it has been observed during past lahars that confined diluted flows, still capable of transporting large blocks, can reach towns, such as La Becerrera, San Marcos and Quesería, where they have been responsible for some damage such as bank erosion and undermining of structures near the river. In contrast, the areas that could be exposed to large post-eruptive lahars are more extensive, with flows potentially reaching greater distances (up to 20 km) than the primary pyroclastic flows from which they

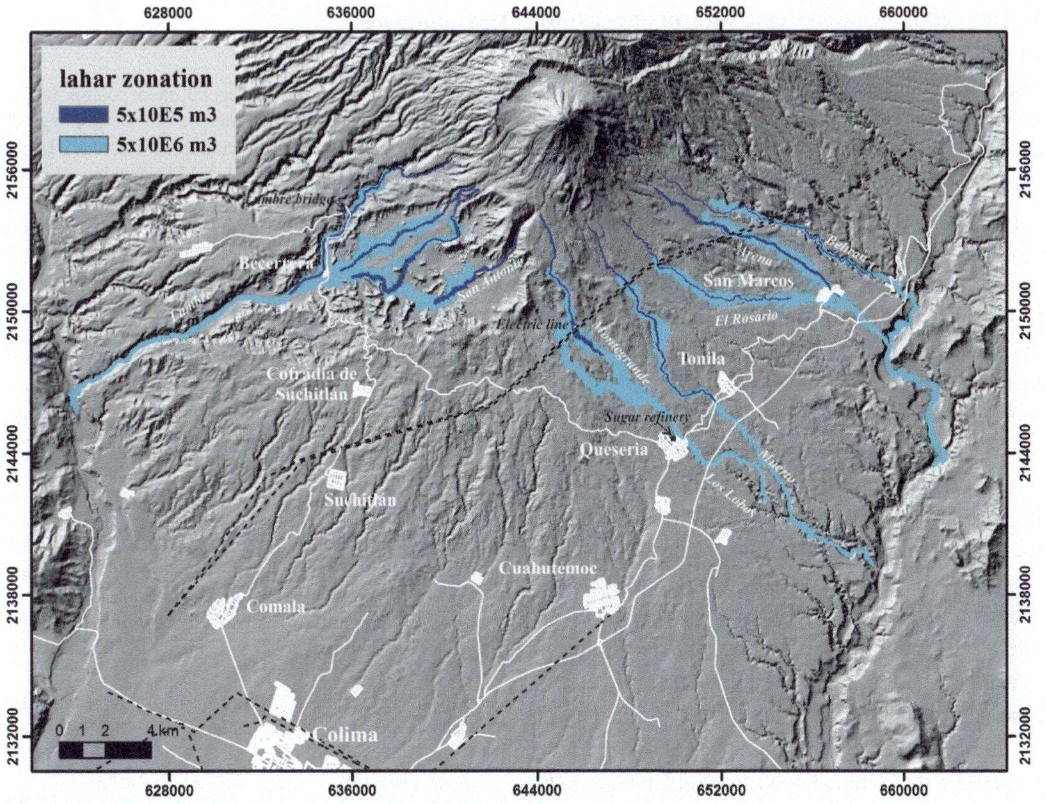

Fig. 8 Hazard zonation for lahars at Volcán de Colima using the LaharZ code

originate, as was observed during the 1913 eruption (Saucedo et al. 2010).

The growing human occupation of lahar hazard areas has important implications for future damage and loss. In 1913 the lahar inundation area was not as developed as it is today, so damage would be much greater if that 1913 Plinian eruption and associated post-eruptive lahars occurred today. In fact, all the plains surrounding the volcano are utilized for extensive cultivation of sugarcane, avocado, corn, coffee, blackberries, and agave for tequila production, with a total of 300,000 inhabitants including Colima city. So, in addition to an impact to various human settlements, the economy of the region could be drastically affected. In particular, an important sugarcane refinery is located at Quesería, very close to the Los Lobos (Montegrande) ravine, which could be affected by a lahar (Fig. 8).

7 Discussion and Conclusion

Rain-triggered lahars represent a very common phenomenon at Volcán de Colima. During the last century, the most voluminous events were associated with the 1913 Plinian eruption, being still hot several months after the eruption, and causing one reported death (Saucedo et al. 2010), several cases of damage to coffee-houses within the SW sector of the volcano, and some loss of animals. An inter-eruptive lahar in 1955 at Atenquique was an extraordinary rain-triggered debris flow that resulted from the liquefaction and coalescence of several landslides along a narrow and deep valley and caused the deaths of 23 inhabitants (Saucedo et al. 2008). An additional risk factor could be the occurrence of a strong tectonic earthquake occurring during a period of heavy rain. One such event occurred in

January 2003 with the epicentre of the 7.6 M earthquake being in Tecoman (Colima). This event triggered several small landslides in the volcanic highlands, and resulted in the accumulation of a large number of rocks at the foot of various slopes (Keefer et al. 2006). Had this happened during the rainy season, lahars could have been generated. Our hazard zonation indicates that annually occurring lahars will continue to affect infrastructure to some degree, but they do not pose significant danger to human settlements. During the period 2006–2012 the volcano presented a very low level of activity. A new dome started to grow in January 2013, and during May 2013 evolved into a lava flow with frequent rockfalls and small pyroclastic flows heading down the west flank (Chap. 7). As a consequence, the 2013–2014 lahar seasons was characterized by several lahars, varying in magnitude depending on the accumulated rainfall (Vazquez et al. 2016a). The 10–11 July 2015 block-and-ash flows generated from a catastrophic dome collapse (Capra et al. 2016; Reyes-Dávila et al. 2016), engulfed and drastically changed the morphology of the Montegrande ravine. During the next rainy seasons the frequency and magnitude of lahars will increase substantially, as was observed just after the 2004–2005 eruptive activity. The occurrence of a Plinian eruption could result in a large volume of material being emplaced by column-collapse pyroclastic flows, completely filling the main ravines. In this case, large lahars could occur in response to heavy rains for several years after the eruption, potentially affecting the main towns within a radius of 20 km from the volcano with catastrophic effects.

Acknowledgements This work was supported by projects DGAPA-UNAM (IN-103107 and 106710), CONACyT (99486, 360) and SRE-CONACYT 146324 to Lucia Capra. We appreciate support during field work by Pablo Davila, Victor Vargas and several students from the Science Faculty of Colima University. This chapter benefits from comments by Thomas Pierson and Frank Lavigne.

References

Alanís, R.M.: Regeneración vegetal en depósitos laháricos de la barranca Monte Grande, Volcán de Fuego de Colima. Honor's Thesis, Facultad de Filosofia y Letras e Instituto de Geografía, UNAM, Mexico, 178 pp (2011)

Capra, L., Macías, J.L.: The cohesive Naranjo debris flow deposit (10 km^3): a dam breakout flow derived from the pleistocene debris-avalanche deposit of Nevado de Colima volcano (México). J. Volcanol. Geoth. Res. **117**, 213–235 (2002)

Capra, L., Borselli, L., Varley, N., Norini, G., Gavilanes, J.C., Sarocchi, D., Caballero, L.: Rainfall-triggered lahars at Volcán de Colima, Mexico: surface hydro-repellency as initiation process. J. Volcanol. Geoth. Res. **189**(1–2), 105–117 (2010)

Capra, L., Macias, J.L., Cortes, A., Saucedo, S., Osorio-Ocampo, S., Davila, N., Arce, J.L., Gavilanes-Ruíz, J.C., Corona-Chávez, P., García-Sánchez, L., Sosa-Ceballos, G., Vázquez, R.: Preliminary report on the July 10–11, 2015 eruption at Volcán de Colima: pyroclastic density currents with exceptional runouts and volumes. J. Volcanol. Geoth. Res. **310**, 39–49 (2016)

Carn, S.A., Watts, R.B., Thompson, G., Norton, G.E.: Anatomy of a lava dome collapse: the 20 March 2000 event at Soufrière Hills Volcano, Montserrat. J. Volcanol. Geoth. Res. **131**(3–4), 241–264 (2004)

Cortés, A., Macías, J.L., Capra, L., Garduño-Monroy, V. H.: Sector collapse of the SW flank of Volcán de Colima, México. The 3600 yr BP La Lumbre-Los Ganchos debris avalanche and associated debris flows. J. Volcanol. Geoth. Res. **197**, 52–66 (2010)

Dávila, N., Capra, L., Gavilanes, J.C., Varley, N., Norini, G., Gomez, A.: Recent lahars at Volcán de Colima (México): drainage variation and spectral classification. J. Volcanol. Geoth. Res. **165**, 127–141 (2007)

De Belizal, E., Lavigne, F., Robin, A.K., Sri Hadmoko, D., Cholik, N., Thouret, J.C., Sawudi, D.S., Muzani, M., Sartohadi, J., Vidal, C.: Lahar hazards and risks following the 2010 eruption of Merapi volcano, Indonesia. In: Jousset, P., Pallister, J., Surono (eds.) Merapi 2010 centennial eruption. J. Volcanol. Geoth. Res. Spec. Issue **261**, 330–347 (2013)

Doyle, E.E., Cronin, S.J., Cole, S.E., Thouret, J.C.: The coalescence and organization of lahars at Semeru volcano, Indonesia. Bull. Volcanol. **72**(8), 961–970 (2010)

Instituto Mexicano de Tecnología del Agua: ERIC II: Extractor Rápido de Información Meteorológica. CD-room (2008)

Iverson, R.M., Schilling, S.P., Vallance, J.W.: Objective delineation of lahar-inundation hazard zones. Geol. Soc. Am. Bull. **110**(8), 972–984 (1998)

Jakob, M.: A size classification for debris flows. Eng. Geol. **79**, 151–161 (2005)

Keefer, D.K., Wartman, J., Navarro, C., Rodriguez-Marek, A., Wieczorek, G.F.: Landslides caused by the M 7.6 Tecomán, Mexico earthquake of January 21, 2003. Eng. Geol. **86**(2–3), 183–197 (2006)

Lavigne, F., Thouret, J.C., Voight, B., Suwa, H., Sumaryono, A.: Lahars at Merapi volcano, Central Java: an overview. J. Volcanol. Geoth. Res. **100**, 423–456 (2000)

Lavigne, F., Thouret, J.C.: Sediment transport and deposition by rain-triggered lahars at Merapi Volcano, Central Java, Indonesia. Geomophology **49**, 45–69 (2002)

Lavigne, F., Suwa, H.: Contrasts between debris flows, hyperconcentrated flows and stream flows at a channel of Mount Semeru, East Java, Indonesia. Geomorphology **61**(1–2), 31–58 (2004)

Manville, V., Németh, K., Kano, K.: Source to sink: a review of three decades of progress in the understanding of volcaniclastic processes, deposits, and hazards. Sed. Geol. **220**, 136–161 (2009)

Naranjo, J.A., Sigurdsson, H., Carey, S.N., Fritz, W.J.: Eruption of the Nevado del Ruiz volcano, Colombia, on 13 November 1985: tephra fall and lahars. Science **233**, 961–963 (1986)

Reyes-Dávila, G., Arámbula-Mendoza, R., Espinasa-Pereña, R., Pankhurst, M.J., Navarro-Ochoa, C., Savov, I., Vargas-Bracamontes, D.M., Cortés-Cortés, A., Gutiérrez-Martínez, C., Valdés-González, C., Domínguez-Reyes, T., González-Amezcua, M., Martínez-Fierros, A., Ramírez-Vázquez, C.A., Cárdenas-González, L., Castañeda-Bastida, E., Vázquez Espinoza de los Monteros, D., Nieto-Torres, A., Campion, R., Courtois, L., Lee, P.: Volcán de Colima dome collapse of July, 2015 and associated pyroclastic density currents. J. Volcanol. Geoth. Res. **320**, 100–106 (2016)

Rodolfo, K.S., Arguden, A.T.: Rain-lahar generation and sedimentation-delivery systems at Mayon Volcano, Philippines. In: Fisher, R.V., Smith, G.A. (eds.) Sedimentation in volcanic settings. SEPM Special Publication, pp. 71–87 (1991)

Saucedo, R., Macias, J.L., Sarocchi, D., Bursik, M.I., Rupp, B.: The rain-triggered Atenquique volcaniclastic debris flow of October 16, 1955 at Nevado de Colima Volcano, Mexico. J. Volcanol. Geoth. Res. **132**, 69–83 (2008)

Saucedo, R., Macias, J.L., Gavilanes, J.C., Arce, J.L., Komorowski, J.C., Gardner, J.E., Valdez, G.: Eyewitness, stratigraphy, chemistry, and eruptive dynamics of the 1913 Plinian eruption of Volcan de Colima, Mexico. J. Volcanol. Geoth. Res. **191**, 149–166 (2010)

Scott, K.M., Macias, J.L., Vallance, J.W., Naranjo, J.A., Rodriguez-Elizarrarás, S.R., McGeehin, J.P.: Catastrophic debris flows transformed from landslides in volcanic terrains: mobility, hazard assessment, and mitigation strategies. U.S. Geological Survey Professional Paper 1630, 61 pp (2001)

Scott, K.M., Vallance, J.V., Kerle, N., Macias, J.L., Strauch, W., Devoli, G.: Catastrophic precipitation-triggered lahars at Casita Volcano, Nicaragua: occurrence, bulking and transformation. Earth Surf. Proc. Land. **30**, 59–79 (2005)

Torres, R., Mouginis-Mark, P., Self, S., Garbeil, H., Kallianpur, J., Quiambao, R.: Monitoring the evolution of the Pasig-Potrero alluvial fan, Pinatubo Volcano, using a decade of remote sensing data. J. Volcanol. Geoth. Res. **138**, 371–392 (2004)

Van Wyk, Vries B., Kerle, N., Petley, D.: Sector collapse forming at Casita volcano, Nicaragua. Geology **28**(2), 167–170 (2000)

Vazquez, R., Capra, L., Caballero-García, L., Arambula-Mendoza, R., Reyes, G.: The Anatomy of a lahar: deciphering the 15th September 2012 lahar at Volcán de Colima. J. Volcanol. Geoth. Res. **272**, 126–136 (2014)

Vazquez, R., Capra, L., Coviello, V.: Factors controlling erosion/deposition phenomena related to lahars at Volcán de Colima, Mexico. Nat. Hazards Earth Syst. Sci. **16**, 1881–1895 (2016a)

Vazquez, R., Suriñach, E., Capra, L., Arámbula-Mendoza, R., Reyes-Dávila, G.: Seismic characterisation of lahars at Volcán de Colima, Mexico. Bull. Volcanol. **78**(2), 8 (2016b)

Williams, S.N.: Nevado del Ruiz volcano, Colombia: the November 1985 eruption and related events. J. Volcanol. Geoth. Res. **1987**, 355–360 (1987)

Zobin, V.M., Placencia, I., Reyes, G., Navarro, C.: The characteristics of seismic signal produced by lahars and pyroclastic flows: Volcán de Colima, Mexico. J. Volcanol. Geoth. Res. **179**, 157–167 (2009)

Monitoring the Recent Activity: Understanding a Complex System

N. R. Varley

Abstract

The recent history (from 1998) of Volcán de Colima witnessed the most intense period of activity since the last catastrophic eruption in 1913. During this period, monitoring efforts were increased, resulting in a large quantity of data, whose integration is enabling significant advances in the understanding of the eruptive mechanism of the volcano. New insights have been possible, like the definition of what drives the multi-time-scale cyclicity. In this chapter we summarize the different components of the monitoring network on Volcán de Colima and give an overview of some of the most important data that has been generated in recent years. Details of the activity during this important period are reviewed, together with the hazards associated with the different styles of eruption.

Keywords

Monitoring · Precursors · Explosive activity
Effusive activity · SO_2 flux

N. R. Varley (✉)
Facultad de Ciencias, Universidad de Colima,
Colima, Mexico
e-mail: nick@ucol.mx

1 Introduction

Volcán de Colima has produced many periods of eruptive activity during historic times. These have varied from large Plinian (or sub-Plinian) events, the last one being in 1913, to slow effusive emplacement of lava domes, last observed in 2016. The complex produced even larger magnitude events going back a few thousand years (see Crummy et al., this volume). Each type of activity has associated hazards, and for such an active volcano, it is vital that probabilistic evaluations are undertaken to optimise risk mitigation in the form of emergency plans, and land-use planning. Geological investigation of the surviving and accessible deposits has been augmented with numerical models, allowing the simulation of hazards, such as tephra fall or pyroclastic density currents, with the final product being zonation maps.

Activity since 1998 has been almost continuous, with frequent transitions between eruption styles. Detailed accounts of each eruption are important for the definition of new scenarios, which in turn should be incorporated into any hazard evaluation. Recently, the available maps have been greatly improved, providing a valuable resource for the Civil Protection Authorities. The large eruption in July 2015 illustrated how new scenarios are always possible, underlining the need to frequently update the tools used as part of hazard mitigation, such as zonation maps.

N. Varley et al. (eds.), *Volcán de Colima*, Active Volcanoes of the World,
https://doi.org/10.1007/978-3-642-25911-1_12

To forecast an eruptive scenario a volcano needs to be monitored, the goal essentially being the detection and characterization of magma bodies beneath the volcano. How much magma is there? Where is it and how fast it is ascending, if at all? This is done using a combination of techniques, with much of the work involving the analysis of time series of data, and identifying and interpreting anomalies. The interpretation of monitoring data requires a conceptual model of the conduit processes. Certain factors can of course be generalized between volcanoes, but their inherent individuality emphasises the importance of well documented chronologies of past activity, in order to establish the context of the monitoring data, allowing its incorporation into any conceptual model.

The most recent activity of Volcán de Colima can be divided into two periods, separated by one and a half years of almost complete calm. The first eruptive period lasted from late 1998 until mid-2011, and the second started in 2013 and continues until the present (September 2016). These periods easily represent the most intense phase since the 1913 sub-Plinian eruption, and therefore their importance cannot be overstated. This chapter summarizes the characteristics of the recent activity at Volcán de Colima and how it has been monitored.

2 Monitoring Volcán de Colima

To mitigate volcanic risk it is imperative to ascertain how the activity of a volcano is evolving. Some volcanoes tend to exhibit extended periods with little or no signs of activity, the aim of monitoring is to detect signs of reawakening. In the case of persistently erupting volcanoes, like Volcán de Colima, monitoring has the objective of detecting variations in eruptive style. El Chichón, in SE Mexico had no eruptions for over 500 years prior to the large Plinian event in 1982 (Macías et al. 2008); unfortunately there was no monitoring at that time and over 2000 perished. To produce this type of eruption, large volumes of magma have to climb rapidly to the surface, and in doing so will inevitably produce clear and large seismic and geodesic signals that can be detected by a monitoring network. If the ascent is rapid with high concentrations of volatiles in the magma, the scenario is set for a devastating Plinian eruption, like that which occurred at El Chichón.

It is more of a challenge when the volcano is more continuously active. In this case, the activity might be evolving between different regimes, and making an interpretation from the signals generated from the monitoring network is not trivial. A modern monitoring network consists of various systems, which produce signals over a wide range of sampling rates. The aim of the network is to provide warning of a change in the eruptive scenario. Scientists interpret the warnings and then transmit their definitions of most likely scenarios to the Civil Protection authorities who make decisions to reduce the risk, such as evacuations.

Monitoring of Volcán de Colima started in 1985 with the establishment of the RESCO (*Red Sismológica Telemétrica del Estado de Colima*) local seismic network at the University of Colima, located in the state capital Colima, 35 km from the volcano. Since then, other systems have been installed in two main initiatives fuelled by increasing activity: firstly by the onset of a new eruptive period in 1998 and secondly by the large explosions of 2005, in an attempt to further our understanding of the system and look for precursors to future increases in activity. Figure 1 includes the field locations for both permanent monitoring stations and sites used for periodic measurements. Table 1 summarizes the different monitoring systems installed as of 2016. These are supported by other tools, such as a network of cameras and rain gauges for understanding lahar generation. In this section, a summary of the current monitoring network is presented, and with a view to the future, some of its deficiencies are identified. Currently, as in many parts of the world, financial limitations hamper the efforts of the local scientists to improve the mitigation of volcanic risk.

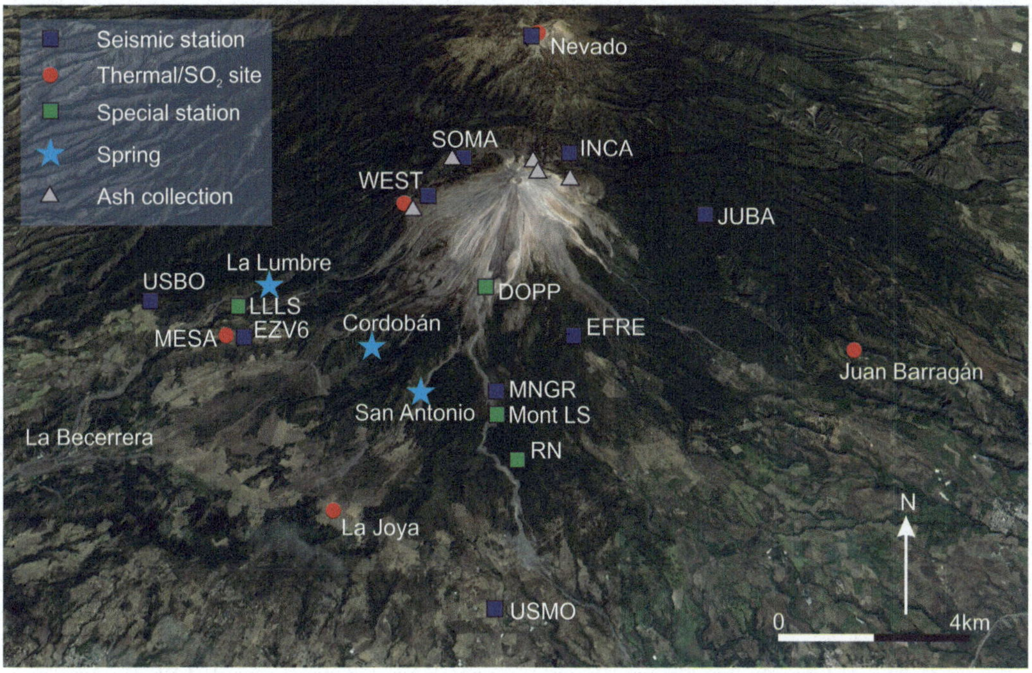

Fig. 1 Map showing the main locations for monitoring activites, unless otherwise mentioned, all stations are maintained and measurements performed by the Universidad de Colima. Seismic stations: USBO—El Borbollon; EFRE—Fresnal; JUBA—Juan Barragán; MNGR—Montegrande; USMO—Montitlán; Nevado—USNE; EZV6—Yerbabuena (see Arámbula et al., this volume for further details). Special Stations: DOPP—Doppler Radar (destroyed in July 2015 eruption; installed by Hamburg University); RN—Radon station; LLLS—La Lumbre lahar station (installed and maintained by UNAM); Mont LS—Montegrande lahar station (installed and maintained by UNAM). Thermal/SO_2 refers to sites where thermal and SO_2 flux measurements are taken periodically. Nevado is the location for the USNE seismic station and for Thermal/SO_2 monitoring. Additionally, various cameras are installed around the volcano

Table 1 Summary of monitoring network at Volcán de Colima

Technique	Stations	Sampling frequency	Notes
Seismic	13	100 Hz	3 broadband stations co-located with 3 short period instruments
Gas flux	N/A	Up to weekly	Location depends upon current wind direction
Geodesic – EDM	N/A	Variable	Recent activity destroyed prisms
Geodesic – tiltmeters	4	2 min.	
Spring geochemistry	4	Monthly	
Thermal monitoring	N/A	Up to weekly	Imaging carried out from flights, flank locations and Nevado de Colima
Acoustic	4	100 Hz	Original stations (2 arrays and 2 single sensor) no longer functioning. 4 new single sensor stations installed during 2017
Doppler radar	1	~ 15 Hz	Destroyed in July 2015
Radon gas	2	15 min.	
Lahar monitoring	2	Variable	Include geophones, remote cameras and rain gauges

2.1 Seismic Network

The seismic network started with short-period instruments, but has since been expanded to include nine broadband 3-component seismometers. As is the case at many monitored volcanoes, it has been the mainstay of the network, benefitting from a much longer history and level of investment than other monitoring options. Chapter 8 covers in detail the seismicity associated with Volcán de Colima. Attempts have been made to correlate the seismicity with other types of observations, such as the thermal properties of explosions (Webb et al. 2014), thermal emission from satellite-based sensors (Fig. 2; Sorge2011); or petrological information on the crystallisation depth of minerals (Reubi et al. 2013).

Seismicity has played a key role in attempts to decipher the mechanism that generates the Vulcanian activity of different magnitudes (Varley et al. 2010a; b) with the application of a variety of analysis techniques to try and unravel the complexities that characterize this type of system (Arámbula-Mendoza et al. 2011).

Recently advanced statistical methods have been increasingly applied to extract the intricate insights that often lay hidden within the data streams (Bebbington 2012). In the case of Volcán de Colima; Varley et al. (2006) applied statistical techniques to the time series of explosion occurrence, which confirmed an independent generation mechanism for two distinct types of explosion, distinguished by their seismic waveform. More recently techniques such as detrended fluctuation analysis (Lachowycz et al. 2013) and the quantification of the entropy within the system (Lachowycz et al. 2016) have proved to be useful prospects for the detection of changes, and make a useful addition to the arsenal of analytical techniques applied to monitoring data. Lamb et al. (2014) specifically looked for evidence of cyclicity within the seismic data. Two types of explosion signature were separated (emergent or impulsive waveforms) and examined along with long-period (LP) events as time series. Some intriguing results emerged, with cycles of 50 and 100 days being highlighted within the explosion data. Examination of variations in the probabilistic distribution of the data

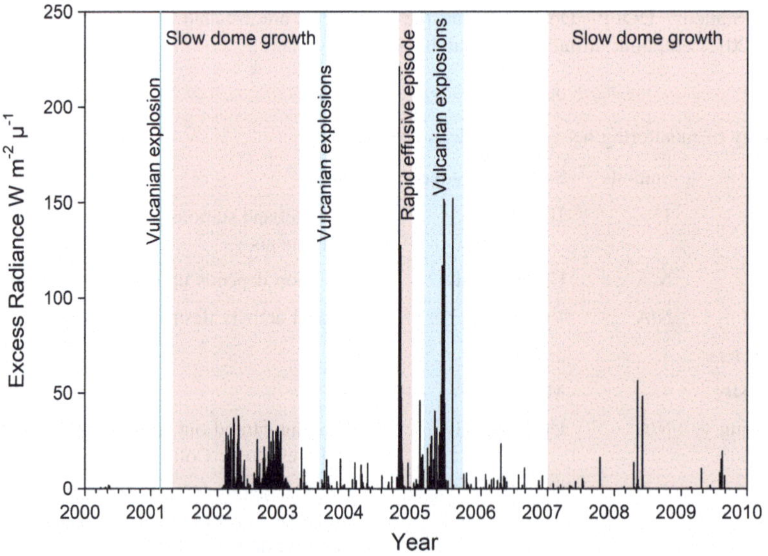

Fig. 2 Combined thermal data from GOES and MODIS satellite systems (hourly excess radiance) from 2000 to 2009. Prolific cloud cover during the rainy season (June–October) greatly reduced data availability. Periods of higher thermal emission from the crater can clearly be seen, i.e., periods of dome growth. Thermal anomalies can also be seen during explosive episodes when a satellite pass coincides with the period following the generation of a pyroclastic density current (PDC). Data analysed by Jolane Sorge, Simon Fraser University

revealed periodic changes in the dominating processes as was first shown with earlier data by Varley et al. (2006). Lamb et al. (2014) were unable to clearly relate many of their results to specific processes, due to the lack of supporting observations or complimentary time series of other data. This highlights the need for multiple data sets for reliable interpretation of monitoring data.

Chapter 8 discusses many of the ways seismicity is being used to assess the activity at Volcán de Colima. With recent expansions of the network, there are abundant opportunities to expand our awareness of some of the complexities of magma ascent and eruption, and improve our detection and interpretation of precursors.

2.2 Geochemical Monitoring

Magma ascent, volcano explosivity, the generation of pyroclastic density currents (PDCs) and the production of tephra are all determined by the gases contained within the magma. Understandably, monitoring the concentrations dissolved in the magma, the generation of a separate gas phase and the processes that occur during decompression once the magma is making its way towards the surface is critical for the appreciation of the inner workings of any volcanic system. Many advances have been taking place recently, with new methodologies finally allowing the generation of time series of data with sufficient sample rates to permit the critical changes within the system to be identified and quantified (e.g. Boichu et al. 2010).

Geochemical monitoring might include direct measurement of gas fluxes or composition from volcanic fumaroles (e.g. de Moor et al. 2016) or indirect approaches, such as observing variations in different parameters in local aquifers (Lowenstern et al. 2006). Monitoring diffuse gas emission also has proven successful, whereby carbon dioxide, or a tracer gas such as radon, is detected above permeable pathways, such as faults, within the volcanic edifice (Cardellini et al. 2017). Each of these methods has the potential to detect precursors to an eruption.

The flux of sulphur dioxide has been measured at Volcán de Colima with a variable frequency, largely dependent upon funding. Early monitoring was carried out using the COSPEC instrument, usually flown under the plume. Precursory increases in the flux were detected prior to the eruption onset in 1998 (Taran et al. 2000). Prior to the 2003 shift in activity style to multiple daily Vulcanian eruptions, Volcán de Colima produced a fairly steady 0.58 kg s^{-1} (50 t d^{-1}) of SO_2. A visible plume could be seen on clear days extending across the sky in a direction depending upon the season. With the new regime of daily explosions, a continuous plume could no longer be seen from the volcano; instead periodic plumes were produced by explosions, their colour controlled by the amount of ash emitted.

More recently, the more modern Flyspec is being used, usually in ground-based scanning mode. This device is built around a small-sized UV spectrometer and includes a rotating mirror to enable scanning through the plume (Horton et al. 2006). More frequent measurements have been possible, ideally weekly or greater, permitting a clearer picture of how the SO_2 flux compares to other parameters. Figure 3 shows an example of SO_2 flux data between 2014 and June 2016. The relationship between flux and activity is not always simple; there was a short-lived peak in flux during the emplacement of the SW lava flow in September 2014, but it dropped again, only to increase after the large explosion on 21 November. The largest flux in this period (43.8 kg s^{-1}) was measured a few days after the explosion in January 2015, although a precursor was also recorded prior to this event. The flux also increased prior to the explosion on 9 February 2016 and the large eruption on 10 July 2015.

Vulcanian explosions result from pressure increasing as a result of permeability decreases at the top of the ascending magma column. It is a complex mechanism due to variations in a variety of factors related to magmatic processes, such as its crystallisation, ascent rate, and its decompression and degassing. The consequence are deviations in the gas flux, possibly further complicated by a flux of deeper sourced gas. The

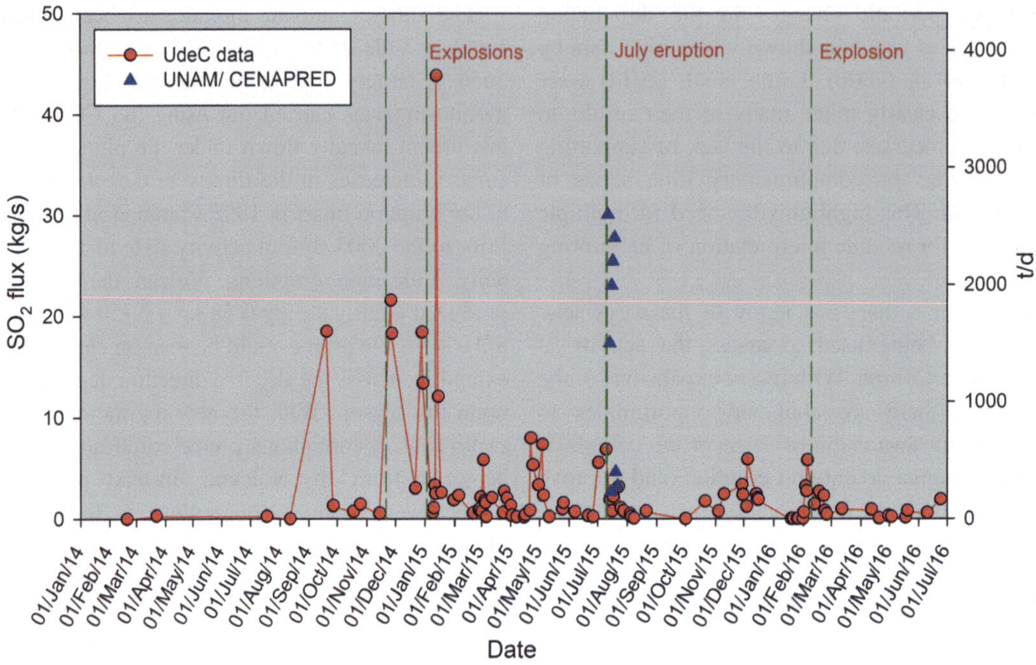

Fig. 3 SO_2 flux time series from early 2014 to October 2016. Peaks relating to larger Vulcanian explosions can be observed (21 November 2014, 03 January 2015 and 09 February 2016) as well as the large eruption of 10–11 July 2015. UdeC data (in red) was recorded using a Flyspec UV spectrometer and processed in the Universidad de Colima. The data after the July 2015 eruption, represented in blue, were collected using COSPEC or a UV spectrometer and processed using differential optical absorption spectroscopy (DOAS). Thanks to Ramón Espinasa (CENAPRED) and Robin Campion (UNAM)

relationship between the gas flux and the generation of explosions can also be inconsistent. Interpretation is often hampered due to a poor temporal resolution of the data. Often it is assumed that the gas flux is proportional to the magma output, thus giving a convenient indirect method to estimate it (e.g. Hirabayashi et al. 1995 for Unzen, Japan; Badalamenti et al. 2004 at Mt. Etna, Italy). Indications at Volcán de Colima are that there is often a poor relationship with much variation in the volatile-contents, possibly related to multiple magma sources.

Attempts have been made to quantify the gas emitted during explosions, but one major problem is the reduced transmission of the ultra-violet light through the ash-laden plumes. All spectrometric methods measure the absorption of UV to determine a column concentration of SO_2. The assumption of a direct path through the volcanic plume has been showed to be somewhat false (Kern et al. 2010).

Four springs are regularly monitored, each located at a distance of between 5 and 6 km from the crater. In the field, basic properties like temperature, pH and conductance are measured with samples returned to the laboratory for analysis of major ions. The isotopic analyses of helium and carbon dioxide in the dissolved gases have shown that there is a magmatic component to the waters, which elevates the temperature in three of the four springs. The clearest precursor signal has been presented by the volatile element boron. An anomalous peak was first observed prior to the effusive episode in 1998 in three of the springs (Taran et al. 2000). Since, peaks have been observed preceding almost all the effusive episodes, usually between two and four months before magma first emerges at the summit.

Diffuse degassing has been measured with maps generated of both carbon dioxide and radon (Varley & Taran 2003). Generally low values were encountered, but there was a large anomaly in La Lumbre ravine, probably indicating a fault. A radon probe has been installed and is continuously measuring its emission. It is hoped that any large magma intrusion will be accompanied by a peak in mass gas flow though the volcano's flanks. This should register as a radon anomaly; however, no significant anomalies have been detected during the past few years.

2.3 Geodesy

Factors related to the regional and local tectonics combine to shape the topography, and have given rise to the volcanism of the Colima Volcanic Complex (see Norini et al., this volume). To further our understanding of any volcanic system, and for the detection of critical inflation precursors, a permanent geodesic network is an invaluable component of a monitoring system. Unfortunately this continues to be a weakness of the Volcán de Colima network due to a lack of funding.

Geodetic monitoring in the form of Electronic Distance Meters, or EDM, revealed a precursory signal with positive deformation detected prior to the eruption in 1998 (Zobin et al. 2002). Unfortunately, however, this system has not been maintained during recent years. A telemetered tiltmeter network has shown inflation-deflation cycles of the volcano, however, confidence in the data has been undermined by variations observed that can be more easily related to changes in meteorological conditions, rather than the volcano. Zobin et al. (2013) present a summary of results. Some campaign geodesic monitoring has been performed, e.g. repeated occupation of a levelling line crossing the area to the north of the volcano (Playón) has shown that the volcano suffers slumping towards the south (Murray et al. 2002).

InSAR (interferometric synthetic aperture radar) is a technique that uses radar data obtained from a satellite platform to obtain quantified differences in altitude between consecutive passes. Two studies have been carried out at Volcán de Colima, both concluding that significant inflation did not occur during either 2002–6 (Pinel et al. 2011) or 2007–11 (Chaussard et al. 2013), despite the emplacement of considerable volumes of magma. This was interpreted as a testament to an open-system, with unrestricted ascent of magma within the upper edifice. In fact Pinel et al. (2011) registered subsidence of around 1 cm per year of the summit, thought to represent a deflating magma source. More recently, short-term variations were detected prior to the explosions in January 2013 (Salzer et al. 2014). These deformation results were modelled, which resulted in the suggestion of two sources for the magma.

2.4 Thermal Monitoring

Thermal monitoring is useful to gain an insight into how energy is released during both effusive eruptions, from the dome, and explosive eruptions, from the eruption column. The nearby Nevado de Colima peak affords a view at the same altitude as the active peak. Regular images taken from there using an infrared camera have enabled the temperature of the fumaroles to be monitored on a regular basis (Stevenson and Varley 2008). A dramatic decrease in temperature was registered between 2005 and 2008 (Fig. 4a). The higher values in 2005 were related to much higher gas fluxes with a hotter magma body close to the surface. Slow dome growth commenced in 2007 but the magma was much more degassed than the batch that generated large Vulcanian explosions during 2005. Anomalous temperatures might be related to variations in effusion rate and the evolution of the near-surface plumbing system. Figure 4b shows a later period, during which two periods of effusion in 2015 and 2016 generated elevated temperatures in the monitored fumaroles. Earlier dome growth and lava flows, emplaced with a lower effusion rate in 2013 and 2014 were not associated with increased temperatures.

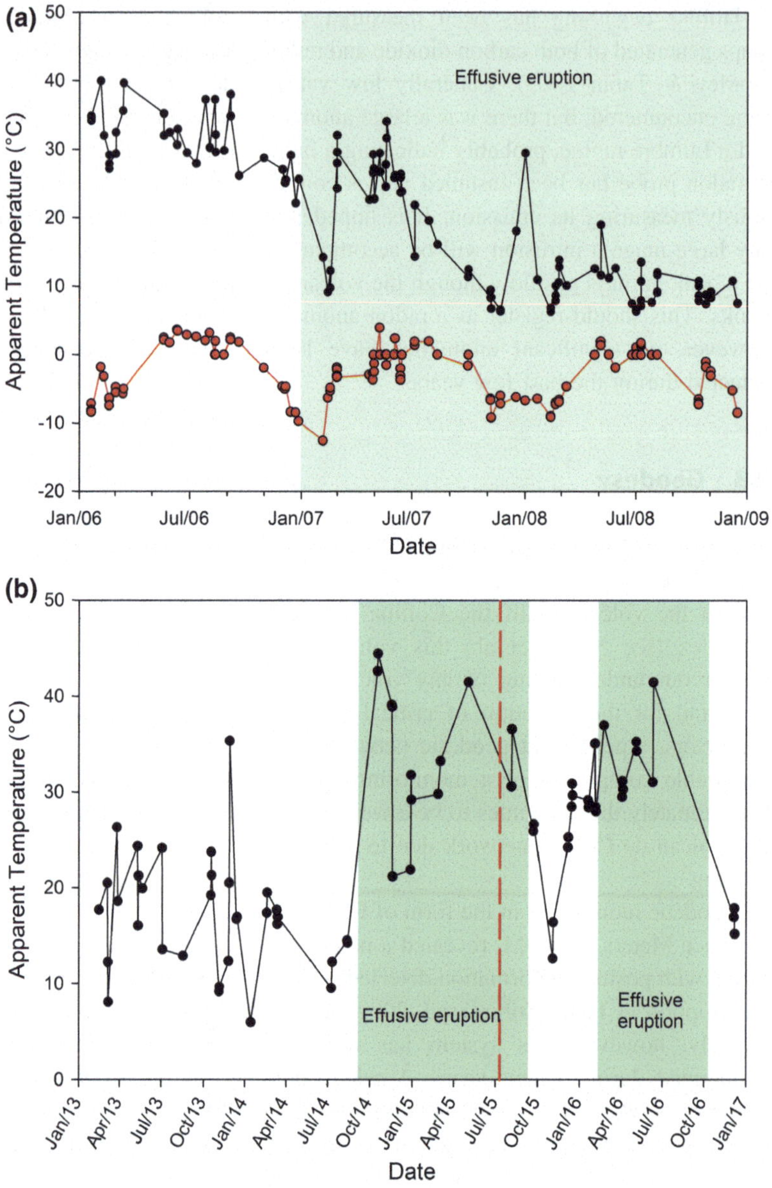

Fig. 4 Fumarole temperatures on N upper flank measured remotely using an infrared camera located on Nevado de Colima. **a** Decreasing temperatures over central fumaroles group from 2006 until the end of 2008 (black line). This was associated with decreasing activity after the 2005 explosive episode. The red line shows a control area on the lower flank of the volcano, where temperature fluctuations correspond to seasonal variation. Some interesting anomalies occur, e.g. at the start of the dome emplacement in early 2007 there is a negative anomaly. **b** Graph shows a later period (2013–2016) during which the moderately large eruption occurred during July 2015 (shown as a dashed red line). Two periods of effusive activity are shown during 2015 and 2016, both associated with higher fumarole temperatures

Regular flights were carried out to maintain thermal monitoring of dome growth during 2007–2011 (Fig. 5 shows a thermal image of an early stage dome). Analysis of the images determined that the emplacement changed from exogenic to endogenic after the accumulation of significant volumes of talus around the hot core, and then back to exogenic with the emplacement of a new lobe on the west side (Figs. 6 and 7; Hutchison et al.2013). It is clear that the accumulation of talus played an important role by buttressing the enlarging dome and thus creating a high aspect ratio (height/radius; Bernstein & Varley 2009). This could have implications to the risk of partial dome collapse.

Recent advances in photogrammetry software have made it far easier to combine photographs taken from multiple angles to obtain a three-dimensional model. James & Varley (2012) applied the technique, known as structure-from-motion and

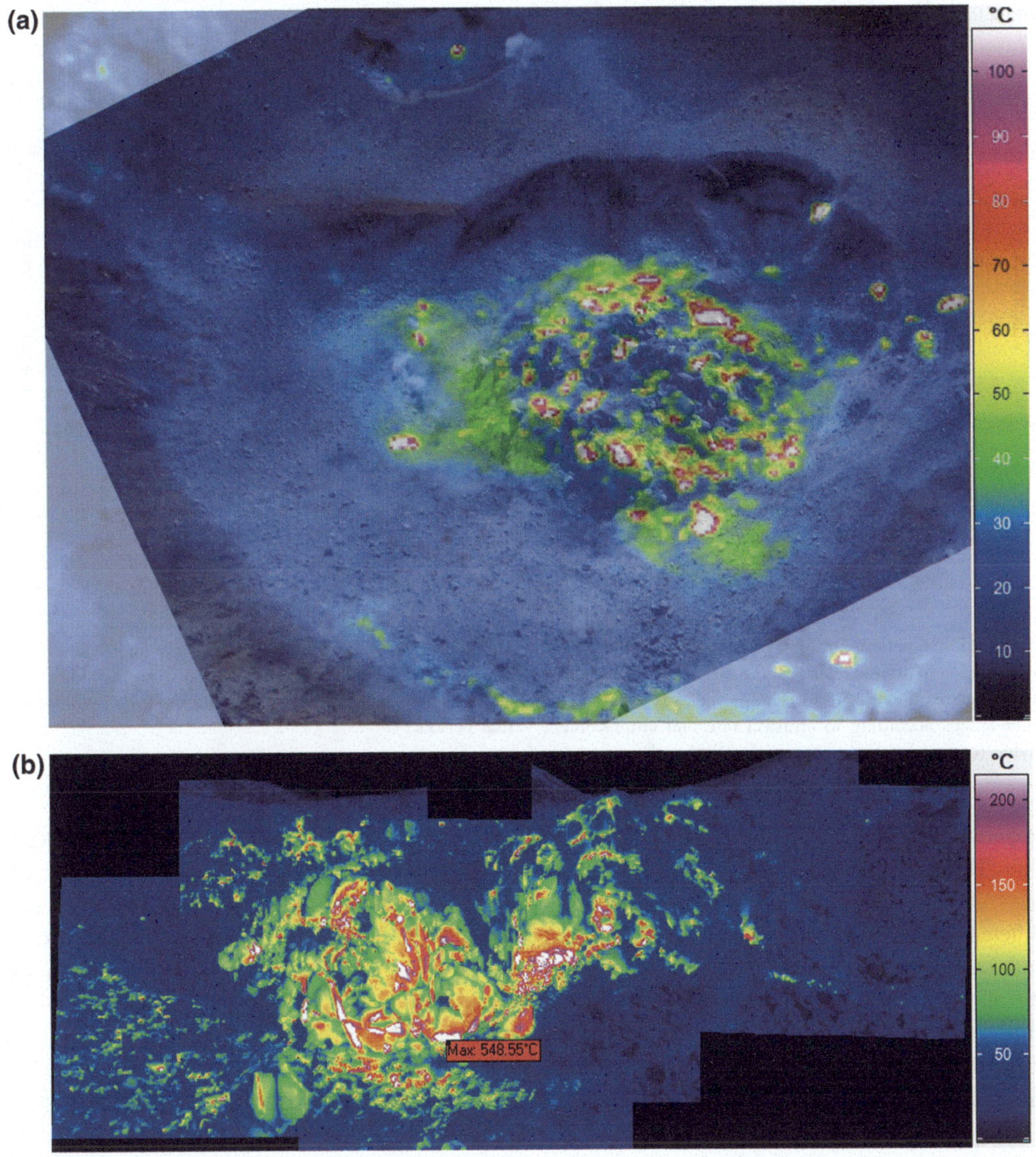

Fig. 5 **a** Thermal image of small dome superimposed on top of photograph taken from flight on 09 February 2007 first evidence of the new episode of effusion that was to continue for 4.5 years. A thermal anomaly also can be seen within the explosion crater in the NE of the main crater. The white areas indicate apparent pixel temperatures of >200 ° C. The diameter of the hot dome is about 68 m. **b** Thermal image of growing lava dome taken from crater rim on 11 November 2007 with lobe being emplaced towards the south

multi-view stereo (SfM-MVS) to demonstrate deformation during the end of the lifetime of the 2007–2011 dome (see Fig. 8b). Regular flights since 2013 have permitted the application of photogrammetry to reproduce 3D models of the growing lava domes and flows. When elevated gas fluxes or meteorological clouds obscured the dome for regular photographs, thermal images were used to construct the model (Thiele et al. 2017).

During the 2013–2015 dome growth, thermal images taken from flights have provided an insight into magma extrusion processes, with

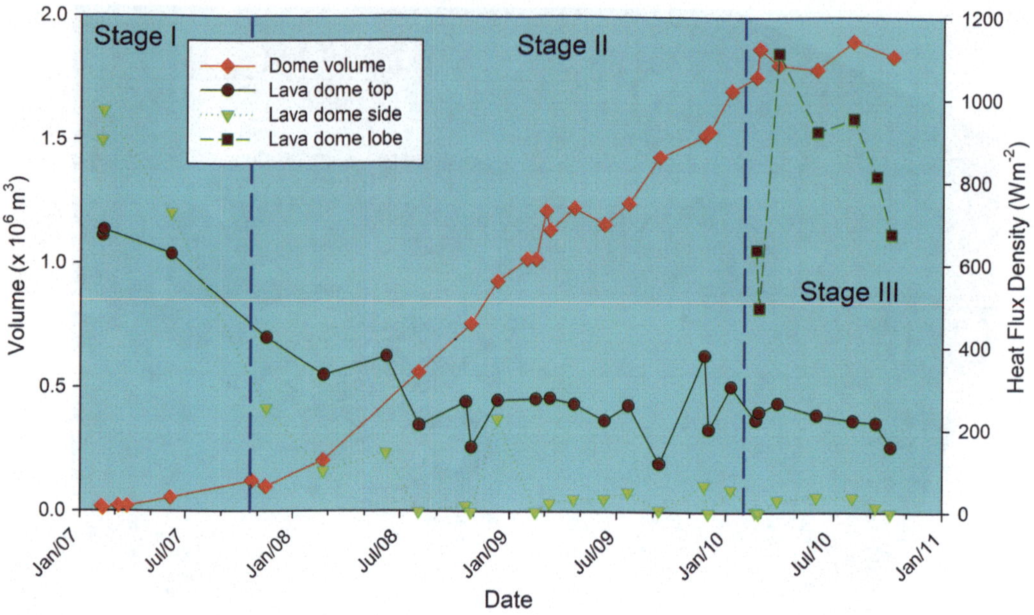

Fig. 6 Dome volume and heat flux density (from dome top, sides and lobe) during 2007-2010 period (adapted from Hutchison et al. 2013). The period was divided into three stages according to effusion rate and emplacement style (exogenic or endogenic). Stage I represents endogenic growth when the volume is small, stage II is exogenic with much less heat emitted from the sides, and stage III includes the high flux from the lobe extruded in that period

important implications for the July 2015 eruption. Thermal models created using photogrammetry have greatly assisted in the visualization of different anomalies both on the interior and exterior of the crater. In May 2015, migration of the effusion source occurred, with simultaneous emission from two points, clearly shown in the thermal model (Fig. 9). This preceded the large eruption on 10–11 July.

Satellite-based sensors are gaining an ever-increasing importance for volcano monitoring. An example is shown in Fig. 2, where increased thermal emission can be explained by either magma effusion, with the growth of lava domes and later flows, or from the deposition of hot PDCs associated with explosions and column collapse. The largest peaks were associated with the rapid onset of effusion in 2004 and the large explosions of 2005. The MODVOLC (Koeppen et al. 2011) and MIROVA (Coppola et al. 2016) systems both are designed to quantify thermal emissions from erupting volcanoes and are contributing to the monitoring efforts at Volcán de Colima. It is often not possible to carry out overflights with the infrared camera, and the satellite data can contribute to the data pool, facilitating interpretation and the definition of eruptive scenarios.

2.5 Lahar Monitoring

Lahars represent the most frequently occurring hazard at Volcán de Colima, and have resulted in damage to infrastructure in 2000, 2007 and 2015. Two lahar monitoring stations have been installed in La Lumbre and Montegrande ravines, which generate photographic images of the flows, record seismic data and include a rainfall gauge. Vázquez et al. (2014) report on the observation of one particular lahar event within Montegrande in 2012. Advances in our understanding of lahar generation and transport will assist in improving plans for risk reduction.

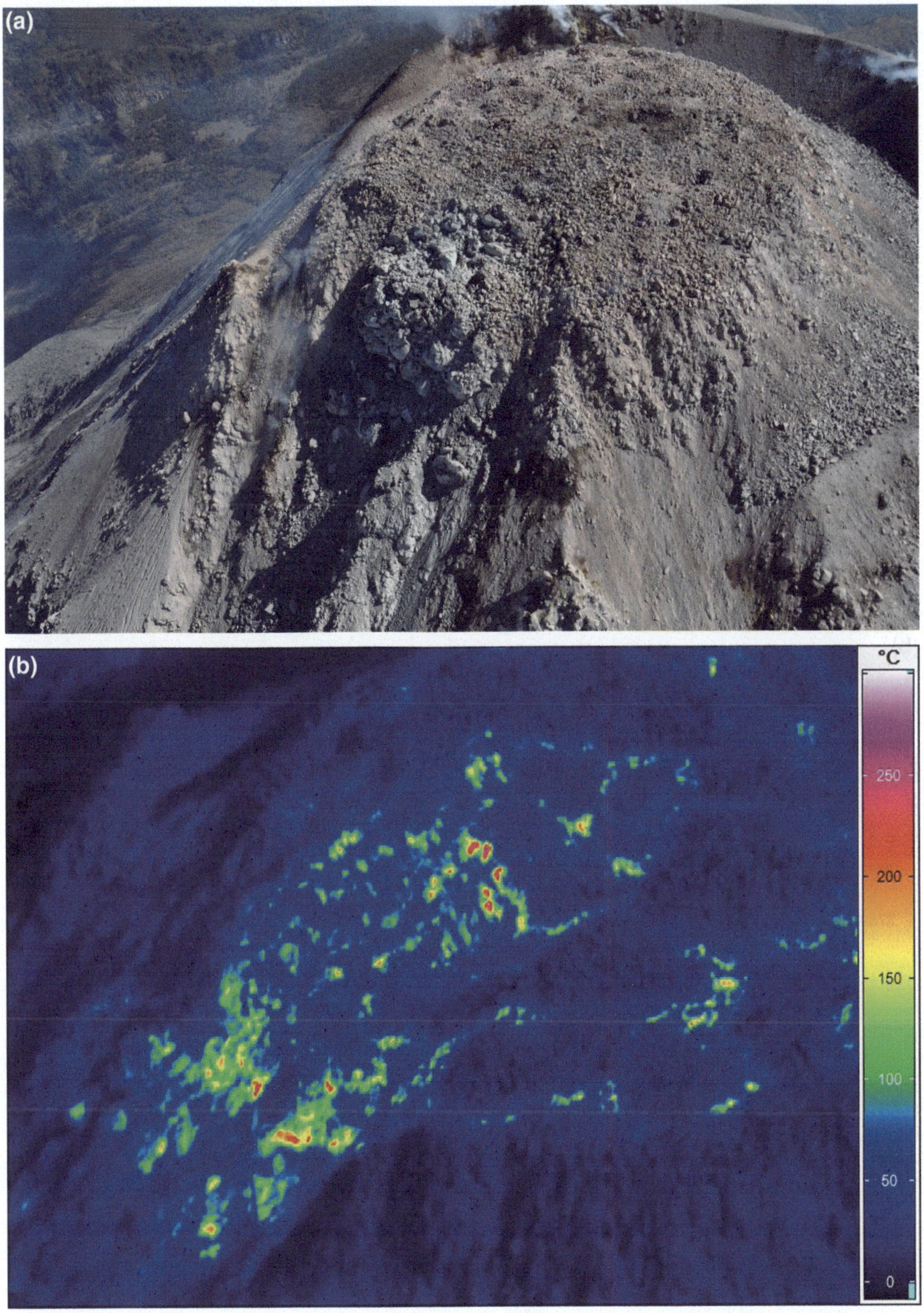

Fig. 7 **a** The Western lobe formed as dome growth returned to an exogenic emplacement style during 2010 (photo taken during flight on 26 December 2010). Frequent rockfalls were produced down the W flank. Diameter of upper flat dome surface was about 150 m. **b** Thermal image also taken on 26 December 2010. Maximum apparent temperature recorded was 525 °C

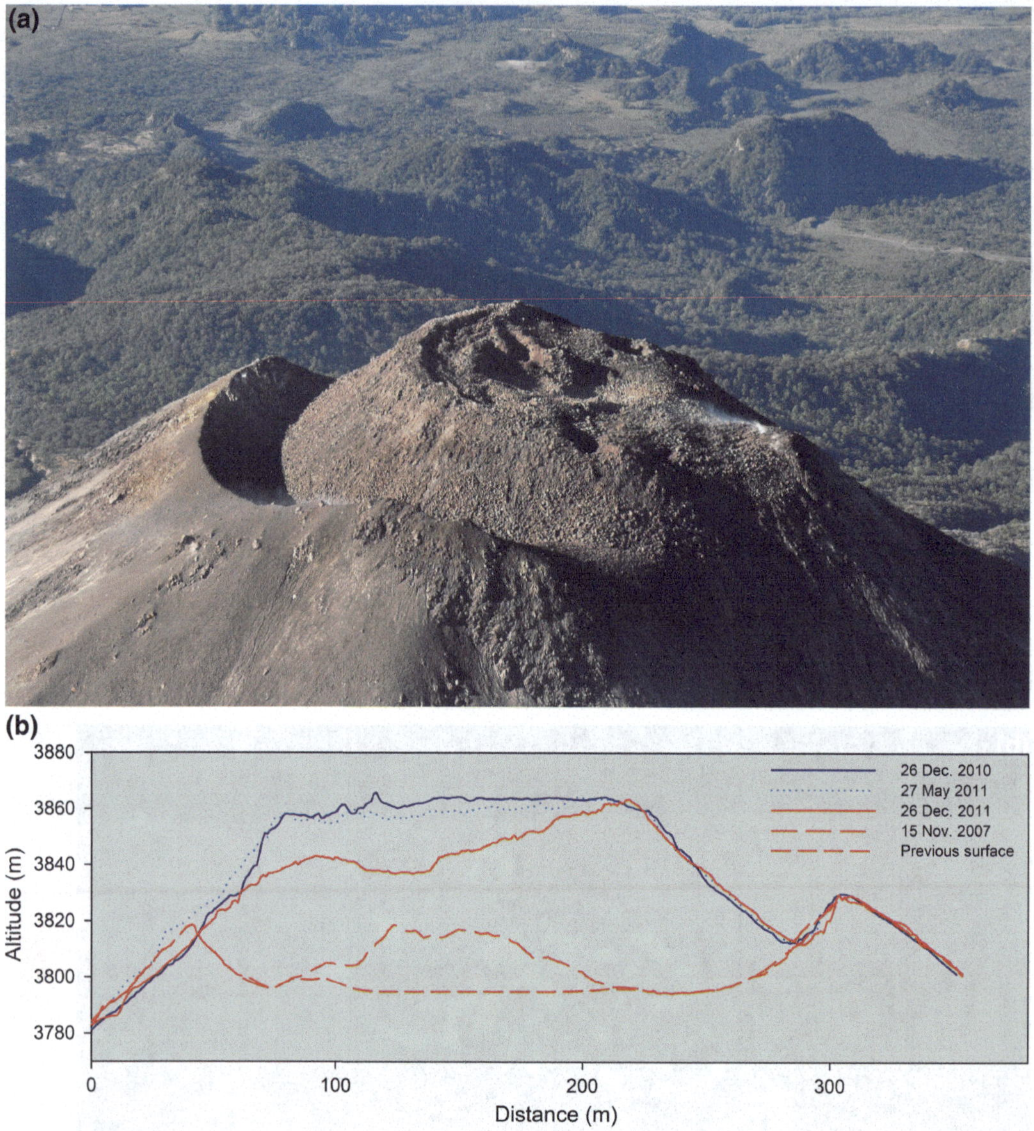

Fig. 8 a Dome after June 2011 explosion. Photo taken on 15 November 2011 showing view towards S with various hummocks visible in the background that resulted from Holocene debris avalanches (see Cortés et al., this volume). Dome no longer has horizontal flat upper surface. b Variation in profile obtained from photogrammetry. A difference in the dome profile can be noted between December 2010 and May 2011, which is due to subsidence. The large reduction in height that followed was due to the explosion on 21 June 2011 (see James & Varley 2012 for further details)

Figure 10 shows how the activity in 2005 increased the potential for lahars on the flanks. Increased instrumental observation is enabling the possibility of forecasts given the rainfall intensity and duration (Vázquez et al. 2016).

2.6 Other Monitoring Techniques

One critical monitoring parameter is the effusion rate, this is particularly clear when considering the most common scenario that can be termed as

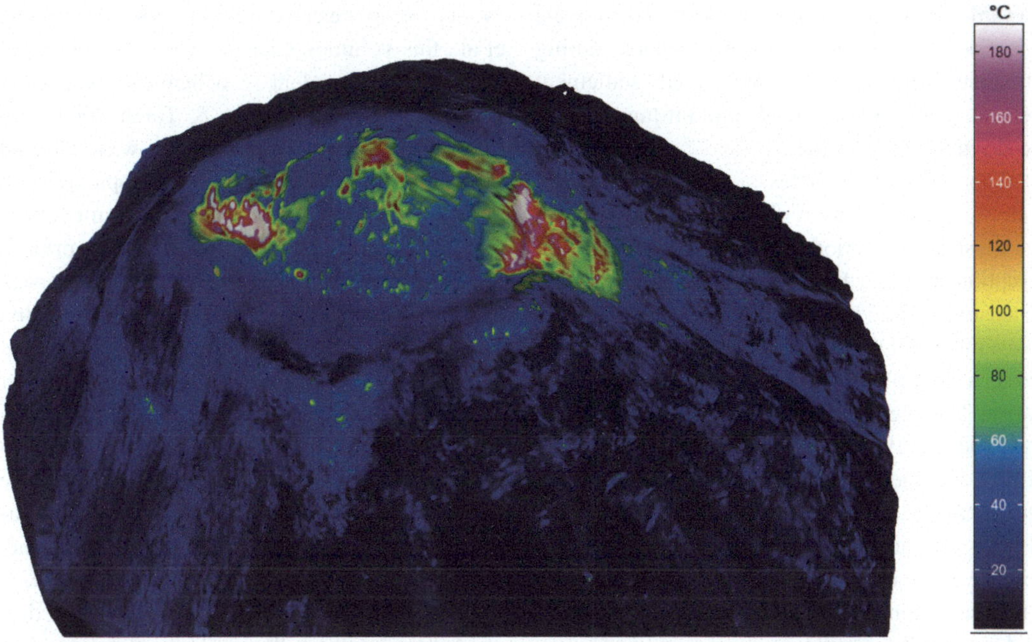

Fig. 9 3D thermal model constructed using structure-from-motion techniques showing thermal anomalies during the flight on 07 June 2015. The view is from the WSW. Two thermal anomalies can be observed, indicating the emission of magma both to the north and south of the crater. Two actively extruding vents had not previously been observed at Volcán de Colima

dangerous: a dome collapse. Large PDCs were generated in 2004 and 2015, largely due to collapses. An accelerating effusion rate, specifically the integrated volumetric effusion rate, can result in the collapse of a dome, either due the sheer size of the magma body, or due to the accumulation of pressure within. A gravity initiated partial collapse can result in rapid decompression and fragmentation within the dome, with resulting mobile PDCs (Sulpizio et al. 2010). It is clearly important to obtain data that allows the estimation of the effusion rate as frequently as possible.

As previously mentioned, aerial photographs or thermal images can be combined using photogrammetric techniques, which permits a reasonably accurate volume estimation of domes, flows or a crater after explosive activity. An example of results obtained using this technique are shown in Fig. 8. The time series of dome profiles permit detailed analysis of growth or subsidence, with the potential for identifying critical processes in the evolution of activity.

Another addition to the monitoring network was a Doppler radar system installed on the south flank in 2007. Although the installation was dogged by technical issues, some useful results were obtained of particle velocities during Vulcanian explosions (Scharff et al. 2011). This technique permits a detailed view of the complex variation in velocity and revealed the presence of pulses, separated by a few seconds (Scharff et al. 2015). Unfortunately the system was a major casualty during the July 2015 eruption (Fig. 23c).

3 Activity 1998–2011

Recent activity at Volcán de Colima has been divided into two eruptive periods, the first lasting 12.5 years and the second which is ongoing at the time of writing (September 2016), having so far lasted three years and nine months. The periods have included some intense eruptions, prompting fears of a near-future escalation towards the next cataclysmic event. To mitigate the risk, several villages were evacuated during different periods, with one small settlement relocated in controversial circumstances (see Rodríguez, this volume).

This section includes a summary of the activity presented by Volcán de Colima during its most active period since its last catastrophic eruption in 1913, both in terms of duration and intensity. The eruptive period lasted from November 1998 until June 2011. Previously, Volcán de Colima had been quiet for over four years. The preceding effusion episode emplaced a dome in 1991 (Rodríguez-Elizarrarás et al. 1991) and there was an isolated and somewhat enigmatic explosion in 1994 (Jiménez et al. 1995). Seismicity occurring prior to this explosive event (see Arámbula et al., this volume) suggests magma ascent was occurring, though it did not reach the surface.

3.1 Reactivation in 1998

The first signs of a new reactivation of the system were increases in the seismicity with swarms of events being detected during 1997 (Arámbula et al., this volume). Changes were also observed in three independent geochemical precursors (Taran et al. 2000; Varley & Taran 2003). An increasingly magmatic signature was detected from July 1997 by analysing the isotopic ratios in water samples obtained from the summit fumarole discharge. In August 1998, 3 months prior to the first appearance of the new lava dome, a peak in boron was measured in spring waters. Finally, three weeks before extrusion, at the end of October 1997 an increase in the flux of sulphur dioxide was recorded (Taran et al. 2000). In November 1998, the emplacement of a new lava dome was the first appearance of magma at the surface since 1991. The initial effusion rate was estimated to be 4.4 $m^3\ s^{-1}$ (Navarro-Ochoa et al. 2002) filling the 1994 crater relatively quickly, and heralding the start of lava flows which descended the southern flanks. These flows reached a maximum distance of 3.8 km before the eruption terminated in February 1999.

Figure 11 shows the different lava flows that have been emplaced and have remained visible on Volcán de Colima since 1961, the date of the first flow since the 1913 eruption. Prior to this,

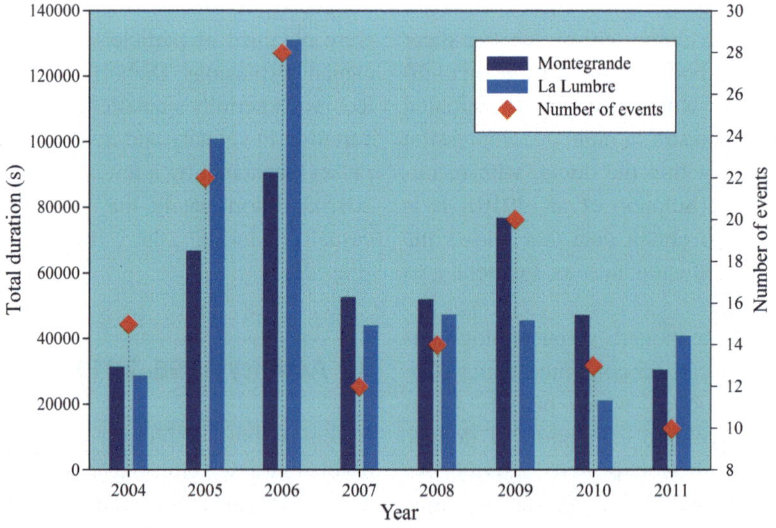

Fig. 10 Total duration and number of lahar events within La Lumbre and Montegrande ravines each year (2004–2011) as recorded by the associated seismicity. As the volume of material in the upper ravines available to be remobilized progressively decreased after the 2005 explosive activity, so did the number and magnitude of lahars

the historical records are sparse, resulting in many lava flows having an unknown age. Important morphological features include the dome of Volcancito formed with its associated flow from a flank eruption in 1869, and the complex of three domes to the south, known as Los Hijos (both shown in Fig. 11). These are of unknown age; however, given the scant vegetation cover on two of the domes, an age of a few hundred years seems likely.

In 1999, three Vulcanian events occurred, with eruption columns reaching 11 km in the case of the largest, on 17 July (Savov et al. 2008). This sequence of events left a large crater in the summit with dimensions of 230–261 m diameter and 6–35 m in depth. These explosions had been the largest since monitoring of the volcano had started in the mid-1980s.

3.2 2001–2003 Effusive Phase

Following the 1999 explosions, the volcano was relatively quiet until February 2001, when a further major Vulcanian explosion occurred, signalling the start of a new phase in the eruption. This was somewhat atypical, since on this occasion the explosion preceded dome growth, which commenced in May 2001. Figure 12a shows the early dome growing in the west side of the crater. No precursors were identified and the effusion rate was slow at the beginning (<0.1 $m^3\ s^{-1}$), decreasing further and possibly stagnating in September, before a shift in the effusive centre occurred within the summit crater. The second phase of dome growth started with a more crystalline magma batch, resulting in the emplacement of a lava spine, in the NE sector of the crater (Fig. 12b). After its collapse, an acceleration of the effusion rate occurred (up to 0.7 $m^3\ s^{-1}$) with the crater being filled by early February 2002.

This effusive phase lasted 22 months, finally finishing during March 2003. Table 2 compares the effusion rate for the seven dome-building phases of the recent eruptive period. During the 2001–3 period no significantly long lava flows were formed due to the low rate the magma was

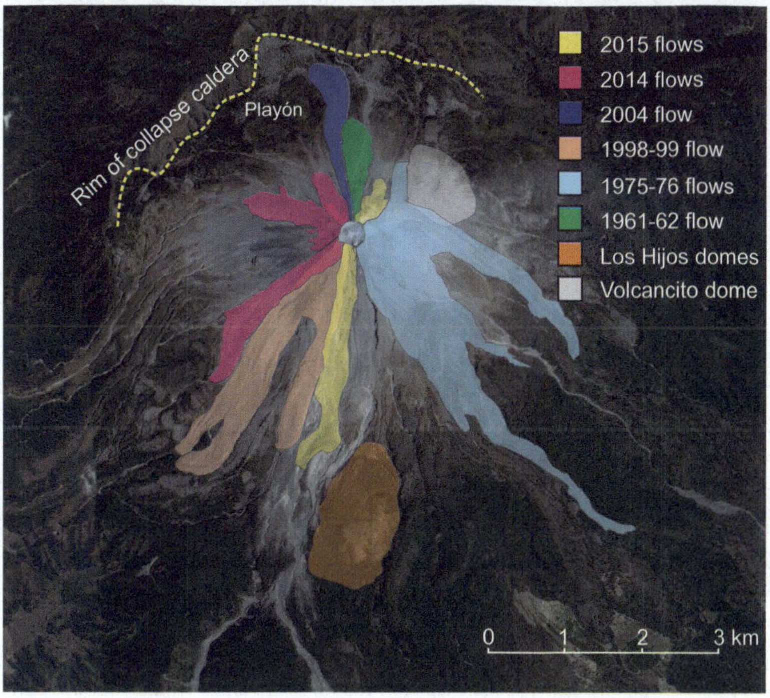

Fig. 11 Historic lava flows (from 1961–62 until the flows emplaced in 2015). Other lava flows subsequently buried by more recent events were emplaced during the eruptive phases of 1981, 1991, 2001–03 and 2007–10

Fig. 12 2001–3 dome. **a** May 2001 dome – first phase of growth originating from the west central vent (26 May 2001). Photo taken from NE rim of crater. The main crater diameter was 230–260 m, with the inner crater diameter estimated at 127 m. **b** Spine that formed in October 2001. A change in the magma rheology resulted in it becoming brittle at deeper levels within the conduit. Photo taken from E rim of crater looking towards the SW. Diameter of dome along N-S axis was 98 m

Table 2 Effusive phases of recent eruption

Start date	End date	Period	Peak effusion rate ($m^3 s^{-1}$)	Max. lava flow length (km)
November 1998	February 1999	2.5 months	4.4[a]	3.6
May 2001	March 2003	22 months	0.07	1.2
September 2004	December 2004	2.5 months	6–8	2.15
May 2005	June 2005	Hours	1–5	N/A
January 2007	June 2011	54 months	0.02	0.43
January 2013	April 2015	27 months	9.7	2.61
May 2015	September 2015	5 months	> 20	2.88
February 2016	October 2016	8 months	4–6	1.84

[a]Navarro-Ochoa et al. (2002). The case of 2005 was exceptional with larger magnitude Vulcanian explosions destroying some small emplaced domes only hours after their first observation. The peak effusion rate for the May–September 2015 period is a minimum estimate. There was exceptionally fast dome growth and collapse to form large PDCs in July 2015

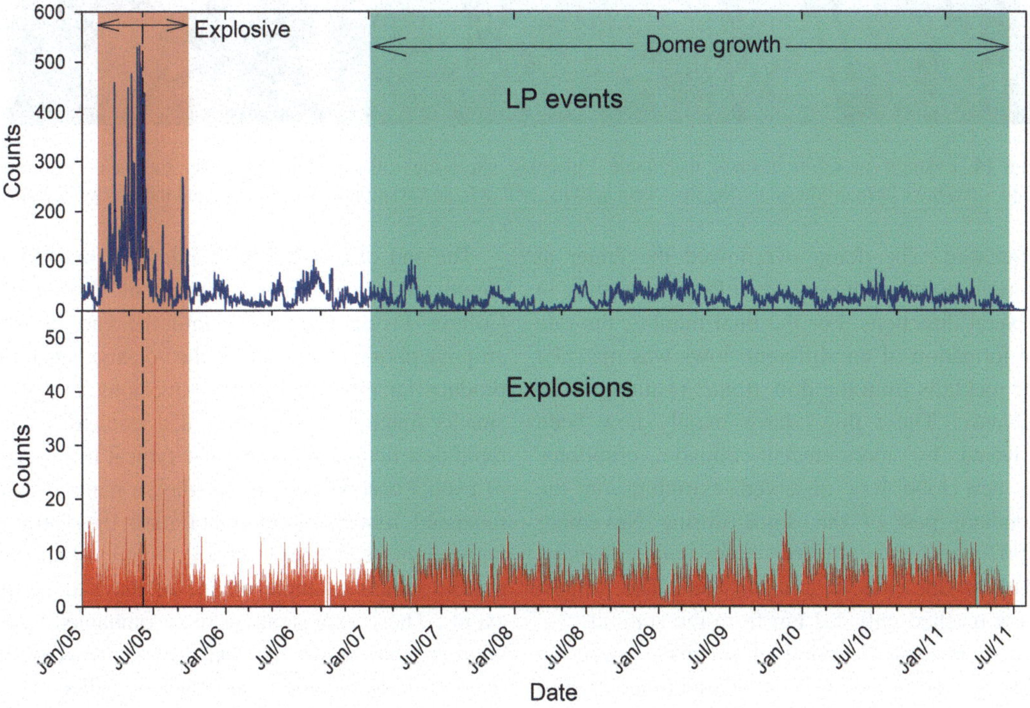

Fig. 13 Number of long period and explosions per day during January 2005 to June 2011. Explosions have been manually identified from the seismic records using a complex series of criteria to differentiate between different event types. The green period represents an episode of dome growth with a very low effusion rate. Variations in LP counts might be related to variations in effusion rate (Varley et al. 2010a). The orange period represents the most energetic explosive period at Volcán de Colima in the last 100 years, lasting from February to September 2005 with the peak shown by the vertical dashed line on 05 June. The swarms of LP events that were associated with each larger explosion can be distinguished. The number of explosions per day typically varied between 5 and 10 during the same period

Fig. 14 Example of ash-rich small magnitude Vulcanian explosion (30 April 2016). Many thousands of these explosions have been produced during the recent activity, with a large variation in ash-contents and the energy liberated

emplaced. The dome overflowed the crater at various points with very small flows extending in several directions. For the final months, the rate of formation of the different flows was matched by rockfalls from the flow fronts, resulting in no advance. These flows have mainly now been covered by more recent effusive eruptions. A new lobe was observed extruding on the southern part of the dome during November 2002; the shift in effusion centre again coincided with an increase in effusion rate. The longest flow reached only 1.2 km from the summit.

An interesting period of seismicity occurred during this phase with large amplitude tremor detected in May and June 2002. Tremor of this magnitude had not been recorded prior to this period on the RESCO network and has not been recorded since. This precedent resulted in the evacuation of several close communities. The mechanism that generated this tremor remains to be identified, and although no increase in effusion rate was observed, there were higher fluxes of sulphur dioxide detected in the volcanic plume.

The end of effusion in March 2003 marked a significant change in the activity of Volcán de Colima. From this period until the end of the eruptive period in June 2011, the volcano began to produce far more Vulcanian explosions, but of a smaller magnitude than those that resulted in the rapid destruction of domes. The typical frequency of events was around six per day, a number that remained remarkably constant until they terminated eight years later (Fig. 13). Figure 14 shows a more recent example of one of the Vulcanian events. These explosions were accompanied by a varying volume of ash and were generated in several locations within the crater. Lavallée et al. (2011) proposed that the locations of the effusive and explosive centres within the crater are controlled by a denser block of lava, which remained in the central conduit after the 1913 eruption. Observations during both the recent eruptive periods continue to reveal the presence of multiple points of emission within the crater.

The 2001–3 lava dome was destroyed by three larger magnitude explosions that occurred on the

Fig. 15 **a**, **b** Examples of craters formed to the north of the volcano as a result of ballistic impacts from explosions in July and August 2003. Distance from the summit is about 1.5–2 km. **c** Amazing highly fractured breadcrust bomb, observed to north of volcano in August 2003

17 July, 02 and 28 August 2003. An extensive study has revealed a greater density of impacts to the north/north-east of the volcano. They reached up to 2.5 km from the summit and the impact craters had diameters of up to 8 m. Figure 15 shows some results of impacts from ballistics of the 2003 explosions. In some cases, branches of trees were cut, which allowed an estimate of the trajectory of the volcanic bomb on its way to impact the ground. Many bombs showed the characteristic bread-crust textures with a glassy outer layer, testament to rapid cooling during its flight.

3.3 2004—Large Lava Flow to the North

In July 2004, another increase in boron in the spring waters was detected, and a swarm of long-period (LP) seismic events occurred (Arámbula-Mendoza et al. 2011). Two months later, during September 2004, magma was again observed, with a lava dome filling the summit crater in less than 2 days. Thermal images recorded during a flight in early October confirmed that the effusion was centred on the north side of the crater and subsequently this was the side where lava overtopped the crater rim and descended in two flows, one to the north, and the other to the northwest (Table 2). Figure 16a shows the growing dome, with intense gas discharge from a zone lower on the East flank. A faster effusion rate meant that the dome and flows frequently suffered collapses, which resulted in frequent small PDCs descending in the same directions as the two flows, but not reaching further than 2.5 km (Fig. 16b). Often called block-and-ash flows, these often represent a major volcanic hazard associated with dome emplacement.

Fig. 16 **a** Dome growing in the summit crater (16 October 2004). Note the intense fumarolic activity on the eastern flank. Photo taken looking towards the NW; dome diameter is about 280 m. **b** Small pyroclastic density current descending NW flank (02 October 2004). **c** Flow front of the October 2004 pyroclastic density current, which was produced from a major collapse of the growing lava dome. **d** 2004 lava flow from the air, flowing towards the north in the Playón. Contrast in rheology with the 1961–62 flow to the left is very clear. The darker colouration in the centre of the flow reflects a greater vesicularity of the blocks in this zone

One major collapse occurred from the northwest flow on 06 October 2004 sending an estimated 900,000 m^3 of material down the La Lumbre ravine to a distance of 6.1 km from the volcano (Fig. 17). Although still far short of the village of La Becerrera, which is located at 14.7 km along the ravine, the event served as an important reminder of the hazards associated with dome collapse. Figure 16c shows the PDC front with large blocks having been transported from the crater or remobilized from some point in its trajectory. To assess the possible impact on the local population computer simulations are used to estimate where the flows might reach. For the case of Volcán de Colima, using the TITAN 2D code, it was shown that no large community would be threatened by a major dome collapse, since the maximum distance they might reach is 7 km (Sulpizio et al. 2010). Subsequent activity in 2015 has, however, demonstrated the limitation of these models (see Sect. 4.1). Simplifications do not account for certain factors that can greatly increase the mobility of this type of flow.

The north lava flow advanced at a faster velocity than the northwest flow, mainly due to differences in slope angles. The less inclined

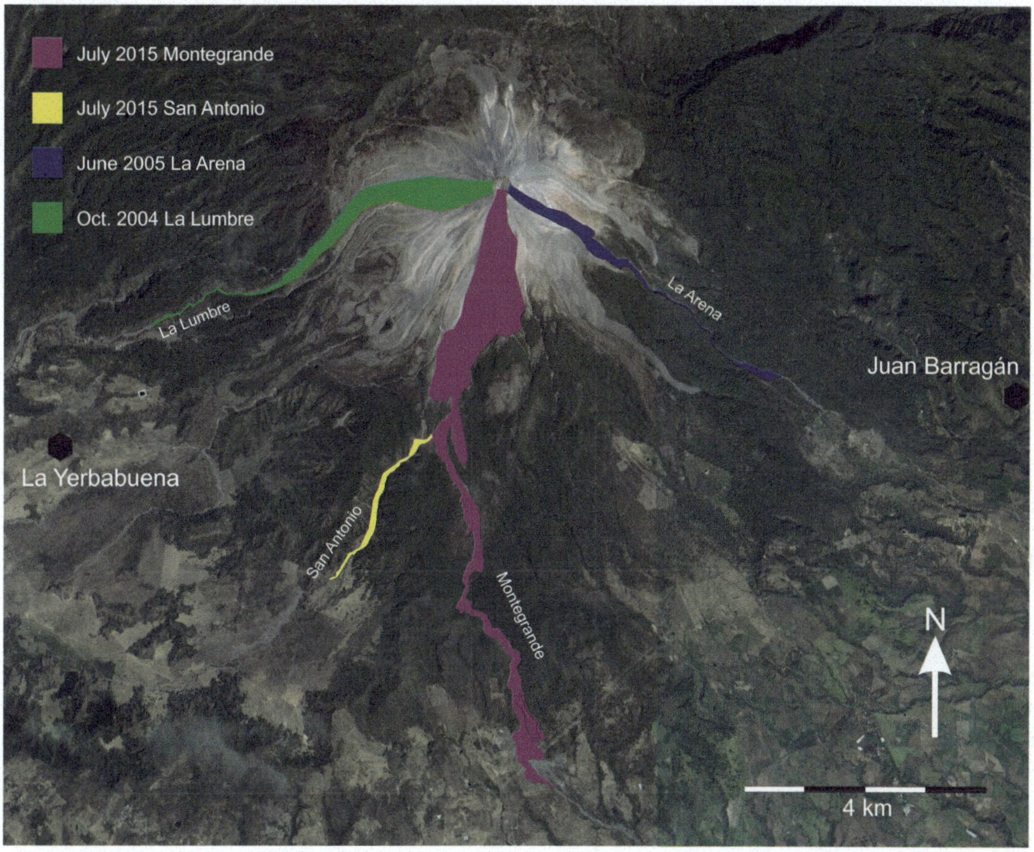

Fig. 17 Map showing most important pyroclastic density currents during the 1998–2011 and 2013–2017 eruptive periods. The flow in La Lumbre was from a partial dome collapse in 2004 reaching 6.1 km, whilst the largest runout during the 2005 Vulcanian explosion period was in La Arena with 5.4 km. Finally, the July 2015 complex event is shown with 9.7 km runout to the south in Montegrande, whilst the overbank diverted flow that entered San Antonio reached 6.8 km (all distances are straight-line along the ground distances). The two most vulnerable communities are also shown: La Yerbabuena in Colima and Juan Barragán in Jalisco

northern slopes gave this flow front more stability, whereas the northwest flow descended the steepest part of the whole edifice and was almost continuously collapsing. After one month, the north flow reached the relatively flat floor of the area known as the Playón, located between the current edifice and the wall, which marks the boundary of the collapse of the previous edifice (see Fig. 11). Figure 16d shows the flow deviating to the west to follow the gentle slope in this direction. The height of the flow front was 18 m and the final volume was estimated at 7.45 x 10^6 m^3. The effusion rate showed two peaks, with the highest at 6–8 m^3 s^{-1}. Towards the end there was a gradual decrease in its advance, with the eruptive phase terminating during December 2004.

3.4 Multiple Vulcanian Explosions in 2005

During 2005 the largest explosions occurred at Volcán de Colima since the last major sub-Plinian eruption in 1913. At least 30 explosive eruptions occurred during the period 17 February–27 September, each producing a PDC from the collapse of the dense columns. Their run-out distances varied between 2.5 and 5.4 km and they descended the ravines to the SW, S and

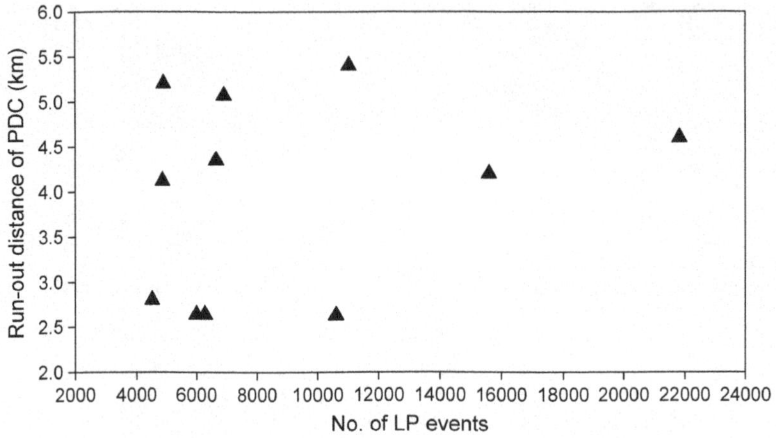

Fig. 18 The run-out distance of pyroclastic density currents generated from column collapse during the explosions of 2005 plotted against the count of LP events in each swarm that pre-empted the explosion. No correlation is present, reflecting the complexity of the physical processes controlling the generation of the LP events during magma ascent, explosion generation, column formation and its collapse, and finally the transport and deposition of the resulting PDC

SE of the volcano. Often evidence for multiple pulses was observed in the field. The seismic signal gives an insight into the Vulcanian process, with the formation of an impermeable cap, pressurization and then breaching, with fragmentation generating the ash fraction expelled along with large ballistic bombs and blocks (Varley et al. 2010a). Each explosion was preceded by a swarm of low magnitude LP seismic events, lasting up to three days (Fig. 13). They represented a remarkably reliable precursor for the explosions, with no false positive occurring, that is, every swarm was followed by a large explosion. However, no correlation was observed between swarm characteristics and the magnitude of the explosion. Various parameters can be applied to define the magnitude of an explosion, including the magnitude of the related seismicity, which also did not correlate with swarm characteristics (Varley et al. 2010a). Figure 18 shows no correlation between the magnitudes of these swarms (determined in this case by the number of LP seismic events constituting the swarm) with the recorded maximum distance reached by the PDCs, for those observed in the field. The runout distance achieved by the PDCs represents an appropriate measure when considering the hazard associated with these events. The maximum distance of 5.4 km was reached by the PDC that followed La Arena ravine on the 09 June 2005 (Fig. 17). There had been a switch in direction with previous PDCs reaching their maximum runout along the Montegrande ravine to the south.

Varley et al. (2010b) examined in detail the relationship between the seismic swarms and the explosion characteristics and proposed a generation mechanism based on brittle fracture associated with the enhanced strain-rate found along the conduit margins. Clearly there was a complex relationship between magma ascent, the generation mechanism for the Vulcanian explosion and the factors controlling PDC generation.

Magma was ascending during the entire period from February to September, but only made it appreciably above the surface during the end of May to the beginning of June 2005 period, when small lava domes were observed to have formed in the crater. They were then destroyed by the Vulcanian explosions. No overflights were taking place, but these domes reached sufficient height to be observed above the crater rim from the observation station on Nevado de Colima. Although effusion rates were similar during 2004 and 2005, the latter episode featured large explosions, absent in 2004. This was most likely

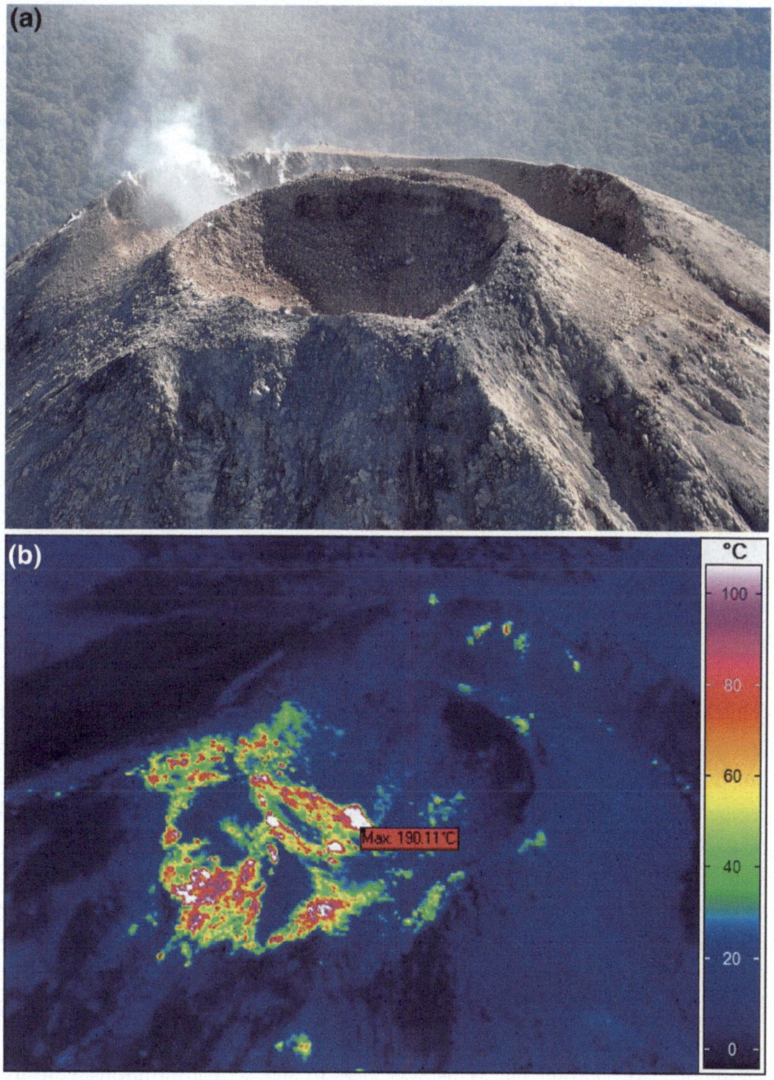

Fig. 19 **a** New crater formed within the 2007-11 dome from explosions on 06 January 2013 (diameter about 150 m). The photo is looking towards the east, showing the 2005 crater rim behind the dome. Vigorous fumarolic gas emission can be seen in the NE corner of the crater. **b** New extrusion shown in thermal image of crater recorded during flight on 11th January 2013

due to increased volatile-contents. Permeability decreased within the ascending magma due to crystallization, creating a seal. When pressurization beneath such a seal reached critical levels, Vulcanian explosions destroyed the embryonic domes.

Large volumes of material, due to the PDCs, were deposited in the ravines, which run south from the volcano, principally Montegrande. This loose unconsolidated material was readily remobilized to form lahars during the 2005 and subsequent rainy seasons, as reported by Dávila et al. (2007) and discussed in Chap. 6. Lahars, or mud-flows represent the most damaging phenomena associated with the volcano during recent years. Electric pylons were damaged and a house was inundated during July 2000. In Fig. 10, the correlation between the available volume of material in the ravines and the incidence of lahars is shown.

3.5 Long-Lived Dome Extrusion 2007–2011

During 2005, the daily smaller Vulcanian events continued with a similar occurrence rate. These explosions continued into 2006 before a new effusive phase commenced in early January 2007. A small dome was spotted during a helicopter flight with subsequent thermal imaging confirming that it was an active emplacement of hot lava (Fig. 5a). This phase was the longest lived, with the lowest effusion rate of the 1998–2011 eruptive period. Photogrammetry was used to calculate dome volumes and hence the effusion rate between flights.

Analysis of thermal images taken during overflights proved to be enlightening during the 2007–2011 episode. Thermal images taken from flights permitted not only the observation of the dome, but also the calculation of heat flux, and variations in the temperatures of the fumarole fields located on the inner and outer crater walls (Hutchison et al. 2013). Based on this, the effusive episode could be divided into three phases. Initial emplacement (stage 1) was exogenic with high temperatures recorded, due to the exposed magma (Fig. 5). Towards the end of 2007, emplacement switched to endogenic, due to the accumulation of talus from material falling from the slowly growing dome, which had reached 1×10^5 m^3 (Fig. 6). This marked the start of stage II, which continued until early 2010, with the surface temperatures of the dome remaining relatively low during its slow expansion. This was due to the absence of hot material breaking through the cooled dome exterior. The dome grew from a vent located on the west side of the crater, which meant that the first overflow occurred on that same west side. Once the dome grew over the rim, rockfalls became frequent, with material unloading that side of the dome. This resulted in the deviation of the magma flow direction. Rather than the dome being fed in the centre, a lobe formed on the western side, with a return to exogenic growth in early 2010, shown as stage III in Fig. 6, with the lobe shown in Fig. 7.

Rockfalls then became the principal fate for any new magma, with the volume of the dome not increasing appreciably. To continue quantifying the effusion rate, it was necessary to estimate the volume of the different size rockfalls descending the steep western flank. The seismic signal was compared with the thermal signature of the rockfalls, as well as an estimate made of the volume of material lost from the dome when possible (Mueller et al. 2013). From an analysis of the rockfalls it was determined that the extrusion rate was not constant: it increased from 0.008 to 0.02 m^3 s^{-1} during 2010 and dropped down to 0.008 m^3 s^{-1} again in March 2011. Typical dome emplacements on stratovolcanoes grow at rates faster by 2 or 3 orders of magnitude. This episode of activity at Volcán de Colima was notable for its long duration and exceptionally slow effusion rate. The morphology of the 2007–11 dome was also different to those emplaced in 2001–3 and 2004 (comparing Fig. 12 with Figs. 16 and 19) with a higher aspect ratio (height/diameter) indicating a different rheology.

The 1998–2011 eruption terminated in a surprisingly abrupt fashion. With no warning a larger magnitude Vulcanian explosion occurred on 21 June 2011. This event destroyed only a part of the dome, but signalled a remarkable and unexpected cessation of activity. The seismicity dropped to almost nothing, the daily smaller magnitude explosions stopped after eight years of continuous activity, and only low values of SO_2 flux were then measured (<0.2 kg s^{-1}). This was a dramatic transformation, and the first time activity had dropped to these low levels since before the onset of this eruptive period in 1998. Photographs taken during flights were processed using structure-from-motion (Sfm) to calculate volumes and reveal changes in the dome morphology (James & Varley 2012). It was discovered that the dome had started to subside prior to the large explosion, possibly indicating the start of magma withdrawal once the effusion rate had dropped (Fig. 6).

Fig. 20 11 January 2013 PDC to west of Volcán de Colima deposited in area known as Playón. Photo taken at 1.7 km from crater. This section of deposit comprised of relatively small clasts with larger clasts concentrated in levées. Unburnt vegetation is evidence of low emplacement temperatures. Here the PDC deposit was about 20 m in width

4 2013–2016 Eruption[1]

Following the abrupt cessation of activity in 2011, an 18 month period of quiet followed. Because of this, the 2013–2016 activity is considered as a separate eruption. Comparing the 1998–2011 and 2013–2016 eruptive periods, similarities can be observed, such as the resumption of the small explosions several times per day; however, important differences can also be highlighted. The eruption is ongoing at the time of writing (September 2016); it is proving to be the most complex during recent times, with multiple batches of magma reaching the surface at different rates, initiating lava flows to multiple directions (west, south-west, south, north-northeast and north); a greater variation in the frequency and magnitude of the Vulcanian explosions (from one or two per day up to in excess of 40); and the most important eruptive event since the 1913 sub-Plinian eruption, the July 2015 episode.

The volcano reawakened somewhat unexpectedly with an explosion on 06 January 2013. It was preceded by three days of precursory seismicity (see Arámbula et al., this volume). Initial reports defined the eruption as phreatomagmatic, considering an influence of rainfall given the coincidence with the short rains that typically occur in January/February in this region. However, the relatively low ash-content of the eruption column, apparent from examination of photos, as well as limited ashfall failed to support this idea. Phreatomagmatic activity is characterized by more efficient fragmentation of the magma and large volumes of ash. The explosion was Vulcanian, resulting from the ascent and decompression of a relatively volatile-rich batch of magma.

[1]The eruption is ongoing at the time of writing: September 2016.

Fig. 21 Blocky dome and young lava flow growing towards west with small Vulcanian eruption. Large block can be seen initiating a rockfall. Dome width as shown in photo was about 135 m, taken on 15 March 2013

Figure 19a shows the crater that was excavated within the 2007–11 dome, indicating a common emission point within the crater for both eruptions. The first sign of a new lava emplacement was identified in the depths of the crater on 11 January. The thermal image in Fig. 19b confirmed elevated temperatures.

A further three larger explosions occurred during January, with PDCs generated from column collapse travelling to the west on the 11 January, and to the south on the 29 January, with the latter having the maximum runout of these events at 2.4 km. Figure 20 shows the remarkably smooth convex surface of the western PDC. Larger rocks can be seen in small levées running down each side of the flow. Emplacement was at relatively low velocity and temperature, evidenced by surviving unburnt vegetation on either side of the deposit. The explosions expelled ballistics within a radius of 2.5 km, mainly to the north/north-east of the volcano.

Effusion continued with magma filling the crater and forming a dome composed of particularly large blocks, many measuring several metres in diameter. In early March, the dome overflowed the rim to initiate a new lava flow towards the west, following a similar path to flows during 2010. The flow continued advancing at a slow rate through 2013 and 2014. Observations of rockfall activity revealed that the magma was ascending in pulses. Small volume ($<10{,}000$ m^3), slow-moving (<10 m s^{-1}) PDCs were sometimes observed descending the western flank when decompression in the dome after a rockfall was sufficient to generate fragmentation. Figure 21 shows an early advancing lava flow with the initiation of a rockfall.

An increase in effusion rate in September 2014 heralded the onset of a much faster advancing lava flow that descended to the south-west. Similarly to what was observed in 2004, a later magma pulse, with a significant increase in the effusion rate, initiated a second flow emplaced on top of the earlier flow rather than this flow increasing its flow-rate. The second flow reached about 0.8 km, whilst the first flow came to rest above the El Zarco ravine at a distance of 2.61 km from the crater.

The next larger magnitude Vulcanian explosion occurred on 21 November 2014, reflecting the faster ascent of a gas-rich magma batch. Again a PDC was generated from column collapse, which reached 2.9 km in the Montegrande and 3.0 km in the neighbouring San Antonio ravine towards the south. There were a large number of distinctly vesicular clasts found within the deposit, unlike any observed within the 2005 or 2013 PDCs. It was assumed that they represented juvenile material and were evidence of higher volatile-contents. A substantial dilute cloud was associated with the eruption, a result of more efficient fragmentation within the conduit. The cloud coated large parts of the forest on the flanks of the two ravines, as can be observed in Fig. 22.

Two larger magnitude Vulcanian explosions occurred during January 2015, the first on 03 January. Observations of the PDC deposit again implied a high volatile presence in the magma. Small degassing vents were abundant within the large volume of fine material. The PDC ($\sim$100,000 m^3) reached 1.6 km to the north in the Playón, with negligible deposition to the south. The second explosion occurred on 29 January and was followed by magma extrusion, with a slowly forming dome extruding from a vent in the southern section of the crater. This dome had a much flatter morphology than previous domes in 2001–3, 2004 or 2007–11 (compare Fig. 23a, left panel, with Figs. 8, 12 and 16). This suggests a lower viscosity, which most likely reflects higher temperatures and volatile-contents of the magma. Further work is in progress to constrain these variables. For the

Fig. 22 Deposition from PDC on 21 November 2014. There was a greater proportion of fine material generated compared with previous PDCs in 2005 or 2013. Trees can be seen coated outside of the main ravine

Fig. 23 **a** Comparison of the crater as viewed from the south before and after the July 2015 eruptive episode (photos from 07 June and 03 August 2015). A proportion of the PDC deposit was composed of remobilised material that collapsed forming the scar visible in the second photo. **b** Still steaming PDC deposit within the Montegrande ravine in July 2015. Trees on the ridges have been stripped of branches, much of their bark, and left charred by the hot flow. Incredibly, many of these trees actually survived and recovered with fresh leaves growing the following year. **c** Destroyed Doppler radar station located on a ridge ~50 m above the base of the Montegrande ravine and 2.8 km from the summit. The station was destroyed by the PDCs generated in July 2015. Two dishes were installed, one was sheared off its steel base and carried for 25 m, whilst the other in the photo was ripped from the ground, the concrete base for the legs proving to be insufficient to sustain it

first time in recent eruptive history at Volcán de Colima, effusion was observed originating simultaneously from two vents: one to the south, the other to the north. Both emissions eventually overflowed the crater rim to form short lava flows during June 2015.

4.1 July 2015 Eruption

The eruption that occurred over two days during July 2015 can be described as enigmatic, with a complex combination of fast magma ascent with the collapse of newly emplaced material, along with a portion of the previous crater rim. The eruption generated a series of PDCs, concentrated within two periods separated by 15 h on 10 and 11 July. The furthest distance reached was 10.7 km from the summit, travelling down the Montegrande ravine (9.7 km considering a straight line from the collapse site). This is an incredible distance for a dome collapse, and it was highly fortunate that there were no fatalities. The longest run-out previously recorded for a dome collapse was the October 2004 event, which reached 6.1 km in the La Lumbre ravine. As discussed in Sect. 3.3, its volume was estimated at 9 x 10^5 m^3, making it considerably smaller. The deposits of major PDCs emplaced since 2004 are shown in Fig. 17. These represent the farthest reaching events recorded since the sub-Plinian 1913 eruption, however, it is important to note that little data exists regarding PDC deposition prior to the 1990s. The enhanced mobility of the 2015 flow points to a high gas-content, with high temperatures also demonstrated by the burning of vegetation (Fig. 23b) and the impact upon a monitoring station that was located 2.8 km from the summit (Doppler radar station; Fig. 23c).

Capra et al. (2016) and Reyes-Dávila et al. (2016) both describe the main events, but without fully elucidating to the mechanism. The eruption followed an acceleration in the emplacement of the summit dome. Inspection of the PDC deposit revealed that at certain locations along its length, the deposit is finer than would be expected from a block-and-ash flow originating from a simple dome collapse. This suggests extensive and efficient fragmentation associated with some of the pulses. Finer-grained PDCs have greater capacity to retain gas, which creates fluidization within the flow and gives them greater mobility (Druitt et al. 2007). The topography and restrictions in the channel width may also have combined to increase the PDC mobility. Further evidence to the mobile nature of the flow was demonstrated at two bends in the ravine at 3.8 and 4.1 km from the summit, where the PDC continued in a straight line, leaping out of the ravine. In the latter case, the flow made the jump into the San Antonio ravine and continued to a distance of 6.8 km (Fig. 17). Accumulated deposits from previous PDC pulses meant the flow no longer had a channel to follow in Montegrande.

Observation before and after the events show how a proportion of the crater rim collapsed (Fig. 23a), but the volume is small compared to the complete flow. Hydrothermal alteration of this south-east section of the crater rim would have contributed to its collapse. Figure 16a shows intense degassing in this zone during the 2004 eruption. The volume of the July 2015 PDC deposit was relatively large, when compared to other Volcán de Colima PDCs within the last 25 years (current estimates range from 2–8×10^6 m^3). It cannot be totally accounted for by a collapse of the growing lava dome along with this section of the crater rim. Besides, the suggestion of a high proportion of highly fragmented magma implies an explosion, although no eruption column was observed. An initial collapse would have decompressed the upper conduit volume, sending a fragmentation front down into the magma column. The absence of an eruption column is due to the high density of the expulsed material. There is a paradox, however, since the high-mobility has been explained by a large volatile component of the erupted mass, whereby this would reduce the density, making the development of an eruption column likely.

Reyes-Dávila et al. (2016) refer to a boiling-over mechanism for the July PDCs. This term has been used to describe PDCs that have been generated with only a minimal column height, and have been explained as resulting from a high volume flux coupled with low exit velocity (Dufek & Bergantz 2007). Another problem is presented by the typical characteristics of this type of PDC: due to their density they do not have high mobility, creating short and thick deposits. A large proportion of vesicular clasts would also be expected with this mechanism, this was not observed in the deposit. Numerical modelling has shown that with increased leakage of volatiles from the conduit through surrounding country rock, the boiling-over mechanism can be produced (Clarke et al. 2002), a condition not compatible with this event.

Clearly further work is required with careful analysis of deposits and other field evidence to determine the mechanism of the July 2015 eruptive episode. Furthermore, eruptive scenarios considered for the recent updating of hazards maps for the volcano do not contemplate such a mobile PDC (Varley 2017). This is a clear example of the necessity to regularly update this type of map.

5 Cyclicity & Transitions Between Eruptive Regimes

At Volcán de Colima during recent years the variations between eruptions styles have been frequent, and at times, dramatic. Cyclicity is being explored at one extreme to obtain a probabilistic risk assessment of the larger eruptive scenarios (Varley 2017; De la Cruz et al., this volume), whilst at the other end of the spectrum, cycles of just minutes or hours are being considered in the quest to understand the Vulcanian explosion process (Kendrick et al. 2016). Statistics have also been applied to the repose period between explosions in an attempt to highlight variations in behaviour related to conduit processes (Varley et al. 2006; Lachowycz et al. 2013; Lamb et al. 2014).

The transition between different eruptive regimes is a complex phenomenon: what are the factors whose variation can lead to a transition? First, we need to define what is meant by an eruptive regime, since it is possible to consider transitions across a large range of scales. In terms of mitigating volcanic risk, the most important phenomena are explosive eruptions of a large magnitude, which produce hazardous emissions with the potential to impact the local population and infrastructure. During recent activity at Volcán de Colima, various periods of both effusive and explosive activity have been experienced, with clear variation between regimes. These regimes can be defined, for example, by considering the differences between fast and slow emission of magma, or through variation in the characteristics of explosions, such as the ash-contents of the eruption column. One of the factors that makes Volcán de Colima an exciting prospect for study is its rapid variability in eruptive characteristics. Herein lies the challenge of explaining these transitions, which may be achieved through modelling the conduit processes. For the 2010 Merapi eruption in Indonesia, for example, samples of different eruptive products, along with monitoring data, were used to create a schematic progression for the eruption, explaining the transition between effusive and explosive phases (Preece et al. 2016). This is a great example of how petrological and geochemical data can provide much insight into the eruptive processes, however, the time required for analyses often restricts its use as a monitoring tool.

The relationship between effusive and explosive activity can show many patterns. The most important factors that determine the style of an eruption are the gas-content of the magma and the rate at which it ascends the conduit: a faster ascent rate meaning accelerated decompression and a more likely explosive scenario. During recent times, Volcán de Colima has shown a pattern of dome-building, followed by their destruction, with the delay between the end of effusion and the explosivity varying from a matter of days to several months. This sequence of events occurred in 1999 and was repeated in 2003 and 2004–5. Partial dome destruction occurred in 2011, which marked the end of that eruptive period. The cause of this transition has yet to be firmly identified, but was most likely due to a late gas-rich batch of magma encountering an impermeable cap, which resulted in pressurization and a Vulcanian explosion.

The magnitude-frequency relationship of volcanic eruptions can be complex. Correctly analysing the occurrence of rare large events can be especially difficult. Deligne et al. (2010) underlined the problem of underreporting of eruptions in the records, and used extreme value theory to characterize the issue. Based on the observational records of Volcán de Colima, the well-known Plinian or sub-Plinian eruption cycle of around 100 years was highlighted some time ago (Robin et al. 1991; Luhr 2002). At the time of writing, the 100th anniversary of the most recent eruption with a magnitude greater than VEI 4 (Volcano Explosivity Index; Newhall & Self 1982) was recently passed in January 2013. Given the fact that only three cycles of Plinian or sub-Plinian eruptions have been registered (1622–1690; 1690–1818; 1818–1913), it would not be prudent to consider this historical cycle alone when pondering over the probability of the next catastrophic eruption. The sub-tropical climate at Colima makes the identification of eruptions in the geological record challenging.

One approach to fill in the gaps is the application of statistical methods. Mendoza-Rosas & De la Cruz-Reyna (2008) and De la Cruz-Reyna (this volume) have combined historic observational records with information from the geologic record, to greatly extend the record of eruptive activity at Colima. The Colima Volcanic Complex has produced a large number of debris avalanches during the Holocene and Pleistocene

(Cortés et al., this volume). It is not always clear whether these collapse events are associated with a large eruption ($\geq$ VEI 5) or if they might have occurred during periods of calm, however, some evidence points to a connection, and through dating the deposits, a record of these large eruptions is obtained. De la Cruz-Reyna et al. (this volume) obtain recurrence intervals of 170 years for a VEI 4 eruption, 1283 for VEI 5, and about 10,000 years for VEI 6.

The most generally accepted model for an abrupt acceleration in activity to a Plinian event is the injection of mafic, therefore hotter, more buoyant magma into the magma chamber (Caricchi et al. 2014). The transition between other eruptive regimes, which often occur in cycles, is somewhat harder to define. Viscosity plays a key role, with the efficiency of the degassing process largely being responsible for style of the eruption (Eichelberger 1995; Dingwell 1996). The physical properties of the magma, before and after crystallisation, plays a crucial role in determining how different monitoring signals will vary reflecting changes in the conduit or surface (Lavallée et al., this volume). Modelling of the conduit conditions for silicic magma has shown that an effusive and explosive solution are both possible for similar magma conditions (Kozono and Koyaguchi 2009).

Since the 1913 eruption, activity at Volcán de Colima has produced further cycles on a several year scale. Periods of effusive activity have been followed by several years of quiet. Little is known of the activity between 1913 and 1961, often no activity is acknowledged in the literature; however, the 350 m deep crater formed during the sub-Plinian event was being slowly filled (Saucedo et al., this volume; Connor et al., this volume). Once the crater rim was overtopped in 1961, a new lava flow descending the N flank clearly evidenced a new eruption. Between 1961 and 1998 there were five effusive episodes with a mean repose period of nine years between them. Saucedo et al. (this volume) gives further details of these eruptions.

Vulcanian eruptions are common at volcanoes producing intermediate lavas (andesite and dacite; definition based upon silica-content) and often occur regularly over extended periods of time (e.g. Santa Maria, Guatemala, Johnson et al. 2014; Sakurajima, Japan, Semeru, Indonesia, Iguchi et al. 2008; Tungurahua, Ecuador, Hall et al. 2015). They are often, though not exclusively, generated during dome emplacement. A simple model has been used to explain their generation, whereby a permeability reduction is a consequence of the crystallization of an ascending magma body at shallow depth (<1 km; Self et al. 1979). This leads to gas accumulation, due the continued decompression of the ascending magma, subsequently pressurization, then failure, producing the Vulcanian explosion. Inter-event time is largely determined by complex variations of physical magmatic properties that create multiple feedback loops, which modify the magma viscosity, permeability and buoyancy (Gottsmann et al. 2011; Dominguez et al. 2016).

Seismicity can also be used to gain an insight into the Vulcanian explosion process (e.g. Inza et al. 2014). In the case of Volcán de Colima, different models have been presented to account for the signals detected, including the long-period (LP) swarms observed as a precursor (Varley et al. 2010b) and the low frequency onset of the majority of explosions (Zobin et al. 2009; Varley et al. 2010a; Varley 2016). Cole et al. (2014) have divided events at Soufrière Hills Volcano, Montserrat into two types: Vulcanian explosions and ash-venting events, based upon seismicity and the duration. Shear-induced fragmentation along the conduit wall (Gonnermann & Manga 2003) is cited as the mechanism for the ash-venting, such as that identified at Volcán de Colima during the 2005 activity (Varley et al. 2010b). Explosions have been divide into two categories at Volcán de Colima using both the associated seismicity (Varley et al. 2006) and the thermal signal from the eruption column (Webb et al. 2014).

6 Modelling the Eruptive Mechanism

Various questions arise from an examination of the activity at Volcán de Colima during recent years. What was the mechanism behind the large explosions of 2005? Why does the explosivity decrease during effusive episodes? Why did eruptive activity stop during June 2011 to December 2012?

It is becoming increasingly clear that volcanic systems are a far leap from the simplistic ideas presented in older textbooks. Volcanoes are likely to have more than one magma reservoir, each with distinct magma composition and volatile-contents. Evidence for this can be from geochemical or geophysical data (e.g. Moretti et al. 2013). At Volcán de Colima there has been widespread evidence of the heterogeneity of the magma, largely associated with variations in degassing (Reubi et al. 2013). There is an increasing body of evidence pointing to a complex system of magma storage and ascent. A study of disequilibria between members of the U decay chain concluded that multiple magma batches ascend via different pathways and at different velocities in a complex system, possibly stalling for periods of up to 10 years (Reubi et al. 2015). Multiple magma reservoirs, with a contrast in volatile-contents, has been used to explain some observed inconsistencies, such as the variation in the porosity of ash particles and the eruption of magma batches with contrasting ascent rates (Cassidy et al. 2015); results of satellite-based InSAR deformation studies suggest multiple sources for magma that ascended in early 2013 (Salzer et al. 2014) and variations in the characteristics of melt inclusions captured in pyroxene crystals (Reubi et al. 2013).

Data from the monitoring network needs to be processed quickly to provide insight into the evolution of activity, and to forecast changes in the eruptive scenario, which might necessitate actions to reduce the risk related to the different volcanic hazards. Interpretation of multiple data sets is often a challenge and needs to be grounded in models explaining the internal processes. As summarized in Chap. 10, experimental work is significantly expanding the potential for creating models whose divergence from the real situation is increasingly negligible. Work at Volcán de Colima is at the forefront of this revolution in our understanding of volcanic systems. The importance of the relationship between permeability of magma and other parameters is ever clearer, since it is these relationships that govern how the gas can escape from the magma, which defines the eruption style (Heap et al. 2014; Farquharson et al. 2015, 2016).

The close examination of eruption products and experimental determination of critical magma parameters has become as critical as the new technologies being applied in volcano monitoring. The objective is to improve our capacity to not only generate, but successfully interpret observational data, and ultimately refine our eruption forecasts.

Acknowledgements I wish to thank the many CIIV students who have devoted their time in Colima to obtain and process data. In particular Sam Thiele for the creation of the 3D thermal and photographic models, and Will Hutchison for the thermal analysis during 2007–2010. Thanks to RESCO (Gabriel Reyes & Raúl Arámbula) for the dedication required to maintain a telemetered seismic network and for access to data. The University of Hamburg (Matthias Hort & Lea Scharff) installed the Doppler radar system. Thanks to John Stevenson for helping me launch many of the monitoring efforts. Some SO_2 flux data presented was collected and processed by CENAPRED and Robin Campion (UNAM).

Bibliography

Arámbula-Mendoza, R., Lesage, P., Valdés-González, C., Varley, N.R., Reyes-Dávila, G., Navarro, C.: Seismic activity that accompanied the effusive and explosive eruptions during the 2004-2005 period at Volcán de Colima, Mexico. J. Volcanol. Geoth. Res. **205**(1–2), 30–46 (2011)

Badalamenti, B., Bruno, N., Caltabiano, T., Di Gangi, F., Giammanco, S. and Salerno, G.: Continuous soil CO_2 and discrete plume SO_2 measurements at Mt. Etna (Italy) during 1997–2000: a contribution to volcano monitoring. Bull Volcanol. **66**(1), 80–89 (2004)

Bebbington, M.: Models for temporal volcanic hazard. Stat. Volcanol. **1**, 1–24 (2012)

Bernstein, M.L., Varley, N.R.: Growth processes at the Colima dome 2007–2008: results of thermal imaging (abstract). In: Proceedings of Volcán de Colima XI International Conference. Universidad de Colima, Mexico, 14, 3–6 Feb 2009

Boichu, M., Oppenheimer, C., Tsanev, V., Kyle, P.R.: High temporal resolution SO_2 flux measurements at Erebus volcano, Antarctica. J. Volcanol. Geoth. Res. **190**(3–4), 325–336 (2010)

Capra, L., Macías, J.L., Cortés, A., Dávila, N., Saucedo, R., Osorio-Ocampo, S., Arce, J.L., Gavilanes-Ruiz, J.

C., Corona-Chávez, P., García-Sánchez, L., Sosa-Ceballos, G., Vázquez, R.: Preliminary report on the July 10–11, 2015 eruption at Volcán de Colima: Pyroclastic density currents with exceptional runouts and volume. J. Volcanol. Geoth. Res. **310**, 39–49 (2016)

Cardellini, C., Chiodini, G., Frondini, F., Avino, R., Bagnato, E., Caliro, S., Lelli, M., Rosiello, A.: Monitoring diffuse volcanic degassing during volcanic unrests: the case of Campi Flegrei (Italy). Sci. Rep. **7** (1), 6757 (2017)

Caricchi, L., Annen, C., Blundy, J., Simpson, G., Pinel, V.: Frequency and magnitude of volcanic eruptions controlled by magma injection and buoyancy. Nature Geosci **7**(2), 126–130 (2014)

Cassidy, M., Cole, P.D., Hicks, K.E., Varley, N.R., Peters, N., Lerner, A.H.: Rapid and slow: Varying magma ascent rates as a mechanism for Vulcanian explosions. Earth Planet. Sci. Lett. **420**, 73–84 (2015)

Chaussard, E., Amelung, F., Aoki, Y.: Characterization of open and closed volcanic systems in Indonesia and Mexico using InSAR time series. J. Geophys. Res. Solid Earth **118**(8), 3957–3969 (2013)

Clarke, A.B., Voight, B., Neri, A., Macedonio, G.: Transient dynamics of vulcanian explosions and column collapse. Nature **415**, 897–901 (2002)

Cole, P.D., Smith, P., Komorowski, J.-C., Alfano, F., Bonadonna, C., Stinton, A., Christopher, T., Odbert, H. M. and Loughlin, S., 2014. Ash venting occurring both prior to and during lava extrusion at Soufrière Hills Volcano, Montserrat, from 2005 to 2010. In: Wadge, G., Robertson, R.E.A., Voight, B. (eds) The Eruption of Soufrière Hills Volcano, Montserrat from 2000 to 2010; Geological Society, London, Memoirs, pp. 71–92

Coppola, D., Laiolo, M., Cigolini, C., Donne, D.D., Ripepe, M.: Enhanced volcanic hot-spot detection using MODIS IR data: Results from the MIROVA system. Geological Society, London, Special Publications **426**(1), 181–205 (2016)

Dávila, N., Capra, L., Gavilanes-Ruiz, J.C., Varley, N., Norini, G., Gómez-Vázquez, Á.: Recent lahars at Volcán de Colima (Mexico): Drainage variation and spectral classification. J. Volcanol. Geoth. Res. **165**(3–4), 127–141 (2007)

de Moor, J.M., Aiuppa, A., Pacheco, J., Avard, G., Kern, C., Liuzzo, M., Martínez, M., Giudice, G., Fischer, T. P.: Short-period volcanic gas precursors to phreatic eruptions: Insights from Poás Volcano, Costa Rica. Earth Planet. Sci. Lett. **442**, 218–227 (2016)

Deligne, N.I., Coles, S.G., Sparks, R.S.J.: Recurrence rates of large explosive volcanic eruptions. J. Geophys. Res. **115**(B6), B06203 (2010)

Dingwell, D.B.: Volcanic dilemma: flow or blow? Science **273**, 1054–1055 (1996)

Dominguez, L., Pioli, L., Bonadonna, C., Connor, C.B., Andronico, D., Harris, A.J.L., Ripepe, M.: Quantifying unsteadiness and dynamics of pulsatory volcanic activity. Earth Planet. Sci. Lett. **444**, 160–168 (2016)

Druitt, T.H., Avard, G., Bruni, G., Lettieri, P., Maez, F.: Gas retention in fine-grained pyroclastic flow materials at high temperatures. Bull. Volcanol. **69**, 881–901 (2007)

Dufek, J., Bergantz, G.W.: Dynamics and deposits generated by the Kos Plateau Tuff eruption: Controls of basal particle loss on pyroclastic flow transport. Geochem. Geophys. Geosyst. **8**(12), Q12007 (2007)

Eichelberger, J.C.: Silicic volcanism: ascent of viscous magmas from crustal reservoirs. Annu. Rev. Earth Planet Sci. **23**, 41–63 (1995)

Farquharson, J., Heap, M.J., Varley, N.R., Baud, P., Reuschlé, T.: Permeability and porosity relationships of edifice-forming andesites: a combined field and laboratory study. J. Volcanol. Geoth. Res. **297**, 52–68 (2015)

Farquharson, J.I., Heap, M.J., Lavallée, Y., Varley, N.R., Baud, P.: Evidence for the development of permeability anisotropy in lava domes and volcanic conduits. J. Volcanol. Geoth. Res. **323**, 163–185 (2016)

Gonnermann, H.M., Manga, M.: Explosive volcanism may not be an inevitable consequence of magma fragmentation. Nature **426**, 432–435 (2003)

Gottsmann, J., De Angelis, S., Fournier, N., Van Camp, M., Sacks, S., Linde, A., Ripepe, M.: On the geophysical fingerprint of Vulcanian explosions. Earth Planet. Sci. Lett. **306**(1–2), 98–104 (2011)

Hall, M.L., Steele, A.L., Bernard, B., Mothes, P.A., Vallejo, S.X., Douillet, G.A., Ramón, P.A., Aguaiza, S.X., Ruiz, M.C.: Sequential plug formation, disintegration by Vulcanian explosions, and the generation of granular Pyroclastic Density Currents at Tungurahua volcano (2013–2014), Ecuador. J. Volcanol. Geoth. Res. **306**, 90–103 (2015)

Heap, M.J., Lavallée, Y., Petrakova, L., Baud, P., Reuschlé, T., Varley, N.R. and Dingwell, D.B.: Microstructural controls on the physical and mechanical properties of edifice-forming andesites at Volcán de Colima, Mexico. J. Geophys. Res. Solid Earth. **119** (4). 2013JB010521 (2014)

Hirabayashi, J., Ohba, T., Nogami, K.: Discharge rate of SO_2 from Unzen volcano, Kyushu. Jpn. Geophys. Res. Lett. **22**(13), 1709–1712 (1995)

Horton, K., Williams-Jones, G., Garbeil, H., Elias, T., Sutton, A., Mouginis-Mark, P., Porter, J., Clegg, S.: Real-time measurement of volcanic SO_2 emissions: validation of a new UV correlation spectrometer (FLYSPEC). Bull. Volcanol. **68**(4), 323–327 (2006)

Hutchison, W., Varley, N., Pyle, D.M, Mather, T.A.: Airborne thermal remote Sensing of the Volcán de Colima (Mexico) Lava Dome from 2007 to 2010 In: Pyle, D.L., Mather, T.A., (eds) Remote sensing of volcanoes & volcanic processes: integrating observation & modelling. Geol. Soc. Lond. (2013)

Iguchi, M., Yakiwara, H., Tameguri, T., Hendrasto, M., Hirabayashi, J.-I.: Mechanism of explosive eruption revealed by geophysical observations at the Sakurajima, Suwanosejima and Semeru volcanoes. J. Volcanol. Geoth. Res. **178**(1), 1–9 (2008)

Inza, L.A., Métaxian, J.P., Mars, J.I., Bean, C.J., O'Brien, G.S., Macedo, O., Zandomeneghi, D.: Analysis of dynamics of vulcanian activity of Ubinas volcano,

using multicomponent seismic antennas. J. Volcanol. Geoth. Res. **270**, 35–52 (2014)
James, M.R., Varley, N.: Identification of structural controls in an active lava dome with high resolution DEMs: Volcán de Colima. Mexico. Geophys. Res. Lett. **39**(22), L22303 (2012)
Jiménez, Z., Reyes, G., Espíndola, J.M.: The July 1994 episode of seismic activity at Colima Volcano, Mexico. J. Volcanol. Geoth. Res. **64**, 321–326 (1995)
Johnson, J.B., Lyons, J.J., Andrews, B.J., Lees, J.M.: Explosive dome eruptions modulated by periodic gas-driven inflation. Geophys. Res. Lett. **41**(19), 6689–6697 (2014)
Kendrick, J.E., Lavallée, Y., Varley, N., Wadsworth, F. B., Lamb, O.D, Vasseur, J.: Blowing off steam: Tuffisite formation as a regulator for lava dome eruptions. Front. Earth Sci.vol. 4 https://doi.org/10.3389/feart.2016.00041(2016)
Kern, C., Deutschmann, T., Vogel, L., Wöhrbach, M., Wagner, T., Platt, U.: Radiative transfer corrections for accurate spectroscopic measurements of volcanic gas emissions. Bull. Volc. **72**(2), 233–247 (2010)
Koeppen, W., Pilger, E., Wright, R.: Time series analysis of infrared satellite data for detecting thermal anomalies: a hybrid approach. Bull. Volc. **73**(5), 577–593 (2011)
Kozono, T., Koyaguchi, T.: Effects of relative motion between gas and liquid on 1-dimensional steady flow in silicic volcanic conduits: 2. Origin of diversity of eruption styles. J. Volcanol. Geoth. Res. **180**(1), 37–49 (2009)
Lachowycz, S.M., Pyle, D.M., Mather, T.A., Varley, N. R., Odbert, H.M., Cole, P.D., Reyes-Dávila, G.A.: Long-range correlations identified in time-series of volcano seismicity during dome-forming eruptions using detrended fluctuation analysis. J. Volcanol. Geoth. Res. **264**, 197–209 (2013)
Lachowycz, S.M., Cosma, I.A., Pyle, D.M., Mather, T.A., Rodgers, M., Varley, N.R., Odbert, H.M., Cole, P.D., Reyes-Dávila, G.A.:. Analysis of uncertainty in time-series of volcano seismicity during dome-forming eruptions using Shannon entropy. Uncertainty in Geoscience workshop, Buffalo. http://geohazards.buffalo.edu/wp-content/uploads/sites/39/2016/03/Combined.pdf (2016)
Lamb, O.D., Varley, N.R., Mather, T.A., Pyle, D.M., Smith, P.J., Liu, E.J.: Multiple timescales of cyclical behaviour observed at two dome-forming eruptions. J. Volcanol. Geoth. Res. **284**, 106–121 (2014)
Lavallée, Y., Varley, N., Alatorre-Ibarguengoitia, M., Hess, K.-U., Kueppers, U., Mueller, S., Richard, D., Scheu, B., Spieler, O., Dingwell, D.: Magmatic architecture of dome-building eruptions at Volcán de Colima. Mexico. Bull Volcanol **73**(8), 1–12 (2011)
Lowenstern, J.B., Smith, R.B., Hill, D.P.: Monitoring super-volcanoes: geophysical and geochemical signals at Yellowstone and other large caldera systems. Philos. Trans. Roy. Soc. A Math. Phys. Eng. Sci. **364**(1845), 2055–2072 (2006)
Luhr, J.F.: Petrology and geochemistry of the 1991 and 1998–1999 lava flows from Volcán de Colima, Mexico: implications for the end of the current eruptive cycle. J. Volcanol. Geoth. Res. **117**(1–2), 169–194 (2002)
Macías, J.L., Capra, L., Arce, J.L., Espíndola, J.M., García-Palomo, A., Sheridan, M.F.: Hazard map of El Chichón volcano, Chiapas, México: Constraints posed by eruptive history and computer simulations. J. Volcanol. Geoth. Res. **175**(4), 444–458 (2008)
Mendoza-Rosas, A.T., De la Cruz-Reyna, S.: A statistical method linking geological and historical eruption time series for volcanic hazard estimations: applications to active polygenetic volcanoes. J. Volcanol. Geoth. Res. **176**(2), 277–290 (2008)
Moretti, R., Arienzo, I., Civetta, L., Orsi, G., Papale, P.: Multiple magma degassing sources at an explosive volcano. Earth Plan. Sci. Lett. **367**, 95–104 (2013)
Mueller, S.B., Varley, N.R., Kueppers, U., Lesage, P., Reyes Davila, G.Á., Dingwell, D.B.: Quantification of magma ascent rate through rockfall monitoring at the growing/collapsing lava dome of Volcán de Colima, Mexico. Solid Earth **4**(2), 201–213 (2013)
Murray, J.B., Ramirez Ruiz, J.J.: Long-term predictions of the time of eruptions using remote distance measurement at Volcán de Colima, Mexico. J. Volcanol. Geoth. Res. 117(1–2), 79–89 (2002)
Navarro-Ochoa, C., Gavilanes-Ruiz, J.C., Cortes-Cortes, A.: Movement and emplacement of lava flows at Volcán de Colima, Mexico: November 1998-February 1999. J. Volcanol. Geoth. Res. **117**(1–2), 155–167 (2002)
Newhall, C.G., Self, S.: The volcanic explosivity index (VEI) an estimate of explosive magnitude for historical volcanism. J. Geophys. Res. Oceans **87**(C2), 1231–1238 (1982)
Pinel, V., Hooper, A., De la Cruz-Reyna, S., Reyes-Davila, G., Doin, M.P., Bascou, P.: The challenging retrieval of the displacement field from InSAR data for andesitic stratovolcanoes: Case study of Popocatepetl and Colima Volcano, Mexico. J. Volcanol. Geoth. Res. **200**(1–2), 49–61 (2011)
Preece, K., Gertisser, R., Barclay, J., Charbonnier, S.J., Komorowski, J.-C., Herd, R.A.: Transitions between explosive and effusive phases during the cataclysmic 2010 eruption of Merapi volcano, Java. Indonesia. Bull. Volcanol. **78**(8), 1–16 (2016)
Reubi, O., Blundy, J., Varley, N.R.: Volatiles contents, degassing and crystallisation of intermediate magmas at Volcán de Colima, Mexico, inferred from melt inclusions. Contrib. Miner. Petrol. **165**(6), 1087–1106 (2013)
Reubi, O., Sims, K.W.W., Varley, N., Reagan, M.K., Eikenberg, J.: Timescales of degassing and conduit dynamics inferred from 210Pb–226Ra disequilibria in Volcán de Colima 1998–2010 andesitic magmas. In: Caricchi, L., Blundy, J.D. (eds.) Chemical, Physical and Temporal Evolution of Magmatic Systems, pp. 189–206. Geological Society, Special Publications, London (2015)
Reyes-Dávila, G.A., Arámbula-Mendoza, R., Espinasa-Pereña, R., Pankhurst, M.J., Navarro-Ochoa, C.,

Savov, I., Vargas-Bracamontes, D.M., Cortés-Cortés, A., Gutiérrez-Martínez, C., Valdés-González, C., Domínguez-Reyes, T., González-Amezcua, M., Martínez-Fierros, A., Ramírez-Vázquez, C.A., Cárdenas-González, L., Castañeda-Bastida, E., Vázquez Espinoza de los Monteros, D.M., Nieto-Torres, A., Campion, R., Courtois, L., Lee, P.D.: Volcán de Colima dome collapse of July, 2015 and associated pyroclastic density currents. J. Volcanol. Geoth. Res. vol. 320, 100–106 (2016)

Robin, C., Camus, G., Gourgaud, A.: Eruptive and magmatic cycles at Fuego de Colima volcano (Mexico). J. Volcan. & Geotherm. Res. **45**, 209–225 (1991)

Rodríguez-Elizarrarás, S., Siebe, C., Komorowski, J.-C., Espíndola, J.M., Saucedo, R.: Field observations of pristine block-and ash-flow deposits emplaced April 16–17, 1991 at Volcán de Colima. Mexico. J. Volcan. & Geotherm. Res. **48**, 399–412 (1991)

Salzer, J.T., Nikkhoo, M., Walter, T.R., Sudhaus, H., Reyes-Dávila, G., Bretón, M. and Arámbula, R.: Satellite radar data reveal short-term pre-explosive displacements and a complex conduit system at Volcán de Colima, Mexico. Front. Earth Sci. **2** (2014)

Savov, I.P., Luhr, J.F., Navarro-Ochoa, C.: Petrology and geochemistry of lava and ash erupted from Volcán Colima, Mexico, during 1998–2005. J. Volcanol. Geoth. Res. **174**(4), 241–256 (2008)

Scharff, L., Hort, M.K., Varley, N. R. and Herzog, M., Installation of a permanent doppler radar monitoring system at Colima Volcano, Mexico, and its use for eruption cloud modelling. American Geophysical Union, Fall Meeting 2011, abstract V33A-2611 (2011)

Scharff, L., Hort, M. and Varley, N.R.: Pulsed vulcanian explosions: a characterization of eruption dynamics using Doppler radar. Geology. G36705.36701 (2015)

Self, S., Wilson, L., Nairn, I.: Vulcanian eruption mechanisms. Nature **277**, 440–443 (1979)

Sorge, J.: Characterizing volcanic behaviour using thermal remote sensing and other time series data, 2000–2009, Volcán de Colima, Mexico. MSc Thesis, Department of Earth Sciences, Simon Fraser University, BC, Canada (2011)

Stevenson, J.A., Varley, N.: Fumarole monitoring with a handheld infrared camera: Volcán de Colima, Mexico, 2006–2007. J. Volcanol. Geoth. Res. **177**(4), 911–924 (2008)

Sulpizio, R., Capra, L., Sarocchi, D., Saucedo, R., Gavilanes-Ruiz, J.C., Varley, N.R.: Predicting the block-and-ash flow inundation areas at Volcán de Colima (Colima, Mexico) based on the present day (February 2010) status. J. Volcanol. Geoth. Res. **193** (1–2), 49–66 (2010)

Taran, Y.A., Gavilanes, J.C., Cortes, A., Armienta, M.A.: Chemical precursors to the 1998–1999 eruption of Colima volcano. Mexico. Revista Mexicana de Ciencias Geológicas **17**(2), 112–126 (2000)

Thiele, S.T., Varley, N., James, M.R.: Thermal photogrammetric imaging: A new technique for monitoring dome eruptions. J. Volcanol. Geoth. Res. **337**, 140–145 (2017)

Varley, N.R., Taran, Y.A.: Degassing processes of Popocatépetl and Volcán de Colima, Mexico. In: Oppenheimer, C., Pyle, D.M., Barclay, J. (eds.) Volcanic Degassing, pp. 263–280. Geological Society, London (2003)

Varley, N.R., Johnson, J., Ruiz, M., Reyes-Dávila, G.A., Martin, K., Applying statistical analysis to understanding the dynamics of volcanic explosions. In: Mader,H.M., Coles, S.G., Connor, C.B., Connor, L.J., (eds) Statistics in Volcanology. Special publication of IAVCEI, pp. 57–76 (2006)

Varley, N., Arámbula-Mendoza, R., Reyes-Dávila, G., Sanderson, R., Stevenson, J.: Generation of Vulcanian activity and long-period seismicity at Volcán de Colima, Mexico. J. Volcanol. Geoth. Res. **198**(1–2), 45–56 (2010a)

Varley, N., Arámbula-Mendoza, R., Reyes-Dávila, G., Stevenson, J., Harwood, R.: Long-period seismicity during magma movement at Volcán de Colima. Bull. Volcanol. **72**(9), 1093–1107 (2010b)

Varley, N.: Comment on "Field and seismic evaluation of the block-and-ash flows emplaced from eruption columns of the 2005 Vulcanian explosions at Volcán de Colima, Mexico" by Zobin et al. Bull Volcanol 78:27. **78**(10), 1–4 (2016)

Varley, N.: Riesgos Volcánicos. Universidad de Colima, Protección Civil del Estado de Colima, Atlas de Riesgos del Estado de Colima (2017)

Vázquez, R., Capra, L., Caballero, L., Arámbula-Mendoza, R., Reyes-Dávila, G.: The anatomy of a lahar: Deciphering the 15th September 2012 lahar at Volcán de Colima, Mexico. J. Volcanol. Geoth. Res. **272**, 126–136 (2014)

Vázquez, R., Capra, L., Coviello, V.: Factors controlling erosion/deposition phenomena related to lahars at Volcán de Colima, Mexico. Nat. Hazards Earth Syst. Sci. **16**(8), 1881–1895 (2016)

Webb, E.B., Varley, N.R., Pyle, D.M., Mather, T.A.: Thermal imaging and analysis of short-lived Vulcanian explosions at Volcán de Colima, Mexico. J. Volcanol. Geoth. Res. **278–279**, 132–145 (2014)

Zobin, V.M., Luhr, J.F., Taran, Y.A., Breton, M., Cortes, A., De La Cruz-Reyna, S., Dominguez, T., Galindo, I., Gavilanes, J.C., Muniz, J.J.: Overview of the 1997–2000 activity of Volcán de Colima, Mexico. J. Volcanol. Geoth. Res. **117**(1–2), 1–19 (2002)

Zobin, V.M., Reyes, G.A., Guevara, E., Bretón, M.: Scaling relationship for Vulcanian explosions derived from broadband seismic signals. J Geophys Res, **114**: B03203, doi:03210.01029/02008JB005983 (2009)

Zobin, V.M., Ramírez, J.J., Santiago, H., Alatorre, E., Navarro, C. Bretón, M.: Resolution of different-scale deformation data measured at volcanoes and its importance for eruption monitoring: Volcán de Colima, Mexico. In: Zobin, V.M., (ed) Complex Monitoring of Volcanic Activity. Nova Science Publishers, Inc., pp. 137–151 (2013)

Seismic Activity Associated with Volcán de Colima

Raúl Arámbula-Mendoza, Gabriel Reyes-Dávila, Tonatiuh Domínguez-Reyes, Dulce Vargas-Bracamontes, Miguel González-Amezcua, Alejandro Martínez-Fierros and Ariel Ramírez-Vázquez

Abstract

Many different types of seismic events are registered simultaneously at Volcán de Colima, reflecting the variety of physical processes that occur at the same time. Formal recording of seismic activity began in 1989, when seismic short-period sensors were installed. This has been expanded to a network of broadband sensors currently working on the volcano. Seismicity associated with the most recent volcanic activity has consisted of Volcano-Tectonic events (VTs) and/or High Frequency events (HFs), Long-Period events (LPs), tremor, explosions with associated Very Long- Period events (VLPs), rockfalls, pyroclastic flows and lahars. The time evolution of these signals varies along with the depth and its characteristics. The seismicity and related energy release have been a good indicator of the magnitude of volcanic activity. Combined with other observations, seismic monitoring has been a very efficient tool for forecasting volcanic activity and helping to reduce volcanic risk at Volcán de Colima.

Keywords

Volcán de Colima · Volcano seismicity
Seismic monitoring · Seismic events
Forecast

1 Introduction

The objective of this chapter is to give an overview of the seismicity at Volcán de Colima and its relationship with volcanic activity. The history of the seismic network is reviewed, along with the different types of seismic signal and how their presence is being interpreted in the context of eruptive activity.

1.1 Seismic Monitoring Network

The Colima seismic network (*Red Sísmica Telemétrica del Estado de Colima* or RESCO) was established at the Universidad de Colima and has been monitoring seismic activity of Volcán de Colima since 1989, when five telemetered seismic stations on and around the volcano were installed. Each station was equipped with a short-period (Ts = 1 s) vertical seismometer,

R. Arámbula-Mendoza (✉) · G. Reyes-Dávila
T. Domínguez-Reyes · M. González-Amezcua
A. Martínez-Fierros · A. Ramírez-Vázquez
Centro Universitario de Estudios Vulcanológicos (CUEV), Universidad de Colima, Av. Bernal Díaz del Castillo no. 340, Col. Villas San Sebastián, 28045 Colima, C.P., Mexico
e-mail: raul_arambula@ucol.mx

D. Vargas-Bracamontes
CONACYT-CUEIV, Universidad de Colima. Av, Bernal Díaz del Castillo no, 340, Col. Villas San Sebastián, 28045 Colima, C.P., Mexico

N. Varley et al. (eds.), *Volcán de Colima*, Active Volcanoes of the World,
https://doi.org/10.1007/978-3-642-25911-1_1

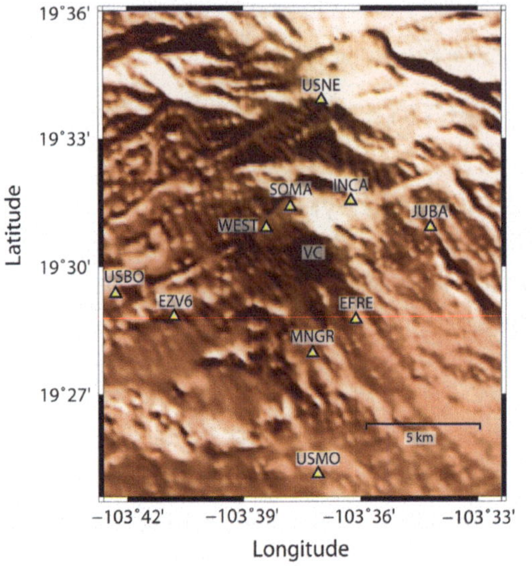

Fig. 1 Current seismic monitoring network at Volcán de Colima operated by RESCO. The EZV6 is a short-period station. INCA, JUBA, MNGR and WEST are broadband stations. At USNE, SOMA and EFRE there are both short-period (EZV3, EZV4, EZV5) and broadband sensors. All the signals are sent by radio to RESCO-Universidad de Colima. Additionally, RESCO have others seismic stations for monitoring tectonic seismicity. 'VC = Volcán de Colima'

with analogue recording. The first episode of lava dome growth recorded by RESCO was in 1991.

During the 1998–1999 effusive episode, RESCO was able to record seismicity digitally, with continuous recording achieved since March 1999. The data generated has proved to be a powerful tool for monitoring and studying the volcanic activity. In recent years RESCO has been upgraded with more stations and broadband sensors. Figure 1, shows the current seismic network: four short-period SS-1 Ranger vertical seismometers, and five broadband seismometers Guralp CMG-6TD and one broadband Guralp CMG-40TD. Due to their location, stations MNGR, WEST and EZV6 also record the occurrence of lahars.

2 Seismicity Associated with Volcanic Activity

2.1 Volcano-Tectonics Events (VTs)

Volcano-Tectonic events (VTs) generally take place during the first stage of eruptive activity at volcanoes (White and McCausland 2015). At Volcán de Colima they appear in swarms, usually after extended periods of quiescence. Since the start of seismic recording of volcanic activity, the following episodes have produced this type of event: 1991 (Núñez-Cornú et al. 1994), 1994 (Jiménez et al. 1995), and 1997–1998 (Domínguez et al. 2001; Zobin et al. 2002a, b, c; Zamora-Camacho et al. 2007). Some of these events have been linked with the Tamazula fault (Domínguez et al. 2001; Zobin et al. 2002b). The tensional tectonic regimen associated with the Colima Graben can also produce VTs without a clear relationship with the ascent of magma. As is typical for this type of event, they have well-defined P and S phases (Fig. 2a), meaning their origin can be located. Their frequency content is mainly between 1 and 15 Hz (Fig. 2b, c), and their origin is typically at depths of up to 12 km below the crater (Zobin et al. 2002a, b, c). During the recent eruptive period from 2013 to present, a small number of isolated VTs have also been observed, probably associated with an open system, whereby the magma ascends without significant resistance. Also, with the more recently added broadband seismometers, high frequency events (HFs) without a clear S phase have been registered by the closest stations to the crater. These events are likely to be associated with lava dome growth as a brittle rupture process at shallow depth.

During the 1991 episode, VTs were located to the north of the summit, some of them relatively deep (7–11 km below the summit), underneath the saddle between Volcán de Colima and Nevado de Colima (Núñez-Cornú et al. 1994).

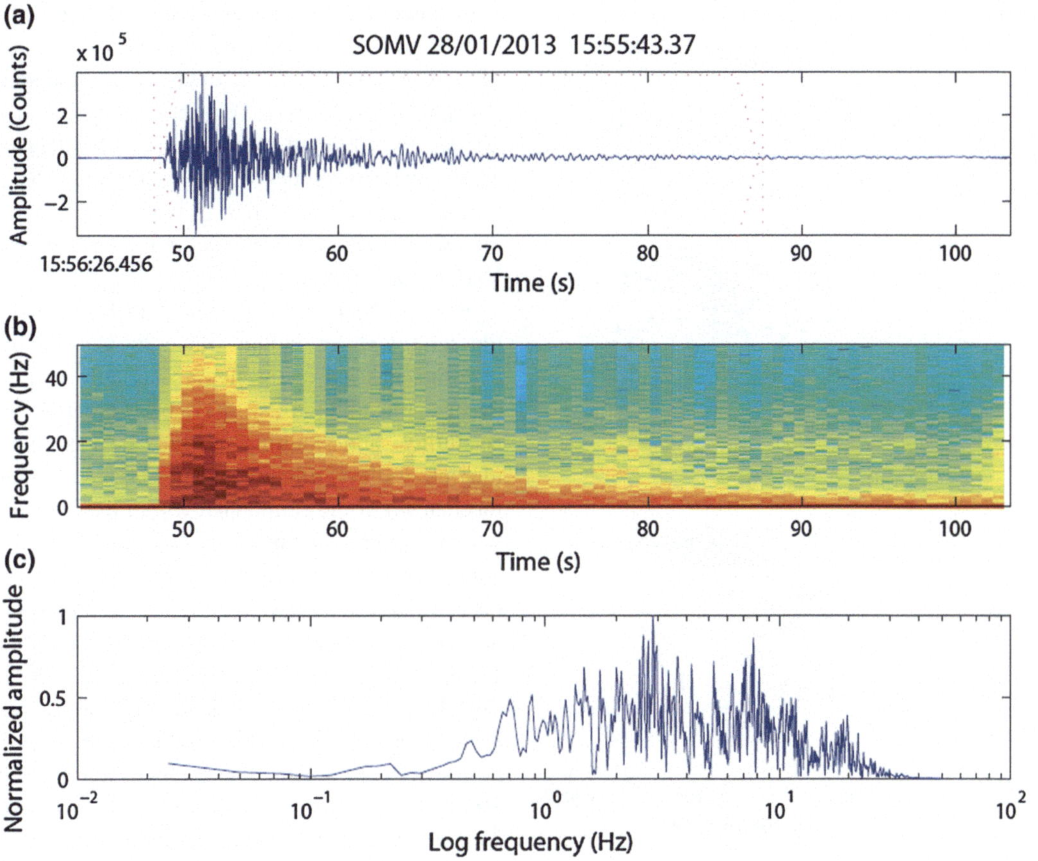

Fig. 2 Volcano-Tectonic event (VT) at SOMA (vertical component), on 28 January 2013. **a** Velocity signal, **b** spectrogram, where the exponential decay of the amplitude can be observed and **c** normalized amplitude spectrum of the signal

In 1994 the earthquakes had hypocentres towards the SSW of the summit and under the crater, with depths of up to 6 km (Jiménez et al. 1995). The location of the earthquakes during the years 1997–1998 lie within a 50 km^2 area including the active crater of Volcán de Colima and the region that extends towards Nevado de Colima, 5.5 km to the north (Fig. 3). During 1997–1998 the majority of hypocentres within this area did not exceed a depth of 5 km below sea level, and were mainly distributed within the volcanic edifices of Volcán de Colima and Nevado de Colima (Domínguez et al. 2001; Zobin et al. 2002a, b, c).

2.2 Long-Period Events (LPs)

Long-period events (LPs), are the events that most frequently occur at Volcán de Colima. They are seismic events of short duration (<1 min in most cases), with an emergent first arrival P phase, but no observable second S phase (Fig. 4a). Their frequency is mainly below 5 Hz (Fig. 4b, c), and they are commonly located at depths of less than 3 km (Jiménez et al. 1995; Arámbula-Mendoza 2011a; Petrosino et al. 2011).

LP events have been observed mainly prior to effusive activity and preceding explosions.

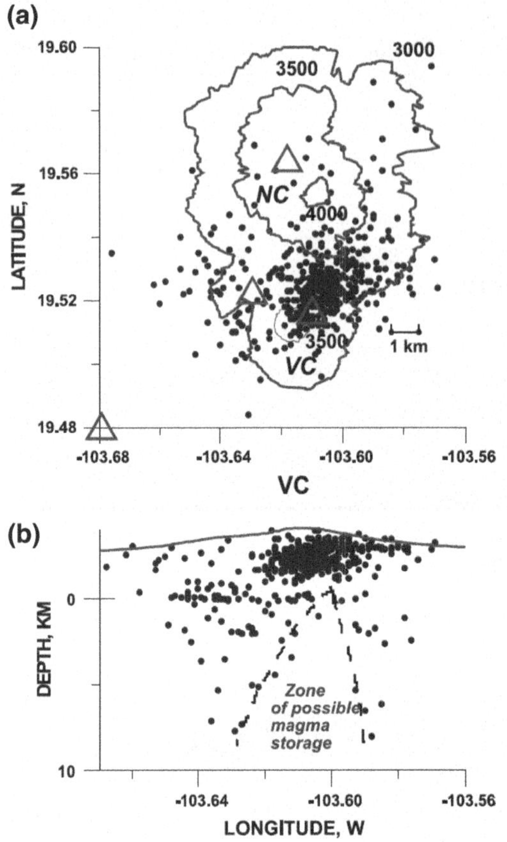

Fig. 3 **a** Location of epicentres for VTs located during the seismic crisis of 1997–1998. **b** Distribution of hypocentres for the same interval projected onto an E-W cross section through the summit of Volcán de Colima. An earthquake-free zone is outlined as a possible zone of magma storage. The error is about 0.6 km in depth (after Zobin et al. 2002a, b, c)

A number were detected during the third seismic crisis of the 1991 volcanic activity, coincident with an increase in fumarolic activity (Núñez-Cornú et al. 1994). Similarly they were detected during the seismic swarm that preceded the explosion of 1994 (Jiménez et al. 1995), and within the swarm of earthquakes that preceded the effusion of 1998. LPs preceded the 1999 Vulcanian explosions and were observed during the 2001–2003 effusive activity. During September 2004, an LP swarm announced the beginning of an effusive stage (Arámbula-Mendoza et al. 2011b). Between February and September of 2005, low amplitude LPs known in Colima as '*pulgas*' occurred preceding episodes of small lava dome growth and moderate Vulcanian explosions (Fig. 5, Varley et al. 2010a, b; Arámbula-Mendoza et al. 2011b). They appeared frequently during the 2007–2011 volcanic episode, as well as in January 2013 with the reactivation and fresh lava dome growth (see Varley, this volume).

Recent observations of abundant brittle–ductile fault textures within silicic lavas, such as andesite, has led to the hypothesis that seismic events, including LPs, may be triggered by fracturing and faulting within the erupting magma itself (Tuffen et al. 2008). A trigger mechanism of the earthquake within the conduit, which could be the breaking of magma along the conduit walls or within the dome itself, as has been proposed by various authors as an explanation for the generation of LP events (Goto 1999; Tuffen et al. 2003; Neuberg et al. 2006). The LPs observed at Volcán de Colima are probably generated by the fracturing of magma during magma ascent. Observations during the 2005 eruptive period indicated that repetitive sources are active for long durations in swarms lasting hours to days (Varley et al. 2010a, b; Arámbula-Mendoza et al. 2011b), generating LP families, as is the case with other volcanoes producing silicic lavas (e.g. Green and Neuberg et al. 2006; Neuberg et al. 2006). Twelve families were identified during the 2004–2005 episode (Arámbula-Mendoza et al. 2011b).

During the 2004–2005 effusive and explosive stages, another type of LP event was observed: monochromatic LP events or '*tornillos*' (Arámbula-Mendoza et al. 2011b). These events were characterised by having only one fundamental frequency and a slowly diminishing coda, and likely to be due to very strong pressure differences within the conduit. A popular model that explains the existence of this type of event is the resonance of fractures associated with the pressurization processes that precede Vulcanian explosions in some cases, for example at Galeras volcano (Stix et al. 1997; Kumagai and Chouet 1999).

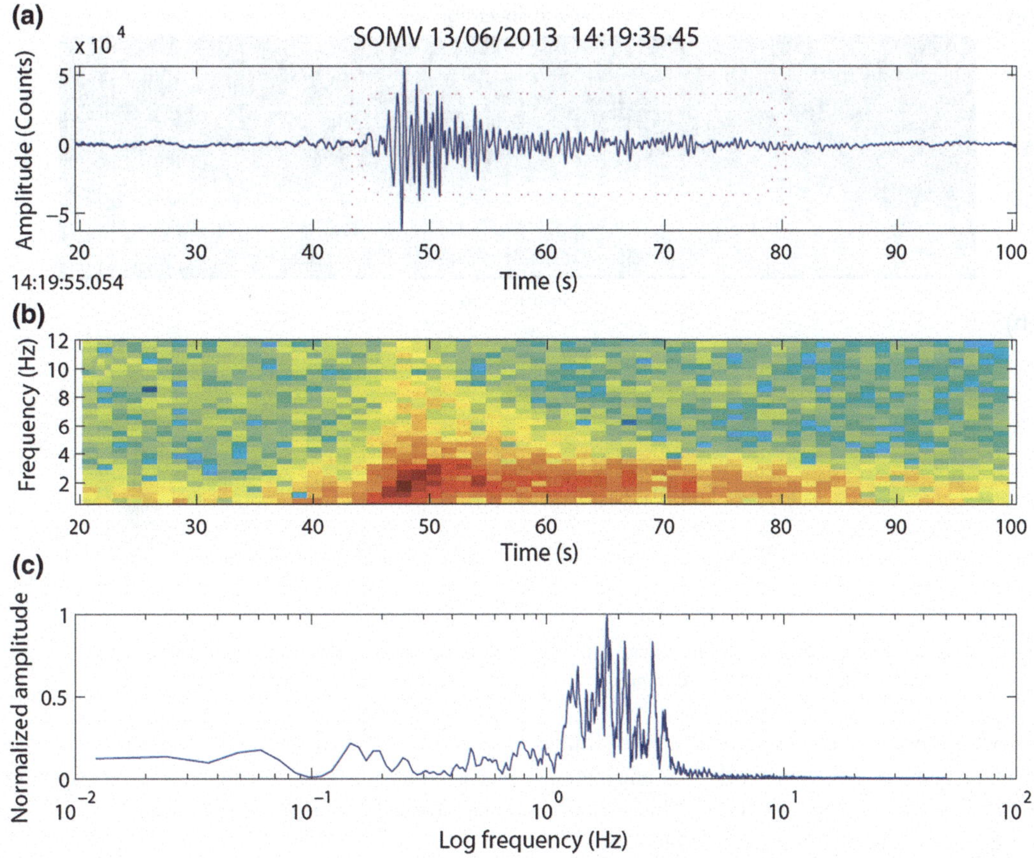

Fig. 4 Long-Period event (LP) at SOMA (vertical component) on 13 June 2013. **a** Velocity signal, **b** spectrogram and **c** normalized amplitude spectrum. These kinds of events have energy mainly between 1 and 5 Hz

2.3 Volcanic Tremor

Another type of event frequently observed at Volcán de Colima is volcanic tremor. It is a continuous signal that can last seconds to hours and sometimes days (Fig. 6a). The frequency content is mainly between 1 and 5 Hz, although this can be higher, up to 8 Hz, or as low as 0.5 Hz (Fig. 6b, c). At Volcán de Colima, three types of tremor have been observed: harmonic (with a clear fundamental frequency and harmonics; Fig. 6), spasmodic (with large variation in both the amplitude and frequency; Fig. 7), and pulsating (formed of a series of discrete events closely spaced in time; Fig. 8). For further information on the classification of tremor see Konstantinou and Schlindwein (2003).

Volcanic tremor at Volcán de Colima can be associated with every observed episode of volcanic activity, but variations in its characteristics have been registered. Typically, tremor is observed as a precursor, e.g. during the explosive episode of 1999, harmonic tremor appeared a few hours before the occurrence of large explosions. During many episodes, the tremor has continued up to the onset of the explosion. At times occurrences do not coincide with clear variations in eruptive activity, e.g. during the 2002 effusive period, an isolated period of medium to small amplitude monochromatic tremor was observed, which was likely a consequence of the degassing process or an interaction between the magma and groundwater. Some episodes were continuous with durations in excess of 24 h. During this

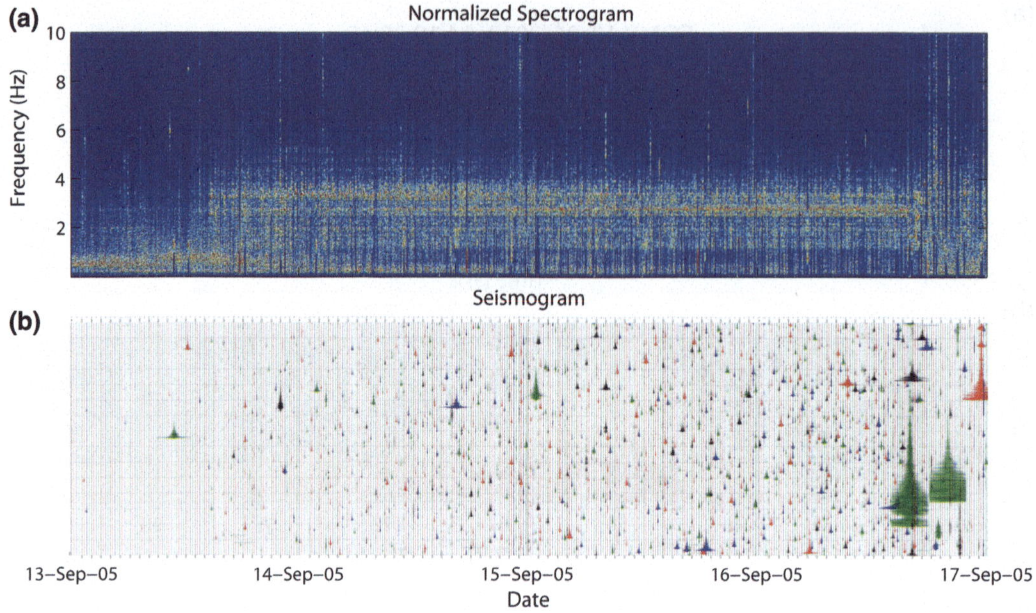

Fig. 5 **a** Normalized spectrogram and **b** seismogram from 13 to 16 September 2005, from short-period station (EZV4). Here the occurrence of low amplitude LPs ('*pulgas*') before an explosion can be observed

period, gliding frequencies were observed similar to those seen at Soufrière Hills volcano, Monserrat (Powell and Neuberg 2003). This variation in frequency is probably generated by changes in the pressure within the conduit and/or an opening and sealing processes. The source location of the tremor is identical to that of the LPs and explosions, mainly between 0 and 3 km below the crater (Arámbula-Mendoza 2011a), although some episodes of harmonic tremor in 2015 have been located at a greater depth.

There are various models that aim to explain the existence of tremor at volcanoes: (a) fluid-flow-induced oscillations of conduits transporting magmatic fluids; (b) excitation and resonance of fluid-filled cracks; (c) bubble growth or collapse due to hydrothermal boiling of groundwater; (d) a variety of models involving the oscillations of magma bodies with different geometries (summarized in Konstantinou and Schlindwein (2003)). Also, the similar frequency content of LPs and tremor suggest they are physically related, indeed tremor can be considered a continuous manifestation of LP events as suggested for other volcanoes (e.g. Neuberg et al. 2000). At Volcán of Colima, it is models (a), (b) and (c) that are the most likely processes to occur, but further research is necessary to understand the generation of tremor.

2.4 Explosion Events

Since the major eruption of 1913, volcanic activity has largely consisted of periods of effusive activity followed by explosive stages. The characteristics of the seismic signals associated with the explosive events vary depending on the style and size of the eruption. Commonly for Volcán de Colima, the seismic signals begin with an emergent low frequency phase (LF, Fig. 9a), followed by an impulsive high amplitude phase (Zobin et al. 2002a, b, c); this last phase can be attributed to magma fragmentation. The energy is concentrated at low frequencies (mainly 1–5 Hz) although we have observed high frequency content related to the occurrence of rockfalls and/or pyroclastic density currents related to the explosions (Fig. 9b, c). Sometimes the shockwave or airwave generates a high-frequency pulse in the seismogram (Fig. 9a).

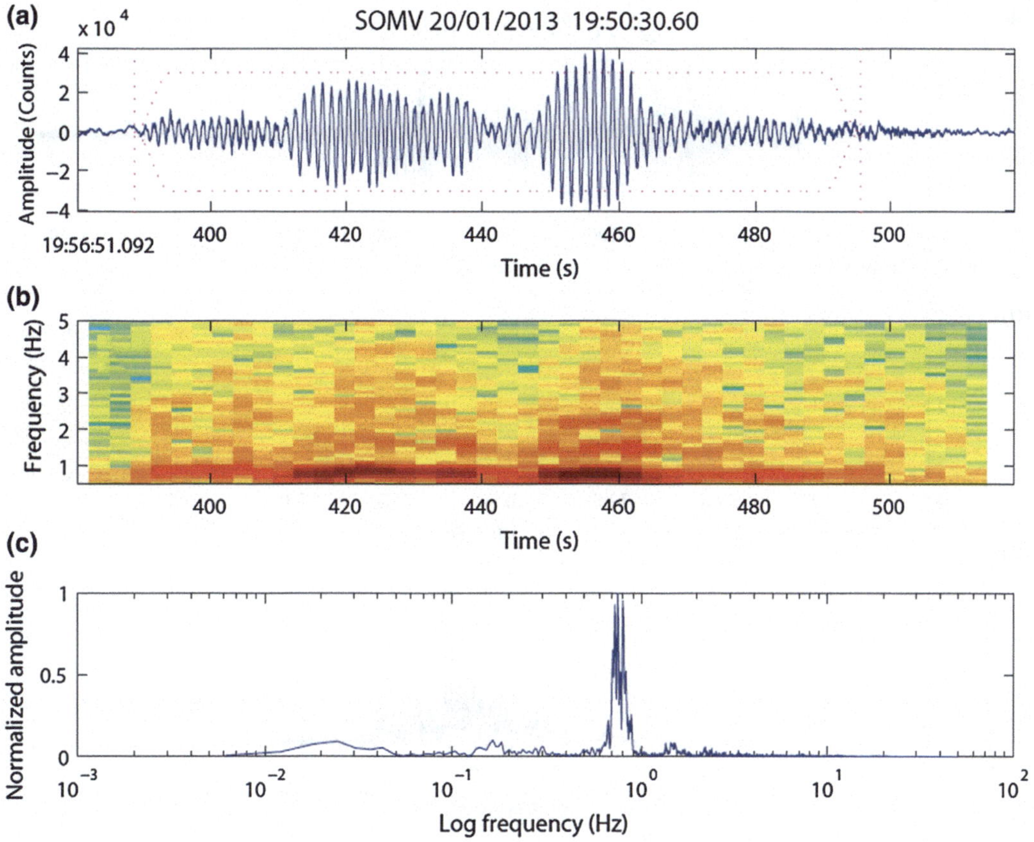

Fig. 6 Harmonic tremor at SOMA (vertical component) on 20 January 2013. **a** Velocity signal, **b** spectrogram and **c** normalized amplitude spectrum. In this case, this tremor is practically monochromatic

Explosive events at Volcán de Colima occur frequently when a fresh dome is emplaced in the crater. During cooling of either the lava dome or the magma column within the conduit there is an acceleration of crystallization, resulting in the formation of an impermeable cap within the vent at a relatively shallow level. Gas pressure then increases in the pore spaces beneath the cap until the lithostatic pressure above is exceeded, the cap fails, and a Vulcanian eruption occurs (Varley et al. 2010a). The first arrival of the seismic signal could reflect the breaking of the seal, but the explanation of the LF phase is still open to debate. Zobin et al. (2009) consider that it is due to fragmented magma movement within the conduit, whilst Varley et al. (2010b) suggest that after the initial rupture of the plug there is a rapid pressure drop and degassing of the magma, which generates the LF phase.

The majority of the larger explosive events tend to be associated with lava dome destruction. The largest explosions in the most recent periods of activity occurred in 1994, 1999, 2003, 2005, 2011, 2013, 2014 and 2015. In Table 1, the main Vulcanian explosions can be observed with the corresponding seismic energy calculated with the methodology proposed by Boatwright (1980), and subsequently used by Johnson and Aster (2005). The energy was calculated using only the vertical component of the EFRE station.

The largest explosions registered to date in 2005, with more than 30 significant events, the majority with pyroclastic flows generated by column collapse (Zobin et al. 2006b; Varley et al. 2010a, b; Varley, this volume). In a dense seismic array installed at Volcán de Colima, an apparent slowness variation of the first onset of the LF phase associated with the explosions was

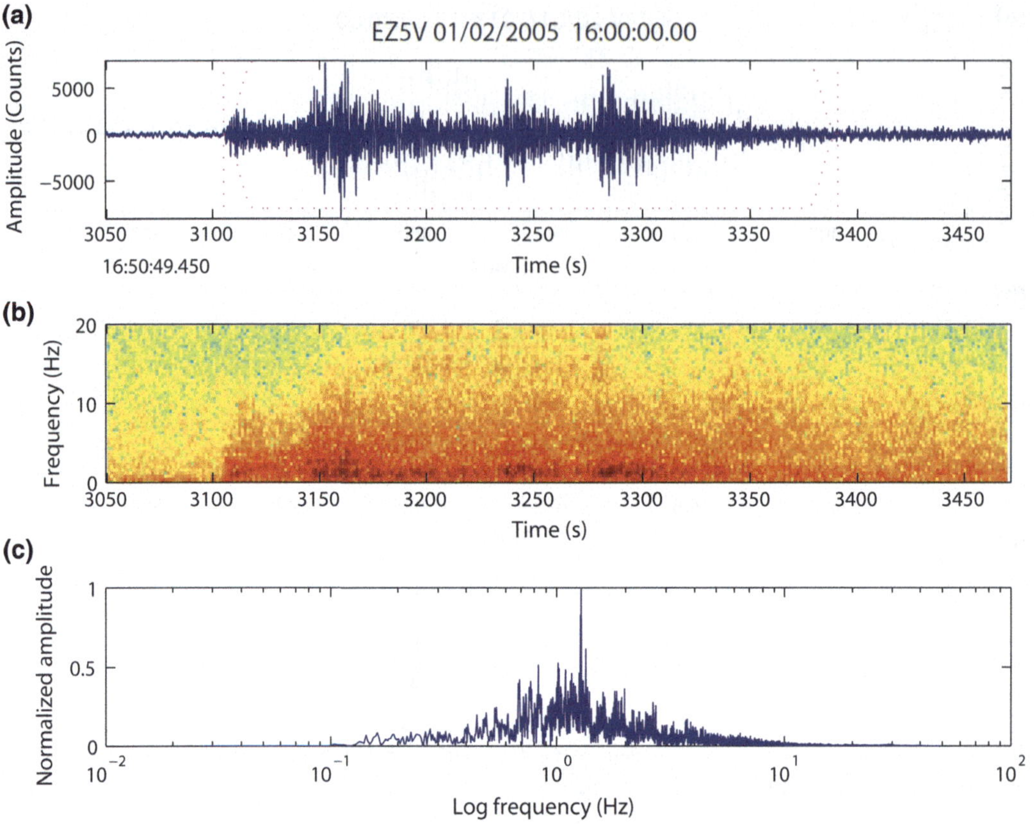

Fig. 7 Spasmodic tremor at EFRE (vertical component), on 01 February 2005, associated with an ash and gas emission. **a** Velocity signal, **b** spectrogram and **c** normalized amplitude spectrum

observed, with a possible upward migration of the source, the depth of which has been identified within a range of between 2.6 and 3.3 km below the crater (Palo et al. 2009). The location of these events in the last years, has been determined using a method based on seismic amplitudes (Battaglia and Aki 2003), whereby the attenuation of the seismic energy with distance is considered, the source being defined when the error between the theoretical model of attenuation and the observed data is a minimum for all the seismic stations. During recent activity, the start of the larger explosions (first 6 s of signal) have been localized between 3 and 0 km beneath the crater (Fig. 10).

Very-Long-Period events (VLPs) have also been recorded, associated with moderate-large Vulcanian explosions in 2003, 2005, 2011, 2013, 2014 and 2015. These events have been observed once the signal is filtered, leaving just low frequencies. The VLPs at Volcán de Colima have periods of between 10 and 30 s, whereas in other volcanoes these events have been seen to be characterized by larger periods, e.g. at Popocatépetl volcano, VLPs associated with Vulcanian explosions were observed with periods between 25 and 100 s (Arciniega-Ceballos et al. 1999). Other examples of VLPs associated with Vulcanian explosions have been observed at Stromboli, Italy (D’Auria et al. 2006), Tungurahua, Ecuador (Kumagai et al. 2010, 2011), Augustine, USA (Dawson et al. 2011) and Redoubt, USA (Haney et al. 2012).

In Fig. 11, seventeen of these events can be observed; regrettably they were only observed with a single broadband sensor installed in 2005 (EFRE). The similar waveforms represent the process of pressurization and depressurization

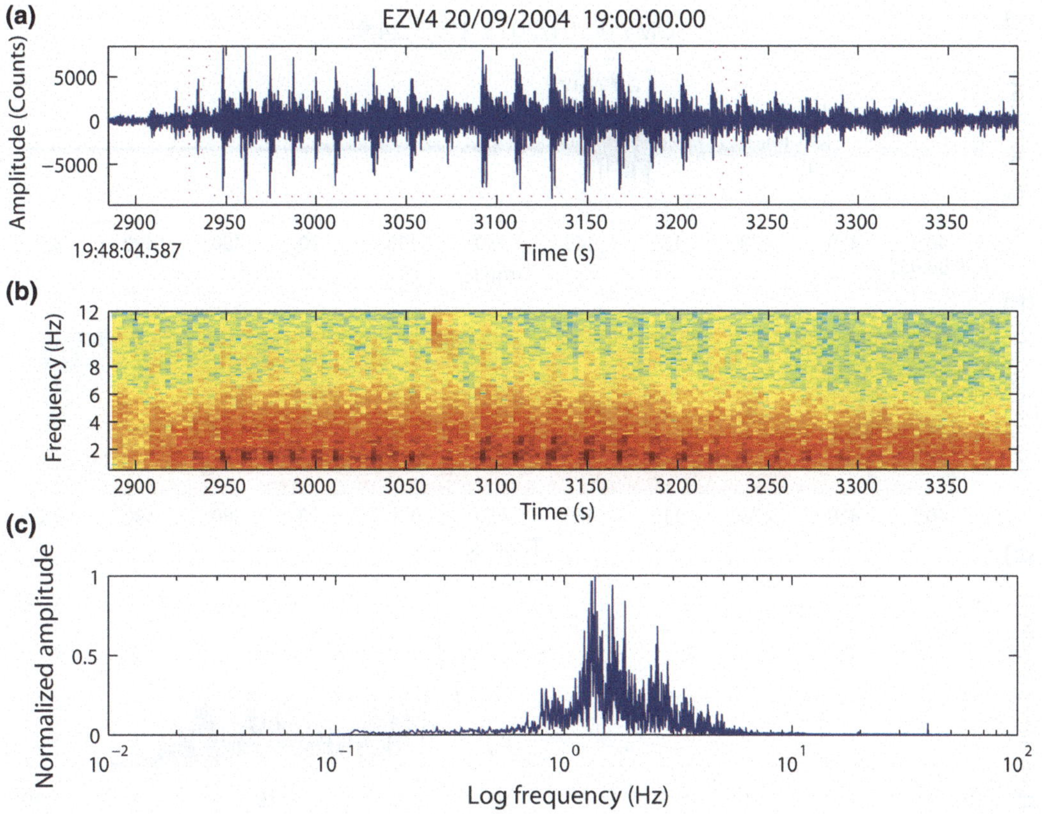

Fig. 8 Pulsating tremor at short-period seismometer EZV4 on 20 September 2004. **a** Velocity signal, **b** spectrogram and **c** normalized amplitude spectrum

during the explosions. In 2013, 2014 and 2015, VLPs were recognized at all the broadband stations.

The source mechanism of VLPs observed at many volcanoes is generally well reproduced by inversion, assuming six moment tensor components and three single force components in a volumetric source (Chouet 2003), although De Barros et al. (2013) mentioned that errors in both source location and velocity modelling can lead to strong spurious single forces. The moment tensor represents expansion or contraction of rock around a source, and a vertical single force can be generated as a consequence of the release of gravitational energy within the source volume (Takei and Kumazawa 1994). In the case of Vulcanian explosions, the recoil force generated by a volcanic jet during an eruption can generate one single force (Uhira and Takeo 1994; Chouet 2003), although this and other models are still being debated. A more detailed study is necessary to determine the origin of these events at Volcán de Colima.

2.5 Rockfalls and Pyroclastic Density Currents

In effusive episodes such as those observed at Volcán de Colima, the lava dome expands within the crater until it reaches the edges when it begins to flow down the flanks of the volcano. Due to the relatively high viscosity of andesite, lava flows are blocky, which generate rockfalls and possibly pyroclastic flows, particularly when the slope is steep. This kind of activity has been recorded in every episode of effusive activity at Volcán de Colima since 1991 (Fig. 12).

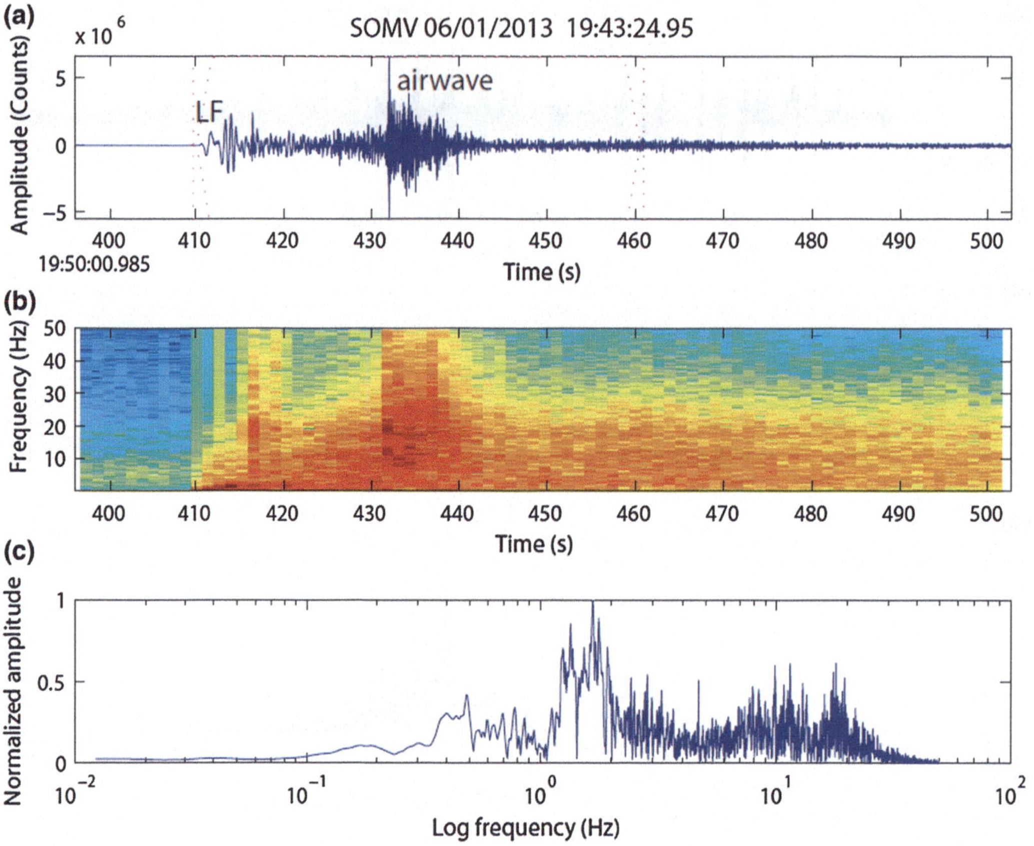

Fig. 9 Explosion event on SOMA (vertical component), on 06 January 2013. **a** Velocity signal; LF phase at the beginning and a shockwave that generates a high frequency pulse. **b** Spectrogram and **c** normalized amplitude spectrum for entire window

The seismic response to these events is an emergent high frequency signal with a duration of seconds to minutes. The generally symmetrical seismic waveform starts as an emergent signal, reaching its maximum amplitude near the centre, then finally decaying (Fig. 14a). The waveforms are of high frequency (mainly 1–15 Hz), but what is recorded is influenced by the distance between the seismic station and the source. Due to the generation of mainly surface waves and being of high frequency, they attenuate rapidly with distance (Fig. 13b, c). Zobin et al. (2009b) observed that seismic records of pyroclastic density currents generated by either the collapse of a lava dome or of an eruptive column, have a similar spectral content. The number and duration of rockfalls per day are directly related to the emission rate of the magma: when the rate increases, the number and duration of rockfalls also increase (Mueller et al. 2013).

2.6 Lahars

The eruption rate of Volcán de Colima is one of North America's highest. During the last twenty years, there have been more than ten episodes of volcanic activity (see Varley, this volume). Frequently, there are significant amounts of non-consolidated material waiting to be remobilised during the rainy seasons. Each year there are around ten to fifteen lahars during the rainy season but after a year of volcanic activity this

Table 1 Details of the seismicity associated with the main Vulcanian explosions at Volcán de Colima during recent activity (derived from the vertical amplitude of EFRE)

Date (GMT)	Comment	Energy × 10^9 (J)
10-02-1999	Explosion with incandescent blocks that showered the flanks out to 5 km, forming impact craters and triggering numerous forest fires (Zobin et al. 2002a, b, c)	Short-period seismometers saturated
10-05-1999	Explosion was felt and heard in Colima city; it was accompanied by two small pyroclastic flows that mainly descended the SW flank, where they entered into the ravines La Lumbre and Cordobán 04/1999 (BGVN 24:04)	Short-period seismometers saturated
17-07-1999	Large explosion, column reached ~11 km altitude. Pyroclastic flows had 4–4.5 km runout distances 08/1999 (BGVN 24:08)	Short-period seismometers saturated
22-02-2001	Column reached ~7 km altitude and produced ashfall at a distance of ~25 km from the volcano in Tuxpan, Jalisco. Small pyroclastic flows formed in the SW sector and descended to 3000 m. 04/2001 (BGVN 26:04)	Short-period seismometers saturated
17-07-2003	Explosion sent incandescent volcanic material to 500 m above the volcano, and produced a SW-drifting ash cloud to a height of ~7 km asl. At least five pyroclastic flows travelled down the volcano's flanks to a maximum distance of ~2 km. 08/2003 (BGVN 28:08)	0.27
02-08-2003	Large explosion was recorded on 2 August at 15:41. The Washington VAAC reported a plume at ~7.6 km altitude. 08/2003 (BGVN 28:08)	2.00
29-08-2003	The explosion produced an ash column at 7 km altitude with ashfall up to 60 km W-NW of the volcano and was accompanied by pyroclastic flows out to 2.5 km. 08/2003 (BGVN 28:08)	0.98
10-03-2005	Major explosion whose eruptive column reached up to 9.1 km asl and its pyroclastic flows travelled about 3 km into the Montegrande ravine[a]	0.48
13-03-2005	Column reached 8.9 km asl, with pyroclastic flows following principally the Muerto, Montegrande, Cordobán and San Antonio ravines. Ashfall in Los Mazos, Jalisco, located 12.5 km from the volcano[a]	1.50
26-03-2005	Moderate explosion[a]	0.38
12-04-2005	Moderate explosion[a]	0.09
20-04-2005	Moderate explosion, aviation authorities report an ash cloud reaching a maximum height of 6.1 km asl[a]	0.73
10-05-2005	Major explosion occurs with an ash column of 7.6 km asl[a]	0.99
16-05-2005	Major explosion with pyroclastic flows covering all the volcano's slopes[a]	2.14
24-05-2005	Major explosion occurred, whose ballistics reached up to 2.5 km and whose pyroclastic flows mainly travelled down the San Antonio and Montegrande ravines, causing some forest fires[a]	5.38
30-05-2005	A major explosion occurred. The column reached 8.5 km asl with pyroclastic flows mainly to the west[a]	7.13
02-06-2005	One major explosion occurred, causing ashfall in Colima city[a]	3.98
05-06-2005	Major explosion with a column of 8.9 km asl. Reports of broken glass were received from Juan Barragan 9 km away and in San Marcos, Jalisco, at 14 km. The Buenavista National Airport located to SSE of the volcano, remained closed for more than 12 h due to ashfall[a]	14.43
07-06-2005	Column reached a height of 8.5 km asl, the wind carried some of the ash to Colima city[a]	1.47
10-06-2005	Major explosive event occurred at night, this explosion was audible within a radius of 50 km, with probable pyroclastic flows up to 5 km[a]	0.94

(continued)

Table 1 (continued)

Date (GMT)	Comment	Energy × 10^9 (J)
05-07-2005	Major explosion with pyroclastic flows to 4.8 km to SE; there were reports of ashfall in the Becerrera and Yerbabuena villages[a]	0.89
27-07-2005	Major explosion, eruptive column reaching 9.1 km asl[a]	0.22
16-09-2005	Major explosion with an eruptive column of 9.8 km asl, with ashfall in some villages to the NW. Pyroclastic flows in all directions[a]	1.17
27-09-2005	Explosion with an eruptive column that reached 7.6 km asl, depositing ash in Comala, Villa de Alvarez and Colima city[a]	0.71
21-06-2011	Explosion with pyroclastic flows mainly to the west, this event destroyed a small part of the lava dome formed during 2007–2011	0.11
06-01-2013	Explosion with pyroclastic flows mainly to west, the ash plume rose 2 km above the crater. Ash fell on Ciudad Guzman	0.08
11-01-2013	Explosion with pyroclastic flows mainly to west; this event destroyed another part of lava dome formed during 2007–2011	0.06
13-01-2013	Moderate explosion	0.02
29-01-2013	Explosion with ballistics covering the entire edifice, residents up to 20 km away reported a loud noise, shaking ground, and rattling windows. The ash plume rose to 6.5 km asl. 04/2013 (BGVN 38:04)	0.15
21-11-2014	Explosion with pyroclastic flows to the south and southwest	0.06
30-11-2014	Largest explosion during the period 2013–2015	0.23
03-01-2015	Explosion with pyroclastic flows to the north	0.04

[a]Taken from Arámbula-Mendoza et al. (2011b)

can increase to as many as sixteen (see Capra et al., and Varley, this volume). Lahars can last from minutes to hours (Fig. 14a), and have signals with a broadband frequency range, from low frequencies to 50 Hz or higher (Fig. 14b, c). Similar to rockfalls, frequencies observed depend on the distance to the source. Sometimes lahars have many pulses or surges, associated with variations in the intensity of rain. Recent studies have observed that the amplitude and duration of the seismic signal is proportional to the volume of solid material and water (Vázquez et al. 2014).

3 Forecasting of Volcanic Eruption Using Seismicity

Real-time Seismic Energy Measurements (RSEM; De la Cruz-Reyna and Reyes-Dávila 2001) is the variable that represents the volcano seismic activity in a simple and effective way, although sometimes the RSEM can be affected by wind, noise, tides, etc. RSEM represents the RMS or Root Mean Square of the seismic signal in a defined period of time. It is proportional to the seismic energy, and is only slightly different from the RSAM (Real-time Seismic-Amplitude Measurement) that is traditionally used at many volcanoes around the world, and is the mean of the seismic signal amplitude. The RSEM is defined as:

$$RSEM = \sqrt{\frac{1}{N}\sum_{i=1}^{N} A_i^2} \quad (1)$$

where A is the amplitude of the velocity of the ground (m/s), and N is the number of samples of the seismic signal. In Fig. 15, the accumulative monthly RSEM can be observed, highlighting the main periods of activity, 1998–1999, 2001–2003, 2004–2005, 2007–2011 and 2013–present (December 2016). In many cases, first lava dome growth was observed (effusive stage), then moderate-large explosions destroyed the lava

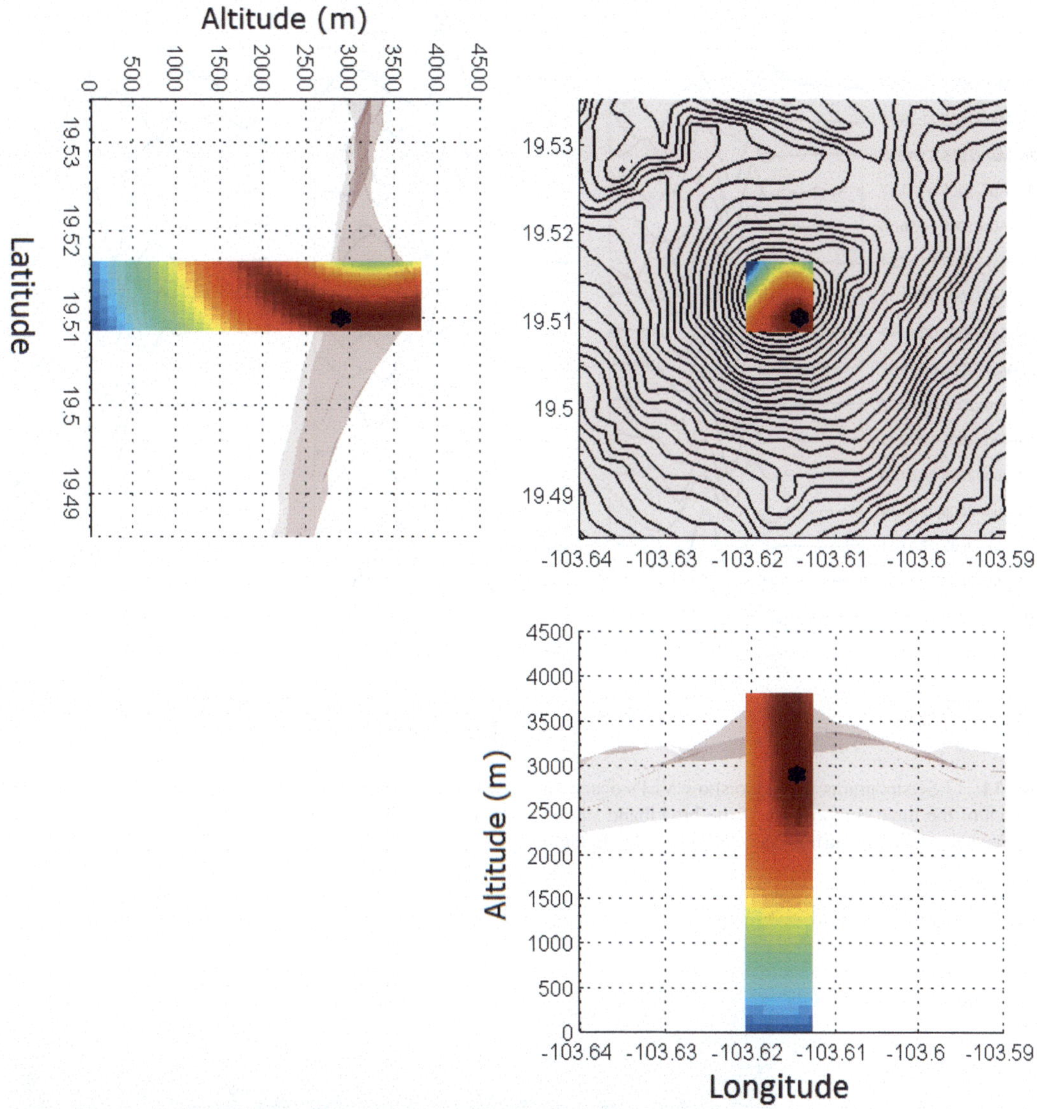

Fig. 10 Location of the start (first 6 s) of the explosion on 29 January 2013, using the seismic amplitudes method; the star is the location with a minimum misfit (difference between theoretical model and the data observed). The dark red colours indicate a better fit or less error. Further details of the method are available in Battaglia and Aki (2003)

dome (explosive stage). Only in 2001 and 2013, has the period of activity started with one or more explosions, followed by effusion with lava dome growth then flows.

During July 2015, a major episode of lava dome growth and two partial dome collapses generated large deposits from pyroclastic density currents within the Montegrande ravine up to distances of 10 km (Reyes-Dávila et al. 2016; Varley, this volume). This episode of activity was preceded by an increase in rockfalls and LP activity about two days before; only few VTs were observed, indicating that the system was "open" and the magma was easily able to reach the surface.

The seismicity has been used for forecasting eruptions, De la Cruz-Reyna and Reyes-Dávila (2001), and Reyes-Dávila and De la Cruz-Reyna

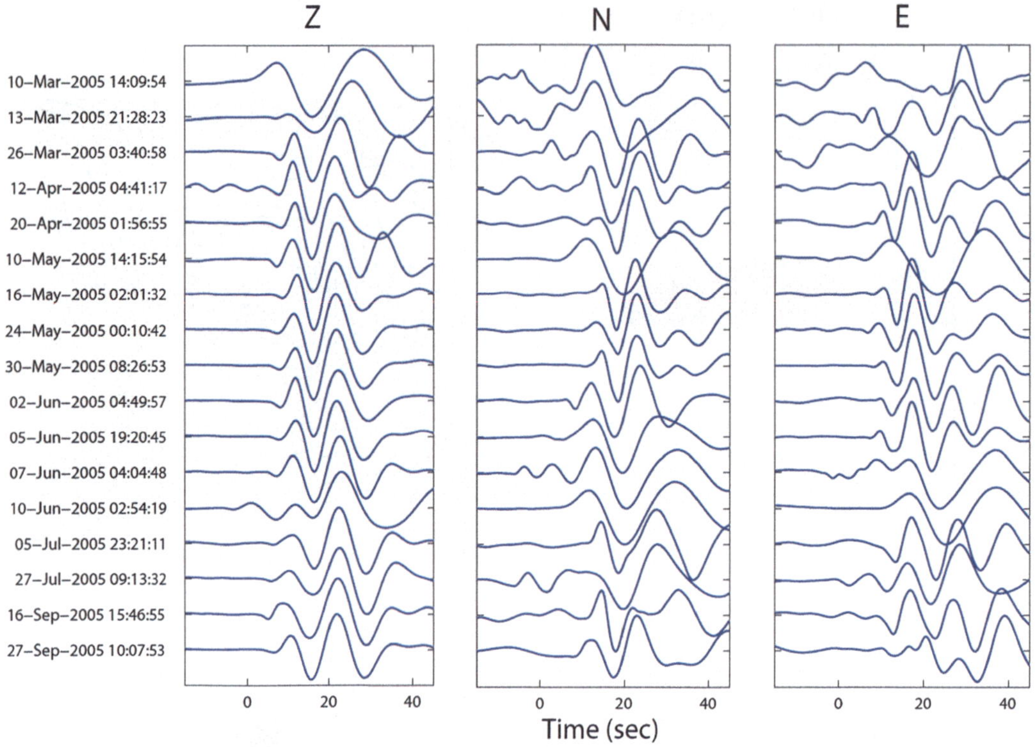

Fig. 11 17 Seismograms filtered to show VLPs observed in each of the three components of the broadband station of Fresnal (EFRE) during the 2005 moderate-large Vulcanian explosions. The signals are filtered between 30 and 10 s. The instrument response has been removed

Fig. 12 Example of small pyroclastic density current down southern flank of Volcán de Colima on 11 July 2015

(2002) used a continuum-mechanical model that reproduces the observed increase of strain rate preceding eruptions. The model is based on the strain response of a generalized Kelvin-Voight viscoelastic body with an exponential retardation spectrum to forecast the time of eruptions at

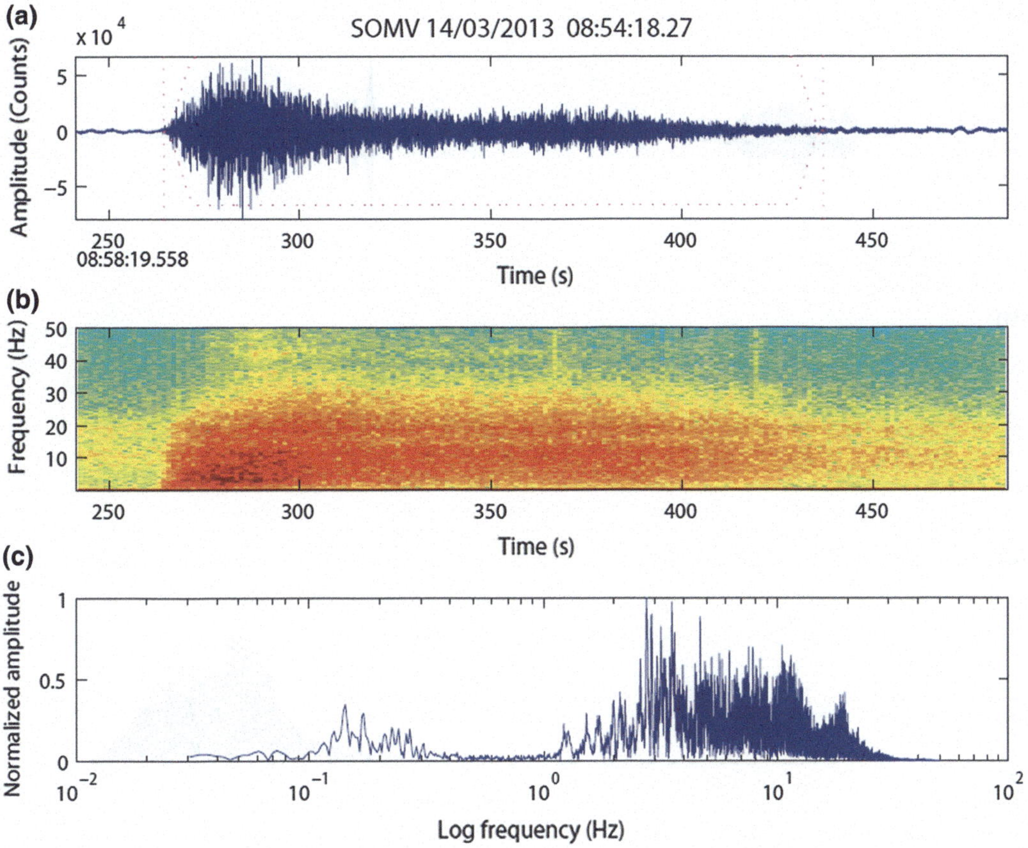

Fig. 13 Rockfall event at SOMV on 14 March 2013. **a** Velocity signal, **b** spectrogram. High frequencies can be observed here, **c** normalized amplitude spectrum

Volcán de Colima. This is a representation of how a material can undergo failure in a rigid state when it is subjected to pressure. When the volcano generates an explosion or a new emission of lava, this model can be applied. To model the process they used RSEM, which is a quantity directly proportional to the rate of strain release (De la Cruz-Reyna and Reyes-Dávila 2001). The pattern of RSEM evolution preceding eruptions at Volcán de Colima may be classified into three groups: type A, no precursory patterns can be recognized, type B precursory patterns are difficult to recognize or appear over time scales too short to allow warning or response, and type C precursory activity with patterns that can be recognized several days before an eruption (Fig. 16, Reyes-Dávila and De la Cruz-Reyna 2002). Seismicity during the precursory eruptive periods analysed (1994–2001) was composed mainly of HFs and LPs. Some of these eruptions were predicted before of the event, e.g. the extrusion in November 1998; others were analysed a posteriori as is the case for the explosion in July 1994 (De la Cruz-Reyna and Reyes-Dávila 2001). In these two cases the forecast was successful.

In a posteriori study Arámbula-Mendoza et al. (2011b) also observed an increase in the rate of liberation of seismic energy, mainly comprising of LPs, before each major explosion that occurred in 2005. The time of the explosion was estimated by applying the material failure forecasting method (FFM, Voight 1988, 1989). The FFM helps to forecast the time of failure in a process of rupture, and is observed with the growth of the accumulated seismic energy. In

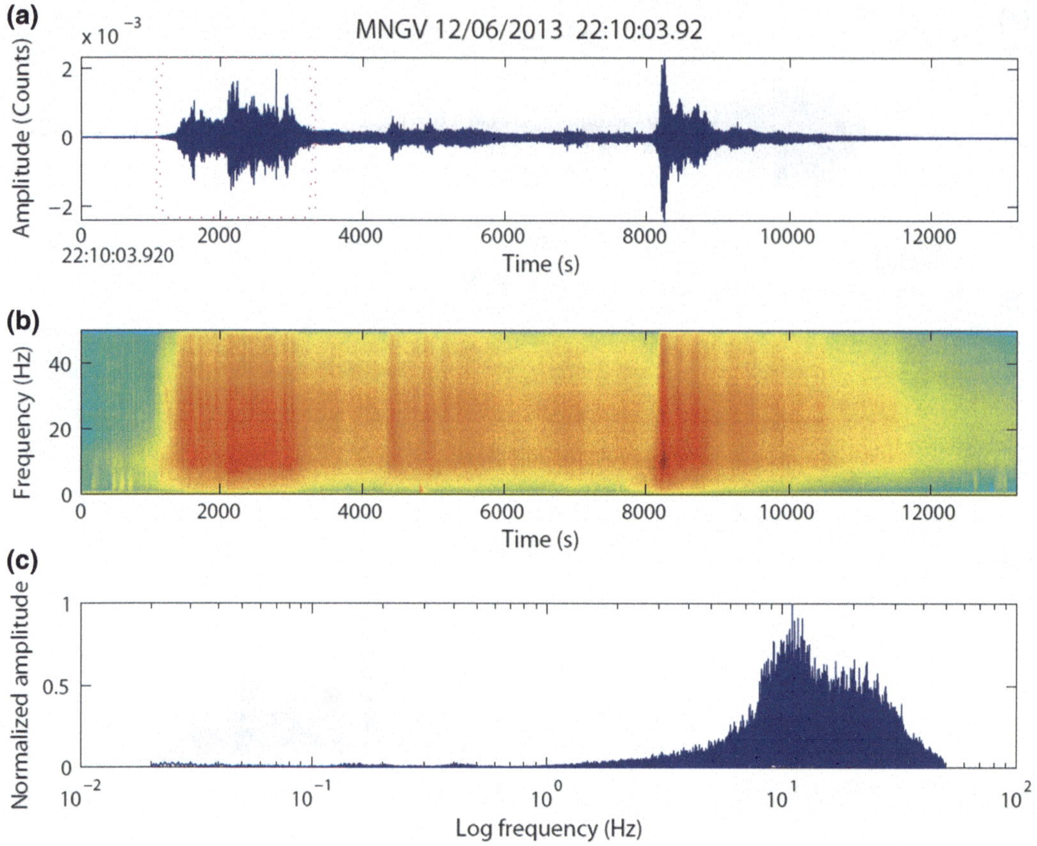

Fig. 14 Lahar recorded at MNGR (vertical component), on 12 June 2013; **a** velocity signal. Many pulses can be observed reflecting the variation of mass flow rate within the Montegrande ravine; **b** spectrogram; **c** normalized amplitude spectrum

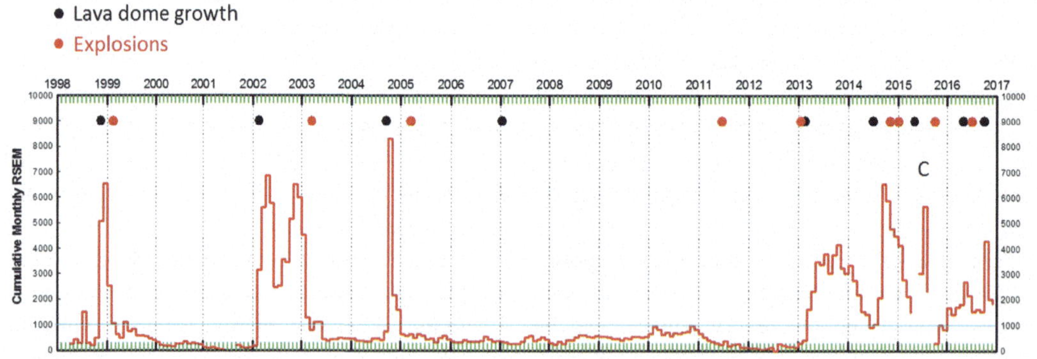

Fig. 15 Seismic activity observed at the short-period station EZV4 with RSEM, from 1998 to 2016. The black dots represent lava dome growth with lava flows (effusive stages), and the red dots represent destruction of lava domes (explosive stages). 'C' refers to lava dome collapses in July 2015

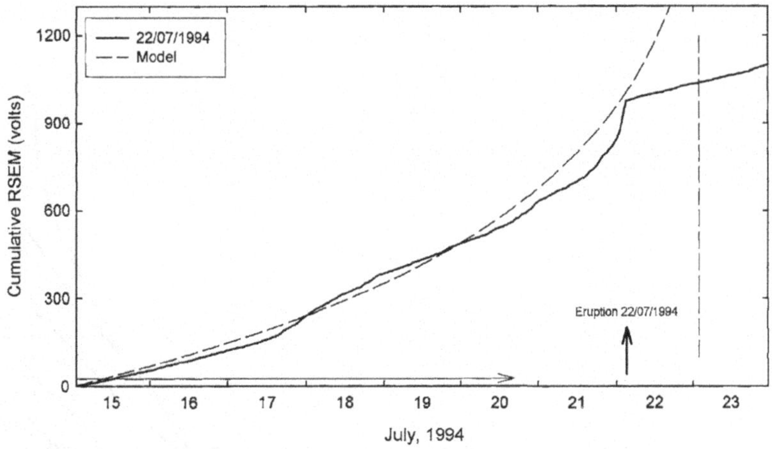

Fig. 16 The heavy curve represents cumulative RSEM data in volts for the period 15-23 July 1994, at Volcán de Colima. A 'forecast' was made in hindsight from part of the observed cumulative curve by trial fitting. The dashed curve is the prediction of the model, calculated from the cumulative data of the first 5.76 days as indicated by the horizontal arrow, counted from the beginning of the swarm on 15 July 1994. The dashed vertical line is the time of failure given by the model prediction from the data in that interval (After Reyes-Dávila and De la Cruz-Reyna 2002)

this case the Seismic Spectral Energy Measurement (SSEM), was utilized, which is the RSEM filtered between 1 and 3 Hz. The inverse SSEM was calculated, giving a straight line. This was extrapolated to a zero value, which provided a posteriori forecast of the time of the explosion that could be compared with that observed. For the 2005 data, a systematic delay of a few hours between the actual time of occurrence of the explosions and the estimated time was observed, which leads to the suggestion that more complex processes than a simple disruption of a volcanic plug occur prior to each event. For the 2005 activity, 53% of the major explosions could be forecast. This methodology has been utilised at many volcanoes around the world, with some degree of success for forecasting explosions (Ortiz et al. 2003; De la Cruz-Reyna et al. 2010). It seems that forecasting can be improved by first classifying the seismicity (Boué et al. 2015).

In Fig. 17, the accumulated RSEM of the precursory seismicity can be observed in the principal explosions of 1999, 2001, 2003, 2005, 2011 and 2013. This seismicity can be intense with many LPs, HFs or tremor, or discrete with few events, and can last a few hours or up to several days. In many cases, a notable increase in the seismic energy and number of events was seen a short time prior to the explosions, sometimes with harmonic or spasmodic tremor. This behaviour indicates that the associated seismicity can be complex, due to multiple magma pathways, leading to different vents, or variations in any combination of the following factors: gas contents, permeability, or rheology (Lavallée et al. 2008; Lavallée et al., this volume).

New techniques of seismic monitoring are being tested at Volcán de Colima. Automatic recognition of volcanic seismic signals based on Hidden Markov Models (HMMs; Benitez et al. 2009; Cortés et al. 2009) is one such technique. The classification of each seismic signal is done automatically, the program based on HMMs can recognize the signals without help from the operator. First, however, the system must be trained with many examples of each seismic event, then, a process of recognition and evaluation of the system is performed, until the program classifies with a high level of accuracy. The HMMs have the capacity to work in a continuous way, unlike other algorithms. The use of automatic recognition systems is very useful during

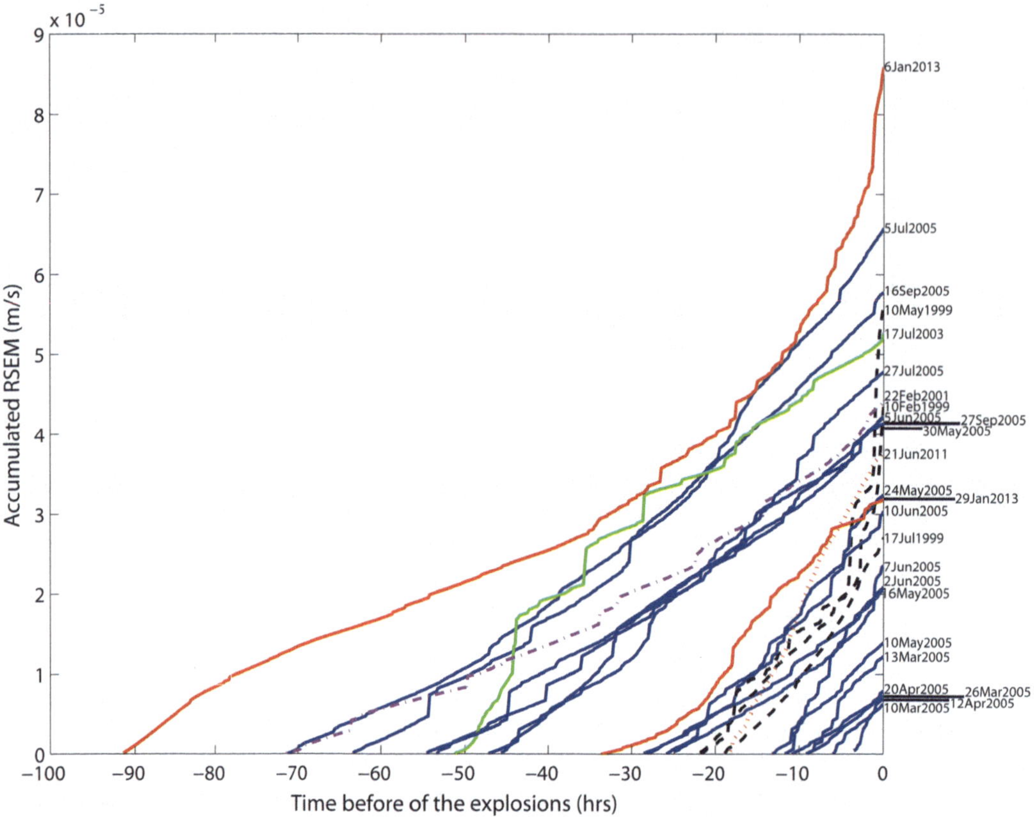

Fig. 17 Precursory seismicity (RSEM calculated each 10 min) before each of the main explosions observed at the short-period station EZV4 during the most recent explosion episodes. Zero is the onset of each explosion, the start time being taken as the occurrence of a notable increase in the RSEM associated with the initiation of a swarm. The explosions observed were: 10 Feb, 10 May, 17 Jul. 1999; 22 Feb. 2001; 17 July 2003; 10 Mar., 13 Mar., 26 Mar., 12 Apr., 20 Apr., 10 May, 16 May, 24 May, 30 May, 2 June, 5 June, 7 June, 10 June, 5 July, 27 July, 16 Sep., 27 Sep. 2005; 21 June 2011; 6 Jan., 29 Jan. 2013

times of crisis. It allows a fast evaluation of the activity, thus allowing time to be spent on other analysis.

Currently the system is run continuously at Volcán of Colima, whereby the software generates a seismogram with various colours representing different types of events (Fig. 18). The implementation of the software in RESCO has allowed the characterization of the last two periods of volcanic activity: 2007–2011 (Fig. 19) and the ongoing period of activity that began in January 2013. This system is still in development and currently its accuracy is about 82% (Arámbula-Mendoza 2011a), when compared with manual event picking.

The level of seismicity, as illustrated by the RSEM, at Volcán de Colima can be variable and depends mainly on the effusion rate. Eruptive periods with a low effusion rate generate low numbers of rockfalls per hour, and low explosive activity as in 2001–2003 and 2007–2011. The maximum extrusion rate during recent activity, prior to July 2015 was between 6 and 8 $m^3 s^{-1}$ for the last days of September and first few days of October, 2004 (Varley et al. 2010a). With an effusion rate at this level, large volume rockfalls and small pyroclastic density currents are commonplace. For the most recent eruptive period (2013–present), the maximum daily number of rockfalls has been between 250 and 350 events,

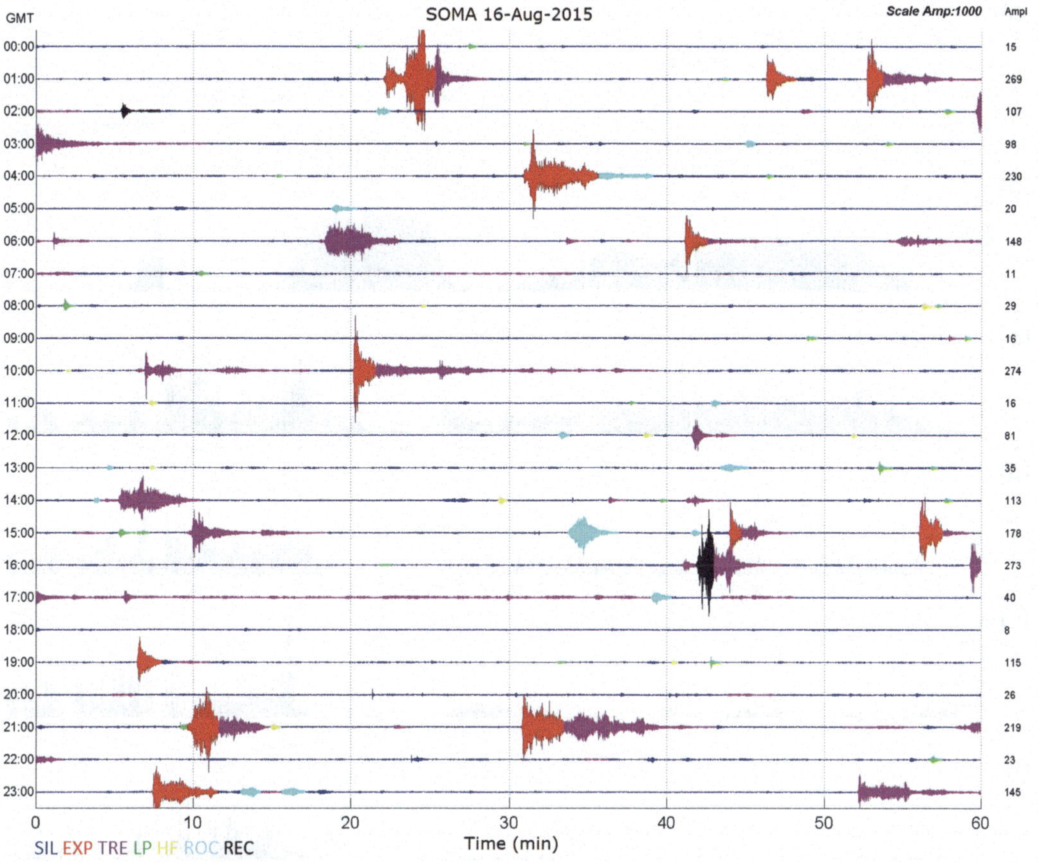

Fig. 18 Automated seismogram classified with Hidden Markov Models (HMMs), the colours indicate the type of seismic event; explosions (red), LPs (green), HFs (yellow), rockfalls (cyan), tremor (magenta), regional tectonic event (black) and silence (blue). The station used is a broadband sensor SOMA, located at 1.7 km from the crater; the seismic signal is filtered between 0.5 and 50 Hz. Taken from Arámbula-Mendoza et al. (2017), submitted

with an extrusion rate within the range 0.1–1 m^3 s^{-1}. The mean number of explosions was 8.5 per day up to December 2014, with some days having up to 30 events. Considering the activity over the past century, the most important period of explosive activity was between February and September 2005 (see Varley, this volume). During that time, the number of explosive events per day was actually lower than certain other periods, but the mean energy release by the explosions was greater, indicating a variation in the generation mechanism of the explosions, which remains undefined.

4 Discussion and Conclusions

Seismic monitoring is vital in the surveillance of a volcano. Since 1989, when the first seismic station was installed at Volcán de Colima, many thousands of signals (VTs, LPs, explosions, tremors, rockfalls and lahars) have been recorded by the seismic network. These signals represent different physical and geological processes, which can happen both inside and outside the volcano.

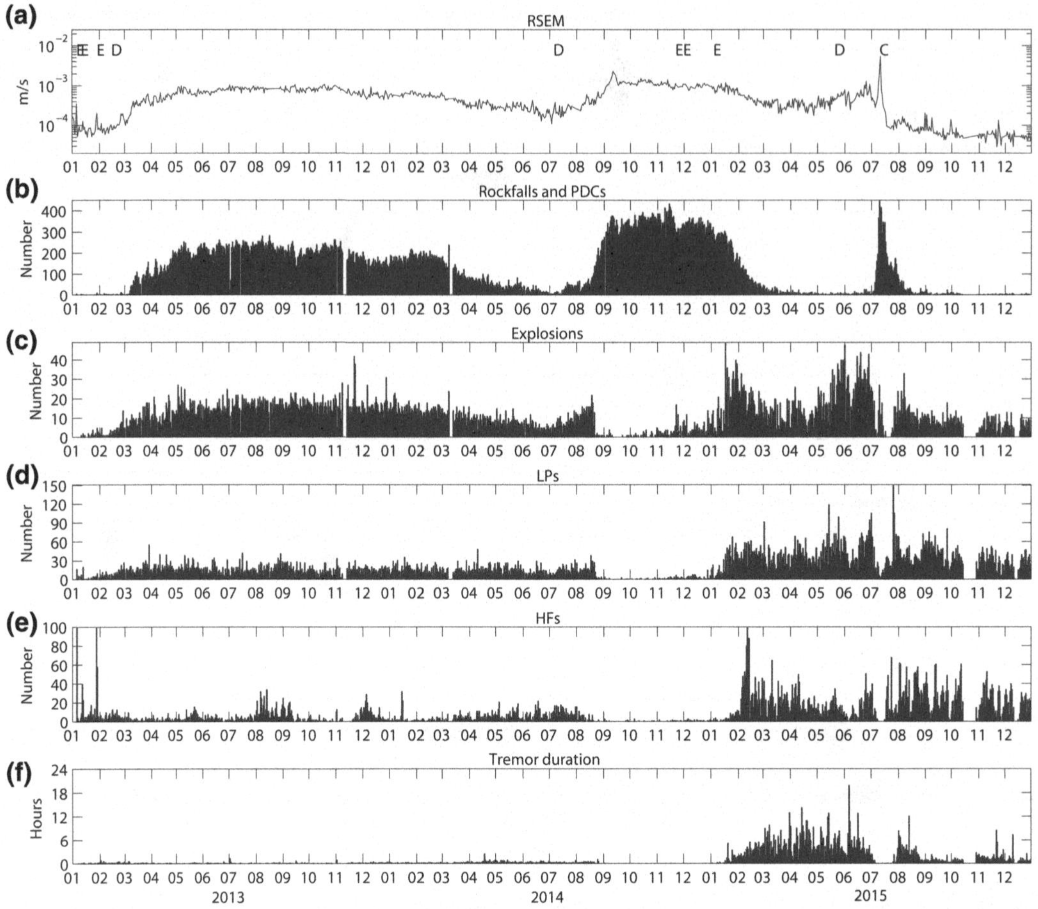

Fig. 19 **a** Daily RSEM, daily number of **b** rockfalls and pyroclastic density currents (PDCs), **c** explosions, **d** LPs, and **e** HFs, **f** tremor, duration in hours. 'E = moderate explosion', 'D = start of lava dome growth' and 'C = lava dome collapse'. Taken from Arámbula-Mendoza et al. (2017), submitted

VTs are usually the first seismic signals that show up, indicating the beginning of a new period of volcanic activity, for example prior to the 1991, 1994 and 1998 eruptions. During these periods, the system is considered as closed or partially closed after a long period of quiescence. The origin of the VTs is related to changes in the stresses within the volcano, as a consequence of the migration of magma to the surface. At Volcán de Colima, the tensional tectonic regimen associated to the Colima Graben can also produce VTs without a clear relationship to the ascent of magma.

Usually, the next type of event to occur are LPs. However, for the 2001, 2004, 2007 and 2013 periods, LPs were the first seismic signals of reactivation. Evidence exists to indicate that LPs are generated by the fracturing of magma during its ascent to the surface. The occurrence of LP events in swarms can be a precursor of the emplacement of a lava dome and/or major explosions. 2005 was a year with extensive LP activity associated with magma ascent that ended with moderate-large Vulcanian explosions (Varley et al. 2010a, b; Arámbula-Mendoza et al. 2011b). Civil Protection authorities were warned a few hours before several explosions occurred, with the information based on the occurrence of LPs.

Volcanic tremor is another seismic signal that is often observed at Volcán de Colima. During explosive episodes in 1999, spasmodic tremor appeared a few hours before the occurrence of large explosions. During the 2001–2003 effusive period, we observed a lengthy period of harmonic tremor in 2002, probably associated with degassing processes or interaction of groundwater. This tremor was of low amplitude, which could be associated with phreatic eruptions (Benoit et al. 2013), a reflection of a small diameter volcanic conduit (McNutt and Nishimura 2008), or a low gas content of the magma (Nadeau et al. 2011).

Rockfalls and pyroclastic flows are also common events at Volcán de Colima, occurring after the crater is filled with new material and begins to overflow. The size and number of rockfalls are both related to the emission rate. Explosions can be accompanied by a large energy release, with major events generating pyroclastic flows due to column collapse. The seismicity associated with explosions can be complex. With the increase in the number of broadband seismometers, VLPs have also been observed during explosions, most likely resulting from the liberation of gravitational energy during a volcanic jet. Finally, during the rainy season, lahars can occur due to remobilization of the non-consolidated material on the volcano's slopes. The size of these events depends of the amount of rain and the available material on the volcano (see Capra et al. this volume). As the occurrence of this type of event endangers the people that live close to, or even within, the ravines, seismic monitoring is a vital tool in reducing the risk. The moment that a lahar event is identified, state civil protection authorities are notified. The message is then relayed to the municipal civil protection offices who will advise the people that live near to the rivers or ravines and if necessary, carry out evacuations of the most vulnerable population.

At Volcán de Colima many successful forecasts of volcanic activity have been carried out. Seismic monitoring has been the vanguard of the network since its introduction 27 years ago. However, improvements in the observations of others seismic parameters in real time such as Q value (attenuation of the seismic waves for changes in the medium; Domínguez et al. 2003), b value (statistical relationship between the magnitude and occurrence of VTs; Zobin et al. 2002a, b, c, anisotropy (variations in the fractures associated with magmatic activity; Johnson and Savage 2012), seismic noise (variations in the velocity of the medium; Lesage et al. 2014), etc., could help to better our knowledge about the process that accompanies the eruptions.

Acknowledgements We appreciate the comments of three anonymous referees and the editor Nick Varley who improved the manuscript. RESCO has been maintained by the University of Colima, *Fideicomiso Preventivo* (FIPREDEN) of the Federal Government, and the Civil Protection Authorities of Colima and Jalisco states. Also many thanks to VDAP (USGS) for the donation of four broadband seismic stations. Finally, thanks to CONACYT who provided funding via a *Proyecto por Problemas Nacionales* 2015, #916, which supported a part of this project.

References

Arámbula-Mendoza, R.: Clasificación automática de eventos sísmicos volcánicos y análisis de la actividad reciente en el Volcán de Colima. Ph.D. Thesis. Instituto de Geofísica, UNAM, México, 176 pp (2011a)

Arámbula-Mendoza, R., Lesage, P., Valdés-González, C., Varley, N.R., Reyes-Davila, G., Navarro, C.: Seismic activity that accompanied the effusive and explosive eruptions during the 2004–2005 period at Volcán de Colima, Mexico. J. Volcanol. Geoth. Res. **205**, 30–46 (2011)

Arámbula-Mendoza, R., Reyes, D.G., Vargas, B.D., González, A.M., Navarro, O.C., Velásquez, M.A., Ariel, R.C.: Seismic monitoring of effusive-explosive activity and large lava dome collapses during 2013–2015 at Volcán de Colima, Mexico. Submitted to J. Volcanol. Geoth. Res. (2017)

Arciniega-Ceballos, A., Chouet, B., Dawson, P.: Very long-period signals associated with vulcanian explosións at Popocatepetl Volcano, Mexico. Geophysical. Res. Lett. **26**, 3013–3016 (1999)

Battaglia, J., Aki, K.: Location of seismic events and eruptive fissures on the Piton de la Fournaise volcano using seismic amplitudes. J. Geophys. Res. **108**(B8). https://doi.org/10.1029/2002jb002193 (2003a)

Benitez, M.C., Lesage, P., Cortés, G., Segura, J.C., Ibáñez, J.M., De la Torre, A.: Automatic recognition of volcanic-seismic events based on continuous hidden markov models. In: Bean, C.J., Braiden, A.K., Lokmer, I., Martini, F., O'Brien, G.S. (eds.) The Volume Project, Volcanoes: understanding Subsurface Mass Movement, pp. 130–139 (2009)

Benoit, J.P., McNutt, S.R., Barboza, V.: Duration-amplitude distribution of volcanic tremor. J. Geophys. Res. **108**(B3), 2146. https://doi.org/10.1029/2001jb001520 (2003)

Boatwright, J.: A spectral theory for circular seismic sources: simple estimates of source dimension, dynamic stress drop, and radiated seismic energy. Bull. Seismol. Soc. Am. **70**(1), 1–27 (1980)

Boué, A., Lesage, P., Cortés, G., Valette, B., Reyes-Dávila, G.: Real-time eruption forecasting using the material failure forecast method with a Bayesian approach. J. Geophys. Res: Solid Earth **120** (4), 2143–2161 (2015)

Cortés, G., Arámbula, R., Álvarez, I., Benítez, C.M., Ibáñez, J.M., Lesage, P., González, A M., Reyes, D G.: Analysis of Colima, Popocatépetl, and Arenal volcanic seismicity using an automatic Continuos Hidden markov Model-based recognition system. In: Bean, C.J., Braiden, A.K., Lokmer, I., Martini, F., O'Brien, G.S. (eds.) The Volume Project—Volcanoes: understanding Subsurface Mass Movement, pp. 150–160 (2009)

Chouet, B.: Volcano seismology. Pure. Appl. Geophys. **160**, 739–788 (2003)

D'Auria, L., Giudicepietro, F., Martini, M., Peluso, R.: Seismological insight into the kinematics of the 5 April 2003 vulcanian explosión at Stromboli volcano (southern Italy). Geophys. Res. Lett. **33**, L08308 (2006). https://doi.org/10.1029/2006GL026018

Dawson, P.B.A., Chouet, B.A., Power, J.: Determining the seismic source mechanism and location for an explosive eruption with limited observational data: Augustine Volcano, Alaska. Geophys. Res. Lett. **38**, L03302. https://doi.org/10.1029/2010gl045977 (2011)

De Barros, L., Lokmer, I., Bean, C.J.: Origin of spurious single forces in the source mechanism of volcanic seismicity. J. Volcanol. Geotherm. Res. **262**, 1–6 (2013)

De la Cruz-Reyna, S., Reyes-Dávila, G.A.: A model to describe precursory material failure phenomena: application to short-term forecasting at Colima volcano, Mexico. Bull. Volcanol. **63**, 297–308 (2001)

De la Cruz-Reyna, S., Tárraga, M., Ortiz, R., Martínez--Bringas, A.: Tectonic earthquakes triggering volcanic seismicity and eruptions. Case studies at Tungurahua and Popocatépetl volcanoes. J. Volcanol. Geotherm. Res. **193**, 37–48 (2010)

Domínguez, T., Zobin, V.M., Reyes-Dávila, G.A.: The fracturing in volcanic edifice before an eruption: the June-July 1998 high-frequency earthquake swarm at Volcán de Colima, México. J. Volcanol. Geoth. Res. **105**, 65–75 (2001)

Domínguez, R.T., Flores, C.F., Reyes, D.G.: Temporal change in coda wave attenuation observed at Volcán de Colima, México before the 1998 eruption. J. Volcanol. Geoth. Res. **125**, 215–223 (2003)

Haney, M.M., Chouet, B.A., Dawson, P.A., Power, J.A.: Source characterization for an explosion during the 2009 eruption of Redoubt Volcano from very-long-period seismic waves. J. Volcanol. Geotherm. Res. https://doi.org/10.1016/j.jvolgeores.2012.04.018 (2012)

Goto, A.: A new model for volcanic earthquake at Unzen volcano: melt rupture model. Geophys. Res. Lett. **26**, 2541–2544 (1999)

Green, D.N., Neuberg, J.: Waveform classification of volcanic low-frequency earthquake swarms and its implication at Soufriere Hills Volcano, Montserrat. J. Volcanol. Geoth. Res. **153**, 51–63 (2006)

Jiménez, Z., Reyes, G., Espíndola, J.M.: The July 1994 episode of seismic activity at Colima Volcano, Mexico. J. Volcanol. Geoth. Res. **64**, 321–326 (1995)

Johnson, J.B., Aster, R.C.: Relative partitioning of acoustic and seismic energy during Strombolian eruptions. J. Volcanol. Geotherm. Res. **148**, 334–354 (2005)

Johnson, J.H., Savage, M.K.: Tracking volcanic and geothermal activity in the Tongariro Volcanic Centre, New Zealand, with shear wave splitting tomography. J. Volcanol. Geotherm. Res. **223**, 1–10 (2012)

Konstantinou, K.I., Schlindwein, V.: Nature, wavefield properties and source mechanism of volcanic tremor: a review. J. Volcanol. Geoth. Res. **119**, 161–187 (2003)

Kumagai, H., Chouet, B.A.: The complex frequencies of long-period seismic events as probes of fluid composition beneath volcanoes. Geophys. J. Int. **138**, F7–F12 (1999)

Kumagai, H., Nakano, M., Maeda, T., Yepes, H., Palacios, P., Ruiz, M., Arrais, S., Vaca, M., Molina, I., Yamashima, T.: Broadband seismic monitoring of active volcanoes using deterministic and stochastic approaches. J. Geophys. Res. **115**, B08303 (2010). https://doi.org/10.1029/2009JB006889

Kumagai, H., Palacios, P., Ruiz, M., Yepes, H., Kozono, T.: Ascending seismic source during an explosive eruption at Tungurahua volcano. Ecuador. Geophys. Res. Lett. **38**, L01306 (2011). https://doi.org/10.1029/2010GL045944

Lavallée, Y., Meredith, P., Dingwell, D.B., Hess, K.U., Wassermann, J., Cordonnier, B., Gerik, A., Kruhl, J. H.: Seismogenic lavas and explosive eruption forecasting. Nature **453**, 507–510 (2008)

Lesage, P., Reyes-Dávila, G., Arámbula-Mendoza, R.: Large tectonic earthquakes induce sharp temporary decreases in seismic velocity in Volcán de Colima, Mexico. J. Geophys. Res. Solid Earth **119**, 4360–4376. https://doi.org/10.1002/2013jb010884 (2014)

McNutt, S.R., Nishimura, T.: Volcanic tremor during eruptions: temporal characteristics, scaling and constraints on conduit size and processes. J. Volcanol. Geoth. Res. **178**(1), 10–18 (2008)

Mueller, S.B., Varley, N.R., Kueppers, U., Lesage, P., Reyes-Davila, G., Dingwell, D.B.: Quantification of magma ascent rate through rockfall monitoring at the growing/collapsing lava dome of Volcán de Colima, Mexico. Solid Earth **4**, 201–213 (2013)

Nadeau, P.A., Palma, J.L., Waite, G.P.: Linking volcanic tremor, degassing, and eruption dynamics via SO_2 imaging. Geophys. Res. Lett. **38**(1), L01304 (2011)

Neuberg, J., Luckett, R., Baptie, B., Olsen, K.: Models of tremor and low frequency earthquake swarms on Montserrat. J. Volcanol. Geoth. Res. **101**, 83–104 (2000)

Neuberg, J.W., Tuffen, H., Collier, L., Green, D., Powell, T., Dingwell, D.: The trigger mechanism of low-frequency earthquakes on Montserrat. J. Volcanol. Geoth. Res. **153**, 37–50 (2006)

Núñez-Cornú, F., Nava, F. A, De la Cruz-Reyna, S., Jiménez, Z., Valencia, C., García-Arthur, R.: Seismic activity related to the 1991 eruption of Colima Volcano, Mexico. Bull. Volcanol. **56**, 228–237 (1994)

Ortiz, R., Moreno, H., García, A., Fuentealba, G., Astiz, M., Peña, P., Sánchez, N., Tarraga, M.: Villarica volcano (Chile): characteristics of the volcanic tremor and forecasting of small explosions by means of a material failure method. J. Volcanol. Geotherm. Res. **128**, 247–259 (2003)

Palo, M., Ibáñez, J.M., Cisneros, M., Bretón, M., Del Pezzo, E., Ocaña, E., Orozco-Rojas, J., Posadas, A. M.: Analysis of the seismic wavefield properties of volcanic explosións at Volcán de Colima, México: insights into the source mechanism. Geophys. J. Int. **177**, 1383–1398 (2009)

Petrosino, S., Cusano, P., La Rocca, M., Galluzzo, D., Orozco-Rojas, J., Breton, M., Del Ibáñez, J., Pezzo, E.: Source location of long period seismicity at Volcán de Colima, México. Bull. Volc. **73**, 887–898 (2011)

Powell, T., Neuberg, J.: Time dependent features in tremor spectra. J. Volcanol. Geoth. Res. **128**, 177–185 (2003)

Reyes-Dávila, G.A., De la Cruz-Reyna, S.: Experience in the short-term eruption forecasting at Volcán de Colima, México, and public response to forecasts. J. Volcanol. Geoth. Res. **117**, 121–127 (2002)

Reyes-Dávila, G.A., Arámbula-Mendoza, R., Espinasa-Pereña, R., Pankhurst, M.J., Navarro-Ochoa, C., Savov, I., Vargas-Bracamontes, D.M., Cortés-Cortés, A., Gutiérrez-Martínez, C., Valdés-González, C., Domínguez-Reyes, T., González-Amezcua, M., Martínez-Fierros, A., Ramírez-Vázquez, A., Cárdenas-González, L., Castañeda-Bastida, E., Vázquez-Espinoza de los Monteros, D.M., Nieto-Torres, A., Campion, R., Courtois, L., Lee, P. D.: Volcán de Colima dome collapse of July, 2015 and associated pyroclastic density currents. J. Volcanol. Geotherm. Res. **320**, 100–106 (2016).

Stix, J., Torres, R.C., Narváez, M.L., Cortés, J.G.P., Raigosa, J.A., Gómez, M.D., Castonguay, R.: A model of vulcanian eruptions at Galeras volcano, Colombia. J. Volcanol. Geoth. Res. **77**, 285–303 (1997)

Takei, Y., Kumazawa, M.: Why have the single force and torque been excluded from seismic source models? Geophys. J. Int. **118**, 20–30 (1994)

Tuffen, H., Dingwell, D., Pinkerton, H.: Repeated fracture and healing of silicic magma generates flow banding and earthquakes? Geology **31**, 1089–1092 (2003)

Tuffen, H., Smith, R., Sammonds, P.R.: Evidence for seismogenic fracture of silicic magma. Nature **453**, 511–514 (2008)

Uhira, K., Takeo, M.: The source of explosive eruptions of Sakurajima volcano, Japan. J. Geophys. Res. **99** (B9), 17775–17789 (1994)

Varley, N., Arámbula-Mendoza, R., Reyes-Dávila, G., Stevenson, J., Harwood, R.: Long-period seismicity during magma movement at Volcán de Colima. Bull. Volc. **72**, 1093–1107 (2010a)

Varley, N., Arámbula-Mendoza, R., Reyes-Dávila, G., Sanderson, R., Stevenson, J.: Generation of Vulcanian activity and long-period seismicity at Volcán de Colima, Mexico. J. Volcanol. Geoth. Res. **198**, 45–56 (2010b)

Vázquez, R., Capra, L., Caballero, L., Arámbula-Mendoza, R., Reyes-Dávila, G.: The anatomy of a lahar: Deciphering the 15th September 2012 lahar at Volcán de Colima, Mexico. J. Volcanol. Geoth. Res. **272**, 126–136 (2014)

Voight, B.: A method for prediction of volcanic eruptions. Nature **332**, 125–130 (1988)

Voight, B.: A relation to describe rate-dependent material failure. Science **243**, 200–203 (1989)

White, R., McCausland, W.: Volcano-tectonic earthquakes: a new tool for estimating intrusive volumes and forecasting eruptions. J. Volcanol. Geoth. Res. (2015). https://doi.org/10.1016/j.jvolgeores.2015.10.020

Zamora-Camacho, A., Espíndola, J.M. Reyes-Dávila, G.: The 1997–1998 Activity of Volcán de Colima, Western Mexico: some aspects of the associated seismic activity. Pure Appl. Geophys. **164**, 39–52 (2007)

Zobin, V.M., Luhr, J.F., Taran, Y.A., Bretón, M., Cortés, A., De La Cruz-Reyna, S., Domínguez, T., Galindo, I., Gavilanes, J.C., Muñíz, J.J., Navarro, C., Ramírez, J. J., Reyes, G.A., Ursúa, M., Velasco, J., Alatorre, E., Santiago, H.: Overview of the 1997–2000 activity of Volcán de Colima, México. J. Volcanol. Geotherm. Res. **117**(1–2), 1–19 (2002a)

Zobin, V.M., González-Amezcua, M., Reyes-Dávila, G. A., Domínguez, T., Cerda-Chacón, J.C., Chávez Alvarez, J.M.: Comparative characteristics of the 1997–1998 seismic swarms preceding the November 1998 eruption of Volcán de Colima, México. J. Volcanol. Geoth. Res. **117**, 47–60 (2002a)

Zobin, V.M., González-Amezcua, M., Reyes-Dávila, G. A.: Seismotectonic deformation of the volcanic edifice prior to the 1998 lava eruption of Volcán de Colima, México. Bull. Volcanol. **64**, 349–355 (2002b)

Zobin, V.M., Navarro-Ochoa, C.J., Reyes-Dávila, G.A.: Seismic quantification of the explosions that destroyed the dome of Volcán de Colima, Mexico, in July-August 2003. Bull. Volcanol. **69**, 141–147 (2006a)

Zobin, V.M., Navarro, C., Reyes-Dávila, G., Orozco, J., Bretón, M., Téllez, A., Reyes-Alfaro, G., Vázquez, H.: The methodology of quantification of volcanic explosións from broadband seismic signals and its application to the 2004–2005 explosións at Volcán de Colima, México. Geophys. J. Int. **167**, 467–478 (2006b)

Zobin, V.M., Reyes, G.A., Guevara, E., Bretón, M.: Scaling relationship for Vulcanian explosións derived from broadband seismic signals. J. Geophys. Res. **114**, B03203 (2009). https://doi.org/10.1029/2008JB005983

Zobin, V.M., Plascencia, I., Reyes, G., Navarro, C.: The characteristics of seismic signals produced by lahars and pyroclastic flows: Volcán de Colima, México. J. Volcanol. Geoth. Res. **179**, 157–167 (2009b)

Petrological Monitoring of Volcán de Colima Magmatic System: The 1998 to 2011 Activity

Olivier Reubi, Jonathan Blundy and Joe Pickles

Abstract

Volcán de Colima's nearly continuous activity since 1998 has shifted repeatedly from effusive to explosive Vulcanian eruptions. Strong explosions occurred in 1999 and 2005, which raised concerns that the current eruptive phase may lead to a major Plinian eruption, as in 1913. Understanding the processes controlling the eruptive style and intensity, and identifying the associated precursory signals, are essential for hazard mitigation. Here we present an overview of the petrology of the magmas erupted since 1998 and use these data to constrain the storage conditions and pre-eruptive evolution of the magmas. The potential of research strategies linking underground magmatic processes with geophysical and geochemical surface records, an essential step to identify key precursory signals, is established by comparing the melt inclusion and volcanic earthquakes record. The significance of the 1998–2011 activity in the context of historical eruption cycles is addressed by comparing the petrology of the 1998–2011 magmas with magmas from the 1913 Plinian eruption. This demonstrates the importance of mafic recharges before the major cycle-ending Plinian eruptions.

Keywords

Petrology · Geochemistry · Magmatic system Volatiles

1 Introduction

Over the last 200 years, Volcán de Colima's eruptive behaviour has encompassed highly explosive Plinian eruptions, moderately explosive Vulcanian eruptions, and gentle effusion of lava domes and flows (Varley, this volume). The large range of eruption styles and intensities involved and the rapid transitions between them cause major challenges for volcanic hazards mitigation. Understanding the processes controlling these transitions and identifying the associated precursory signals is essential to developing predictive models.

Factors that control eruption intensity include physical factors such as magma temperature, crystallinity, ascent and degassing rates, and chemical factors such as dissolved volatile content and melt composition (e.g. Eichelberger et al. 1986; Sparks 1997; Villemant and Boudon 1998). The minerals crystallizing in the magmas and the droplets of quenched melt trapped within the crystals during growth (melt inclusions) record

O. Reubi (✉)
Institut des Sciences de la Terre, Université de Lausanne, Lausanne, Switzerland
e-mail: olivier.reubi@unil.ch

J. Blundy
School of Earth Sciences, University of Bristol, Bristol BS8 1 RJ, UK

J. Pickles
Camborne School of Mines, University of Exeter, Exeter TR10 9FE, UK

N. Varley et al. (eds.), *Volcán de Colima*, Active Volcanoes of the World,
https://doi.org/10.1007/978-3-642-25911-1_9

these factors as the magmas are ascending through the Earth's crust and evolve in response to changes in pressure and temperature. Advances in petrological tools enable quantitative determination of the melt composition, volatile contents and crystallinity, the depth and temperature of magma storage, the nature of magmatic processes, and the rates at which magmas evolve and ascend (Putirka 2008a; Blundy and Cashman 2008). The presence and composition of coexisting minerals can constrain critical parameters (temperature, pressure, and oxygen fugacity) (Putirka 2008b) during magmatic evolution (crystallization, magma mixing or mingling, etc.). Changes in critical parameters occurring during crystallization can also be inferred from the compositional and textural zoning of minerals (e.g. Ginibre et al. 2002a, b; Streck 2008). Additional constraints on the pre-eruptive conditions and evolution of the magmas can be obtained from the chemistry of melt inclusions. Melt inclusions are small (<50 μm) droplets of melt trapped in minerals during crystallization. Once sealed, melt inclusion become physically disconnected from the surrounding melt that continues to evolve in response to changes in temperature and/or pressure, and magmatic processes such magma mixing. Melt inclusions, therefore, provide snapshots of the chemical evolution of the melt while it is crystallizing in the subvolcanic magmatic system (Roedder 1984). Melt inclusions provide particularly valuable information on the pre-eruptive volatile contents of magmas, and on the degassing and crystallization paths followed during their ascent through the upper crust (e.g. Blundy and Cashman 2005; Lowenstern 2003; Wallace 2005). Petrological studies, therefore, provide an efficient means to document the pre-eruptive evolution of the magmas. By comparing the products of successive eruptions covering a range of styles and intensities, the factors and processes controlling the eruptive behaviour may be indentified. The drawback is that petrologists can only interrogate volcanic products retrospectively, i.e. after an eruption has occurred.

Monitoring of active volcanoes involves a broad range of methodologies such as seismology, gas fluxes and chemistry, ground-deformation, microgravity, and thermal imaging (McNutt et al. 2000; Sparks 2003). These techniques effectively record changes occurring in the magmatic or hydrothermal systems beneath the volcano. Relating unequivocally the monitored signals to specific physical processes and evaluating their influences on eruption likelihood, style, and intensity is challenging. Linking the underground magmatic processes identified on the basis of petrological studies with surface manifestations of the type that are routinely monitored, albeit after the event, is essential to successfully interpret the volcano monitoring record and develop eruption forecast models, a critical goal in modern volcanology. Integrated research strategies, combining the petrology and geochemistry of magmas with the geophysical and geochemical surface records, require persistently active volcanoes showing a range of eruption styles and for which a continuous monitoring record capturing these shifts in volcanic activity is available. Volcán de Colima is amongst the few volcanoes that match these criteria and, providing that the volcanic activity continues, significant advances in our understanding of the dynamics of subvolcanic magmatic systems, the processes controlling eruption style and intensity, and the associated surface warning signals may be obtained from multidisciplinary studies of this volcano.

This chapter provides an overview of our understanding of Volcán de Colima's active magmatic system from a petrological point of view, and discusses the link with surface monitoring data. Integrated multidisciplinary research strategies linking underground magmatic processes with geophysical and geochemical surface records remains the goal of future projects. One aim of this chapter is to lay the petrological basis and outline the feasibility and potential of this type of approach. Specific topics that are addressed include: (I) petrology of the magmas erupted at Volcán de Colima between 1998 and 2011; (II) constraints on the pre-eruptive state of the magmas (temperature, volatile content, etc.) and the nature of pre-eruptive magmatic processes (mixing, recharge, degassing, etc.); (III) links between the petrological and monitoring records; (IV) significance of the 1998–2011 activity in the context of historical eruption cycles; (V) outline of future research directions to address the dynamics of the magmatic system.

1.1 Historical Eruptive Cycles of Volcán de Colima

Volcán de Colima has been persistently active over at least the last 500 years flows (Varley, this volume). Luhr and Carmichael (1980) proposed the concept of historical eruptive cycles at Volcán de Colima. Each cycle comprises an initial period of dome growth followed by intermittent eruptions of lava flows, and gas and ash-to-scoria Vulcanian eruptions. These phases are dominated by andesitic lavas with ~61 wt% SiO_2. The cycles terminate with major explosive Plinian eruptions involving relatively mafic andesites with 58 wt% SiO_2, as occurred in 1818 and 1913. The frequency of the major Plinian or sub-Plinian eruptions has been approximately every 100 years, creating concerns amongst volcanologists that the current activity may be ramping up to a major cycle-ending Plinian eruption.

1.2 1988 to 2011 Volcanic Activity at Volcán de Colima

Detailed accounts of the volcanic activity between 1998 and 2011 can be found in Varley (this volume), Savov et al. (2008), or online at Smithsonian Global Volcanism Program (http://www.volcano.si.edu); only a brief outline is given here and illustrated in Fig. 1. Lava extrusion in the summit crater started in late November 1998 following months of precursory seismic and fumarolic activity (Zobin et al. 2002b; Taran et al. 2002) that marked the end of 7 years of repose. Lava flows and occasional block-and-ash flows generated by collapse of the front of the flows lasted until February 1999. An explosive cycle that included several large explosions followed from February until July 1999. From August 1999 to February 2001 the activity was characterized by sporadic low intensity explosions. A new dome appeared in the summit crater in May 2001. This effusive activity lasted until February 2003 and gave way to a new explosive phase with several strong Vulcanian explosions in July, August, and November. Effusive activity accompanied by frequent small explosions resumed in September 2004 and lasted until November 2004. From December 2004 to January 2007 the activity consisted of intermittent

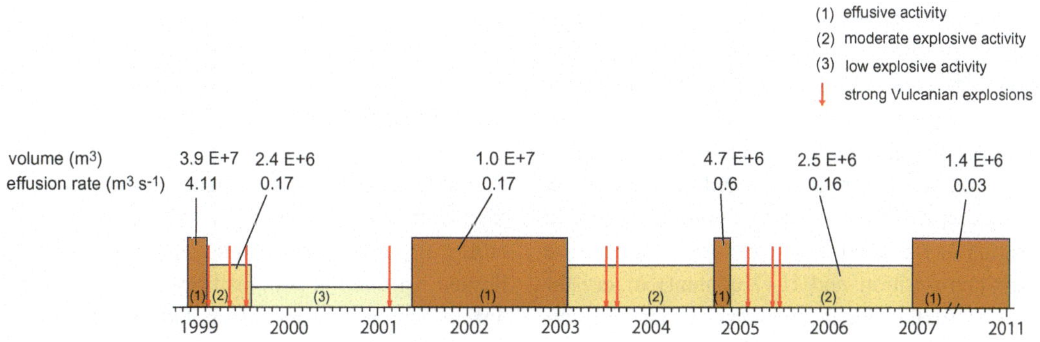

Fig. 1 Schematic depiction of Volcán de Colima volcanic activity between 1998 and 2011. Arrows mark the most powerful Vulcanian explosions. The volumes of magma erupted and the effusion rates for the different periods of activity are also indicated

steam and ash Vulcanian explosions. Large explosions, associated with pyroclastic flows, occurred in March, May, June, and September 2005. These eruptive events represent the most explosive behaviour of the volcano since the last Plinian eruption in 1913. A new dome appeared in the crater in February 2007 and grew at a very low rate until June 2011.

The total volume of magma erupted between 1998 and 2011 is estimated at 6.0×10^7 m^3 or 5.0×10^7 m^3 dense rock equivalent (DRE) assuming an average porosity of 16 vol% (Lavallée et al. 2012). 66% of this volume was erupted during the end of the 1998–early 1999 effusive phase. The average effusion rates for the different periods of activity range from 4.1 to 0.03 m^3/s during the 1998–1999 effusive and 2007–2011 dome growth periods, respectively (Fig. 1).

2 Petrology of the 1998–2011 Magmas

2.1 General Petrology

Descriptions of the post 1998 andesites are presented in several publications (Luhr 2002; Mora et al. 2002; Reubi and Blundy 2008; Reubi et al. 2017; Savov et al. 2008). The lavas are crystal-rich with 30–40 volume percent of phenocryst-sized minerals (i.e. irrespective of their origin) of plagioclase (23–34%), orthopyroxene (6–9%), clinopyroxene (3–8%), titanomagnetite (1.3–2.1%), and rare (<0.5%) ilmenite and resorbed hornblende (Fig. 2). Resorbed hornblende show a rim where they are progressively replaced by pyroxenes and titanomagnetite indicating that they are out of equilibrium in the magma at the conditions (pressure, temperature, melt composition and H_2O content) of crystallization of the others phenocrysts. The groundmass comprises glass and the same minerals, except that amphibole is lacking, and clinopyroxene and ilmenite are present in very small amounts. Olivine crystals surrounded by reaction coronas of pyroxenes and titanomagnetite (Fig. 2a), as well as resorbed quartz crystals are occasionally observed and have been interpreted as xenocrysts owing to their pronounced resorption textures (Luhr 2002; Mora et al. 2002; Reubi and Blundy 2008). Crystal clots, up to few mm in size, and showing a range of textures from aggregates of touching crystals with vesiculated interstitial glass to plutonic (i.e. fully crystallized) fragments are observed in all lavas. Most crystal clots are gabbroic and comprise strongly zoned clinopyroxenes and orthopyroxenes, resorbed olivines, and interstitial plagioclases ± glass (Fig. 2c). These mafic clots represent up to 15% of the total volume in the 1998–2011 magmas. Crystal clots comprising plagioclase, orthopyroxenes, ± clinopyroxenes are also observed.

Like most andesites, Volcán de Colima have, in detail, complex petrography comprising a large range of mineral textures and zoning patterns indicative of complex magmatic histories. Similar general patterns are, nevertheless, observed in all 1998–2011 lavas and variations concern mostly the relative proportions of the different types of mineral textures and subtle variations in chemical zoning across the crystals. The only exception is a sample collected on 8 December 1998, at the onset of the eruption, and presented in Mora et al. (2002) (their sample 1). This sample comprises an unusual amount of strongly resorbed (sieved-textured) plagioclases and granodioritic clasts, showing clear signs of melting with resorbed plagioclase and dark interstitial glasses. Similar textures are not common in subsequent magmas, including later 1998 lavas.

2.2 Plagioclase

Probably the most striking petrological feature of the 1998–2011 magmas is the abundance of euhedral plagioclase crystals showing relatively limited variation in compositions (Figs. 2a, b and 3a), a feature further illustrated by the broad unimodal main compositional population of $An_{60\text{-}40}$ (Fig. 4a). This contrasts with the bimodal populations and strong resorption (sieve) textures often observed in intermediate arc magmas (e.g. Eichelberger 1978; Kent et al. 2010). All samples show similar ranges of crystal interior compositions with a main populations around $An_{60\text{-}40}$ and

Fig. 2 Photomicrograph illustrating the petrographic features of Volcán de Colima 1998–2011 magmas. **a** Euhedral plagioclase, orthopyroxene, and clinopyroxene phenocrysts. A large resorbed olivine crystal mantled by a rim of orthopyroxene can be observed at the bottom left. **b** Large oscillatory zoned plagioclase. **c** Close-up view of gabbroic fragment. **d** Resorbed amphibole mantled by pyroxene, plagioclase and titanomagnetite

subordinate Ca-rich plagioclases (to An_{82}) (Fig. 4a). Rim compositions are more variable in the 1998–2002 samples (An_{63-40}) than in the 2004–2007 samples (An_{54-40}). The predominant type of plagioclase show two superimposed types of zonations, (1) oscillatory variations in An with short wavelength (<10 μm) and amplitudes of 5–10 mol% and (2) long wavelength (>100 μm) variations in An with amplitudes of 10–15 mol% (Fig. 3a). A significant proportion of phenocrysts have a core with slightly lower average An% delimited by a sharp increase in An% associated with a more or less pronounced resorption zone and surrounded by outer zones with higher average An% (Fig. 3b)The Ca-rich plagioclases (<5 vol%) have corroded high-An (An_{82-65}) cores in-filled in by low-An plagioclase (An_{55-45}) and melt inclusions, forming a patchy texture. These resorbed high-An plagioclases have oscillatory-zoned rims with the same range of An as the oscillatory phenocrysts (Fig. 3c). Groundmass microlite plagioclases range from An_{53} to An_{32} (Fig. 4a).

2.3 Pyroxenes

As for plagioclase, orthopyroxene and clinopyroxene show similar ranges of compositions in all samples with a broad main population around Mg-number 76-70 and Mg-rich outliers to

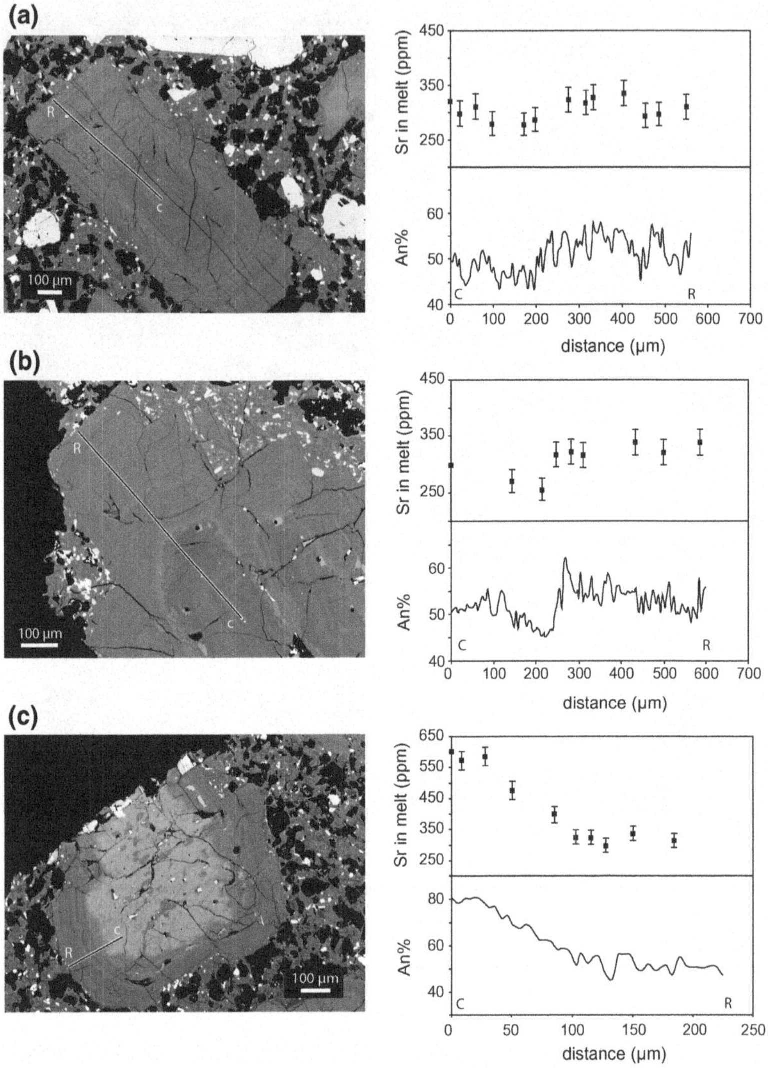

Fig. 3 BSE images and variations in An and calculated Sr_{melt} along traverses for selected plagioclases representative of the textural variability observed in the 1998–2011 magmas. The core (C) to rim (R) traverses are indicated on the BSE images by thick white and narrow black lines. Sr contents of melts were determined from the plagioclase Sr concentrations using the $^{Sr}K_{D\ plag/melt}$ expression from Blundy and Wood (1991) for a temperature of 990 °C. Error bars for calculated melt Sr contents take into account the Sr measurement analytical standard error and an uncertainty of ±40 °C on the temperature estimates. **a** Plagioclase showing weak oscillatory zoning. **b** Crystal showing weak oscillatory zoning with a relatively pronounced resorption zone. **c** Plagioclase with a corroded high-An core filled in by low-An plagioclase ($An_{55\text{-}45}$) and melt inclusions, forming a patchy texture. Zoning style shown in (**a**) is predominant in the 1998–2005 magmas

Mg-number 88 (Fig. 4b, c) [Mg-number = molar Mg/Mg+Fe, which gives an indication of how mafic was the melt in which the crystal formed (higher Mg-number correspond to more mafic compositions)]. The rim compositions are more variable and extend to higher Mg-number (80-66) in the 1998–1999 magmas than in the subsequent lavas (Mg-number 74-61) (Fig. 4b, c). The majority of crystals show weak oscillatory zoning, but pyroxenes displaying complex zoning with

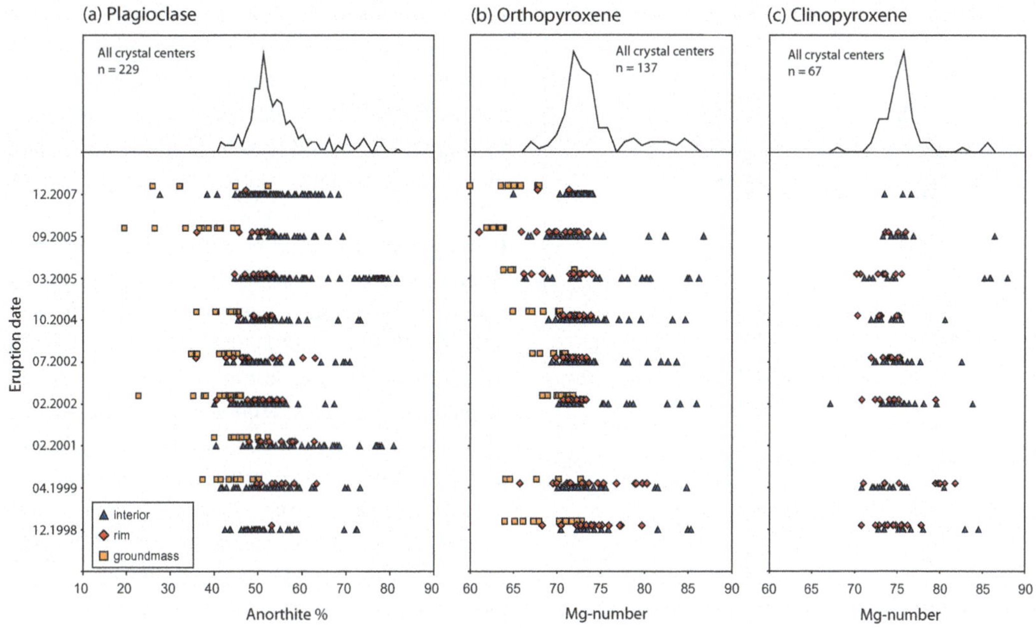

Fig. 4 Ranges of compositions of phenocrysts and groundmass microlites in the 1998–2011 magmas. **a** plagioclases, **b** orthopyroxenes and **c** clinopyroxenes. "Rim" denotes the outer 10 μm of the crystals, whereas "interiors" denotes the rest of the crystal. Histogram at the top show the range of compositions measured at the center of the crystal as seen in thin section for all samples

several successive reverse (to Mg-number 88) and normal zones are also observed. These pyroxenes always have a low-Mg-number rim compositionally similar to the weakly zoned pyroxenes. A small proportion of crystals had a narrow rim with Mg-number higher than the crystal interior (reverse zoning). Pyroxene microlites are orthopyroxenes Mg-number 73-60 (Fig. 4b).

2.4 Amphibole

Amphibole phenocrysts are rare in the 1998–2011 magmas (<0.5 vol%) and always show disequilibrium textures. Two distinct types of resorption are observed: (1) Subhedral crystals with thin (10–50 μm) black, fine-grained "opacite" reaction rims, (2) anhedral amphiboles partially to completely replaced by fine- to medium-grained pyroxene, plagioclase and titanomagnetite forming gabbroic crystal clusters (Fig. 2d). The first type of reaction rims occur in the 1998–1999 lavas (Luhr 2002) but have not been observed in later eruptions. Amphiboles are hornblende with Mg-number 71-63.

3 Chemistry of the 1998–2011 Magmas

3.1 Bulk Rock Chemistry

Bulk rock compositions of post-1998 magmas were published by Luhr (2002), Mora et al. (2002), Reubi and Blundy (2008) and Savov et al. (2008). The 1998–2011 magmas are fairly homogenous calc-alkaline andesites with 59–61.4 wt% SiO_2 (Fig. 5). Broad linear correlations are observed between most major elements. SiO_2 correlates negatively with CaO, MgO, FeO, and positively with K_2O. No clear temporal trend is observed between 1998 and 2011. Individual lava flows and samples from single eruptive phase show variations as large as the complete data set. Wider compositional variations and slightly more mafic magmas with lower SiO_2 are

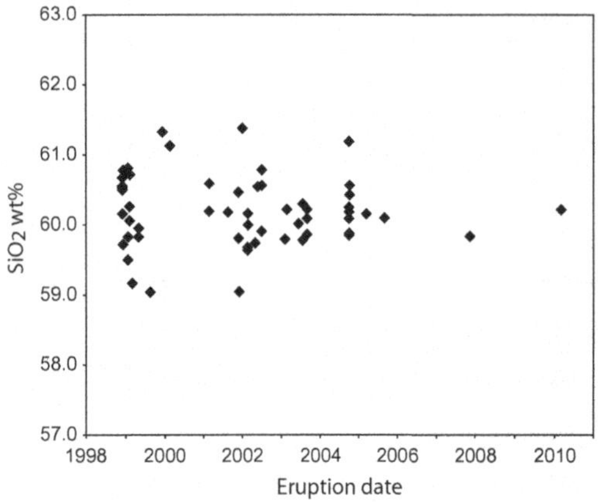

Fig. 5 Bulk-rock SiO_2 contents versus eruption date. Data are from Luhr (2002), Mora et al. (2002), Reubi and Blundy (2008), Savov et al. (2008)

observed in early 1999 and at the end of 2001 to early 2002 (Fig. 5). These more mafic magmas have compositions that approach those of the mafic andesites erupted during the cycle-ending Plinian eruptions, which raised concerns about the imminence of a major explosive eruption (Luhr 2002). This point is discussed later in this chapter.

3.2 Melt Inclusion Chemistry

Detailed studies of melt inclusions (MI) in the 1998–2005 magmas were presented by Reubi and Blundy (2008) and Reubi et al. (2013). Melt inclusions in pyroxenes show a broad range of compositions with SiO_2 contents from 62 to 76 wt% (H_2O-free) and can be divided into two distinct chemical groups on the basis of their K_2O contents: termed "low-K" and "high-K" by Reubi and Blundy (2008) (Fig. 6). Low-K MIs show decreasing MgO, CaO, Al_2O_3 and Na_2O, and increasing K_2O trends with increasing SiO_2. These trends are consistent with crystallization of the phenocrysts assemblage, and overlap with the groundmass glasses (Fig. 6), indicating that the low-K MIs effectively record the composition of the crystallizing melt. There is a compositional gap between the least evolved MI (dacite) and the most evolved bulk-rock (andesite) and the MI and bulk-rock trends are oblique to each other on most major element plots (Figs. 6 and 7). These observations have important implications for the petrogenesis of the magmas and are discussed below. High-K melt inclusions are characterized by higher K_2O and lower CaO contents at a given SiO_2 content compared with the low-K MIs. Their major and trace element (e.g. Ba, Sr, Rb) contents are clearly distinct from the bulk compositions of Volcán de Colima magmas, but are similar to interstitial glasses found within the gabbroic clots (Reubi and Blundy 2008). The high-K MIs are thought to have formed by partial melting of plutonic fragments. Although not representative of the crystallizing melts, this group of inclusions also provides important petrogenetic information.

The ranges of melt inclusion compositions remain essentially constant in the 1998–2005 magmas. H_2O and CO_2 contents obtained by ion probe analyses in the 1998–2005 low-K melt inclusions are presented in Reubi et al. (2013). H_2O contents range from 4.1 to 0.1 wt% and CO_2 range from 236 ppm to below analytical detection limit (12–28 ppm) (Fig. 8). Sulphur contents extend to high values in Volcán de Colima MIs (≤ 1300 ppm) and Cl contents range from 1540 to 2600 ppm [microprobe measurements from Reubi et al. (2013)]. No systematic differences in volatile contents are observed between the effusive and Vulcanian explosive eruptions.

4 Pre-eruption Magma Conditions

4.1 Temperature and Oxygen Fugacity

Estimates of the temperature and oxygen fugacity (fO_2) of magmas during crystallization can be obtained from the compositions of coexisting minerals and melts (glasses). Two-pyroxene geothermometers have been used extensively to constrain the temperature of Volcán de Colima post-1998 magmas. Temperatures between 979 and 1030 °C were obtained using the formulation of Andersen et al. (1993) (Luhr 2002; Reubi and Blundy 2008; Savov et al. 2008). Similar ranges of temperatures were reported for calculations according to the methods of Wood and

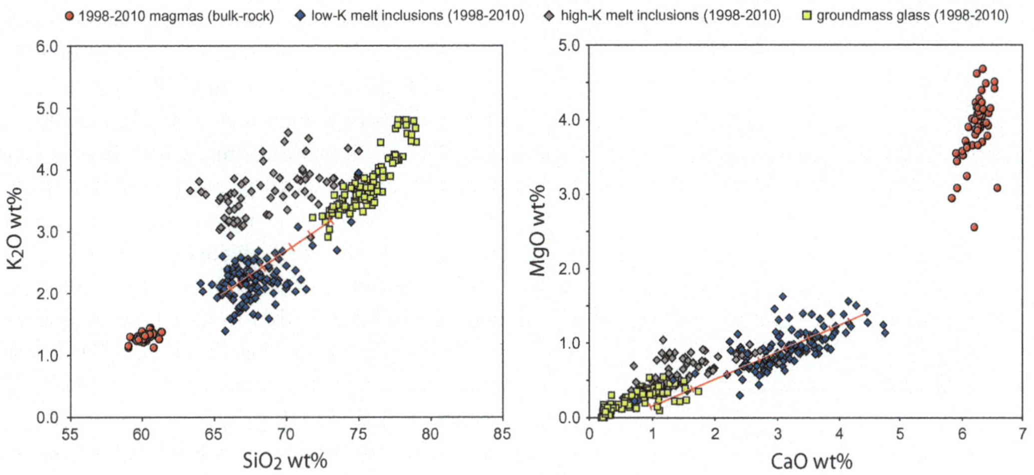

Fig. 6 K_2O versus SiO_2 and MgO versus CaO (wt%) variations in melt inclusions, groundmass glasses and bulk-rocks. All data are plotted H_2O-free. Red lines show the crystallization trend produced by fractionation of the phenocrysts assemblage with 5% increments. Melt inclusion data are from Reubi and Blundy (2008) and Reubi et al. (2013)

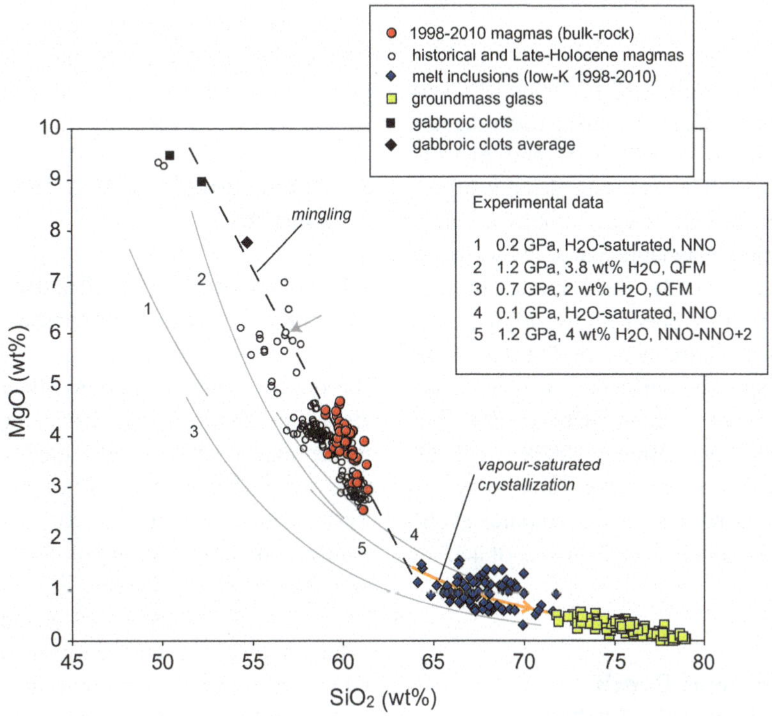

Fig. 7 Chemical variations in melt inclusions, bulk-rock compositions, and groundmass glasses. Curved lines represent the liquid lines of descent based on experimental studies of hydrous basalts over a range of crustal pressures and H_2O contents. Experimental data are from; 1. Sisson and Grove (1993), 2. Muntener et al. (2001), 3. Sisson et al. (2005), 4. Grove et al. (1997), and 5. Alonso-Perez et al. (2009). Compositions of gabbroic clots and melt inclusions are from Reubi and Blundy (2008). Bulk-rock compositions for historical magmas, as in Fig. 5, Late-Holocene bulk-rock compositions are from Luhr et al. (2010). Grey arrow indicates the olivine-andesite used for fractional crystallization model by Luhr and Carmichael (1980), see text for detail

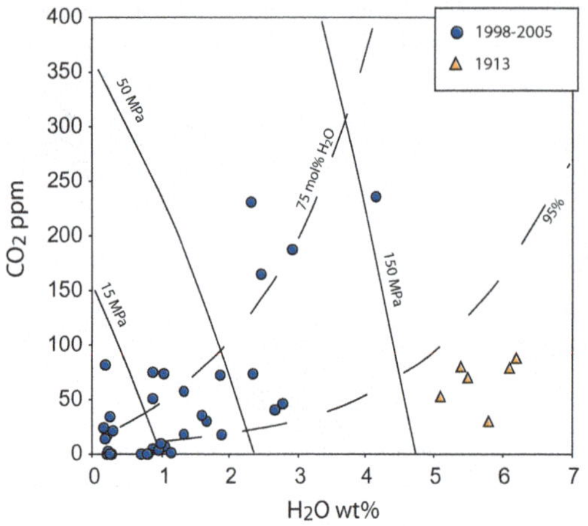

Fig. 8 H_2O and CO_2 contents of melt inclusions from Volcán de Colima. Data for the 1998–2005 magmas are from Reubi and Blundy (2013) and for the 1913 magmas from Luhr (2006). Vapour isopleths (long dash lines) and isobars (short dash lines) are calculated using the solubility model of Papale et al. (2006) for the average 1998–2005 low-K melt inclusion composition

Banno (1973) and Wells (1977). Ilmenite is very rare in Volcán de Colima's historical magmas, but temperatures of 938–993 °C and log(fO_2) values of −9.2 to −9.96 were calculated using the method of Ghiorso and Evans (2008) for three touching ilmenite-magnetite pairs (Mora et al. 2002; Reubi et al. 2013). Temperatures of crystallization were also obtained by applying the plagioclase-melt thermometer of Putirka (2005) to melt inclusions and adjacent host plagioclase. This method gives a range between 959 and 1015 °C (Reubi and Blundy 2008). All the magmas investigated give the same range of temperatures and no systematic temporal evolution has been identified since the onset of activity in 1998.

4.2 Pressure and Depth of the Magmatic System

Constraints on the pressure at which the magmas crystallized can be obtained from the volatile contents of melt inclusions using the known solubility-pressure relations for mixed H_2O–CO_2 in silicate melts. This approach implicitly assumes that the magmas were vapour-saturated, i.e. vapour bubbles were present and in equilibrium with the melt at the time of crystallization and sealing of the melt inclusions. In the case of the 1998–2011 magmas this assumption is supported by experimental petrology results (Moore and Carmichael 1998) and the degassing trends recorded by the melt inclusions (Reubi et al. 2013). Pressures of melt inclusion entrapment between 160 and 3 MPa were obtained using the solubility model of Papale et al. (2006) and assuming magma temperatures of 1000 °C (Reubi et al. 2013) (Fig. 8). These pressures are equivalent to depths of 8.4–0 km (below the summit level) calculated using a lithostatic model taking into account the edifice load (Pinel and Jaupart 2000) (Fig. 9b). Significant and systematic variations in entrapment depths are obtained. The 1998 and 2004–2005 magmas record maximum depths between 3 and 4 km, the 1999 and 2003 Vulcanian eruptions shows depths up to 9 km, whereas the 2001 and 2002 magmas record maximum values <2 km (Fig. 9b).

5 Petrogenesis of the 1998–2011 Magmas

5.1 Fractional Crystallization and Crustal Contamination

The origin of intermediate to silicic magmas in subduction settings has been the subject of debate and is widely attributed to two main processes: differentiation of primary magmas by crystallization within the crust or uppermost mantle, and partial melting of older crustal rocks (see Annen et al. 2006; Gill 1981; Tatsumi 2005). Nd–Sr–Pb isotope studies of Volcán de Colima magmas have shown that assimilation of old crustal rocks did not contribute significantly to the production of the magmas (Luhr 1997; Valdez-Moreno et al. 2006; Verma and Luhr 2010), although limited amounts (<10%) of assimilation of granodiorites from the batholiths (Cretaceous) forming the basement beneath Volcán de Colima occurred (Valdez-Moreno et al. 2006; Reubi et al. 2014). The short lived

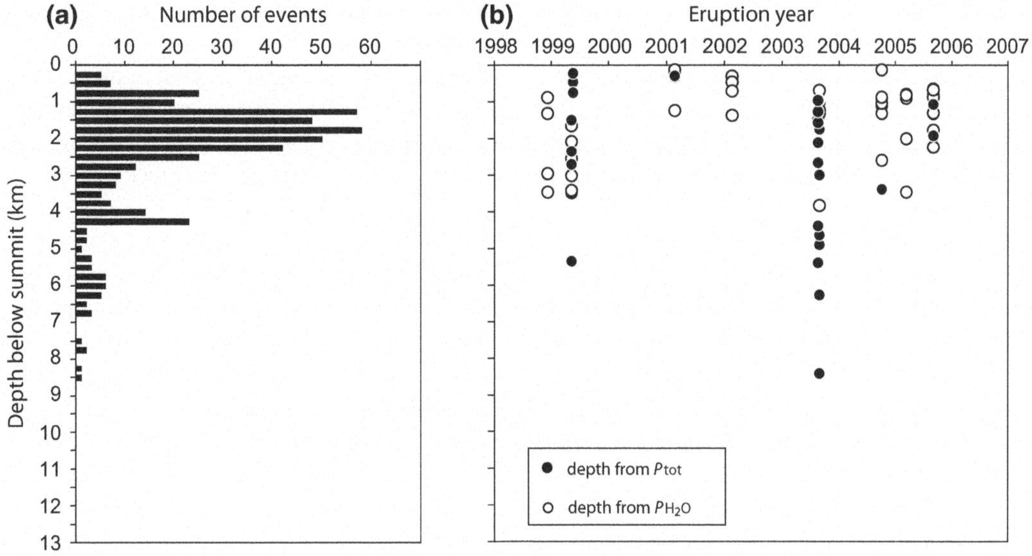

Fig. 9 **a** Histogram showing the depth of volcano-tectonic seismic swarms recorded in 1997–1998 (data from Zobin et al. 2002a, b). **b** Time-series showing the depths of crystallisation estimated on the basis of entrapment pressure of melt inclusions converted to depths according to the method of Pinel and Jaupart (2000). Melt inclusions entrapment pressures are based on dissolved H_2O and CO_2 contents using the solubility model of Papale et al. (2006). Figure modified from Reubi et al. (2013)

U-series isotopes further indicate that the time scales of production of the historical andesites by differentiation of primary magmas resulting from melting of the mantle are short (less than few thousand years) (Reubi et al. 2014). Overall, the geochemical data indicate that crystal fractionation from mafic parent magmas is the primary process that produced the silicic magmas at Volcán de Colima, a process thought to be the predominant producer of silicic magmas in arc setting (e.g. Annen et al. 2006; Reubi and Blundy 2008).

In the context of differentiation dominated by fractional crystallization, the oblique chemical trends exhibited by the bulk rock and melt inclusions (Fig. 7) could reflect a change in the assemblage of crystallizing minerals. Change from high-pressure crystallization of basalt or basaltic andesite parents to generate the dacitic melt, to low-pressure vapour-saturated crystallization of the dacitic melt to form rhyolitic melt is a plausible cause (Annen et al. 2006). Luhr and Carmichael (1980) have shown that crystal fractionation models from prehistoric olivine-andesites (57 wt% SiO_2, 6 wt% MgO) can reproduce the major-element variations in historical andesites, but systematically fail to predict sufficient concentrations of compatible trace elements (Cr, Ni), an indication that magma mixing or mingling and crystal fractionation occurred simultaneously. These authors also noted that the olivine-andesites themselves present petrographic evidence indicative of magma mixing, and that high-Al basalts erupted from late-Quaternary cinder cones surrounding Volcán de Colima likely represent the parental magmas of the historical andesite suites. Additional support for mixing-mingling processes comes from comparisons between the liquid lines of descent produced experimentally for hydrous basaltic melts and the rock compositions. The curved experimental liquid lines of descent reproduce the composition of the silica-rich melt inclusions but fail to reproduce the bulk composition of historical andesites, which lie on linear trends diagnostic of mixing processes (Fig. 7) (Sparks and Marshall 1986; Reubi and Blundy 2008).

The (low-K) melt inclusions with the lowest SiO_2 contents match the required silicic mixing end-member responsible for the observed linear trends (Fig. 7). The bulk composition of the silicic magma that mixed or mingled with a mafic component must, therefore, have been close to these compositions. Starting from such compositions, the melt inclusion trends record up to 40% vapour-saturated crystallization of a crystal assemblage, similar to the phenocryst assemblage (i.e. 20% plagioclase, 3% orthopyroxene, 0.7% clinopyroxene and 0.5% titanomagnetite) (Reubi et al. 2013) (Figs. 6 and 7). Overall, this indicates that the magmas feeding the upper crustal magmatic system are dacitic in composition ($\sim$64 wt% SiO_2) and that these magmas mixed or mingled with a mafic component on their way to the surface and started crystallizing extensively in the subvolcanic system upon reaching vapour-saturation (Fig. 10). The fact that the starting point for crystallization to higher SiO_2 lies close to the proposed silicic end-member for mixing implies that the mixing process did not generate any truly hybrid liquids that subsequently crystallized at low pressure. In that case, we would expect low-pressure crystallization trajectories originating along the mixing array, rather than only from its most SiO_2-rich extremity. This mixing behaviour is suggestive of mixing between evolved (dacitic) melt and crystal residues at low SiO_2, rather than between dacitic and basaltic liquids. This style of mixing is consistent with the petrographic evidence presented above. We explore these possibilities further below.

5.2 The Role of Magma Mixing and Mingling

From a bulk-rock chemistry point of view magma mixing (i.e. complete homogenization of contrasting compositions resulting in magma with intermediate properties) and magma mingling (i.e. mechanical juxtaposition of contrasting compositions with limited chemical homogenization) are equivalent. From a physical (and volcanological) point of view these processes can, however, have very different implications. Transfer of heat and volatiles associated with injection of basaltic magma into silicic magmas is believed to be a common mechanism for triggering explosive eruptions (Sparks et al. 1977). At the other end of the spectrum, entrainment in silicic magmas of partially to fully solidified mafic crystal mush or plutonic fragments formed during previous episodes of magmatism can be regarded as a passive form of magma mingling, because the mafic end member does not provide heat or volatiles. Distinguishing between these processes is essential to understanding the dynamics of the magmatic system, and to assess the impacts in terms of eruption likelihood and intensity.

The compositions of the gabbroic fragments observed in Volcán de Colima 1998–2011 magmas match the composition of the putative mafic mixing end-member (Fig. 7). The gabbroic fragments do not show textural evidence for quenching of the sort typically observed in mafic enclaves. On the contrary, they show evidence for heating and partial melting, the process during which the high-K melt inclusions are inferred to have formed (Reubi and Blundy 2008). Entrainment and heating of gabbroic crystal mush and plutonic fragments from Volcán de Colima's earlier episodes of magmatism, a passive form of magma mingling, is evidently an important process in the evolution of the 1998–2011 magmas. Mingling between these plutonic fragments and a dacitic melt is capable of generating the linear trends observed in Fig. 7. In such case the hybrid magma is a mixture of gabbroic crystals, in various states of disequilibrium, entrained in a dacitic melt. Subsequent, post-mingling crystallization of this melt generates the more evolved melt inclusion and groundmass glasses.

Chemical zoning of the phenocrysts may indicate that mixing occurred concomitantly with mingling. Major-element zoning of plagioclase within the Anorthite-Albite binary system has been widely used as a recorder of magma evolution, although the causes of oscillatory zoning of plagioclase are still a matter of debate (see Ginibre et al. 2002a; Pearce and Kolisnik 1990).

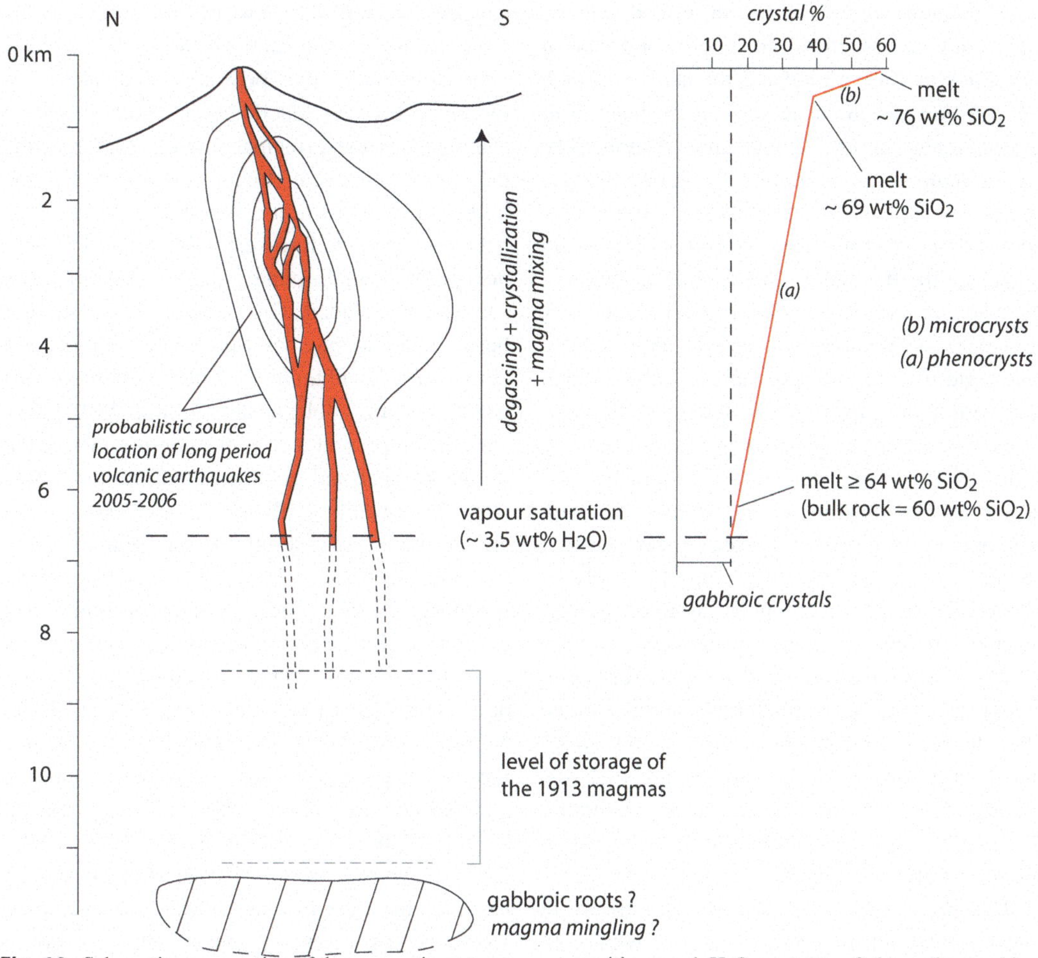

Fig. 10 Schematic cross section of the magmatic system feeding Volcán de Colima 1998–2011 activity. Crystal contents of the magmas are calculated using the melt inclusion compositions. The crystallization path and melt compositions depicted are averages, in detail the compositions and H_2O contents of the melts reaching vapour-saturation and, consequently, the crystallinity and composition of the ascending melts vary slightly. Probabilistic contours of volcanic earthquakes source location are from Petrosino et al. (2011)

The major element composition of plagioclase (An%) is a function of melt composition and H_2O content, temperature, and pressure. Thus, it is difficult to discriminate unequivocally between these rival factors on the basis of binary compositional variations. Minor and trace element analyses have been used to address this issue. For trace elements, for which the partitioning between crystal and melt is reasonably well understood, e.g. Ba and Sr (Blundy and Wood 1991) it is possible to relate trace element variations in the plagioclase to those in the coexisting melt, and thus constrain the role of melt composition in promoting plagioclase zoning (Blundy and Shimizu 1991; Ginibre et al. 2002b; Singer et al. 1995). Sr measurements in plagioclase by ion probe were presented in Reubi et al. (2013).

Two distinct patterns are observed in Volcán de Colima 1998–2011 plagioclases. The oscillatory variations in An% with short wavelength (<10 μm) are not associated with resolvable changes in Sr contents of equilibrium melts (Sr_{melt}) calculated on the basis of plagioclase Sr contents, whereas the long wavelength (>100 μm) variations are associated with changes in Sr_{melt} (Fig. 3a, b). The causes of the first type of oscillations are open to question and may be ascribed either to changes in temperatures or

H_2O content, or to diffusion-limited kinetic effects during crystallization. The second type of variations in An% are related to changes in melt composition and indicate mixing between contrasted melts in the subvolcanic system. The predominant weakly oscillatory zoned plagioclases, as well as the crystals with more pronounced resorption zones, record variations in Sr_{melt} within the range 250–350 ppm (Fig. 3a, b), whereas the melt inclusion Sr contents vary from 180 to 590 ppm. Changes in melt compositions during crystallization are, therefore, small and within the range of the dacitic melt inclusions. The presence of pyroxenes with reversely zoned sectors or rims further indicates the occurrence of magma mixing events. As with plagioclase, changes in pyroxene composition are limited (<5 Mg-number) (Fig. 4), implying small changes in melt composition. Interestingly, reverse zoning is more pronounced (up to 10 Mg-number increase) and the pyroxenes rim compositions extend to higher Mg-number in the 1998 to early 1999 magmas than in the subsequent eruptions (Fig. 4), suggesting that the mixing melts were more mafic and, potentially, that mixing was more important at the onset of the recent eruptive phase.

The high An% plagioclases crystallized in mafic melts with high Sr_{melt} (Fig. 3c) which are also most certainly the source of the resorbed olivines and high Mg-number pyroxenes observed in all samples. Based on compositional and textural similarities, these mafic crystals were interpreted as coming from disaggregation of the gabbroic clots (Reubi and Blundy 2008). They may, nevertheless, also indicate mixing with basaltic magmas. Considering that the mafic crystals are ubiquitous, but that the phenocrysts do not record evidence for extensive mixing with basaltic magmas, mixing would in this case have occurred before crystallization of the phenocrysts. Moreover, if mixing with basaltic magmas was predominant, the least evolved melt inclusions would be expected to match the bulk-rock compositions, as they should be representative of the melt composition at the onset of phenocryst crystallization after mixing. This suggests instead that most of the mafic crystals come from disaggregation of the gabbroic clots.

In summary, the petrology indicates that mingling between dacitic melts and grabbroic fragments is the principal process that controls the linear chemical trends exhibited by the bulk-rock compositions, although an early contribution (i.e. before crystallization of phenocrysts) from mixing with mafic magmas cannot be excluded. On the other hand, mixing between silicic magmas with limited contrasts in composition is ubiquitous in the current subvolcanic system. Volcán de Colima 1998–2011 andesites represent a case of mingling, where the dominant heat and volatiles contributions come from the ascending silicic melt that remobilizes earlier mafic crystal mush or plutons. The opposite scenario, where the ascending mafic magma provides the heat necessary to remobilize partially to fully crystallized silicic crystal mush or plutons, is more commonly envisaged in other andesite systems (e.g. Eichelberger et al. 2006; Murphy et al. 2000). This situation is characterized by the presence of quenched mafic enclaves, and strong resorption textures and chemical zoning in phenocrysts. Volcán de Colima 1998–2011 magmas may, in this regard, represent an end-member scenario in andesite petrogenesis. However, as discussed below, this is not a characteristic of all Volcán de Colima historical magmas, as the relative contribution of mixing varied significantly over time.

5.3 Linking the Petrological and Surface Monitoring Records

5.3.1 Melt Inclusions and Seismicity

In the preceding sections, inferences on the physical conditions and magmatic processes prevailing in the magmatic system beneath Volcán de Colima have been drawn on the basis of petrological and geochemical data. A fundamental assumption subjacent to this approach is that the phenocrysts and melt inclusions are representative of the current state of the system. If the crystals formed hundreds or thousands of years ago, this may not necessarily be the case. Establishing a link between the petrological record and the monitoring signal obtained at the surface prior to the eruption is needed to demonstrate a temporal connection and the feasibility of integrated approaches.

The depths of volcano-tectonic seismic swarms recorded prior to the eruption of the 1998–99 lava flow are in the range 8.5–0.5 km (below the summit) (Zobin et al. 2002a) (Fig. 9 a). The source locations of long-period seismicity recorded by broadband seismic stations between November 2005 and May 2006 are in the same depth range, with maximal probabilistic hypocenter location between 1.5 and 4 km (Petrosino et al. 2011). This is in good agreement with the depth recorded by the melt inclusions for the 1998–2005 period (Fig. 9b) (Reubi et al. 2013). This agreement supports the view that the melt inclusion sealing depths correspond closely to the region in the subvolcanic system, where the magmas were stored before eruption. In other words, both the seismic signals and the melt inclusions record the ascent, degassing, and crystallization of the volatile-saturated magma before eruption. Interestingly, melt inclusions from the 1913 plinian eruption have H_2O contents up to 6.8 wt% (Atlas et al. 2006; Luhr 2006) (Fig. 8) and yield pressures corresponding to depths of 8–11 km, indicating that 100 years ago the magmas crystallized in a region distinct from the currently active seismic zone (Fig. 10). This suggests that there is not only a spatial but also a temporal correspondence between the petrological and seismic records. A similar spatial and temporal correspondence has been established for the Mount St. Helens magmatic system between 1980 and 2006 (Blundy et al. 2008). We note that this does not necessarily imply late crystallization of the phenocrysts, but could reflect predominant sealing of the melt inclusions shortly before eruption. Textural observations indicate that the cavities eventually forming the inclusions are connected to the surrounding melt by thin melt channels or necks. Diffusion-exchange maintain equilibrium until closure of the channel, which may happen long after crystallization of the host and formation of the cavity itself, and may often be caused by magma ascent and decompression prior to eruption (Blundy and Cashman 2005; Blundy et al. 2008).

The close correspondence between the melt inclusion sealing depths and seismic source locations holds when the complete inclusion population is considered. In detail, significant variations are observed with time (Fig. 9b). Most significant are the low melt inclusion pressures recorded in 2001 and 2002. On the basis of petrological arguments, Reubi et al. (2013) argued that these low pressures are melt inclusion artifacts resulting from diffusive loss of H_2O from the inclusions, and that the magmas must have crystallized at the same range of pressures as other 1998–2011 magmas. We suggested that the pressure corresponding to preferential reequilibration of melt inclusions may, nevertheless, indicate magma stalling levels and could potentially provide important information on the flow dynamics in the magma conduit. Unfortunately, locations of seismic sources are not reported for the 2001–2002 period, making it difficult to assess the significance of the melt inclusions pressures of re-equilibration. In addition, it remains to be established over what time scales the melt inclusion and seismic records are correlated. Volcán de Colima data indicates that it is the case for decadal time scales: establishing if this still holds when time scales range from years to months needs to be considered in future research.

5.3.2 Melt Inclusions and Gas Fluxes

Comparisons between gas fluxes from the vent and the fluxes estimated from melt-inclusion volatile contents, combined with the volume of erupted magmas, can provide information on the volume of magma degassing at depth and influx of magmas in the system. Based on melt inclusion S contents (up to 1300 ppm), the petrological estimate of the output between 1998 and 2005 is 2.5×10^8 kg SO_2 (Reubi et al. 2013). This is approximately a quarter of the mass derived from COSPEC measurements for the same period (Varley, this volume). Excess of degassed sulphur relative to petrological estimates is characteristic of arc volcanoes. The excess calculated for Volcán de Colima is small compared to the factor of ten to hundred obtained for many intermediate stratovolcanoes (Shinohara 2008; Wallace 2005). S excess may have several causes: larger volume of magma degassing at depth than the erupted volume, degassing of S before sealing of the melt inclusions, or degassing of S-rich mafic magmas deeper in the system. There are no constraints on the cause at Colima, but assuming that S excess solely results from larger volume of magma degassing at depth and, consequently, reflects the volume of degassing magma between 1998 and 2005, the volume required is about 0.2 km^3 (DRE). Probabilistic hypocenter location analyses of the 2005–2006 seismic swarms indicate a seismic source volume of about 1 km^3 (Petrosino et al. 2011). The density of seismic sources in this volume precludes the presence of a large magma body and rather suggests that the magmas are located in small reservoirs, likely a network of dikes that occupy at most 20% of this volume according to the petrological estimate.

^{210}Pb–^{226}Ra disequilibria in magmas offers the opportunity to constrain the time scale of degassing, since one of the intermediate isotopes, ^{222}Rn, is highly volatile and fractionate into the gas phase (e.g. Berlo and Turner 2010). Using this method, Reubi et al. (2015) obtained time scales of degassing between less than 4 month (i.e. within error of the method) and 11 years for Volcán de Colima 1998 to 2011 magmas. Overall Volcán de Colima ^{210}Pb–^{226}Ra data further suggest a complex subvolcanic magmatic system comprising several conduits and indicates that this system comprises multiple magma batches with variable degassing histories and ascending rates (between >8×10^{-4} and 5×10^{-5} m s^{-1}).

5.4 Will the Current Activity Ramp up to a Plinian Eruption?

Several observations have been taken as indications that the current eruptive phase may terminate with a major Plinian or sub-Plinian eruption. These include; (i) the fact that the 2005 Vulcanian eruptions represent the most explosive behaviour of the volcano since the last Plinian eruption in 1913, (ii) some of the 1998–2011 magmas are approaching the composition of the more mafic 1913 scorias, and (iii) compilations of historical records suggest a 80 to 100 years periodicity for recent Plinian eruptions at Volcán de Colima (Bréton González et al. 2002; Luhr 2002; Luhr and Carmichael 1990). There are strong indications that the eruptive cycles are closely related to magmatic cycles, a link marked by the systematic shifts to more mafic bulk-rock compositions associated with the cycle-ending Plinian and sub-plinian eruptions (Luhr 2002; Luhr and Carmichael 1990; Robin et al. 1991; Savov et al. 2008). The significance of these shifts is a matter of debate. Robin et al. (1991) argued that this results from an influx and mixing of mafic magmas, whereas Luhr (2002) suggested that the mafic magmas represent a mafic layer situated at the bottom of a chemically zoned magmatic chamber, progressively emptied during the previous eruptive cycle. Distinguishing between these scenarios and establishing the significance of the more mafic magmas erupted in 1999 and 2001–2002, that are close in composition to the 1913 scoria, is essential to anticipate the future evolution of the volcanic activity.

Compared to the 1998–2011 magmas, the 1913 scorias are characterized by higher amphibole and lower plagioclase proportions (plagioclase 12–22%, amphibole 3–6%, orthopyroxene 2–3%, clinopyroxene 2–5%, titanomagnetite 0.5–3%). The most significant differences are the presence of euhedral resorption-free amphiboles, higher proportions of resorbed plagioclase with sieve-textures (70–80% of plagioclases), more An-rich and broader compositional distribution of plagioclases, more pronounced reverse zoning of phenocrysts (plagioclase with up to 20 An% increase), and significantly more mafic groundmass glass compositions (Fig. 11) (Pickles 2007). Ion probe measurements of melt inclusions from the 1913 scorias indicate higher H_2O contents (up to 6.8 wt%) and lower CO_2 contents (30–88 ppm) (Atlas et al. 2006; Luhr 2006) than in the 1998–2011 melt inclusions (Fig. 8).

Temperature estimates based on two-pyroxenes and the amphibole composition (Ridolfi et al. 2010) additionally indicate that the temperatures of the 1913 magmas were lower (930–990 °C compared to 980–1030 °C for the 1998–2011 magmas). The volume of magma erupted in 1913 is estimated at 0.64 km^3 (DRE) (Saucedo et al. 2010), which exceeds the inferred volume of magma in the current shallow magmatic system (<0.2 km^3). Similarities between the 1913 and 1998–2011 magmas include; the composition of pyroxenes and melt inclusions (Fig. 11), as well as the presence of gabbroic crystal clots and resorbed olivines.

Several important conclusions can be drawn from these observations. The melt inclusions indicate that initially the 1998–2011 and 1913 magmas had similar compositions, despite the distinct bulk-rock compositions, but the 1913 magmas had higher H_2O contents and where stored a higher pressure and lower temperature (Figs. 8 and 10). Assuming that it will originate

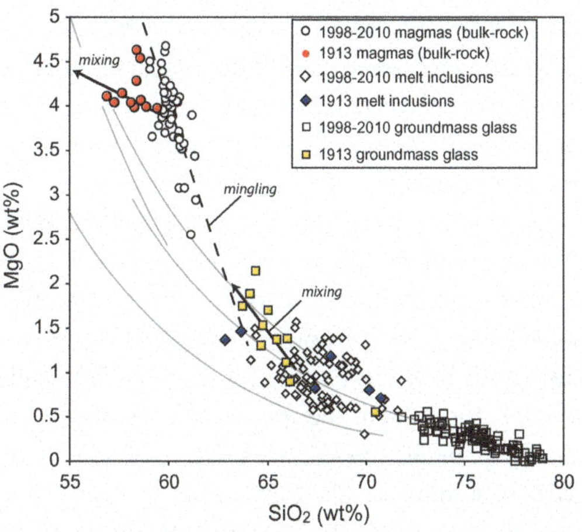

Fig. 11 MgO versus SiO_2 (wt%) diagram showing the differences in compositions between the 1913 and 1998–2011 bulk-rocks, melt inclusions and groundmass glasses. Curves and the dashed line are experimentally produced liquid line of descents and the mingling trend as in Fig. 7. Black arrow represents the effect on the bulk-rock compositions of mixing with a mafic magma. The same mixing process also explains the ground groundmass glasses that are systematically more mafic than the melt inclusion compositions in the 1913 magmas, an obvious difference with the 1998–2011 in which the groundmass glasses are more silicic than the melt inclusions

within the currently active region of the subvolcanic magmatic system (as imaged by the seismicity and petrology), the explosive potential (reduced for magmas with low H_2O content) and the impact of a major explosive eruption terminating the current eruptive cycle may, therefore, be anticipated to be reduced compared to the 1913 eruption as previously suggested by Luhr (2002). In addition, the groundmass compositions that are more mafic than the melt inclusions (Fig. 11) and the pronounced reverse zoning shown by the phenocrysts indicate that mixing with mafic melts was important prior to the 1913 eruption. Similar evidence for mixing with mafic melts is lacking in the 1998–2011 magmas (although there are indications for mixing between silicic melts). The 1999 and 2001 mafic spikes in bulk-rock chemistry are, therefore, unlikely to represent a scenario similar to pre-1913. We note also that the 1998–2011 and 1913 magmas exhibit distinct bulk-rock compositional trends (Fig. 11). As discussed above, the steep 1998–2011 trend corresponds to a mingling trend. The 1913 magmas also contain the mingled gabbroic clots, the displacement of these magmas away from the mingling trend toward the liquid lines of descent (i.e. lower SiO_2) indicates that magma mixing is an additional factor (Fig. 11). Considering that the more mafic 1999 and 2001 lavas fall on the mingling trend and are outliers in a dataset that does not show any temporal evolution (Fig. 5), we suggest instead that these mafic compositions result from higher proportion of mingled mafic fragments.

The petrology and geochemistry of the magmas indicate that not only the conditions of storage, but also the magmatic evolution of the 1913 magmas differed significantly from the evolution of the 1998–2011 magmas. It is also worthwhile mentioning that no significant differences in petrology, volatile-contents, and degassing paths are observed between the 1998–2011 Vulcanian and effusive eruptions, suggesting that, in these cases, the transitions from effusive to explosive result from non-linear dynamics of conduit flow related to the complex relationship between degassing and crystallization (Cassidy et al. 2015; Reubi et al.

2013; Varley et al. 2010). Based on the 1913 scenario, the cycle-ending Plinian and sub-Plinian eruptions are anticipated to follow major perturbations in the subvolcanic system related to the influx of volatile-rich mafic magmas. The strong Vulcanian eruptions that occurred in 2005 are, therefore, most likely not indicative of changes in the subvolcanic system leading to a major Plinian eruption. We stress that if the likelihood of a Plinian eruption emanating from the subvolcanic system currently imaged by the seismicity and petrology can reasonably be estimated to be lower than in 1913, the possibility that such an eruption may be fed by deeper more volatile rich magmas remains open. Seismic signals are induced by movements of magmas or gas (e.g. McNutt 2000) and the absence of seismicity below 8 km does not exclude the presence of magmas at the level of the 1913 reservoir or below, particularly if these magmas are vapour-undersaturated. Investigations using ambient noise tomography have, indeed, revealed the presence of an ellipsoidal low velocity body at a depth of ~15 km beneath Colima volcanic complex (Spica et al. 2017). Evaluating the exact nature, melt content and eruptive potential of this low velocity body is, in this context, of prime importance.

5.5 Open Questions and Future Work

The respectable number of petrological and geochemical investigations of Volcán de Colima historical magmas enables us to define the general characteristics of the magmatic system feeding this hyperactive arc volcano. As always in science, these advances raise as many questions as they give answers. Here, we present a non exhaustive list of some of the issues that in our opinion need consideration.

At Volcán de Colima, like at most intermediate to silicic arc volcanoes, there are strong indications that the phenocrysts crystallized at relatively low pressure (i.e. in the upper crust) and under vapour-saturated conditions (Annen et al. 2006; Reubi and Blundy 2009). Considering that the melts are already dacitic in composition when they reach vapour-saturation, the question arises: where does the initial step of differentiation occur? Lower crustal hot zones are potential candidates (Annen et al. 2006), but mid-crustal reservoirs also need to be considered. Of prime importance, from a volcanological point of view, are the volumes of melt present in these lower reservoirs, that are available to sustain the volcanic activity and potentially feed large explosive eruptions. In this regard, further investigations are required to establish the petrological and physical characteristics of the low velocity body identified at mid-crust depth beneath Volcán de Colima (Spica et al. 2017), to evaluate its role in the evolution of magmas, and to appraise its influence on volcanic activity.

If mixing with mafic magmas is one of the driving forces behind the cycle-ending Plinian eruptions, as inferred for the 1913 eruption, constraining the time lapse between the mixing events, that may in the future be identified by enhanced deeper seismicity, and the eruption itself is essential to establish realistic eruption forecast models (e.g. Saunders et al. 2012). The time scales of mixing between silicic melts, documented for the 1998–2011 magmas, and correlations between the replenishment events and the surface monitoring signal, also need to be established to characterize the dynamics of the active magmatic system and produce a baseline for future eruptions.

Magmas erupted during the sub-Plinian to Plinian eruptions are more mafic than the lavas emitted during effusive activity. This observation holds for the recent eruptive cycles (1818–present) but is also likely to be true for the Late-Holocene activity for which the pyroclastic deposits are relatively mafic (54–61 wt% SiO_2) (Luhr et al. 2010; Crummy et al., this volume). There is a need to investigate the reasons why more mafic magmas, which are expected to have lower viscosities, apparently systematically produce more explosive eruptions. This may be a consequence of magma mixing, as suggested above for the 1913 eruption. Several studies have, however, suggested the opposite scenario, i.e. magma mixing reduces the risk of explosive eruptions as prevailing at intermediate to silicic arc volcanoes (Koleszar et al. 2012; Ruprecht and Bachmann 2010). Determining the processes controlling the apparently systematic shift to more mafic compositions associated with major sub-Plinian to Plinian eruptions at Volcán de Colima and assessing the role and consequences of magma mixing prior to these eruptions is essential to understand the factors controlling eruptive cycles.

The close correspondence between the pressures (depths) recorded by the melt inclusions and the depths of seismic sources indicates that the two records can be matched. This correspondence implies that the melt inclusions and their host-crystals (at least the sections between the melt inclusions and the rims) are representative of the current state of the system, which is an encouraging step toward integrated petrological, geochemical, and geophysical monitoring strategies. A key feature to integrate these records is the application of diffusion chronometry and short-lived U-series isotopes to constrain the timescales of the petrological processes recorded (e.g. Reubi et al. 2017; Costa et al. 2008; Cooper and Reid 2008). Significant advances in our understanding of the architecture and dynamics of magmatic systems can be expected from these multidisciplinary research strategies, and Volcán de Colima represents a fantastic opportunity to do that.

6 Summary

Detailed petrological and geochemical studies of Volcán de Colima historical lavas have provided constraints on the production, storage conditions, and pre-eruptive evolution of the magmas. Lavas emitted between 1998 and 2011 are two-pyroxene andesites showing relatively homogeneous bulk-rock compositions (59.0–61.4 wt% SiO_2). These magmas had pre-eruptive temperatures between 1030 and 940 °C, 4.1 to 0.1 wt% H_2O dissolved in the melt, and were stored at pressures between 1.5 and 0 kbar, corresponding to depths of up to 9 km below the summit of the volcano. The most distinctive petrological features are the preponderance of oscillatory zoned plagioclase with limited variations in An% and the presence of gabbroic crystal clots. Melt inclusions representative of the crystallizing melts are dacite in composition (>64 wt% SiO_2) and exhibit compositional trends consistent with up to 40% crystallization of the phenocryst assemblage under vapour-statured conditions at pressures <1.5 kbar. The bulk-rock compositional trends are oblique to the melt inclusion crystallization trends, and appear to result from entrainment-mingling in the ascending dacite melt of gabbroic crystal clots and mafic crystals formed during previous episodes of magmatism. Compositional zoning of phenocrysts indicates that, in addition to magma mingling, magma mixing was important in the 1998–2011 magmatic system, but involved predominantly dacitic melts with a limited range of compositions. The last major Plinian eruption, in 1913, erupted magmas clearly distinct from those of 1998–2011 in terms of storage conditions and pre-eruptive magmatic evolution. The most distinctive features of the 1913 magmas are the higher H_2O contents (and storage pressures) and evidence for mixing with mafic magmas, indicating major disturbances in the subvolcanic system prior to eruption. Taken together this indicates that the Vulcanian-effusive activity characteristic of the 1998–2011 period and the 1913 Plinian eruption represent clearly distinct pre-eruptive scenarios, rather than a continuum. There is a close correspondence between the depths calculated from the 1998–2005 melt inclusion entrapment pressures and the position of volcanic earthquakes recorded during the same period, indicating that both datasets record the ascent, degassing, and crystallization of the magma before eruption. This correspondence offers potential to integrate the petrological and surface monitoring records to constrain the dynamics of the active magmatic system and identify precursory signals associated with changes in eruptive style.

References

Alonso-Perez, R., Muntener, O., Ulmer, P.: Igneous garnet and amphibole fractionation in the roots of island arcs: experimental constraints on andesitic liquids. Contrib. Miner. Petrol. **157**(4), 541–558 (2009)

Andersen, D.J., Lindsley, D.H., Davidson, P.M.: Quilf—a pascal program to assess equilibria among Fe-Mg-Mn-Ti oxides, pyroxenes, olivine, and quartz. Comput. Geosci. **19**(9), 1333–1350 (1993)

Annen, C., Blundy, J.D., Sparks, R.S.J.: The genesis of intermediate and silicic magmas in deep crustal hot zones. J. Petrol. **47**(3), 505–539 (2006)

Atlas, Z.D., Dixon, J.E., Sen, G., Finny, M., Lillian, A., Pozzo, M.D.: Melt inclusions from Volcan Popocatepetl and Volcán de Colima, Mexico: melt evolution due to vapor-saturated crystallization during. J. Volcanol. Geoth. Res. **153**(3–4), 221–240 (2006)

Berlo, K., Turner, S.: Pb-210-Ra-226 disequilibria in volcanic rocks. Earth Planet. Sci. Lett. **296**(3–4), 155–164 (2010)

Blundy, J., Cashman, K.: Rapid decompression-driven crystallization recorded by melt inclusions from Mount St. Helens volcano. Geology **33**(10), 793–796 (2005)

Blundy, J., Cashman, K.: Petrologic reconstruction of magmatic system variables and processes. Rev. Mineral. Geochem. **69**, 79–240 (2008)

Blundy, J., Cashman, K., Berlo, K.: Evolving magma storage conditions beneath Mount St. Helens inferred from chemical variations in melt inclusions from the 1980–1986 and current eruptions. In: Sherrod, D.R., Scott, W.E., Stauffer, P.H. (eds.) A Volcano Rekindled; The Renewed Eruption of Mount St. Helens, 2004–2006, vol. 1750 (2008)

Blundy, J.D., Shimizu, N.: Trace-element evidence for plagioclase recycling in calc-alkaline magmas. Earth Planet. Sci. Lett. **102**(2), 178–197 (1991)

Blundy, J., Wood, B.: Crystal-chemical controls on the partitioning of Sr and Ba between Plagioclase Feldspar, silicate melts, and hydrothermal solutions. Geochimica Et Cosmochimica Acta **55**, 193–209 (1991)

Bréton González, M.B., Ramirez, J.J., Navarro, C.: Summary of the historical eruptive activity of Volcan de Colima, Mexico 1519–2000. J. Volcanol. Geoth. Res. **117**(1–2), 21–46 (2002)

Cassidy, M., Cole, P.D., Hicks, K.E., Varley, N.R., Peters, N., Lerner, A.H.: Rapid and slow: Varying magma ascent rates as a mechanism for Vulcanian explosions. Earth Planet. Sci. Lett. **420**:73–84 (2015)

Cooper, K.M., Reid, M.R.: Uranium-series crystal ages. Rev. Mineral. Geochem. **69**, 479–544 (2008)

Costa, F., Dohmen, R., Chakraborty, S.: Time scales of magmatic processes from modeling the zoning patterns of crystals. Rev. Mineral. Geochem. **69**, 545–594 (2008)

Crummy et al.:Holocene and Late Pleistocene Eruption History of the Colima Volcanic Complex. This volume

Eichelberger, J.C.: Andesitic volcanism and crustal evolution. Nature **275**(5675), 21–27 (1978)

Eichelberger, J.C., Carrigan, C.R., Westrich, H.R., Price, R.H.: Nonexplosive silicic volcanism. Nature **323** (6089), 598–602 (1986)

Eichelberger, J.C., Izbekov, P.E., Browne, B.L.: Bulk chemical trends at arc volcanoes are not liquid lines of descent. Lithos **87**(1–2), 135–154 (2006)

Ghiorso, M., Evans, B.: Thermodynamics of rhombohedral oxide solid solutions and a revision of the Fe-Ti two-oxide geothermometer and oxygen-barometer. Am. J. Sci. **308**, 957–1039 (2008)

Gill, J.B.: Orogenic Andesites and Plate Tectonics. Springer, Berlin, p. 390 (1981)

Ginibre, C., Kronz, A., Worner, G.: High-resolution quantitative imaging of plagioclase composition using accumulated backscattered electron images: new constraints on oscillatory zoning. Contrib. Miner. Petrol. **142**(4), 436–448 (2002a)

Ginibre, C., Worner, G., Kronz, A.: Minor- and trace-element zoning in plagioclase: implications for magma chamber processes at Parinacota volcano, northern Chile. Contrib. Miner. Petrol. **143**, 300–315 (2002b)

Grove, T.L., DonnellyNolan, J.M., Housh, T.: Magmatic processes that generated the rhyolite of Glass Mountain, Medicine Lake volcano, N California. Contrib. Miner. Petrol. **127**(3), 205–223 (1997)

Kent, A.J.R., Darr, C., Koleszar, A.M., Salisbury, M.J., Cooper, K.M.: Preferential eruption of andesitic magmas through recharge filtering. Nat. Geosci. **3** (9), 631–636 (2010)

Koleszar, A.M., Kent, A.J.R., Wallace, P., Scott, W.E.: Controls on long-term low explosivity at andesitic arc volcanoes: Insights from Mount Hood, Oregon. J. Volcanol. Geoth. Res. **219–220**, 1–14 (2012)

Lavallée, Y., Varley, N.R., Alatorre-Ibargueengoitia, M. A., Hess, K.U., Kueppers, U., Mueller, S., Richard, D., Scheu, B., Spieler, O., Dingwell, D.B.: Magmatic architecture of dome-building eruptions at Volcán de Colima, Mexico. Bull. Volcanol. **74**(1), 249–260 (2012)

Lowenstern, J.B.: Melt inclusions come of age: volatiles, volcanoes and Sorby's legacy. In: De Vivo, B., Bodnar, R.J. (eds.) Melt Inclusions in Volcanic Systems: Methods, Applications And Problems, Developments in Volcanology, vol. 5, pp. 1–22 (2003)

Luhr, J.F.: Extensional tectonics and the diverse primitive volcanic rocks in the western Mexican Volcanic Belt. Can. Mineral. **35**, 473–500 (1997)

Luhr, J.F.: Petrology and geochemistry of the 1991 and 1998–1999 lava flows from Volcan de Colima, Mexico: implications for the end of the current eruptive cycle. J. Volcanol. Geoth. Res. **117**(1–2), 169–194 (2002)

Luhr, J.F.: The 1913 VEI-4 Plinian eruption of Volcán de Colima (Mexico): tephrochronology, petrology and plume modeling. In: AGU Fall Meeting, vol. 87, issue 52. San Fransisco, p. V43B-1786 (2006)

Luhr, J.F., Carmichael, I.S.E.: Colima volcanic complex, Mexico 1. Post-Caldera Andesites from Volcán Colima. Contrib. Mineral. Petrol. **71**(4), 343–372 (1980)

Luhr, J.F., Carmichael, I.S.E.: Petrological monitoring of cyclical eruptive activity at Volcán-Colima, Mexico. J. Volcanol. Geoth. Res. **42**(3), 235–260 (1990)

Luhr, J.F., Navarro-Ochoa, C., Savov, I.P.: Tephrochronology, petrology and geochemistry of Late-Holocene pyroclastic deposits from Volcán de

Colima, Mexico. J. Volcanol. Geoth. Res. **197**(1–4), 1–32 (2010)
McNutt, S.R.: Seismic monitoring. In: Sigurdsson, H. (ed.) Encyclopedia of Volcanoes. Academic Press, pp. 1095–1119 (2000)
McNutt, S.R., Rymer, H., Stix, J.: Synthesis of volcano monitoring. In: Sigurdsson, H. (ed.) Encyclopedia of Volcanoes. Academic press, pp. 1165–1183 (2000)
Moore, G., Carmichael, I.S.E.: The hydrous phase equilibria (to 3 kbar) of an andesite and basaltic andesite from western Mexico: constraints on water content and conditions of phenocryst growth. Contrib. Miner. Petrol. **130**(3–4), 304–319 (1998)
Mora, J.C., Macias, J.L., Saucedo, R., Orlando, A., Manetti, P., Vaselli, O.: Petrology of the 1998–2000 products of Volcán de Colima, Mexico. J. Volcanol. Geoth. Res. **117**(1–2), 195–212 (2002)
Muntener, O., Kelemen, P.B., Grove, T.L.: The role of H_2O during crystallization of primitive arc magmas under uppermost mantle conditions and genesis of igneous pyroxenites: an experimental study. Contrib. Miner. Petrol. **141**(6), 643–658 (2001)
Murphy, M.D., Sparks, R.S.J., Barclay, J., Carroll, M.R., Brewer, T.S.: Remobilization of andesite magma by intrusion of mafic magma at the Soufriere Hills volcano, Montserrat, West Indies. J. Petrol. **41**(1), 21–42 (2000)
Papale, P., Moretti, R., Barbato, D.: The compositional dependence of the saturation surface of $H_2O + CO_2$ fluids in silicate melts. Chem. Geol. **229**(1–3), 78–95 (2006)
Pearce, T.H., Kolisnik, A.M.: Observations of plagioclase zoning using interference imaging. Earth Sci. Rev. **29** (1–4), 9–26 (1990)
Petrosino, S., Cusano, P., La Rocca, M., Galluzzo, M., Orozco-Rojas, J., Bretón, M., Ibáñez, J., Del Pezzo, E.: Source location of long period seismicity at Volcán de Colima, México. Bull. Volcanol. **73**, 887–898 (2011)
Pickles, J.: Magmatic Processes and Eruptive Cyclicity at Hyper-Active Arc Volcanoes: a Perspective from Volcan de Colima, Mexico. M.Sc. thesis, University of Bristol, unpublished (2007)
Pinel, V., Jaupart, C.: The effect of edifice load on magma ascent beneath a volcano. Philos. Trans. Royal Soc. London Ser. A Math. Phys. Eng. Sci. **358**(1770), 1515–1532 (2000)
Putirka, K.A.: Igneous thermometers and barometers based on plagioclase plus liquid equilibria: tests of some existing models and new calibrations. Am. Miner. **90**(2–3), 336–346 (2005)
Putirka, K.D.: Introduction to minerals, inclusions and volcanic processes. Rev. Mineral. Geochem. **69**, 1–8 (2008a)
Putirka, K.D.: Thermometers and barometers for volcanic systems. Rev. Mineral. Geochem. **69**, 61–120 (2008b)
Reubi, O., Blundy, J.: Assimilation of plutonic roots, formation of high-K exotic melt inclusions and genesis of andesitic magmas at Volcán de Colima, Mexico. J. Petrol. **49**, 2221–2243 (2008)
Reubi, O., Blundy, J.: A dearth of intermediate melts at subduction zone volcanoes and the petrogenesis of arc andesites. Nature **461**(7268), 1269–1273 (2009)
Reubi, O., Blundy, J., Varley, N.R.: Volatiles contents, degassing, and crystallisation of intermediate magmas at Volcan de Colima, Mexico: reliability and significance of the melt inclusion record. Contrib. Miner. Petrol. **165**, 1087–1106 (2013)
Reubi, O., Sims, K.W.W., Bourdon, B.: ^{238}U-^{230}Th equilibrium in arc magmas and implications for the time scales of mantle metasomatism. Earth Planet. Sci. Lett. **391**, 146–158 (2014)
Reubi, O., Sims, K.W.W., Varley, N., Reagan, M., Eikenberg, J.: Timescales of degassing and conduit dynamics inferred from ^{210}Pb-^{226}Ra disequilibria in Volcan de Colima 1998–2010 andesitic magmas. In: Caricchi, L., Blundy, J.D. (eds.) Chemical, Physical and Temporal Evolution of Magmatic Systems. Geological Society, London, Special publication. doi: 10.1144/SP422.5 (2015)
Reubi, O., Scott, S.R., Sims, K.W.W.: Evidence of young crystal ages in andesitic magmas from a hyperactive arc volcano - Volcan de Colima, Mexico. J. Petrol. **58** (2), 261–276 (2017)
Ridolfi, F., Renzulli, A., Puerini, M.: Stability and chemical equilibrium of amphibole in calc-alkaline magmas: an overview, new thermobarometric formulations and application to subduction-related volcanoes. Contrib. Miner. Petrol. **160**(1), 45–66 (2010)
Robin, C., Camus, G., Gourgaud, A.: Eruptive and magmatic cycles at Fuego-De-Colima Volcano (Mexico). J. Volcanol. Geoth. Res. **45**(3–4), 209–225 (1991)
Roedder, E.: Fluid Inclusions, vol. 12, p. 644. Mineralogical Society of America, Washington (1984)
Ruprecht, P., Bachmann, O.: Pre-eruptive reheating during magma mixing at Quizapu volcano and the implications for the explosiveness of silicic arc volcanoes. Geology **38**(10), 919–922 (2010)
Saucedo, R., Macias, J.L., Gavilanes, J.C., Arce, J.L., Komorowski, J.C., Gardner, J.E., Valdez-Moreno, G.: Eyewitness, stratigraphy, chemistry, and eruptive dynamics of the 1913 Plinian eruption of Volcan de Colima, Mexico. J. Volcanol. Geoth. Res. **191**, 149–166 (2010)
Saunders, K., Blundy, J., Dohmen, R., Cashman, K.: Linking petrology and seismology at an active volcano. Science **336**(6084), 1023–1027 (2012)
Savov, I.P., Luhr, J.F., Navarro-Ochoa, C.: Petrology and geochemistry of lava and ash erupted from Volcan Colima, Mexico, during 1998–2005. J. Volcanol. Geoth. Res. **174**(4), 241–256 (2008)
Shinohara, H.: Excess degassing from volcanoes and its role on eruptive and intrusive activity. Rev. Geophy. **46**(4) (2008)
Singer, B.S., Dungan, M.A., Layne, G.D.: Textures and Sr, Ba, Mg, Fe, K and Ti compositional profiles in volcanic plagioclase: clues to the dynamics of calc-alkaline magma chambers. Am. Miner. **80**, 776–798 (1995)

Sisson, T.W., Grove, T.L.: Experimental investigations of the role of H_2O in calc-alkaline differentiation and subduction zone magmatism. Contrib. Miner. Petrol. **113**, 143–166 (1993)

Sisson, T.W., Ratajeski, K., Hankins, W.B., Glazner, A. F.: Voluminous granitic magmas from common basaltic sources. Contrib. Miner. Petrol. **148**(6), 635–661 (2005)

Sparks, R.S.J.: Causes and consequences of pressurisation in lava dome eruptions. Earth Planet. Sci. Lett. **150**(3–4), 177–189 (1997)

Sparks, R.S.J.: Forecasting volcanic eruptions. Earth Planet. Sci. Lett. **210**, 1–15 (2003)

Sparks, R.S.J., Marshall, L.A.: Thermal and mechanical constraints on mixing between mafic and silicic magmas. J. Volcanol. Geoth. Res. **29**(1–4), 99–124 (1986)

Sparks, R.S.J., Sigurdsson, H., Wilson, L.: Magma mixing: a mechanism for triggering acid explosive eruptions. Nature **267**, 315–318 (1977)

Spica, Z., Perton, M., Legrand, D.: Anatomy of the Colima volcano magmatic system, Mexico. Earth Planet. Sci. Lett. 459:1–13 (2017)

Streck, J.: Mineral textures and zoning as evidence for open system processes. Rev. Mineral. Geochem. **69**, 595–622 (2008)

Taran, Y., Gavilanes, J.C., Cortes, A.: Chemical and isotopic composition of fumarolic gases and the SO_2 flux from Volcan de Colima, Mexico, between the 1994 and 1998 eruptions. J. Volcanol. Geoth. Res. **117** (1–2), 105–119 (2002)

Tatsumi, Y.: The subduction factory: how it operates in the evolving earth. GSA Today **15**, 4–10 (2005)

Valdez-Moreno, G., Schaaf, P., Schaaf, P., Macias, J.L., Kusakabe, M.: New Sr-Nd-Pb-O isotope data for Colima volcano and evidence for the nature of the local basement. In: Neogene-Quarternary Continental Margin Volcanism: a Perspective from Mexico, pp. 45–63 (2006)

Varley, N., Arambula-Mendoza, R., Reyes-Davila, G., Sanderson, R., Stevenson, J.: Generation of Vulcanian activity and long-period seismicity at Volcan de Colima, Mexico. J. Volcanol. Geoth. Res. **198**(1–2), 45–56 (2010)

Varley, N.: Recent activity - observation & modelling the eruptive mechanism. This volume.

Verma, S.P., Luhr, J.F.: Sr, Nd, and Pb isotopic evidence for the origin and evolution of the Cantaro-Colima volcanic chain, Western Mexican Volcanic Belt. J. Volcanol. Geoth. Res. **197**(1–4), 33–51 (2010)

Villemant, B., Boudon, G.: Transition from dome-forming to plinian eruptive styles controlled by H_2O and Cl degassing. Nature **392**(6671), 65–69 (1998)

Wallace, P.J.: Volatiles in subduction zone magmas: concentrations and fluxes based on melt inclusion and volcanic gas data. J. Volcanol. Geoth. Res. **140**(1–3), 217–240 (2005)

Wells, P.R.A.: Pyroxene thermometry in simple and complex systems. Contrib. Miner. Petrol. **62**(2), 129–139 (1977)

Wood, B.J., Banno, S.: Garnet-orthopyroxene and orthopyroxene-clinopyroxene relationships in simple and complex systems. Contrib. Miner. Petrol. **42**(2), 109–124 (1973)

Zobin, V.M., Amezcua, M.G., Davila, G.A.R., Dominguez, T., Chacon, J.C.C., Alvarez, J.M.C.: Comparative characteristics of the 1997–1998 seismic swarms preceding the November 1998 eruption of Volcan de Colima, Mexico. J. Volcanol. Geoth. Res. **117**(1–2), 47–60 (2002a)

Zobin, V.M., Gonzalez-Amezcua, M., Reyes-Davila, G. A.: Seismotectonic deformation of the volcanic edifice prior to the 1998 lava eruption of Volcan de Colima. Mexico. Bulletin of Volcanology **64**(5), 349–355 (2002b)

The Fragility of Volcán de Colima—A Material Constraint

Yan Lavallée, Michael J. Heap, Jackie E. Kendrick, Ulrich Kueppers and Donald B. Dingwell

Abstract

The accurate assessment of volcanic unrest relies on a precise and timely interpretation of the source mechanisms of the signals monitored. Such an interpretation is facilitated by a comprehensive description of the physico-chemical properties of the geomaterials involved. The current understanding of the properties of the volcanic materials forming Volcán de Colima include (1) the physical properties of the erupted products (e.g., porosity, elastic wave velocities, density, Young's modulus, and permeability), (2) the mechanical properties and failure modes of the rocks forming the volcanic edifice (e.g., compressive and tensile strength, brittle-ductile transition) and, (3) the rheology of recently erupted magmas (e.g., viscosity, viscous-brittle transition). This attempt at a holistic physical description of the volcanic materials involved in eruptions at Volcán de Colima enables an appraisal of the deformation mechanisms and effective material strengths that regulate the volcano edifice stability and the scale of potential hazard scenarios.

Keywords

Volcán de Colima · Andesite · Lava dome Eruption · Rock · Magma · Ductile-brittle transition · Microstructure · Crack density Vesicle size distribution · Uniaxial compressive strength · Elastic moduli Elastic wave velocity · Viscosity · Thermal-stressing

1 Introduction

In the event of volcanic unrest, our ability to mitigate the potential risk imposed during the crisis relies on hazard assessment, defined by our understanding of the source mechanisms of the signals monitored. In such cases, accurate knowledge of the properties of the materials involved is essential. Simply speaking, an eruption includes three fundamental ingredients: rock, magma, and gas. Volcanoes, whether built as an edifice or a depression in the crust, consist of a complex succession of chemically and microstructurally heterogeneous rocks, which may exhibit variable states of physical coherence,

Y. Lavallée (✉) · J. E. Kendrick
Earth, Ocean and Ecological Sciences, University of Liverpool, 4 Brownlow Street, Liverpool, UK
e-mail: Yan.Lavallee@liverpool.ac.uk

M. J. Heap
Géophysique Expérimentale, École et Observatoire des Sciences de la Terre, Institut de Physique de Globe de Strasbourg, Université de Strasbourg (UMR 7516 CNRS), Strasbourg, France

U. Kueppers · D. B. Dingwell
Earth and Environmental Sciences, Ludwig Maximilian University (LMU), Theresienstr. 41, Munich, Germany

N. Varley et al. (eds.), *Volcán de Colima*, Active Volcanoes of the World,
https://doi.org/10.1007/978-3-642-25911-1_7

alteration, and saturation with fluids. Magmas, inherently viscoelastic materials, may equally well exhibit limits to their coherence as, variably charged with volatiles, they range in behaviour from liquid to solid, intact or fragmental, on ascent through the volcanic structure. It is this extremely broad range of magma/rock properties that makes volcanology very much a material science, where the variable material response is accountable for much of the exquisite detail and variety exhibited in volcanic activity and products.

The fragility of volcanic materials may be viewed as their ability to sustain coherence during their deformation in response to the local stress field (e.g., Dingwell 1996; Paterson and Wong 2005). This material characteristic is dependent on the strain rate, temperature, and the timescale of interest. During periods of volcanic unrest, the magnitude of these controlling parameters varies widely, with the result that both magmas and rocks cannot be simply described in a rheological sense. Yet to improve the accuracy of risk assessment it is imperative to understand these changes in real time.

Multi-parameter monitoring arrays are yielding a rapidly improving picture of the physical, chemical, and thermal states of active volcanic systems. These monitoring efforts offer ample evidence that the relationships that exist between the various precursor signals, both temporally and spatially, ultimately hold the key to forecasting eruptive style. Such a goal must rely on a description of the physical, mechanical, and rheological properties of rocks and magmas present at active volcanoes.

2 Volcán de Colima

2.1 Edifice Structure

Volcán de Colima is unstable due to two principal factors: firstly it reaches 3850 m above sea level (i.e., some 1600 m above the base of the volcano; Fig. 1b); and secondly the volcano is frequently subject to a high load provided by an active lava dome. The potential instability of the current edifice, in light of recurrent major eruptions in the last centuries, remains an important component of hazard assessment. The upper part of the edifice raises concerns as it exhibits an arcuate discontinuity focussing outgassing along the contact between the 1913 crater and the overlying lava fill (Fig. 1f). Stochastic flank-collapse probability modelling (Borselli et al. 2011) suggests that Volcán de Colima has reached a state of instability and is either: (1) 110 years overdue for a large-scale flank collapse or, (2) that the next large-scale flank collapse will occur 345 years from the present. It has been argued that the relatively high abundance of VEI 4 eruptions in the 19th century may have helped excavate the summit craters, and in turn removed some of the overburden, whilst not being too large to destabilise lower parts of the edifice (De la Cruz-Reyna et al., this volume).

2.2 Volcanic Activity

Recent volcanic activity at Volcán de Colima has been extensively observed, characterised, and summarised (Fig. 1; Luhr 2002; Zobin et al. 2002; Savov et al. 2008; Varley et al. 2010; Arámbula-Mendoza et al. 2011; Lavallée et al. 2011; James and Varley 2012; Webb et al. 2014; Lamb et al. 2014; Crummy et al., this volume, Varley, this volume), owing to the potential extent of related hazards (Bonasia et al. 2011; Borselli et al. 2011; Capra et al. 2013; De la Cruz-Reyna et al., this volume). The activity has been described as occurring in cycles with intervals of approximately 100 years (Luhr 2002); the last two cycles of which ended in 1818 and 1913 with Volcanic Explosivity Index (V.E.I.) 4 eruptions. The volcano indeed appears to be experiencing the end of this centennial cycle, with unrest resuming with seismicity in 1997 (Lamb et al. 2017), to be followed by protracted dome growth and explosive destruction. Since 1998, the activity has been nearly continuous with only an 18 month period of quiescence from June 2011 to January 2013, and mainly characterised by effusive and endogenic dome growth, although exogenic growth did

Fig. 1 Volcán de Colima. **a** Photograph of Volcán de Colima (looking west), displaying the 1975–76 lava flow (arrow 3), the 2004 lava flow (arrow 2) and the fault scar of an early sector collapse event (arrow 1; see Chapters 2 and 3, this volume). **b** Photograph of the active lava dome and fumarole activity as of 17 February 2009 (looking north). In the background, we can observe the fault scar of the sector collapse, which hosts a dyke that aligns between the summit of Volcán de Colima (foreground) and Nevado de Colima (outside the view, beyond the scar). **c** Photograph of the volcano's south-eastern flank, showing the 1975–76 lava flow as well as aprons of recent pyroclastic deposits. **d** Photograph of the 2004 lava flow (view from the dome looking north). **e** Photograph of a small Vulcanian eruption on 18 February 2009 (looking west); during this dome-eruption period, such explosive events approximately occurred every two hours. **f** Photograph of the dome and the volcano's eastern flank, showing intense fumarolic activity below the current crater rim (16 October 2004). This degasing activity takes place along the discontinuity between the 1913 crater and subsequent lava fill. Photograph provided by Nick Varley. **g** Thermographic image of the dome as of 7 March 2012, showing a relatively hot core in places which were exposed by slumping of blocks from the talus. At the time of this survey, four fumaroles were active along the crater rim. Thermographic image provided by Nick Varley

momentarily erect a spine in September 2001 as well as favouring dome extrusion between 2009 and 2011. Dome growth has generally been accompanied by the daily occurrence of 6 to 25 small, gas-and-ash explosive events (Fig. 1e), but it has also been interrupted by the interspersion of several larger, V.E.I. 2–3 explosive events (see Varley, this volume).

As further discussed by Varley (this volume), the unrest has been monitored extensively via seismicity, gas geochemistry, and thermal imagery. Vulcanian explosive eruptions have often been accompanied by characteristic seismic sequences of low-frequency events, with the main explosive event being characterised by a low-frequency onset associated with outgassing through a ruptured dome carapace, followed by the high-frequency portion, resulting from magmatic fragmentation (Varley et al. 2010). The linking of the long-period events to magma failure processes occurring at variable efficiency offers a view into the level of complexity characterising the mechanistic connections between the dynamics of magma ascent, pressure build-up, and rheological evolution.

Gas emissions have been periodically monitored since the beginning of the recent eruptive phase (Varley and Taran 2003; Varley, this volume). Outgassing has been variable and only occasionally exhibited any characteristic outgassing trends preceding explosive events. It may thus be that outgassing pulses best reflect local fracture outburst events (Kendrick et al. 2016) as witnessed at many active lava domes (e.g., Lavallée et al. 2015; Calder et al. 2015).

Thermal imagery has helped to resolve volcanic activity, both explosive (Webb et al. 2014) and effusive (Walter et al. 2013) events, as well as active outgassing structures. In 2001, fumaroles were monitored to reach up to 820 °C (Varley and Taran 2003). The upper volcanic edifice is frequently characterised by relatively high temperatures, owing to the presence of magma erupting at 960–1031 °C (e.g., Reubi and Blundy 2008; Savov et al. 2008). The temperature of the outer carapace of the lava dome (made of blocks) is dependent on the size of the growing dome and can reach >500 °C when the dome is small and >200 °C when large; thus displaying a crude correlation with effusion rate (Mueller et al. 2013). The core of the dome remains hot, as evidenced by routine thermal imaging of internal dome structures, occasionally exposed by the partial shedding of the talus (Fig. 1g). The prolonged exposure of the crater walls and conduit wallrock to continuous volcanic activity may impact the properties of the rock-forming edifice (Heap et al. 2014); yet, a potential influence on the structural stability of the edifice remains uncertain (Heap et al. 2018).

2.3 Eruptive Products

The erupted products of Volcán de Colima have varied slightly in composition, but have remained monotonously andesitic containing ~58–61 wt % SiO_2 (Luhr 2002; Reubi et al., this volume) and a low volatile content (Reubi et al. 2013) at least for the last 50 years. Since the initiation of dome growth in 1998, several pyroclastic density currents have been shed down the flanks of Volcán de Colima due to dome collapse or relatively large explosive events. This activity has afforded researchers the opportunity to efficiently sample the materials involved in the current eruptive period. The materials erupted are generally pristine, but some show a range of alteration and oxidation (Fig. 2).

Petrographic analysis of the eruptive products reveals rock with a range of crystallinities (phenocrysts: 20–30 vol.% and microlites: 25–50 vol.%), in which the groundmass constitutes as much as 68 vol.% (Luhr 2002) and total glass content generally ranges between 15 and 45 vol.%. The andesites exhibit a porphyritic texture with plagioclase (13–25 vol.%), orthopyroxene (2–4 vol.%), clinopyroxene (3–4 vol.%), and minor hornblende (< 0.5 vol.%) and Fe–Ti oxides (~2 vol.%) (Fig. 3). Rare olivine crystals are present as xenocrysts. In explosive deposits, a fraction (ca. 10 vol.%) of the pyroclasts show a certain degree of oxidation and alteration with a devitrified groundmass and α-cristobalite partly replacing plagioclase; a common feature of vapour-phase altered andesitic rocks (Fang et al. 2003; Horwell et al. 2013).

Fig. 2 Photos of eruptive products. **a** Pristine juvenile dome rock (pen for scale). **b** Oxidised juvenile dome rock. **c** Altered dome rock. **d** Rare flow banded dome rock. **e** Exposed interior of poorly-sintered tuffisite veins, showing a network of tuffisites within a host rock. **f** Well sintered tuffisite veins in the rock of a 2005 block-and-ash flow (sunglasses for scale)

Some of the rocks contain annealed fractures, whilst other rocks show the presence of tuffisite veins that have undergone variable degrees of sintering (i.e., densification as a result of surface tension acting on fragmented particles; Kolzenburg et al. 2012; Kendrick et al. 2016; Farquharson et al. 2016a). The tuffisites present in the pyroclasts can be up to 50 cm long and 0.3–5 cm wide, consisting of coherent, sintered holocrystalline aggregates of fragmented dome material in an equally holocrystalline host rock (Fig. 4a). These features preserve a range of sintering efficiency, suggesting that they can form over variable timescales and at different depths within the conduit. During one of the periods of eruptive quiescence in March 2012, the dome was accessed to examine the rocks exposed at their extrusion point and to assess the state of the dome's carapace. The dome mainly consisted of large angular blocks exceeding

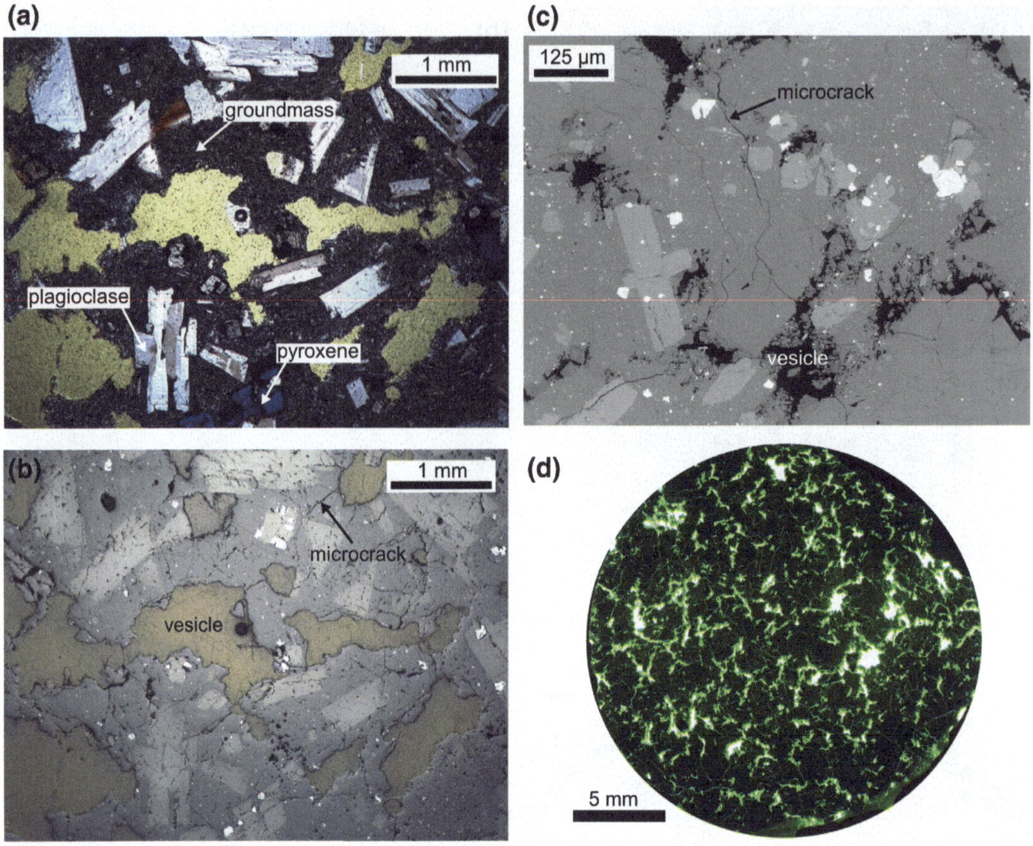

Fig. 3 Microstructures of eruptive products. **a** Photomicrograph of microtextures and microstructures typically present in the andesites from Volcán de Colima (transmitted cross-polarised light). We observe plagioclase and pyroxene phenocrysts in a microlite-bearing groundmass (picture taken from Heap et al. 2014). **b** Photomicrograph of the same area as in panel A, but taken in reflected light. Reflected light allows us to better observe the pervasive microcracked nature of the rock (picture taken from Heap et al. 2014). **c** Scanning electron microscope image showing the complex microstructure (vesicles and microcracks: picture taken from Heap et al. 2014). **d** Photomicrograph map of a sample impregnated with a fluorescent dye to show the complex porosity structure (reflected fluorescent light; picture taken from Heap et al. 2014)

30 cm in diameter (Kendrick et al. 2016). The dome rocks/lavas (Fig. 4b) are very similar to those exposed in the pyroclastic density current deposits, although they are generally more oxidised. The most surprising observation of the dome rocks was the near ubiquity of tuffisites preserved inside or on the surface of the blocks (Kendrick et al. 2016). The veins were generally 0.5–3 cm thick, and occasionally up to ~10 cm, and commonly consisted of coherent ash particles (but not as sintered, nor as dense, as those observed in the pyroclastic deposits; cf. Kolzenburg et al. 2012; Farquharson et al. 2016a) (Fig. 4c), suggesting a shallow and perhaps cooler origin (Kendrick et al. 2016). Cristobalite is commonly present in these shallow tuffisites (Kendrick et al. 2016); cristobalite and tridymite were also found inside fractures in pyroclasts found within the debris-flow tracks (Farquharson et al. 2016a).

Microstructurally, the erupted products contain both pores (vesicles) and microcracks in varying proportions (Heap et al. 2014; Fig. 3); the duality of the porosity microstructure has left the rocks with substantial residual porosities, reaching in a few instances up to 75 vol.%.

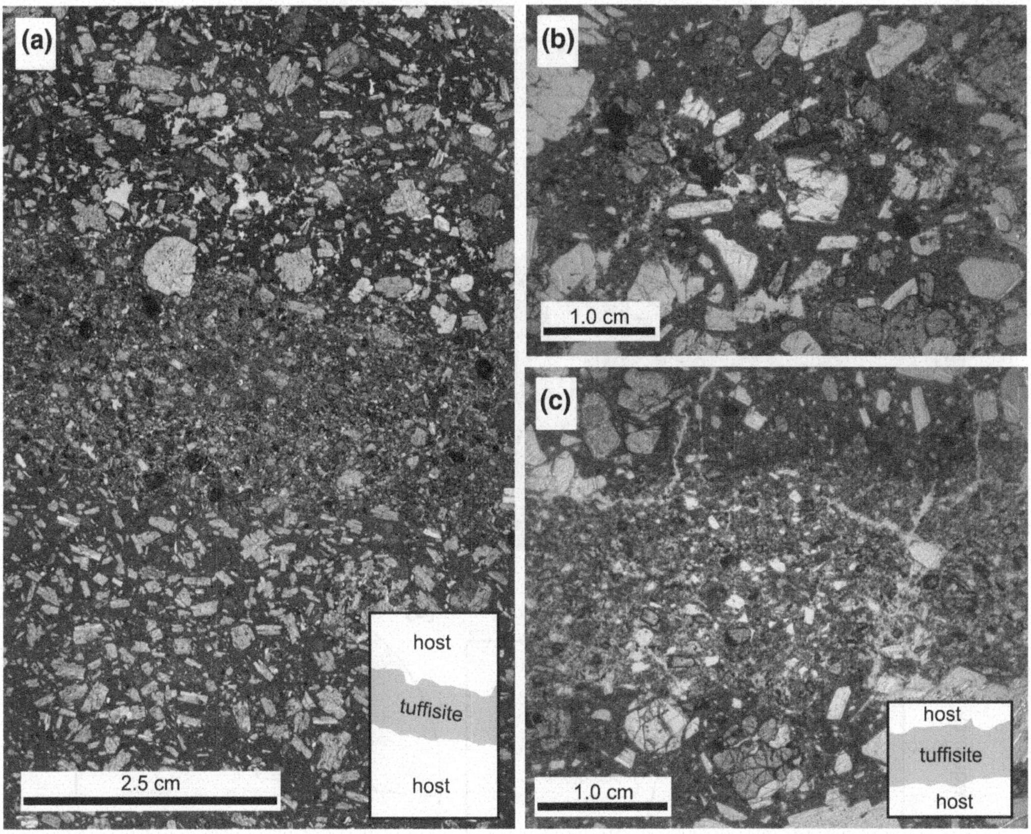

Fig. 4 **a** Photomicrograph showing the textural and structural heterogeneity associated with the presence of a healed tuffisite vein (tuffisite vein is perpendicular to the long axis of the image, in the middle of the image; picture taken from Kolzenburg et al. 2012). **b** Photomicrograph showing the host rock microtextures and microstructures collected from the lava dome during a period of quiescence in 2012. **c** Photomicrograph showing a poorly-sintered tuffisite vein ~10 cm from the area shown in panel **b**

Field-based density studies on 500 blocks in pyroclastic deposits suggested that the porosity of eruptive products exhibit bimodal porosity distribution peaking at approximately 16 and 25 vol.% (Mueller et al. 2011; Lavallée et al. 2012). Another field-based study revealed the porosity of blocks collected from debris-flow tracks at La Lumbre, Montegrande, and El Zarco to range between 2 and 75 vol.% (based on 542 hand samples; Farquharson et al. 2015) and, using the weighted abundance method of Bernard et al. (2015), that the dominant porosity class is between 10 and 25 vol.%. This database was extended by undertaking an extensive campaign of density measurements on some 2635 fist-sized blocks in an attempt to discriminate between dome collapse and dome disruption events (Fig. 5; Kueppers et al. 2009). In this analysis, a density bimodality is not exhibited, although the fact remains that the densest rocks are more altered. Explosive events travelling down Montegrande ravine contain a higher fraction of juveniles. La Lumbre, the ravine that has accommodated most of the dome collapse deposits during recent eruptions, hosts pyroclastic deposits with a wide density distribution as well as larger fraction of exotic clasts. Pyroclastic deposits in La Arena ravine travelled the farthest from their source. Their density is relatively high and they appear to contain the largest fraction of exotic lithics, presumably picked up during flow. Although the density distribution does not

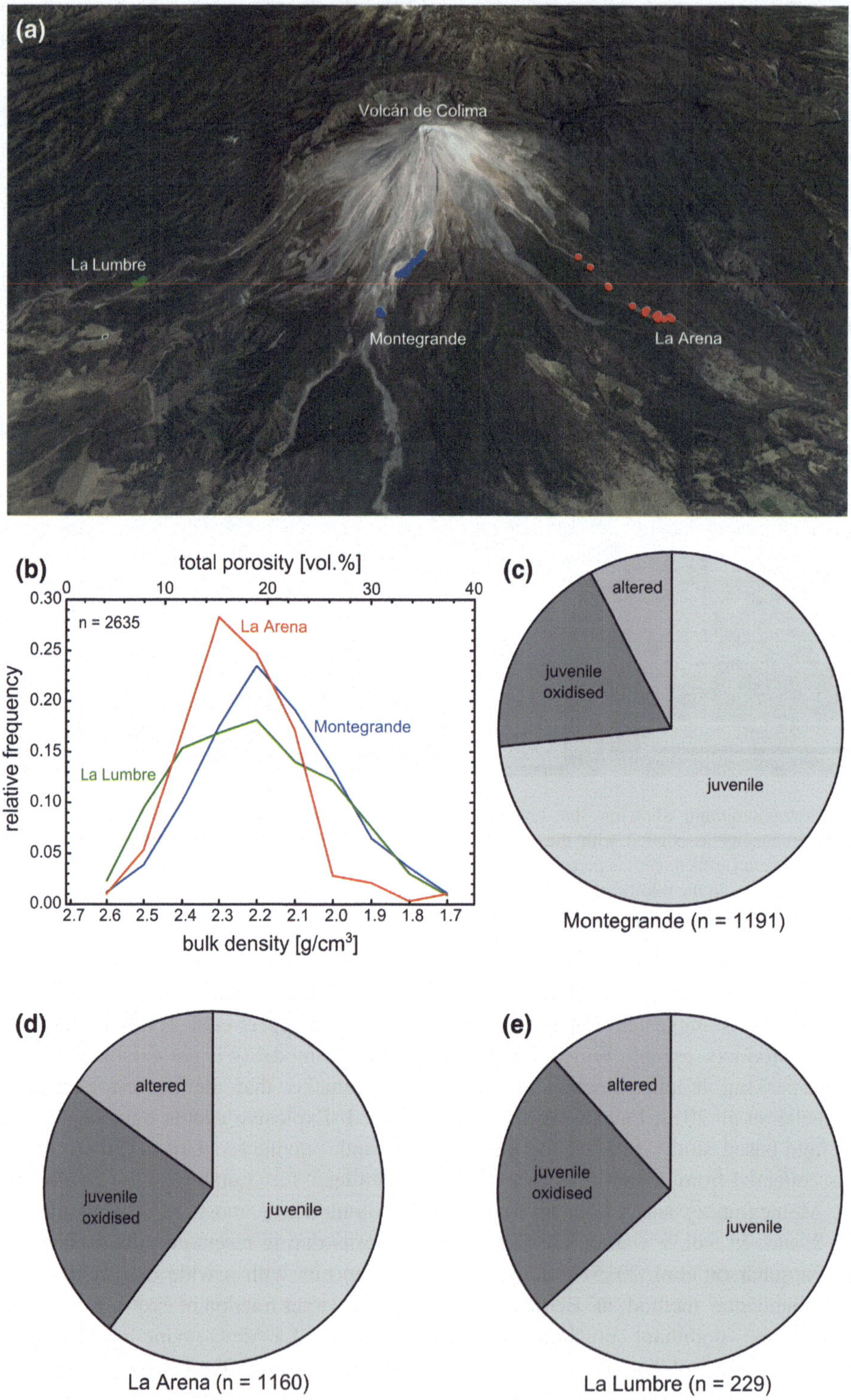

Fig. 5 Field-based density distribution survey. **a** Locations of the three main ravines (La Lumbre, Montegrande, and La Arena) in which the density of clasts in pyroclastic density current deposits was measured. **b** Density/porosity distribution for each valley (porosity was calculated using a skeletal density of 2.84 g/cm^3, the average skeletal density of five powdered samples; measurements were made using a helium pycnometer). The survey was complemented by an analysis of the clasts (juvenile versus oxidised juvenile versus altered as shown in Fig. 2a–c) for **c** Montegrande, **d** La Arena and, **e** La Lumbre. Data in Fig. 5 are unique to this study

provide a conclusive distinction between the mechanisms causing pyroclastic density currents, it does provide a fruitful reconstruction of the magmatic architecture involved in the dome eruption.

3 Rock Mechanics

3.1 Rock Physical Properties

The heterogeneous nature of dome rocks results in highly variable rock physical properties (Table 1)—parameters key to model seismic activity, deformation, outgassing and pore pressure augmentation, and volcano stability. Elastic wave velocity, one such physical property, is a useful tool to probe the physical state of rocks (a low elastic wave velocity is indicative of high porosities and/or high densities of microscopic cracks), and provide valuable input data for models designed to, for example, locate volcano-tectonic earthquakes (e.g., Lesage et al. 2014). Room pressure and temperature elastic wave velocities have been measured for a range of samples. The P- and S-wave velocities range between 2.34–3.11 and 1.09–1.45 km/s, respectively. Vp/Vs ratios range between 1.90 and 2.39. Elastic wave velocity generally decreases (whilst Vp/Vs ratio increases) as the porosity increases (Table 1; Heap et al. 2014). In cases where the samples were water-saturated, the

Table 1 Average physical properties of dome rocks (data from Heap et al. 2014): connected porosity, dry bulk sample density, dry elastic wave velocities, dry Vp/Vs ratio, dry dynamic Young's modulus and Poisson's ratio, wet P-wave velocity, and connected porosity and dry P-wave velocity after exposure to 450 °C

connected porosity [vol.%]	dry bulk sample density [g/cm^3]	dry P-wave velocity [kms^{-1}]	dry S-wave velocity [kms^{-1}]	dry Vp/Vs ratio	dry dynamic Young's modulus [GPa]	dry dynamic Poisson's ratio	wet P-wave velocity [kms^{-1}]	connected porosity after 450°C [vol.%]	dry P-wave velocity after 450°C [kms^{-1}]
7.9	2.48	3.11	1.30	2.39	11.8	0.39	4.84	8.1	2.71
8.6	2.48	2.69	1.42	1.90	13.0	0.31	4.98	8.9	2.52
11.2	2.40	3.11	1.45	2.21	13.3	0.33	4.97	11.6	2.85
17.6	2.13	2.56	1.21	2.07	8.8	0.35	4.30	17.7	2.31
25	1.96	2.34	1.09	2.15	6.4	0.36	3.91	25.2	1.89

"Dry" and "wet" refer to measurements on samples dried in a vacuum oven at 40 °C and samples saturated with distilled water in a vacuum, respectively. Porosities were measured at ambient pressure and elastic wave velocities were determined under an axial stress of 1.9 MPa

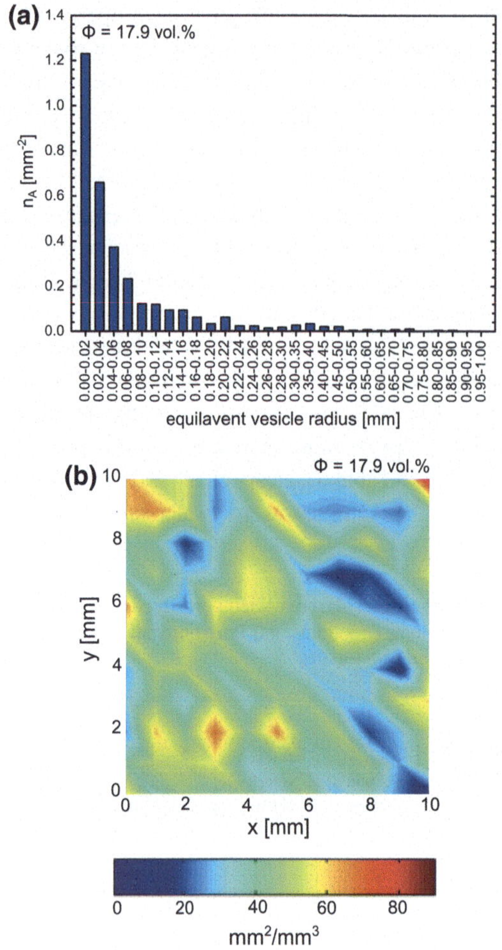

Fig. 6 Quantification of the porous network of a dome rock with 17.9% porosity: **a** vesicle size distribution (n_A is the vesicle density) and, **b** crack area density calculated using standard stereological techniques (after Underwood 1970). Data from Heap et al. (2014)

P-wave velocity shifted to higher values (Table 1; Heap et al. 2014), thereby influencing the resultant dynamic Young's modulus and Poisson's ratio; this effect requires due consideration when deciphering velocity profiles in any natural system, such as the location of dry and wet zones in the edifice rock, necessary in order to improve a description of the 3D velocity structure.

The porosity of dome rocks consists of vesicles (Fig. 6a) and microcracks (Fig. 6b). The porosity of magma and the conduit wallrock comprises an efficient permeable network, which assists the outgassing of the magmatic column (Mueller et al. 2008; Kolzenburg et al. 2012; Kendrick et al. 2013a; Heap et al. 2014, 2015; Farquharson et al. 2015, 2016a, b). The permeability of rocks from Volcán de Colima has been found to span a large range, from $\sim 10^{-17}$ to $\sim 10^{-11}$ m^2 (Fig. 7a), a range typical for volcanic materials at lava domes (Mueller et al. 2005; Ashwell et al. 2015; Farquharson et al. 2015). It should be noted that the permeability (i.e., the measure of the ability of a material to transmit fluids) generally increases with porosity (e.g., Farquharson et al. 2015, 2016a; Kendrick et al. 2016), as previously demonstrated for dome rocks from many volcanoes worldwide (e.g., Mueller et al. 2005). Porosity-permeability relationships in volcanic rocks have typically been described using a single power law model (e.g., Mueller et al. 2005). However, the permeability of the rocks at Volcán de Colima has been found to fall within two distinct families separated by a statistically-relevant "porosity changepoint" at about 14 vol.%, a consequence of a change in the dominant microstructural element controlling fluid flow: i.e., from crack- (<14 vol.%) to pore-dominated (>14 vol.%) flow path geometries (Heap et al. 2014; Farquharson et al. 2015, 2016a). Statistical analysis on the compiled data (Fig. 7a) using the modified Bayesian Information Criterion (BIC)—which penalises models of increased complexity and determines whether or not the data are better described by one power law or two (see Main et al. 1999; Farquharson et al. 2015)—suggests that the data are best described by power laws, with a changepoint porosity (x^*) of 21 vol.% (Fig. 7a). This changepoint is much higher than those previously estimated (Heap et al. 2014; Farquharson et al. 2015, 2016a) and is likely due to the relative abundance of high-porosity, high-permeability data (Fig. 7a). Therefore, although these compiled data can be statistically better described by a two power-law model, the changepoint porosity value of 21 vol.% may not represent a microstructural boundary, as discussed in Heap et al. (2014) and Farquharson et al. (2015, 2016a).

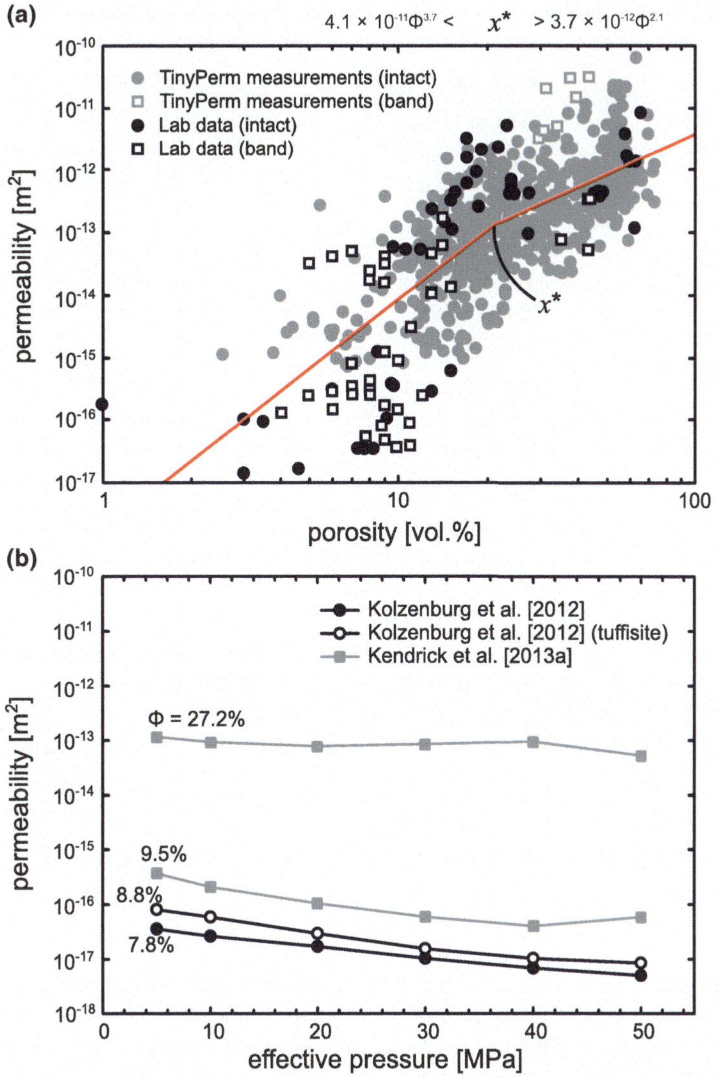

Fig. 7 Permeability of dome rocks from Volcán de Colima at room temperature. **a** Permeability as a function of porosity on log-log axes. All TinyPerm measurements (grey symbols) were measured using a TinyPerm II (Farquharson et al. 2015; Kendrick et al. 2016). The intact laboratory measurements include data from Mueller (2007), Kolzenburg et al. (2012), Richard et al. (2013), Kendrick et al. (2013a, b), Heap et al. (2014, 2015), and Farquharson et al. (2015, 2016a, b, c, 2017). The Heap et al. (2014) data are water permeabilities measured under a confining pressure of 2 MPa and a pore fluid pressure gradient of 0.5 MPa. The Kolzenburg et al. (2012) and the Kendrick et al. (2013a) data are water permeabilities measured under a confining pressure of 5 MPa and a pore fluid pressure gradient of 1 MPa. The Mueller (2007) data are gas (argon) permeabilities measured under a confining pressure of 0 MPa using the pulse decay method (2.5 MPa on one side and 0 MPa on the other). The low porosity samples of Mueller (2007) were performed on lava samples from the 1999 block-and-ash flow deposit, and the high porosity samples were pumice samples from the 1913 eruption. The Richard et al. (2013) datum is for a sample collected from the 1999 block-and-ash flow deposits; gas (argon) permeability was measured under a confining pressure of 0 MPa using the pulse decay method (4 MPa on one side and 0 MPa on the other). The Heap et al. (2015) and Farquharson et al. (2015, 2016a, b, c, 2017) data are gas (nitrogen) permeabilities measured under a confining pressure of 1 MPa. All of the permeability data from samples containing discontinuities (unfilled black squares) are from Farquharson et al. (2016a, b, c, 2017) and are therefore gas (nitrogen) permeabilities measured under a confining pressure of 1 MPa. The red line is the two-slope power law model, as defined by the modified Bayesian Information Criterion (BIC), see text for details. The power law for both slopes are given at the top of the panel. **b** The influence of effective pressure on the permeability of dome rocks. The Kolzenburg et al. (2012) and the Kendrick et al. (2013a) data are water permeabilities measured using a pore fluid pressure gradient of 1 MPa

Table 2 The permeability of dome blocks containing poorly-sintered tuffisites, as measured using a portable hand-held air permeameter (measurements are averages of the values reported in Kendrick et al. 2016)

Block number	Rock type	Permeability (m^2)
1	Host rock	1.1×10^{-13}
	Tuffisite vein 1 Tuffisite vein 2 Tuffisite vein 3	1.5×10^{-11} 8.4×10^{-13} 4.3×10^{-12}
2	Host rock	1.1×10^{-13}
	Tuffisite vein 1 Tuffisite vein 2	1.4×10^{-11} 2.7×10^{-12}
3	Host rock	4.9×10^{-14}
	Tuffisite vein 1 Tuffisite vein 2	5.8×10^{-13} 7.5×10^{-12}
5	Host rock	3.8×10^{-13}
	Tuffisite vein 1	1.9×10^{-11}
7	Host rock	3.2×10^{-15}
	Tuffisite vein 1	2.8×10^{-13}

Table 3 The permeability of pyroclasts found within debris-flow tracks containing variably sintered fractures/tuffisites, as measured using benchtop steady-state gas (nitrogen) permeameter under a confining pressure of 1 MPa (data taken from Farquharson et al. 2016a)

Sample	Host permeability (m^2)	Fracture/tuffisite permeability (m^2)	Relative change in permeability
COL-V-4	1.58×10^{-16}	1.04×10^{-13}	657.23
COL-L-1	1.36×10^{-17}	3.90×10^{-16}	27.68
COL-L-2	1.77×10^{-16}	1.37×10^{-13}	773.01
COL-L-4	2.40×10^{-13}	6.67×10^{-14}	−0.72

Measurements have been provided for the permeability of poorly-sintered (Kendrick et al. 2016) and variably-sintered tuffisites (Kolzenburg et al. 2012; Farquharson et al. 2016a) from Volcán de Colima. To investigate the influence of poorly-sintered tuffisites (found ubiquitously across the dome, Fig. 2e) on permeability, in-situ permeabilities were measured using a portable, hand-held air permeameter (Kendrick et al. 2016). Permeability measured at an aperture to an internally-hosted tuffisite vein was $\sim$1.5 orders of magnitude higher than in the surrounding rock, thereby promoting strong permeability anisotropy within tuffisite-bearing dome rocks (Table 2). Kolzenburg et al. (2012) and Farquharson et al. (2016a) provide laboratory measurements of permeability for a suite of variably-sintered tuffisites. Farquharson et al. (2016a) find that these features can increase permeability by up to three orders of magnitude, depending on the degree of sintering (Table 3). It was found that the permeability of a tuffisite decreased as the degree of sintering increased (Farquharson et al. 2016a). Indeed, very well sintered tuffisites did not significantly alter sample permeability (Kolzenburg et al. 2012). In one case, the permeability of the sintered fracture was found to be lower than the host rock (Farquharson et al. 2016a; Table 3). The presence of poorly-sintered tuffisites in the shallow conduit and at the dome surface could generate high gas-permeability and significantly affect outgassing efficiency (Kendrick et al. 2016). However, at greater depths, where tuffisites can heal efficiently (Kolzenburg et al. 2012; Farquharson et al. 2016a), the permeability increase due to tuffisite formation is likely transient.

Heap et al. (2015) explored the influence of dilatant (brittle) and compactant (ductile) failure modes on the permeability of porous andesite (porosity = 18 vol.%) from Volcán de Colima. The authors measured a permeability increase of a factor of two following tensile macrocrack formation (permeability measured parallel to the fracture plane) and a permeability decrease of up to an order of magnitude following axial compaction of 1.5% strain under triaxial conditions.

A recent study by Farquharson et al. (2016c) showed that the permeability of samples of low porosity andesite (porosity = 8 vol.%) collected from the El Zarco ravine increases as a function of increasing inelastic strain in the brittle regime. Samples were deformed in compression under triaxial conditions to inelastic strains ranging from 1 to 11% and, following deformation, the samples were unloaded and permeability was measured. Sample permeability increased by three orders of magnitude at the highest inelastic strain (11%), relative to the initial permeability (Farquharson et al. 2016c). Microstructural observations show, with increasing inelastic strain in the brittle regime, that narrow shear fractures evolve into more permeable, thick shear zones containing abundant gouge.

The influence of compactant deformation on permeability evolution was recently extended by Farquharson et al. (2017). High-porosity (porosity = 19–23 vol.%) andesite collected from debris-flow track La Lumbre (locality given in Fig. 5a) was deformed triaxially under different effective pressures within the compactant (ductile) regime. It was found that permeability increased slightly at modest axial strain (<0.05), due to an increase in pore connectivity due to distributed microcracking. Permeability decreased at progressively higher axial strains and, at an axial strain of 0.20, sample permeability was reduced by two orders of magnitude relative to the initial values (Farquharson et al. 2017).

3.2 Uniaxial Compressive and Tensile Strength

The monotonously andesitic nature of the rocks forming the edifice of Volcán de Colima provides the opportunity to assess the mechanical behaviour of the upper structural edifice and the lava dome carapace on representative samples of fresh eruptive products. Uniaxial compressive strength (UCS) tests, employing a constant strain rate of 10^{-5} s^{-1} until sample failure, have been conducted to assess the mechanical response of shallow edifice rocks (Heap et al. 2014; Figs. 8 and 9). When subjected to a constant compressive strain rate, rock first accommodates deformation through microcrack closure (the stress increases non-linearly with increasing strain). This is then followed by an elastic response (where the stress-strain relationship is pseudo-linear) and, finally, by microcrack growth, coalescence, and sample failure (where strain hardening precedes the peak stress which is followed by the stress drop associated with brittle failure; Fig. 8a). During UCS tests, acoustic emissions (AEs, which are high frequency elastic waves generated by the rapid release of strain energy such as during microcracking) help monitor the nucleation, propagation, and coalescence of fractures leading to failure. The onset of AE activity corresponds to the onset of microcracking or dilation (commonly termed C′). Analysis of the frequency distribution of the amplitude of AE (commonly referred to as the *b*-value in seismology) shows a progressive decrease as failure is approached (from 1.6 to 0.6), which reflects an increase in the magnitude (size, energy) of microcracks as the sample approaches macroscopic failure, usually associated with microcrack coalescence and localisation (e.g., Smith et al. 2009; Fig. 8a).

The indirect tensile strength of rocks from Volcán de Colima has been constrained using the Brazilian disc method to range between 2.3–4.9 and 0.4–1.9 MPa for rocks containing porosities of 20 and 30 vol.%, respectively (Fig. 9a), values similar to those measured on rocks collected

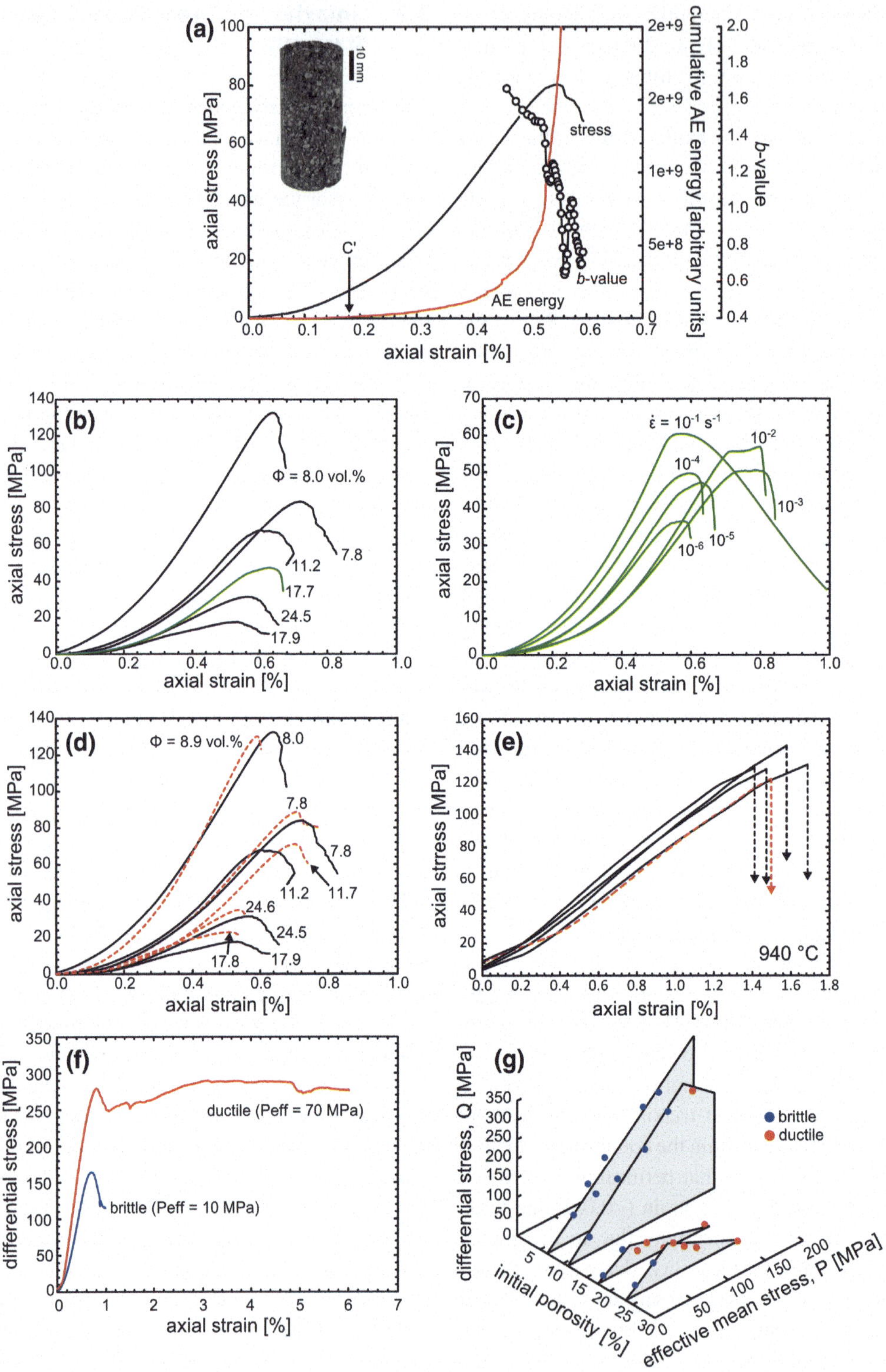
(a)
axial stress [MPa]
cumulative AE energy [arbitrary units]
b-value
10 mm
stress
b-value
AE energy
C'
axial strain [%]
(b)
Φ = 8.0 vol.%
7.8
11.2
17.7
24.5
17.9
axial stress [MPa]
axial strain [%]
(c)
ε̇ = 10⁻¹ s⁻¹
10⁻²
10⁻⁴
10⁻³
10⁻⁶ 10⁻⁵
axial stress [MPa]
axial strain [%]
(d)
Φ = 8.9 vol.%
8.0
7.8
7.8
11.2
11.7
24.6
24.5
17.8
17.9
axial stress [MPa]
axial strain [%]
(e)
940 °C
axial stress [MPa]
axial strain [%]
(f)
ductile (Peff = 70 MPa)
brittle (Peff = 10 MPa)
differential stress [MPa]
axial strain [%]
(g)
differential stress, Q [MPa]
brittle
ductile
initial porosity [%]
effective mean stress, P [MPa]

◀ **Fig. 8** Rock Mechanics. **a** Typical mechanical dataset (black line) and the output of AE energy (red line) associated with a uniaxial compressive strength (UCS) test at a constant strain rate of 10^{-5} s^{-1} (dataset from Heap et al. 2014). Seismic *b*-values shown as open circles. The experiment was performed at room temperature. The inset is a photograph of the post-failure sample. **b** The influence of porosity on the UCS of dome rocks. All dataset was collected using a constant strain rate of 10^{-5} s^{-1} at room temperature (the black curves are taken from Heap et al. 2014; the green curve is unique to this study). **c** The influence of strain rate (from 10^{-6} to 10^{-1} s^{-1}) on the UCS of dome rocks under room temperature conditions. All of these dataset are unique to this study (see also Table 4). **d** The influence of thermal stressing (to 450 °C) on the UCS of dome rocks. All dataset was collected using a constant strain rate of 10^{-5} s^{-1}. **e** The influence of (healed) tuffisites on the UCS of andesitic magmas at 940 °C. The black curves are samples that contain tuffisites and the dashed red curve is for a tuffisite-free sample. All dataset was collected using a constant stressing rate of 2 MPa min^{-1} (dataset from Kolzenburg et al. 2012). **f** Stress-strain curves from triaxial experiments performed at a constant strain rate of 10^{-5} s^{-1} and effective pressures of 10 and 70 MPa. Experiment performed at 10 MPa was brittle and the experiment performed at 70 MPa was ductile (dataset from Heap et al. 2015). **g** 3D Failure envelopes for andesitic lava from Volcán de Colima. Differential stress (Q) at failure (peak stress in the brittle regime and the onset of inelastic compaction in the ductile regime) and the effective mean stress (P) are plotted alongside initial connected porosity. Brittle and ductile experiments are represented by blue and red symbols, respectively (dataset from Heap et al. 2015)

from other volcanoes (e.g., Rocchi et al. 2004). These measurements are therefore in agreement with the expected ratio of compressive to tensile strength (approximately 12:1; Jaeger et al. 2007). Tensile fracturing has been observed to decrease P-wave velocity in the orthogonal direction with respect to the fracture generated (Lamb et al. 2017). Similar seismic wave velocity decreases across a volcanic edifice have been observed to precede the onset of eruptions at many volcanoes (e.g. Duputel et al. 2009; Anggono et al. 2012), but analysis of the seismic record monitored during eruption onset in 1998 at Volcán de Colima shows no such velocity variations (Lamb et al. 2017), suggesting that this forecasting method may not be universal.

The rocks from Volcán de Colima show a general increase in strength with decreasing porosity (Figs. 8b and 9a)—an observation typical to all rock types (Paterson and Wong 2005). The increased presence of microcracks and vesicles (i.e., higher porosities) further affects the elastic portion of deformation, the gradient of which is referred to as the static Young's modulus (or stiffness) of the rocks. The mechanical data (Fig. 8b) shows that the static young's modulus increases with decreasing porosity; an observation also made for the dynamic Young's modulus estimated from ultrasonic velocity (Table 1). Micromechanical modelling of the mechanical behaviour of the rocks has shown that the abundance of microcracks and vesicles in dome rocks prevents their simulation with current models such as the pore-emanated crack model (e.g., Sammis and Ashby 1986; Vasseur et al. 2013) or the sliding wing-crack model (e.g., Ashby and Sammis 1990), demonstrating the need for more refined models to constrain volcanic rock mechanics (as discussed in Heap et al. 2014). Although a recent micromechanical model attempts to address the role of crystallinity on the mechanical behaviour of volcanic rocks (Heap et al. 2016), models that describe the behaviour of volcanic materials must now consider the influence of vesicle size, shape, and size distribution.

3.3 Influence of Confining Pressure (Depth)

Recent experiments have shown that increasing the confining pressure (or depth) on porous andesite from Volcán de Colima can result in a switch in failure mode, from one dominated by dilatancy (shear fracturing) to one dominated by compaction (cataclastic pore collapse) (Heap et al. 2015). Examples of the stress-strain curves for deformation in each regime are provided in Fig. 8f. The curve for an effective pressure (P_{eff}, where P_{eff} = confining pressure (P_c) minus the pore fluid pressure (P_p)) of 10 MPa has a similar form to the UCS tests described in the previous section. However, the P_{eff} = 30 MPa curve does not exhibit a strain softening phase and stress drop associated with the formation of a shear fracture. Deformation in this regime is purely compactant and is manifest as subplanar surfaces of crushed pores (Heap et al. 2015). It is further noted that the switch from dilatant to compactant mode of failure occurs at lower pressure or depth for higher porosity andesites; this is best depicted on a three-axis graph of differential stress (*Q*), initial porosity, and effective mean stress (*P*, where $P = ((\sigma 1 + 2Pc)/3) - Pp$) (Fig. 8b; Heap et al. 2015). The amplitude of the failure envelope (the grey zone) is lower when the porosity is higher, meaning that rock containing lower porosity is intact (or pre-failure) over a much larger stress space and requires higher confining pressure to switch into the compactant regime. As discussed in Heap et al. (2015), compaction of the deep edifice may destabilize the volcano and may impede outgassing of the volcanic conduit.

3.4 Strain Rate Effects

Varying strain rate—a variable seldom employed in rock mechanics, but of utmost importance in volcanic environments as large fluctuations in strain rates are expected to result from the highly variable style of unrest (e.g., edifice inflation, sector collapse, small versus large eruptions, conduit versus dyke fed eruptions, faulting)—can cause strong changes in rock strength and in the deformation mechanisms dictating its behaviour (Rutter 1986; Paterson and Wong 2005). From the new tests conducted (in the Volcanology Laboratory at the University of Liverpool), an increase in strength is observed with an increase in applied strain rate (Figs. 8c and 9a). As strain rate is decreased, time-dependent deformation mechanisms, namely subcritical crack growth (Atkinson 1984; Brantut et al. 2013), appears to become increasingly important, resulting in a lowering of the compressive strength. Under the conditions of these experiments (room temperature and ambient confining pressure) stress corrosion cracking (a process that invokes fluid-rock reactions between a chemically-active pore fluid and the strained atomic bonds close to the crack tips) is likely to be the dominant subcritical crack growth mechanism, even under ambient humidity conditions. Stress corrosion cracking has previously been discussed in relation to time-dependent cracking and failure in basalt (Heap et al. 2011). At higher strain rates, not only is the influence of time-dependent processes diminished, but the stress on the rock is applied faster than the microstructure can evolve; as a result, the damage becomes increasingly localised as strain rate is increased. An apparent increase in static Young's modulus is also induced by an increase in strain rate (Table 4), which could be a result of increasing damage build-up at slow rates, that have been shown to systematically reduce elastic moduli (Ma and Daemen 2004). This preliminary work illustrates the need for more experiments at a slower strain rate where time-dependent behaviour (e.g., stress corrosion cracking, Brantut et al. 2013) may affect the rock response that controls volcano stability and magma transport.

3.5 Thermal Effects

Active volcanic environments harbour temperatures that are vastly higher than regional averages of the Earth's geotherm. It is well known that the stress exerted by the expansion/contraction of phases during thermal stressing may induce thermal microcracks, which alter the physical and mechanical properties of the material (e.g., David et al. 1999; Vinciguerra et al. 2005). Thermally stressing dome rocks from Volcán de Colima to an intermediate temperature of 450 °C (Heap et al. 2014) has yielded the observations that the mineralogical assemblage remains stable and that thermal stressing does not significantly impact the most deleterious microcracks, which play an important role in compressive failure. As a result, uniaxial compressive strength remains unchanged upon thermal stressing to 450 °C (Fig. 8d), although it imparts a small increase in the number of microcracks, reflected by small changes in the porosity and P-wave velocity of the material (Table 1). It remains to be seen

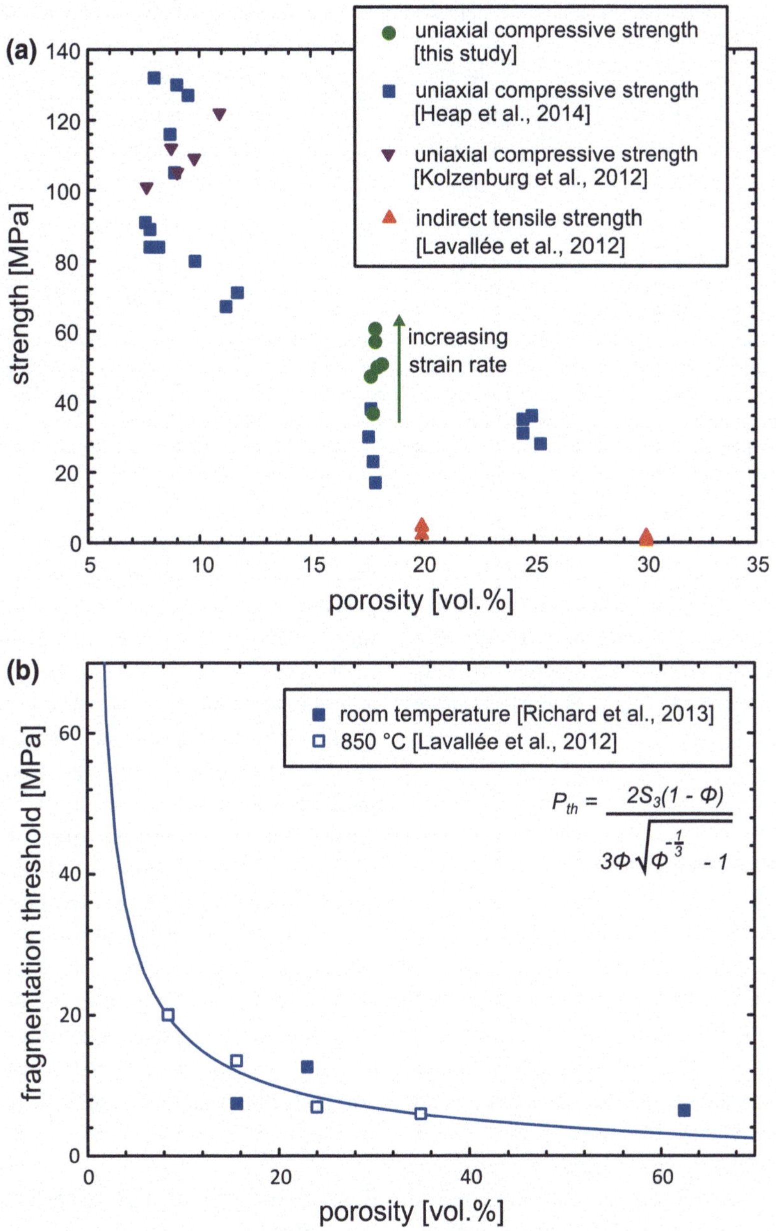

Fig. 9 **a** Compilation of uniaxial compressive and indirect tensile strength as a function of porosity. Data from Heap et al. (2014) were collected using a constant strain rate of 10^{-5} s^{-1} on cylindrical samples (40 mm in length and 20 mm in diameter) at room temperature. Data from Kolzenburg et al. (2012) were collected using a constant stress rate of 2 MPa min^{-1} on cylindrical samples (50 mm in length and 25 mm in diameter) at 940 °C. The data of this study (green circles) were collected using a range of strain rates (10^{-6}–10^{-1} s^{-1}, see Table 4) at room temperature. The indirect tensile strength data were determined on rectangular prisms (160 mm in length and 40 mm in width and thickness) at room temperature. The sample was placed on two static bearing edges (100 mm apart) and loaded in the centre until failure using a hydraulically controlled punching tool (at a rate of 500 MPa s^{-1}; data from Lavallée et al. 2012). **b** Fragmentation threshold (minimum required gas overpressure in pore space to completely fragment a rock by rapid decompression) for samples from Volcán de Colima. Curve shown is the fragmentation threshold modelled using the equation presented in the figure, using an effective tensile strength S = 3.1 MPa (see text and Lavallée et al. (2012) for details)

Table 4 Uniaxial compressive strength (UCS) experiments performed under a variety of constant strain rates (10^{-6}–10^{-1}) on dome rocks until failure

Strain rate (s^{-1})	UCS (MPa)	Static Young's modulus (GPa)	Connected porosity (vol.%)
10^{-6}	36.6	11.992	17.8
10^{-5}	47.1	14.402	17.7
10^{-4}	49.7	14.636	18.0
10^{-3}	50.5	12.867	18.2
10^{-2}	56.9	14.990	17.9
10^{-1}	60.5	19.043	17.9

We also present the static Young's modulus (calculated from the resultant stress-strain curves, see text for details) and the connected porosity for each sample. All experiments were conducted on dry samples at room temperature (in the Volcanology Laboratory at the University of Liverpool). Data in Table 4 are unique to this study. However, the static Young's moduli presented in this table were measured at axial stresses between 20 and 50 MPa, depending on the strength of the sample. Therefore, direct comparison with the dynamic Young's moduli of Table 1 (that are measured at an axial stress of 1.9 MPa) is ill-advised (see Heap et al. (2014) for further details)

whether thermal stressing to higher temperatures, such as those experienced by conduit wall rocks, would induce a strength reduction (e.g., Heap et al. 2018), especially in conjunction with other varying stress conditions imparted by the movement of magma (e.g., Kendrick et al. 2013b).

3.6 Pore Pressure Effects

Pore pressures are likely to fluctuate within an active volcanic edifice. To study the influence of increasing pore pressure in andesites collected from the 1998 block-and-ash flow deposits, Farquharson et al. (2016b) performed triaxial experiments on samples of andesite from Volcán de Colima in which the pore pressure was increased at different rates until macroscopic failure. These authors found that increases in pore pressure over a range of timescales (from minutes to days) can cumulate in the brittle failure of otherwise intact rock. Further, rock deforming in a compactant (ductile) manner can be shunted into the brittle regime through pore pressure augmentation.

Heap and Wadsworth (2016) studied the influence of closed system ("undrained") deformation on the mechanical behaviour of low and high porosity andesites (porosities = 16.5 and 7.5 vol.%, respectively) from Volcán de Colima. Pore pressure decreases during closed system in the brittle regime, although the decrease in pore pressure was insufficient to promote strengthening ("dilatancy hardening"), due, in part, to the high pre-existing microcrack density. The authors find that pore pressure increase during ductile deformation of a rock subjected to a close system. However, pore pressure does not increase sufficiently to embrittle the rock: the emergence of microcracking as the rock approaches the compactant–dilatant transition acts as a negative feedback and prevents the rock from transiting into the dilatant field (Heap and Wadsworth 2016). They employed a dimensionless Darcy number that defines the conditions (strain rate and permeability) required for drained or undrained deformation for the edifice-forming andesites of Volcán de Colima.

3.7 Strength of Tuffisite-Bearing Rocks

The presence of fractures, and their potential infilling by fragmented material, may influence the response of rocks to stress and deformation. A mechanical study of tuffisite-bearing rocks found in pyroclastic density current deposits has shown that, if strongly healed, these structures may not necessarily have a lower strength (Fig. 8e) or a higher permeability (Fig. 7c) than the tuffisite-free host rock (Kolzenburg et al.

2012). Kolzenburg et al. (2012) conclude that healing may be efficient at depth (where temperature enhancing viscous sintering and the load driving compaction are higher) within the conduit and that the mechanical significance of the presence of tuffisites may be restricted to the window in which the particles heal a fracture. The poorly-sintered tuffisites observed in-situ in the dome (e.g., Kendrick et al. 2016) however stand in stark contrast to those described in distal pyroclastic deposits by Kolzenburg et al. (2012) and Farquharson et al. (2016a). Their lower degree of sintering means that the presence of the tuffisite reduces their strength drastically and the rocks are easily broken by hand, which has contributed to the fragmented nature of the dome carapace (Kendrick et al. 2016).

4 Magma Rheology

The rheology of erupted dome lavas at Volcán de Colima has been the focus of several experimental studies (Lavallée et al. 2007, 2008, 2012, 2013; Kendrick et al. 2013a, 2017). The rheology was assessed via compressive deformation experiments under constant uniaxial stresses at magmatic temperatures (900–980 °C). Dome lavas such as those erupted at Volcán de Colima have been described to exhibit a non-Newtonian rheology with a component of shear thinning (Caricchi et al. 2007; Lavallée et al. 2007). A rheological law for the dome lava has been developed (Lavallée et al. 2012) where the stress, σ, can be related to the strain rate, $\dot{\epsilon}$, via:

$$Log\sigma = 0.5Log\dot{\epsilon} + \frac{606.9}{T - T_g} + 0.29 \quad (1)$$

valid for a range of temperature, T, down to the glass transition, Tg, estimated at 710 °C (Fig. 10). Such a rheology has a consequence that the apparent viscosity of dome lava decreases with applied stress or resultant strain rate. The relatively small range of crystallinities and porosities of dome lavas (in contrast with other types of eruptive products) restricts viscosity variation and it has been suggested that the

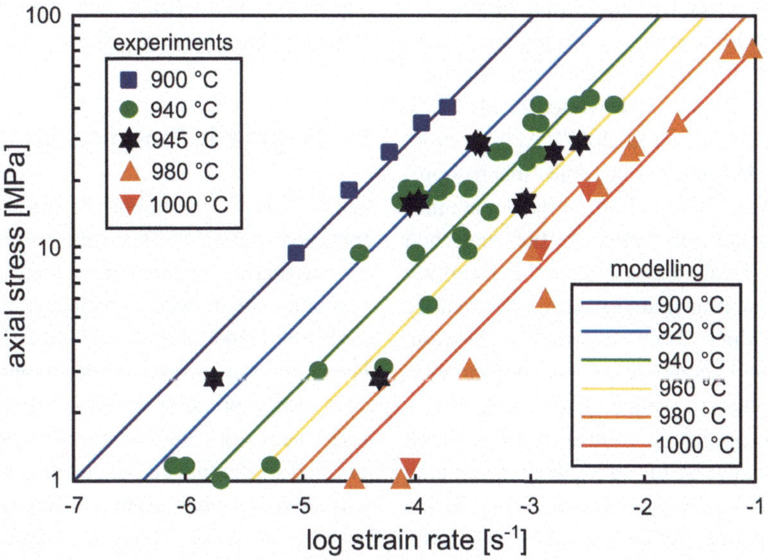

Fig. 10 Lava rheology. Rheological data showing the relationship between stress and strain rate (on log-log axes) as a function of temperature. The data denoted by coloured symbols were used to model flow curves, as described by Eq. 1 (data from Lavallée et al. 2012). The data denoted by black stars highlight the influence of porosity, with higher strain rates resulting from a given applied stress to more porous lavas (27.2% porosity) than to denser lavas (9.5% porosity) (Kendrick et al. 2013a)

rheological contribution of the interstitial liquid is reduced to the extent of a pseudo-plastic rheology with important contributions from the crystal fraction (Lavallée et al. 2012; Kendrick et al. 2013a, 2017). The presence of crystals and bubbles, which can deform, rotate, break, and align during flow, add a further strain-weakening effect (i.e., the viscosity decreases with strain). At higher rates or loads these flaws nucleate fractures that release seismic energy (Lavallée et al. 2012), producing a complex evolution of physical properties (Kendrick et al. 2013a).

Magma flowing at increasing strain rates may undergo changes in its rheology (Fig. 11). In crystal-bearing, bubbly magmas such as those that drive dome eruptions, the mechanical contribution of each phase induces important rheological complexities which complicate the direct application of the concept of the glass transition as a brittle limit and fragmentation criterion (Lavallée et al. 2007; Cordonnier et al. 2009, 2012). Instead, at low strain rates, magma deformation is macroscopically ductile (i.e., pervasive deformation with no strain localisation) and is characterised by viscous deformation of interstitial melt, physical reorganisation of its phases, crystal plasticity, nucleation of microcracks in crystals leading to propagation, and coalescence of microcracks distributed throughout the sample (Fig. 11a; Lavallée et al. 2007, 2012; Kendrick et al. 2013a, 2017). At high strain rates, magma deformation becomes macroscopically brittle (Fig. 11b) as high stresses accumulates in, and partition between, the melt and crystal phases (Deubelbeiss et al. 2011; Kendrick et al. 2017) favouring strain localisation (Lavallée et al. 2007), which promotes crack nucleation, propagation and coalescence as deformation evolves to rupture (Lavallée et al. 2008, 2013), which may no longer be controlled by the melt rheology, but rather by the strength of the magma (Kendrick et al. 2013a; Coats et al. 2018). In this deformation regime, failure occurs at low axial strains and the strain rate increases exponentially with strain (or time) at a constant stress (Fig. 11 c). This exponential acceleration is accompanied by a decrease in the seismic *b*-value (as the magnitude (size, energy) of the microcracking increases, usually associated with microcrack coalescence and localisation) and a non-linear release in acoustic emissions (Fig. 11d). Such a precursory signal trend provides the means to potentially forecast the failure of magma (Lavallée et al. 2008) and presumably of explosive eruptions at Volcán de Colima (Arámbula--Mendoza et al. 2011; De la Cruz-Reyna and Reyes-Davila 2001; Boué et al. 2016) and other active volcanoes (e.g. Cornelius and Voight 1996; McNutt 1996). The generation of microcracks has been observed to scale non-linearly with applied stress or strain rate; thus the total strain-to-failure decreases with applied stress (Fig. 11). This accumulation of damage significantly modifies the construction of the permeable network that regulates outgassing of magma to its surroundings, potentially alleviating the explosive potential of extruding magmas (Lavallée et al. 2013). Faulting near the conduit margins may lead to spine formation (as observed in 2001 at Volcán de Colima)—a process resulting from the rheological evolution of magma undergoing rupture. The occurrence of rupture in the conduit divides a magma ascent regime dominated by a viscous control at depth to a fault control ascent regime in the shallow conduit (Kendrick et al. 2014a, b; Hornby et al. 2015).

5 Magmatic Fragmentation

Failure due to gas bubble overpressure has also been investigated to assess the conditions leading to magmatic fragmentation that underlies the generation of tuffisites, small magnitude ash-rich explosions, and catastrophic explosive eruptions. These experiments have been conducted by the rapid decompression of pressurised magma and assessment of the minimum pore pressure required to completely fragment a sample (Alidibirov and Dingwell 1996a, b; Spieler et al. 2004; Scheu et al. 2008; Arciniega-Ceballos et al. 2014 and references therein). We find that the potential energy required to fragment these dome lavas (i.e., the pore pressure, P_{th}) non-linearly decreases with sample porosity, φ (Fig. 9b; Lavallée et al. 2012), following the relationship:

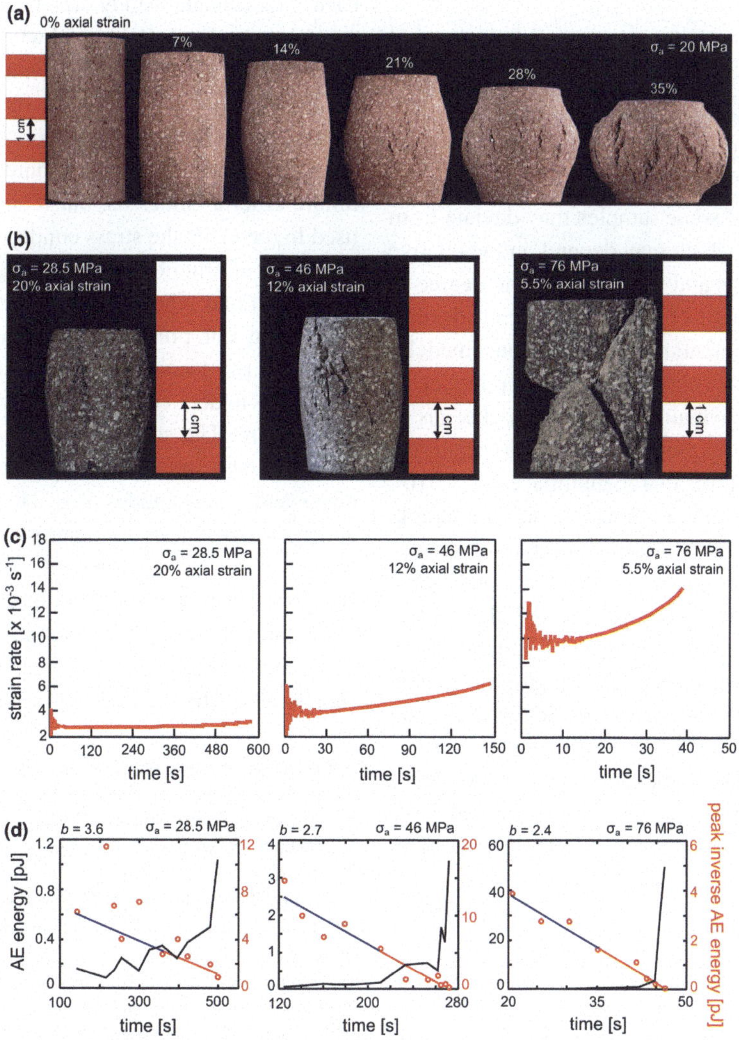

Fig. 11 Stress- and strain-dependent deformation in magmas. **a** Photographs of six post-deformation lava dome samples deformed in the ductile regime under a constant uniaxial stress of 20 MPa at 940 °C. The samples were either undeformed (0% strain) or deformed to axial strains of 7, 14, 21, 28, and 35%. All samples had a starting porosity of about 7 vol.%. We note the development of tensile fractures at the surface of the samples after >20% axial strain. **b** Photographs of three post-deformation lava dome samples deformed in the brittle regime under constant uniaxial stresses, from left to right, of 28.5, 46, and 76 MPa. The final axial strain, indicated on the figure, decreases with increasing constant stress. **c** Strain rate against time for the three experiments shown in panel **b**. We note that the strain rate increases when the sample approaches failure, and is higher at higher constant stress. **d** Acoustic emission (AE) energy and peak inverse energy against time for the three experiments shown in panel **b**. AE energy accelerates as sample failure is approached in each case. An exponential acceleration in released AE can be used as a proxy to forecast failure following the Failure Forecast Method (Voight 1988). Data from Lavallée et al. (2013)

$$P_{th} = \frac{2S_3(1-\varphi)}{3\varphi\sqrt{\varphi^{-1/3}-1}} \quad (2)$$

where S_3 is the effective tensile strength of the material estimated at 3.1 MPa for andesites from Volcán de Colima. We note that the fragmentation threshold of some samples may deviate from the main trend, which may depend on the relative amount of cracks and pores and on the degree of alteration (Richard et al. 2013). During decompression, the potential energy is consumed by variably efficient fragmentation (Kueppers et al. 2006) and the remaining energy of the gas is converted into explosive energy to drive the fragments (Alatorre-Ibargüengoitia et al. 2010). If little kinetic energy remains, it will simply project the fragments into surrounding cracks, producing tuffisites, while the gas, and some fraction of ash, may escape the dome. If sufficient kinetic energy remains, the fragments will be erupted, triggering an explosive event. Field surveys of erupted ballistics and characterisation of impact sites utilising a ballistic model (Alatorre-Ibargüengoitia et al. 2006, 2010) have helped constrain the overpressure of large explosive events to some 20 MPa (Lavallée et al. 2012).

6 Outlook

A thorough characterisation of eruptive products representative of a period of unrest may provide first order constraints on volcanic activity. The experimental data obtained on Volcán de Colima materials enable a description of the volcano and its recent eruptions. This dataset includes a summary of porosity (microcracks and vesicles), crystallinity (phenocrysts and microlites), permeability (with pressure or depth), elastic wave velocity, static and dynamic elastic moduli, compressive strength (with stress, pressure, temperature and heterogeneities), indirect tensile strength, and fragmentation criteria. Although our understanding of the material grows with each experimental study, many aspects remain poorly understood. It is worthwhile pointing out that at this stage in experimental volcanology, our description of multiphase magmas remains somewhat rudimentary. Also, despite many attempts, a general law combining the various failure criteria eludes us, but it should be prioritised to reconcile the stress conditions that lead to the range of eruption styles witnessed at Volcán de Colima and other dome-building volcanoes.

Beyond our present experience is an understanding of mechanical properties of these materials under confining pressure and high temperature. The scarcity of such data prevent a complete description of the deformation mechanisms and failure modes (e.g., Heap et al. 2017), which, for materials as heterogeneous as those found in volcanic environments, are multiple and complex. It is certain that additional efforts need to be made if we wish to constrain the full range of mechanisms underlying changes in monitored signals and better forecast volcanic phenomena.

Acknowledgements We wish to thank Miguel Alatorre-Ibargüengoitia, Soureya Becker, Niklas Berninger, Jonathan Bruce Hanson, Fabian Goldstein, Stephan Kolzenburg, Sebastian Mueller, Sebastian Mueller (not an accidental duplication, but a common Bavarian name) Linda Petrakova, Dominique Richard, and Oliver Spieler for their help during field work, sample collection, and some of the data analysis. M. J. Heap would also like to thank Jamie Farquharson for performing the BIC analysis. Y. Lavallée acknowledges financial support from the DFG grant LA2651/3-1 and ERC Starting Grant on Strain Localisation in Magmas (SLiM, project 306488). D. B. Dingwell acknowledges a research Professorship from the Bundesexzellenzinitiative (LMUexcellent) and the ERC Advanced Grant on Explosive volcanism in the earth system: experimental insights (EVOKES, project 247076). M. J. Heap acknowledges a CNRS INSU grant and an Initiative d'Excellence (IDEX) "Attractivité" grant (VOLPERM). This research was partly funded by the German Research Foundation (DFG; Grant No. 03G0584A), by the DAAD-CONACYT cooperation program between Germany and Mexico, by the Bavarian Elite Network, and by two Hubert Curien Partnership (PHC) PROCOPE grants (grant numbers 27061UE and 332065SG) the Ministry of Foreign and European Affairs (MAE) and the Ministry of Higher Education and Research (MESR), in France, and the Deutscher Akademischer Austauschdienst (DAAD) in Germany.

References

Alatorre-Ibargüengoitia, M.A., Delgado-Granados, H., Farraz-Montes, I.A.: Hazard zoning for ballistic impact during volcanic explosions at Volcán de Fuego de Colima (Mexico). Neogene-Quarternary Cont. Margin Volcanism A Perspect. Mexico **402**, 209–216 (2006)

Alatorre-Ibargüengoitia, M.A., Scheu, B., Dingwell, D.B., Delgado-Granados, H., Taddeucci, J.: Energy consumption by magmatic fragmentation and pyroclast ejection during Vulcanian eruptions. Earth Planet. Sci. Lett. **291**(1–4), 60–69 (2010)

Alidibirov, M., Dingwell, D.B.: An experimental facility for the investigation of magma fragmentation by rapid decompression. Bull. Volc. **58**(5), 411–416 (1996a)

Alidibirov, M., Dingwell, D.B.: Magma fragmentation by rapid decompression. Nature **380**(6570), 146–148 (1996b)

Anggono, T., Nishimura, T., Sato, H., Ueda, H., Ukawa, M.: Spatio-temporal changes in seismic velocity associated with the 2000 activity of Miyakejima volcano as inferred from cross-correlation analyses of ambient noise. J. Volcanol. Geoth. Res. **247–248**, 93–107 (2012). https://doi.org/10.1016/j.jvolgeores.2012.08.001

Arámbula-Mendoza, R., Lesage, P., Valdés-González, C., Varley, N.R., Reyes-Dávila, G., Navarro, C.: Seismic activity that accompanied the effusive and explosive eruptions during the 2004–2005 period at Volcán de Colima, Mexico. J. Volcanol. Geoth. Res. **205**(1–2), 30–46 (2011)

Arciniega-Ceballos, A., Alatorre-Ibargüengoitia, M.A., Scheu, B., Dingwell, D.B., Delgado-Granados, H.: Seismological analysis of conduit dynamics in fragmentation experiments. J. Geophys. Res. **119**(3), 2215–2229 (2014)

Ashby, M.F., Sammis, C.G.: The damage mechanics of brittle solids in compression. Pure Appl. Geophys. **133** (3), 489–521 (1990)

Ashwell, P.A., Kendrick, J.E., Lavallée, Y., Kennedy, B. M., Hess, K.-U., von Aulock, F.W., Wadsworth, F.B., Vasseur, J., Dingwell, D.B.: Permeability of compacting porous lavas. J. Geophys. Res. Solid Earth **120**(3), 1605–1622 (2015)

Atkinson, B.K.: Subcritical crack-growth in geological-materials. J. Geophys. Res. **89**, 4077–4114 (1984)

Bernard, B., Kueppers, U., Ortiz, H.: Revisiting the statistical analysis of pyroclast density and porosity data. Solid Earth. (2015). https://doi.org/10.5194/se-6-869-2015

Bonasia, R., Capra, L., Costa, A., Macedonio, G., Saucedo, R.: Tephra fallout hazard assessment for a Plinian eruption scenario at Volcán de Colima (Mexico). J. Volcanol. Geoth. Res. **203**(1–2), 12–22 (2011)

Borselli, L., Capra, L., Sarocchi, D., De la Cruz-Reyna, S.: Flank collapse scenarios at Volcán de Colima, Mexico: A relative instability analysis. J. Volcanol. Geoth. Res. **208**(1–2), 51–65 (2011)

Boué, A., Lesage, P., Cortés, G., Valette, B., Reyes-Dávila, G., Arámbula-Mendoza, R., Budi-Santoso, A.: Performance of the 'material Failure Forecast Method' in real-time situations: a Bayesian approach applied on effusive and explosive eruptions. J. Volcanol. Geoth. Res. (2016). https://doi.org/10.1016/j.jvolgeores.2016.10.002

Brantut, N., Heap, M.J., Meredith, P.G., Baud, P.: Time-dependent cracking and brittle creep in crustal rocks: a review. J. Struct. Geol. **52**, 17–43 (2013)

Calder, E., Lavallée, Y., Kendrick, J.E., Bernstein, M.: Lava dome eruptions. In: Sigurdsson, H., Hougton, B., McNutt, S.R., Rymer, H., Stix, J. (eds.) Encyclopedia of Volcanoes, 2nd edn. Academic Press 1456 p. (2015)

Capra, L., Bernal, J.P., Carrasco-Nunez, G., Roverato, M.: Climatic fluctuations as a significant contributing factor for volcanic collapses. Evidence from Mexico during the Late Pleistocene. Global Planet. Change **100**, 194–203 (2013)

Caricchi, L., Burlini, L., Ulmer, P., Gerya, T., Vassalli, M., Papale, P.: Non-newtonian rheology of crystal-bearing magmas and implications for magma ascent dynamics. Earth Planet. Sci. Lett. **264**, 402–419 (2007)

Coats, R., Kendrick, J.E., Wallace, P.A., Miwa, T., Hornby, A.J., Matsushima, T., Lavallée, Y.: Failure criteria for porous dome rocks and lavas: a study of Mt. Unzen, Japan. Solid Earth Discuss. 1–42 (2018). https://doi.org/10.5194/se-2018-19

Cordonnier, B., Caricchi, L., Pistone, M., Castro, J., Hess, K.-U., Gottschaller, S., Manga, M., Dingwell, D.B., Burlini, L.: The viscous-brittle transition of crystal-bearing silicic melt: direct observation of magma rupture and healing. Geology **40**, 611–614 (2012). https://doi.org/10.1130/g3914.1

Cordonnier, B., Hess, K.-U., Lavallée, Y., Dingwell, D. B.: Rheological properties of dome lavas: case study of Unzen volcano. Earth Planet. Sci. Lett. **279**, 263–272 (2009)

Cornelius, R.R., Voight, B.: Real-time seismic amplitude measurement (RSAM) and seismic spectral amplitude measurement (SSAM) analyses with the materials failure forecast method (FFM), June 1991 explosive eruption at Mount Pinatubo. In: Fire and mud: Eruptions and lahars of Mount Pinatubo, Philippines, edited by: Punongbayan RS, and Newhall CG, Seattle, University of Washington Press, Quezon City, PHIVOLCS, pp. 249–267 (1996)

Crummy et al. (this issue)

David, C., Menéndez, B., Darot, M.: Influence of stress-induced and thermal cracking on physical properties and microstructure of La Peyratte granite. Int. J. Rock Mech. Min. Sci. **36**(4), 433–448 (1999)

De la Cruz-Reyna, S., Reyes-Davila, G.A.: A model to describe precursory material-failure phenomena: applications to short-term forecasting at Colima volcano, Mexico. Bull. Volcanol. **63**(5), 297–308 (2001)

De la Cruz-Reyna, S., Mendoza-Rosas, A.T., Borselli, L., Sarocchi, D.: Volcanic hazard estimations for Volcán de Colima (this issue)

Deubelbeiss, Y., Kaus, B.J., Connolly, J.A., Caricchi, L.: Potential causes for the non-Newtonian rheology of crystal-bearing magmas. Geochem. Geophys. Geosyst. **12**(5) (2011).

Dingwell, D.B.: Volcanic dilemma: flow or blow? Science **273**(5278), 1054–1055 (1996)

Duputel, Z., Ferrazzini, V., Brenguier, F., Shapiro, N., Campillo, M., Nercessian, A.: Real time monitoring of relative velocity changes using ambient seismic noise at the Piton de la Fournaise volcano (La Réunion) from January 2006 to June 2007. J. Volcanol. Geoth. Res. **184**(1–2), 164–173 (2009). https://doi.org/10.1016/j.jvolgeores.2008.11.024

Fang, J.N., Lo, H.J., Song, S.R., Chung, S.H., Chen, Y.L., Lin, I.C., Yu, B.S., Chen, H.F., Li, L.J., Liu, C.M.: Hydrothermal alteration of andesite in acid solutions: experimental study in 0.05 MH_2SO_4 solution at 110 degrees C. J. Chinese Chem. Soc. **50**(2), 239–244 (2003)

Farquharson, J.I., Heap, M.J., Varley, N.R., Baud, P., Reuschlé, T.: Permeability and porosity relationships of edifice-forming andesites: a combined field and laboratory study. J. Volcanol. Geoth. Res. **297**, 52–68 (2015). https://doi.org/10.1016/j.jvolgeores.2015.03.016

Farquharson, J.I., Heap, M.J., Lavallée, Y., Varley, N.R., Baud, P.: Evidence for the development of permeability anisotropy in lava domes and volcanic conduits. J. Volcanol. Geoth. Res. **323**, 163–185 (2016a). https://doi.org/10.1016/j.jvolgeores.2016.05.007

Farquharson, J.I., Heap, M.J., Baud, P., Reuschlé, T., Varley, N.: Pore pressure embrittlement in a volcanic edifice. Bull. Volcanol. **78**(6) (2016b)

Farquharson, J.I., Heap, M.J., Baud, P.: Strain-induced permeability increase in volcanic rock. Geophys. Res. Lett. **43** (2016c). https://doi.org/10.1002/2016gl071540

Farquharson, J.I., Baud, P., Heap, M.J.: Inelastic compaction and permeability evolution in volcanic rock. Solid Earth **8**(2), 561–581 (2017). https://doi.org/10.5194/se-8-561-2017

Heap, M.J., Baud, P., Meredith, P.G., Vinciguerra, S., Bell, A.F., Main, I.G.: Brittle creep in basalt and its application to time-dependent volcano deformation. Earth Planet. Sci. Lett. **307**, 71–82 (2011). https://doi.org/10.1016/j.epsl.2011.04.035

Heap, M.J., Lavallée, Y., Petrakova, L., Baud, P., Reuschlé, T., Varley, N.R., Dingwell, D.B.: Microstructural controls on the physical and mechanical properties of edifice-forming andesites at Volcán de Colima, Mexico. J. Geophys. Res. Solid Earth. **119**(4), 2925–2963 (2014). https://doi.org/10.1002/2013jb010521

Heap, M.J., Farquharson, J.I., Baud, P., Lavallée, Y., Reuschlé, T.: Fracture and compaction of andesite in a volcanic edifice. Bull. Volcanol. **77**(6), 55 (2015)

Heap, M.J., Wadsworth, F.B.: Closing an open system: pore pressure changes in permeable edifice rock at high strain rates. J. Volcanol. Geoth. Res. **315**, 40–50 (2016)

Heap, M.J., Wadsworth, F.B., Xu, T., Chen, C-f, Tang, C-a: The strength of heterogeneous volcanic rocks: a 2D approximation. J. Volcanol. Geoth. Res. **319**, 1–11 (2016)

Heap, M.J., Violay, M., Wadsworth, F.B., Vasseur, J.: From rock to magma and back again: the evolution of temperature and deformation mechanism in conduit margin zones. Earth Planet. Sci. Lett. **463**, 92–100 (2017). https://doi.org/10.1016/j.epsl.2017.01.021

Heap, M.J., Coats, B., Chen, C.F., Varley, N.R., Lavallée, Y., Kendrick J., Xu, T., Reuschlé, T.: Thermal resilience of microcracked andesitic dome rocks. J. Volcanol. Geoth. Res. (2018). https://doi.org/10.1016/j.jvolgeores.2018.10.021

Hornby, A.J., Kendrick, J.E., Lamb, O., Hirose, T., De Angelis, S., Von Aulock, F.W., Umakoshi, K., Miwa, T., Henton De Angelis, S., Wadsworth, F.B., Hess, K.-U., Dingwell, D.B., Lavallée, Y.: Spine growth and seismogenic faulting at Mt. Unzen, Japan. J. Geophys. Res. **120**, 4034–4054 (2015)

Horwell, C.J., Williamson, B.J., Llewellin, E.W., Damby, D.E., Le Blond, J.S.: The nature and formation of cristobalite at the Soufriere Hills volcano, Montserrat: implications for the petrology and stability of silicic lava domes. Bull. Volcanol. **75**(3) (2013)

James, M.R., Varley, N.: Identification of structural controls in an active lava dome with high resolution DEMs: Volcán de Colima, Mexico. Geophys. Res. Lett. **39** (2012). https://doi.org/10.1029/2012gl054245

Jaeger, J.C., Cook, N.G.W., Zimmerman, R.W.: Fundamentals of Rock Mechanics. Blackwell Publishing Ltd., Victoria (2007)

Kendrick, J.E., Lavallée, Y., Hess, K.U., Heap, M.J., Gaunt, H.E., Meredith, P.G., Dingwell, D.B.: Tracking the permeable porous network during strain-dependent magmatic flow. J. Volcanol. Geoth. Res. **260**, 117–126 (2013a)

Kendrick, J.E., Smith, R., Sammonds, P., Meredith, P., Dainty, M., Pallister, J.: The influence of thermal and cyclic stressing on the strength of rocks from Mount St. Helens, Washington. Bull. Volcanol. **75**, 1–12 (2013b)

Kendrick, J.E., Lavallée, Y., Hess, K.-U., de Angelis, S., Ferk, A., Gaunt, H.E., Dingwell, D.B., Leonhardt, R.: Seismogenic frictional melting in the magmatic column. Solid Earth **5**, 19–208 (2014a)

Kendrick, J.E., Lavallée, Y., Hirose, T., Di Toro, G., Hornby, A.J., De Angelis, S., Dingwell, D.B.: Volcanic drumbeat seismicity caused by stick-slip motion and magmatic frictional melting. Nat. Geosci. **7**, 438–442 (2014b)

Kendrick, J.E., Lavallée, Y., Varley, N.R., Wadsworth, F. B., Lamb, O., Vasseur, J.: Blowing off steam: tuffisite formation as a regulator for lava dome eruptions. Front. Earth Sci. **4**(41) (2016). https://doi.org/10.3389/feart.2016.00041

Kendrick, J.E., Lavallée, Y., Mariani, E., Dingwell, D.B., Wheeler, J., Varley, N.R.: Crystal plasticity as a strain marker of the viscous-brittle transition in magma. Nat. Comm. **8**, 1926 (2017)

Kolzenburg, S., Heap, M.J., Lavallée, Y., Russell, J.K., Meredith, P.G., Dingwell, D.B.: Strength and permeability recovery of tuffisite-bearing andesite. Solid Earth **3**(2), 191–198 (2012)

Kueppers, U., Scheu, B., Spieler, O., Dingwell, D.B.: Fragmentation efficiency of explosive volcanic eruptions: A study of experimentally generated pyroclasts. J. Volcanol. Geoth. Res. **153**(1–2), 125–135 (2006)

Kueppers, U., Varley, N.R., Alatorre-Ibargüengoitia, M. A., Lavallée, Y., Becker, S., Berninger, N., Goldstein, F., Hanson, J.B., Kolzenburg, S., Dingwell, D.B.: 2500 pyroclast puzzle: probing eruptive scenarios at Volcán de Colima, Mexico. In: Annual Assembly of the American Geophysical Union. San Francisco (2009)

Lamb, O.D., Varley, N.R., Mather, T.A., Pyle, D.M., Smith, P.J., Liu, E.J.: Multiple timescales of cyclical behaviour observed at two dome-forming eruptions. J. Volcanol. Geoth. Res. **284**, 106–121 (2014)

Lamb, O.D., De Angelis, S., Wall, R.J., Lamur, A., Varley, N.R., Reyes-Dávila, G., Arámbula-Mendoza, R., Hornby, A.J., Kendrick, J.E., Lavallée, Y.: Seismic and experimental insights into eruption precursors at Volcán de Colima. Geophys. Res. Lett. **44**(12), 6092–6100 (2017). https://doi.org/10.1002/2017GL073350

Lavallée, Y., Hess, K.-U., Cordonnier, B., Dingwell, D. B.: Non-Newtonian rheological law for highly crystalline dome lavas. Geology **35**(9), 843–846 (2007)

Lavallée, Y., Meredith, P.G., Dingwell, D.B., Hess, K.U., Wassermann, J., Cordonnier, B., Gerik, A., Kruhl, J. H.: Seismogenic lavas and explosive eruption forecasting. Nature **453**(7194), 507–510 (2008)

Lavallée, Y., Varley, N., Alatorre-Ibargüengoitia, M., Hess, K.U., Kueppers, U., Mueller, S., Richard, D., Scheu, B., Spieler, O., Dingwell, D.: Magmatic architecture of dome-building eruptions at Volcán de Colima, Mexico. Bull. Volcanol. **74**(1), 249–260 (2011)

Lavallée, Y., Varley, N.R., Alatorre-Ibargüengoitia, M.A., Hess, K.U., Kueppers, U., Mueller, S., Richard, D., Scheu, B., Spieler, O., Dingwell, D.B.: Magmatic architecture of dome-building eruptions at Volcán de Colima, Mexico. Bull. Volcanol. **74**(1): 249–260 (2012)

Lavallée, Y., Benson, P.M., Heap, M.J., Hess, K.-U., Flaws, A., Schillinger, B., Meredith, P.G., Dingwell, D.B.: Reconstructing magma failure and the degassing network of dome-building eruptions. Geology **41**(4), 515–518 (2013)

Lavallée, Y., Dingwell, D.B., Johnson, J.B., Cimarelli, C., Hornby, A.J., Kendrick, J.E., von Aulock, F.W., Kennedy, B.M., Andrews, B.J., Wadsworth, F.B., Rhodes, E., Chigna, G.: Thermal vesiculation during volcanic eruptions. Nature **528**, 544–547 (2015)

Lesage, P., Reyes-Davila, G., Arambula-Mendoza, R.: Large tectonic earthquakes induce sharp temporary decreases in seismic velocity in Volcan de Colima, Mexico. J. Geophys. Res. Solid Earth **119**(5), 4360–4376 (2014)

Luhr, J.F.: Petrology and geochemistry of the 1991 and 1998–1999 lava flows from Volcán de Colima, Mexico: implications for the end of the current eruptive cycle. J. Volcanol. Geoth. Res. **117**(1–2), 169–194 (2002)

Ma, L., Daemen, J.J.K.: Strain rate-dependence of mechanical properties of welded Topopah spring tuff. Int. J. Rock Mech. Mining Sci. **41**(395) (2004)

Main, I.G., Leonard, T., Papasouliotis, O., Hatton, C.G., Meredith, P.G.: One slope or two? Detecting statistically significant breaks of slope in geophysical data, with application to fracture scaling relationships. Geophys. Res. Lett. **26**(18), 2801–2804 (1999)

McNutt, S.R.: Seismic monitoring and eruption forecasting of volcanoes: a review of the state-of-the-art and case histories. In: Scarpa, R., Tilling, R.I. (eds.) Monitoring and Mitigation of Volcano Hazards. Springer, Berlin, pp. 99–146 (1996)

Mueller, S., Melnik, O., Spieler, O., Scheu, B., Dingwell, D.B.: Permeability and degassing of dome lavas undergoing rapid decompression: an experimental determination. Bull. Volcanol. **67**(6), 526–538 (2005)

Mueller, S.: Permeability and porosity as constraints on the explosive eruption of magma: laboratory experiments and field investigations. Ph.D. thesis, Munich (2007). https://edoc.ub.uni-muenchen.de/6909/

Mueller, S., Scheu, B., Spieler, O., Dingwell, D.B.: Permeability control on magma fragmentation. Geology **36**(5), 399–402 (2008)

Mueller, S., Scheu, B., Kueppers, U., Spieler, O., Richard, D., Dingwell, D.B.: The porosity of pyroclasts as an indicator of volcanic explosivity. J. Volcanol. Geoth. Res. **203**(3), 168–174 (2011)

Mueller, S.B., Varley, N.R., Kueppers, U., Lesage, P., Reyes Davila, G.Á., Dingwell, D.B.: Quantification of magma ascent rate through rockfall monitoring at the growing/collapsing lava dome of Volcán de Colima, Mexico. Solid Earth **4**(2), 201–213 (2013)

Paterson, M.S., Wong, T.-F.: Experimental Rock Deformation—The Brittle Field. Science-Technology, 347 p. (2005)

Reubi, O., Blundy, J.: Assimilation of plutonic roots, formation of High-K exotic melt inclusions and genesis of andesitic magmas at Volcán De Colima, Mexico. J. Petrol. **49**(12), 2221–2243 (2008)

Reubi, O., Blundy, J., Varley, N.R.: Volatiles contents, degassing and crystallisation of intermediate magmas at Volcán de Colima, Mexico, inferred from melt

inclusions. Contrib. Mineral. Petrol. **165**(6), 1087–1106 (2013)
Richard, D., Scheu, B., Mueller, S.P., Spieler, O., Dingwell, D.B.: Outgassing: influence on speed of magma fragmentation. J. Geophys. Res. Solid Earth **118**(3), 862–877 (2013)
Rocchi, V., Sammonds, P.R., Kilburn, C.R.J.: Fracturing of Etnean and Vesuvian rocks at high temperatures and low pressures. J. Volcanol. Geoth. Res. **132**(2–3), 137–157 (2004)
Rutter, E.H.: On the nomenclature of mode of failure transitions in rocks. Tectonophysics **122**(3–4), 381–387 (1986)
Sammis, C.G., Ashby, M.F.: The failure of brittle porous solids under compressive stress states. Acta Metall. **34** (3), 511–526 (1986)
Savov, I.P., Luhr, J.F., Navarro-Ochoa, C.: Petrology and geochemistry of lava and ash erupted from Volcán Colima, Mexico, during 1998–2005. J. Volcanol. Geoth. Res. **174**(4), 241–256 (2008)
Scheu, B., Kueppers, U., Mueller, S., Spieler, O., Dingwell, D.B.: Experimental volcanology on eruptive products of Unzen volcano. J. Volcanol. Geoth. Res. **175**(1–2), 110–119 (2008)
Smith, R., Sammonds, P.R., Kilburn, C.R.J.: Fracturing of volcanic systems: experimental insights into pre-eruptive conditions. Earth Planet. Sci. Lett. **280**, 211–219 (2009)
Spieler, O., Kennedy, B., Kueppers, U., Dingwell, D.B., Scheu, B., Taddeucci, J.: The fragmentation threshold of pyroclastic rocks. Earth Planet. Sci. Lett. **226**(1–2), 139–148 (2004)
Underwood, E.E.: Quantitative Stereology. Addison Wesley, Reading (1970)
Varley, N., Arambula-Mendoza, R., Reyes-Davila, G., Sanderson, R., Stevenson, J.: Generation of Vulcanian activity and long-period seismicity at Volcán de Colima, Mexico. J. Volcanol. Geoth. Res. **198**(1–2), 45–56 (2010)
Varley, N.R., Taran, Y.A.: Degassing processes of Popocatépetl and Volcán de Colima, Mexico. In: Oppenheimer, C., Pyle, D.M., Barclay, J. (eds.) Volcanic Degassing, pp. 263–280. Geological Society of London, London (2003)
Varley et al.: Volcán de Colima (this issue)
Vasseur, J., Wadsworth, F.B., Lavallée, Y., Hess, K.-U., Dingwell, D.B.: Volcanic sintering: timescales of viscous densification and strength recovery. Geophys. Res. Lett. (2013). https://doi.org/10.1002/2013gl058105
Vinciguerra, S., Trovato, C., Meredith, P.G., Benson, P. M.: Relating seismic velocities, thermal cracking and permeability in Mt. Etna and Iceland basalts. Int. J. Rock Mech. Min. Sci. **42**, 900–910 (2005)
Voight, B.: A method for prediction of volcanic-eruptions. Nature **332**(6160), 125–130 (1988)
Walter, T.R., Legrand, D., Delgado Granados, H., Reyes, G., Arambula, R.: Volcanic eruption monitoring by thermal image correlation: Pixel offsets show episodic dome growth of the Colima volcano. J. Geophys. Res. Solid Earth **118**(4), 1408–1419 (2013)
Webb, E.B., Varley, N.R., Pyle, D.M., Mather, T.A.: Thermal imaging and analysis of short-lived Vulcanian explosions at Volcán de Colima, Mexico. J. Volcanol. Geoth. Res. **278–279**, 132–145 (2014)
Zobin, V.M., Luhr, J.F., Taran, Y.A., Breton, M., Cortes, A., De la Cruz-Reyna, S., Dominguez, T., Galindo, I., Gavilanes, J.C., Muniz, J.J., Navarro, C., Ramirez Ruiz, J.J., Reyes-Davila, G.A., Ursua, M., Velasco, J., Alatorre, E., Santiago, H.: Overview of the 1997–2000 activity of Volcán de Colima, Mexico. J. Volcanol. Geoth. Res. **117**, 1–19 (2002)

Volcanic Hazard Estimations for Volcán de Colima

Servando De la Cruz-Reyna, Ana Teresa Mendoza-Rosas, Lorenzo Borselli and Damiano Sarocchi

Abstract

Volcán de Colima has one of North America's highest rates of eruption, producing a wide range of eruptions ranging from effusive episodes to Plinian phases and sector collapses. Such activity potentially endangers over 350,000 people dwelling within the reach of volcanic products and runouts. To widen the hazard assessment derived from this activity, a statistical analysis of the volcano past events must include as many historically and geologically reported eruptions and debris avalanches as possible. However, in doing this, factors such as the complex dependence of flank collapse avalanches with the eruptive activity and other external factors, the increased underreporting with age and the uncertainties in the magnitudes, intensities, and dates of past activity, as well as the possible time dependence of the rate of eruptions must be accounted for. This chapter describes the statistical methods used to estimate the probabilities of different hazard scenarios for Volcán de Colima whilst addressing those factors.

Keywords

Colima volcano · Volcanic hazards
Eruption statistics · Volcano flank collapse
Debris avalanches · Extreme volcanic events

1 Introduction

Volcán de Colima rises nearly 4 km above sea level, and currently has one of the highest rates of eruption in North America, as well as an impressive history of flank collapses, generating a potential risk for over 350,000 people dwelling within the reach of its products, either primary or remobilized: pyroclastic falls, pyroclastic flows, lahars, and debris avalanches (Fig. 1). These were some of the reasons that made it eligible as one of the IAVCEI's 15 International Decade for Natural Disaster Reduction (IDNDR) laboratory volcanoes since 1990. The perception of the hazards from Volcán de Colima is not new. In 1893, volcano observatories were installed in

S. De la Cruz-Reyna (✉)
Instituto de Geofísica, Universidad Nacional Autónoma de México, C. Universitaria, Coyoacán, 04510 México D.F., Mexico
e-mail: sdelacrr@geofisica.unam.mx

A. T. Mendoza-Rosas
CONACYT – Centro de Ingeniería y Desarrollo Industrial, Av. Playa Pie de la Cuesta 702, Desarrollo San Pablo, 76125 Querétaro, Qro, Mexico

L. Borselli · D. Sarocchi
Instituto de Geologia/Fac. de Ingeniería, Universidad Autónoma de San Luis Potosí–UASLP, Av. Dr. Manuel Nava 5, C.P., 78240 San Luis Potosí, Mexico

N. Varley et al. (eds.), *Volcán de Colima*, Active Volcanoes of the World,
https://doi.org/10.1007/978-3-642-25911-1_6

Fig. 1 Left: Hummocks or volcanic mounds evidencing the debris avalanche events at Volcán de Colima (Photo courtesy of Raúl Arámbula). Right: Massive avalanche deposit at the Montegrande ravine on the upper part of the vertical wall, at about 7 km south of the Volcán de Colima summit (Photo by Lorenzo Borselli)

Zapotlán (now Ciudad Guzmán) until 1903, and in Colima between 1903 and 1905, as part of a national net of meteorological and astronomical observatories. Daily reports of the activity were made by Arreola (1903, 1915) and Díaz (1906) for about 12 years.

In 1913, this volcano produced its latest major activity: a Plinian eruption that was monitored and reported by those scientists (Arreola 1915). Such an active volcano, surrounded by a large population and infrastructure requires a precise analysis of the hazards that may arise from any possible type of activity. Estimating the current and future hazards at Volcán de Colima poses several problems arising from the postulate inherent to all volcanic hazard analyses: the future activity of the volcano is in some complex way an extension of its past activity, including its randomness, and whatever fluctuations and hiatuses that have occurred in the rate of eruptions, i.e., the number of eruptions per unit time. It is thus essential to construct a complete as possible history of eruptions and collapses, in which the events are well characterized by their dimensions, as well as the time of their occurrences. Clearly, addressing this issue is not always possible. The statistical methods used to estimate the volcanic hazard, defined as the probability of potentially destructive volcanic phenomena occurring in a given time interval, should then account for the small size and possible incompleteness of the database, and for the errors in estimating the size and date of past eruptions and flank collapses.

2 Major-Event History of Volcan de Colima

The analysis of hazard of an eruptive history extending from the present through the Holocene makes it convenient to separate the activity data into two parts according to the nature of the methods used to collect the information, first as an account of the activity in the geological time-scale, in which smaller events are usually missing, dates have a variable degree of uncertainty and the eruption dimensions (magnitude, intensity) are indirectly estimated, or deduced if pertinent data are missing, and secondly, as a detailed score based on the historical database, in which dates are exact, and magnitudes and intensities are more precisely established. However, historical databases (perhaps fortunately) usually lack major or extreme events.

2.1 Geological Time-Scale

Before the current edifice of Volcán de Colima was formed, an ancestral Colima (Paleofuego) volcano grew in the late Pleistocene on the southern flank of Nevado de Colima, the older volcano located to the North of the current edifice. Both, Nevado de Colima and Volcán de Colima, have shown persistent activity during the Holocene and beyond. Such activity has often been characterized by frequent eruptions and large sector collapses generating conspicuous deposits of debris avalanche events (DAE; Borselli et al. 2011). Komorowski et al. (1993, 1997a, b) identify as many as 12 major DAE's from the CVC in the last 45,000 years. More recent studies (Cortés et al., this volume) attribute six of the debris avalanche deposits to Paleofuego, and two to the current edifice. The largest of those avalanches, a massive Bezymianny–St Helens type eruption with a southwards sector collapse, destroyed the ancestral Paleofuego, forming a 5-km-wide horseshoe- shaped caldera, and generated a massive volcanic debris avalanche that blanketed an area of about 1500 km^2. After the massive collapse, a long series of eruptions (Lava dome, lava flows, pyroclastic fall and pyroclastic flows), reconstructed the current volcanic edifice within the caldera (Luhr and Carmichael 1990b). The volume of that debris avalanche deposit has been estimated at 10 km^3. The date of such major event is somewhat controversial, as some authors set it at 4280 $\pm$ 110 yr BP (Luhr and Prestegaard 1988), and others at 9370 $\pm$ 400 yr BP (Robin et al. 1987). Further studies set a likely date for that event at 7040 $\pm$ 160 (Cortés et al., this volume)

Independently from that timing, other evidence shows that the previous Paleofuego edifice showed rather similar behaviour to the current Volcán de Colima, particularly during the last stages before its collapse, with frequent effusive and explosive eruptions and partial sector collapses. In fact, Luhr and Carmichael (1990b) estimate that the average magma production rate of Paleofuego and Volcán de Colima of about 2–3 km^3/kyr was unaffected by the Paleofuego massive sector collapse, signifying that it was an extreme event within a long duration volcanic process of remarkable persistency. For example, Luhr and Carmichael (1990a) estimate the total volume of lava erupted since the 1818 Volcanic Explosivity Index (VEI) 4 eruption at 1.9×10^6 m^3/year, matching the long-term average, which can thus be used to estimate the current rate of cone-building at Volcán de Colima.

Furthermore, sector collapses may be triggered by additional external factors other than eruptions, such as major earthquakes, climatic effects (Capra 2006; Roverato et al. 2011), and others, provided that the volcanic edifice reaches proportions that correspond to a certain level of gravitational potential energy, and possesses an instability condition in its structure.

Numerous studies on the Volcán de Colima products by different authors over the last three or four decades have rendered a large number of eruption deposit ages (see Crummy et al. this volume), which makes the reconstruction of the eruptive history a complex task. Although Crummy et al. (this volume) summarise at least 25 major Plinian eruptions during the last 30,000 years, no detailed account of the probably numerous Holocene eruptions with intermediate magnitudes (VEI 3–4) exists. However, reckoning only major events leaving extensive deposits, five distinct major (VEI $\geq$ 5) Holocene eruptions may be identified. Most of those eruptions have been widely acknowledged by the radiocarbon datings assigned by one or another author, even though the dated samples from the related deposits extend over ample ranges. The left column of Table 1 lists the "nominal" radiometric dates commonly used in the literature to characterize those eruptions.

However, to evaluate volcanic risk, we consider it important to identify those events by their actual ages, and by realistic estimates of their uncertainties. Error ranges reported along with uncalibrated radiocarbon dates usually represent a dispersion of results in repeated readings on a single sample, and they do not represent an actual uncertainty of the deposit age, in the sense that the dispersion of results from different samples of the same deposit, collected and reported by one or different authors does. In addition, uncalibrated radiocarbon ages are calculated on the assumption that the atmospheric radiocarbon concentration has always been the same as it was in 1950, and are always referred to that date.

Table 1 Major eruptions at Volcán de Colima in the Holocene. The events are characterized by their nominal uncalibrated radiocarbon dates (left column)

Radiometric nominal dates of major eruptions and DAE's as commonly addressed in the literature (yr BP)	Calendar date intervals calculated from the radio-carbon datings using the OxCal 4.1 program. Negative numbers refer to dates in years B.C.	Likely calibrated ages referred to the present (yr cal BP 2014) and their uncertainties	VEI	Extensive Debris Avalanche Deposit?	References of radiocarbon datings
2365 ± 55	−752–406	2593 ± 173	>4	No	Navarro and Luhr (2000), Komorowski et al. (1993, 1997a, b)
2400 ± 90					
2420 ± 80					
2455 + 95/−90					
2485 ± 85					
2510 ± 55					
2540 ± 110	−901–798	2864 ± 52	>4	Yes	Luhr and Carmichael (1982), Siebe et al. (1992), Komorowski et al. (1993, 1997a, b), Cortés et al., this volume
2550 + 110/−105					
2690 ± 40					
2870 ± 120					
3060 ± 155	−1680–1451	3580 ± 115	>4	No	Luhr and Carmichael (1982), Komorowski et al. (1997a, b)
3265 ± 55					
3400 ± 95					
3540 ± 60	−2116–1776	3960 ± 170	>4	Yes	Navarro and Luhr (2000), Komorowski et al. (1993, 1997a, b), Cortés et al., this volume
3545 ± 100					
3600 ± 120					
3685 ± 80					
3740 ± 115	−2617–2289	4467 ± 164	>4	No	Kiryanov and Kolosov (1986), Komorowski et al. (1993, 1997a, b)
3785 + 265/−255					
3925 ± 115					
4050 ± 70					
4280 ± 110	−3322–2712	5031 ± 305	>4	Yes	Luhr and Prestegaard (1988), Luhr and Navarro (2002), Robin et al. (1984), Capra and Macías (2002), Capra (2007), Luhr et al. (2010)
4350 ± 100					
4360 ± 140					

(continued)

Table 1 (continued)

Radiometric nominal dates of major eruptions and DAE's as commonly addressed in the literature (yr BP)	Calendar date intervals calculated from the radio-carbon datings using the OxCal 4.1 program. Negative numbers refer to dates in years B.C.	Likely calibrated ages referred to the present (yr cal BP 2014) and their uncertainties	VEI	Extensive Debris Avalanche Deposit?	References of radiocarbon datings
6895 +165/−160	−6056–5772	7928 ± 142	>4	Yes	Luhr and Carmichael (1982), Luhr and Navarro (2002), Komorowski et al. (1993, 1997a, b), Cortés et al. (2005, 2010b), Cortés et al., this volume, Luhr et al. (2010)
6910 ± 190					
6990 ± 130					
7040 ± 160					
7380 ± 160					
9300 ± 420	−9294–8927	11125 ± 184	>4	Yes	Navarro and Luhr (2000), Robin et al. (1984, 1987), Robin et al. (1991), Komorowski et al. (1993, 1997a, b), Cortés et al. (2005, 2010b), Cortés et al., this volume
9370 ± 400					
9671 ± 88					
9770 ± 60					
13275 ± 300	−14893–13663	16292 ± 615	>4	Yes	Komorowski et al. (1997a, b), Cortés et al. (2005)
13370 ± 120					

To illustrate the dispersion of data, interpreted as the uncertainty in the age of the events, the calibrated calendar dates calculated from published data of samples from the eruptions' deposits using the OxCal 4.1 program are listed as pairs of dates representing the error interval in the second column. Estimates of the most likely dates of the events, calculated from the averages of those dates, and their standard deviations are transformed into calibrated ages in years before 2014 and listed in the third column. All of those major eruptions are assumed to have VEI's over 4

We thus calculated the calibrated calendar dates from the radiocarbon datings of Volcán de Colima deposits reported in all of the literature we could find, using the OxCal 4.1 program with the IntCal 04 calibration curve (Bronk Ramsey 2009). This was done pooling together the individual radiocarbon dates and error ranges grouped according to the published stratigraphic information prior to calibration. The program provides a built-in chi-square statistical test for whether the ages are from the same event with a given level of confidence (95.4 as a default). It then yields two dates defining an interval as the most likely time for the event occurrence. In the second column (from the left) of Table 1, we list the intervals of calibrated calendar dates of different groupings representing the Volcán de Colima events. Negative numbers refer to dates BC.

To acquire a more intuitive grasp of these ages, we translated the calibrated calendar ranges into calibrated ages before 2014, using the median of each age range as the likely calendar date of the event, and the half-width of the range as a measure of dispersion. Such calibrated ages, listed in the third column of Table 1 should thus represent the most realistic estimate of the actual ages of the events and their uncertainties.

2.2 Historical Time-Scale

Historical activity is relatively well documented since 1560. Although the available record includes 41 VEI $\geq$ 1 eruptions in that period, we shall not consider in this analysis eruptions in the VEI range $\leq$ 2. The main reasons of this choice are: (a) eruptions in the lowest VEI range are predominantly effusive, and thus difficult to consider as single short-duration events; (b) it is more convenient for hazard analysis purposes to consider the significant explosions that destroyed the products of such effusive processes; (c) minor eruptions may have not been reported, as the completeness of the catalogue usually decreases as the eruption magnitude diminishes. Moreover, to compensate for this, some catalogues assign a default value VEI 2 to eruptions with doubtful information, generating an artefact in that VEI category; (d) it is unlikely that eruptions in the VEI range $\leq$2 represent a serious threat to the population, and omitting them from the analysis may not significantly affect the hazard analysis. In addition, the difficulties of identifying and dating small eruptions far in the past increase the error in the rate of eruption estimations when linking the historical and geological eruptive histories (Mendoza-Rosas and De la Cruz-Reyna 2008).

The historical activity has been characterized by alternate dome-emplacing effusive phases and explosive events that have included major eruptions. Table 2 shows a record of 18 significant eruptions with VEI $\geq$ 3 that have been reported in the past 439 years. The cumulative plot of eruption occurrences shown in Fig. 2 reveals that the overall average eruption rate since 1576 (18/439, thick solid line in the figure) may be composed by two or more age-specific rates of eruption (dashed lines in Fig. 2) suggesting that the volcano shows regimes of more frequent explosive activity alternated with regimes of lower explosive eruption occurrence rate.

After the 1913 Plinian event, the deep crater of the volcano started to slowly fill with lava, showing several eruptive phases after the lava reached the crater rim around 1957 (Mooser 1961). Since then, activity has been characterized, as before, by lava and dome extrusions alternated with dome-destruction explosions and Merapian pyroclastic flows (Saucedo et al. 2005). At times, partial collapses of summit domes and lava flow fronts have produced frequent block-and-ash flows. In the last 23 yr small to moderate eruptive episodes have been reported in 1991, 1994, 1998–1999, 2001–2003, 2004, 2005, 2007–2011 (Rodríguez-Elizarrarás et al. 1991; Saucedo et al. 2005; Sulpizio et al. 2010; Sarocchi et al. 2011; Varley et al., this volume). During heavy rains, which usually occur from June to October at this latitude, these deposits are often remobilized, producing lahars (Capra et al. 2010; Capra et al., this volume). So far, the products emitted during the current post-1913 stage have not significantly affected the surrounding inhabited area. The VEI 4 Plinian phase in 1913 (Saucedo et al. 2010, 2011)

Table 2 Chronology and VEI estimates of the Explosive Historical activity (VEI $\geq$ 3) of Volcán de Colima

Date AD (dd-mm-yyyy)	VEI	References
??-??-1576	3	Tello (1650), Arreola (1903), Medina-Martínez (1983), De la Cruz-Reyna (1993), Bretón et al. (2002)
10-01-1585	4	Ibid.
14-01-1590	3	Ibid.
25-11-1606	4	Ibid.
15-04-1611	3	Ibid.
08-06-1622	4	Ibid.
??-??-1690	3-(4?)	Arreola (1903), Medina-Martínez (1983), De la Cruz-Reyna (1993), Bretón et al. (2002)
10-03-1770	3	Ibid.
15-02-1818	4	Ibid.
12-06-1869	3	Ibid.
26-02-1872	3	Ibid.
06-01-1886	3	Bárcena (1887), Arreola (1903), Medina-Martínez (1983), De la Cruz-Reyna (1993), Bretón et al. (2002)
26-10-1889	3	Arreola (1903), Medina-Martínez (1983), De la Cruz-Reyna (1993), Bretón et al. (2002)
16-02-1890	4	Arreola (1903), Waitz (1932), Medina-Martínez (1983), De la Cruz-Reyna (1993), Bretón et al. (2002)
15-02-1903	3	Arreola (1903), Ordóñez (1903), Waitz (1932), Medina-Martínez (1983), De la Cruz-Reyna (1993), Bretón et al. (2002)
18-12-1908	3	Waitz (1932), Medina-Martínez (1983), De la Cruz-Reyna (1993), Bretón et al. (2002)
20-01-1913	4	Medina-Martínez (1983), De la Cruz-Reyna (1993), Bretón et al. (2002), Saucedo et al. (2010, 2011)
24-05-2005 – 16-09-2005	3	Savov et al. (2008), Smithsonian Institution (2005, 2006), Varley, this volume

How most of the VEI values were assigned is described in De la Cruz-Reyna, 1993

and the VEI 3 (based on column heights exceeding 6 km above the crater) dome-destruction sequence of events in 2005 (Savov et al. 2008; Smithsonian Institution 2005, 2006; Varley, this volume) are represented by the last two steps in Fig. 2.

Was the 2005 event the last of a low age-specific eruption rate, or does it mark the beginning of a high rate episode? What are the probabilities of a significant eruption (VEI $\geq$ 3), a major eruption VEI $\geq$ 4, or a catastrophic VEI $\geq$ 5, eruption occurring in the foreseeable future? There is no evidence of major flank collapses during historical time. However the possibility of such a collapse must also be considered in any hazard assessment (Borselli et al. 2011). Is a flank collapse, related or not to eruptive activity, likely in the near future? Next, we attempt to answer, at least partially, such questions.

3 Statistical Approach to the Assessment of Volcanic Hazard

In the present approach, we shall assume for simplicity that there are only two main sources of hazard from a volcano such as Volcán de Colima: volcanic eruptions and flank collapses. The latter, derived from the volcanic edifice

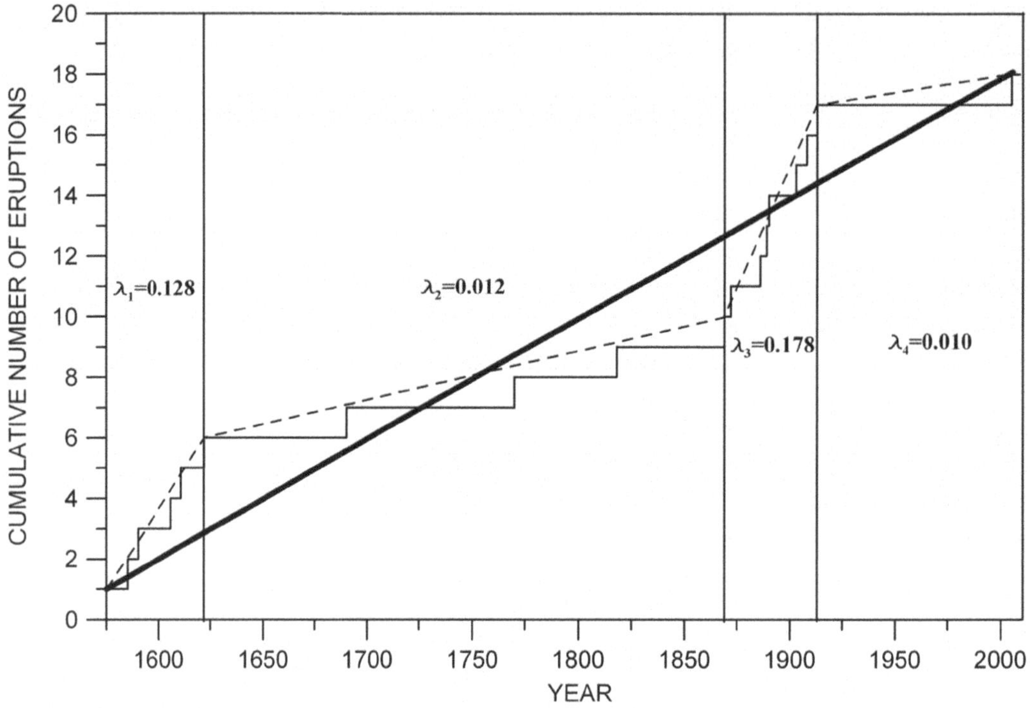

Fig. 2 Cumulative number of eruptions reported in the period 1576–present (2014) in the magnitude class VEI $\geq$ 3 for Volcán de Colima. The solid vertical lines separate four clearly recognizable regimes. The slopes of the dashed lines represent the age-specific eruption rates λ_i of each regime i, with i = 1, …, 4., and the thick line slope represents the mean eruption rate (λ_{global} = 0.0410 er/yr) of the whole historical series

instability, may be induced by eruptive activity, including deposition of voluminous pyroclastic deposits on steep slopes (McGuire 1996), hydromagmatic processes (Dzurisin 1998), and phreatomagmatic activity (Bandai-type activity, Moriya 1980; Siebert et al. 1987), or by other causes that may perturb an unstable volcanic edifice, such as direct magmatic intrusions into the edifice (Bezymianny-type activity, Gorshkov 1963; Siebert et al. 1987) or into the crust under the volcano (Day 1996; Elsworth and Voight 1996). Gravitational failure may also occur in response to external and independent causes such as progressive weakening of an edifice, and discrete triggering mechanisms such as earthquakes (Keefer 1984) and heavy rainfall (Sheridan et al. 1999; Capra 2006; Roverato et al. 2011).

We thus depart from the assumption that the features significant to hazard associated with the future behaviour of a volcano should reflect in some degree the relevant lineaments of its past history and since eruptions and flank collapses may be related or not, we begin addressing the problem by treating both phenomena separately. First, we analyse the statistics of volcanic eruptions, then the statistics of flank collapses. Next, we attempt to obtain an overall estimate of the volcanic hazard.

4 Volcanic Eruption Hazard

The statistical approach to the assessment of volcanic eruption hazards probably begins with the early studies of volcanic time-series by Wickman (1965, 1976) and Reyment (1969). They used stochastic principles to study eruption patterns on specific volcanoes. However, the models presented by Wickman (1976) did not distinguish between eruptions of different types, and were not tested against observed records. Other early studies, analysed specific volcanic eruption series, as was the case of the Hawaiian volcanoes (Klein 1982), the global set of volcanoes (De la Cruz-Reyna 1991, 1996) and Volcán de Colima (De la Cruz-Reyna 1993). Bebbington and Lai (1996a) examined the patterns of Mt. Ruapehu and Mt. Ngauruhoe, and Bebbington and Lai (1996b) also analysed Etna, Kilauea, Yake-Dake, Aso, and St Helens volcanoes. More recently, Bebbington (2010) studied trends and clustering of a group of 14 volcanoes including Volcán de Colima. The tendency of those studies was to use renewal models, i.e., processes in which the eruptions were considered statistically independent events with negligible duration when compared with the repose time between them. If the distribution of the repose times followed a simple exponential distribution the eruption time series was said to be a stationary Poisson process with a constant rate of eruptions (Cox and Lewis 1966). However, although some volcanoes showed stationary patterns, others had a marked time dependence of the rate of eruption occurrences. Subsequent studies have incorporated non-homogeneous behaviour of the eruption sequences, including for instance time-specific eruption rates (De la Cruz-Reyna 1996), transition probabilities of Markov chains (Carta et al. 1981; Aspinall et al. 2006; Bebbington 2007), change-point detection techniques (Mulargia et al. 1987), Rank-order statistics (Pyle 1998), Bayesian analysis of volcanic activity (Ho 1990; Ho et al. 2006; Solow 2001; Newhall and Hoblitt 2002; Marzocchi et al. 2008), non-homogeneous models (Ho 1991; Bebbington and Lai 1996b), a mixture of Weibull distributions (Turner et al. 2008), the mixture of exponentials distribution (Mendoza-Rosas and De la Cruz Reyna 2009, 2010), geostatistical hazard-estimation methods (Jaquet et al. 2000; Jaquet and Carniel 2006), non-homogeneous Poisson processes and extreme value theory (Mendoza-Rosas and De la Cruz-Reyna 2008, 2010; De la Cruz-Reyna and Mendoza-Rosas 2015; Coles and Sparks 2006; Deligne et al. 2010; Sobradelo et al. 2011), long-term hazard models (Marzocchi and Bebbington 2012), statistical analysis of explosions within an eruptive episode to detect contrasting styles of activity (Varley et al. 2006), and many others, that unfairly as it may appear, are impossible to cite here considering the increasing productivity in this field. A recent review of several of those methods may be found in Bebbington (2013).

Different parameters have been used as random variables to characterize the eruptive time series. Among them, the most frequently used are: the duration of eruptions, the repose interval between eruptions, the effusion rate; the volume or mass released, and the intensity of eruptions (Klein 1982; Settle and McGetchin 1980; Walker 1980; Newhall and Self 1982; Fedotov 1985; Mulargia et al. 1987; Carey and Sigurdsson 1989; Pyle 2000; Bebbington 2007, 2013). We prefer to characterize the eruptions by a measure of size that reflects their destructive potential, and assuming that the impact and effects of an eruption are proportional to both, the total mass or energy release (magnitude) and the rate of mass or energy release (intensity). The Volcanic Explosivity Index VEI is the quantity based on those parameters for which the eruption database is most complete (e.g. Newhall and Self 1982; Simkin and Siebert 1994, 2000; Siebert and Simkin 2002), facilitating the comparison among different volcanoes, even though reported VEI's do not always actually reflect both parameters.

For our purpose we shall consider here only significant explosive eruptions, which are usually short-duration events when compared with the repose time between eruptions. The volcanic eruption sequence of Volcán de Colima is thus considered here as a point process developing along the time axis, and both the distribution of eruptions and the repose times between them are analysed for different VEI classes. Long duration effusive processes such as dome emplacements are accounted for by associated dome destructing explosions.

To obtain the most robust volcanic eruption time-series possible we merge the historical (describing more frequent smaller eruptions) and the geological (describing larger, infrequent eruptions) eruptive histories described in the previous section. This is an important factor for a proper estimation of the likelihood of more damaging events (Marzocchi et al. 2004).

To analyse the eruptive time series and its renewal properties we first test the independence between successive eruptions to detect possible memory effects, and the stationarity or time dependence of the eruption sequence to find a possible non-homogeneity of the process. We then use the Mixture of Exponentials Distribution (MOED, Mendoza-Rosas and De la Cruz-Reyna 2009, 2010; Wehrmann and Dzierma 2011) to study the distribution of repose times between successive eruptions, and a Non-Homogeneous Generalized Pareto-Poisson Process (NHGPPP, Mendoza-Rosas and De la Cruz-Reyna 2008) to obtain volcanic hazard estimations. The probabilities of future eruptions of Volcán de Colima within different VEI classes, and over suitable time intervals are estimated using the different distributions, and the results compared with those obtained from a simple Poisson process.

5 Tests for Independence Between Eruptions and Time Dependence of the Eruptive Sequence

One of the simplest independence test for a Poisson process is done by means of a serial correlation scatterplot (Cox and Lewis 1966), in which the duration of each interval T_{i+1} between two successive eruptions is plotted against the previous repose interval durations T_i. This test must be done on a complete time series, which in the present case is the Volcán de Colima historical eruption data with VEI magnitude ≥ 3. Figure 3 shows the diagram with the features of a sequence of independent events: a large dispersion, a concentration of points near the axes, and low correlation coefficients: for each VEI category 2, 3, 4, ≥ 2, and ≥ 3 are: −0.096, 0.256, −0.156, 0.152, and 0.384 respectively, indicating low serial correlations for all VEI categories (Mendoza-Rosas and De la Cruz-Reyna 2008). Therefore, this eruptive series may be characterized as a random process of independent events.

The analysis for time dependence of the eruptive series of Volcán de Colima shows a weak non-stationary behaviour visualized as a succession alternating regimes with age-specific rates of eruption fluctuating around a mean rate in Fig. 2. A simple running mean test (Klein 1982) indicates that such fluctuations may not be attributed to chance and reflect an underlying process of alternating episodes of high and low rates, although the difference in the rates is small (De la Cruz-Reyna 1993, 1996; Mendoza-Rosas and De la Cruz-Reyna 2008, 2009).

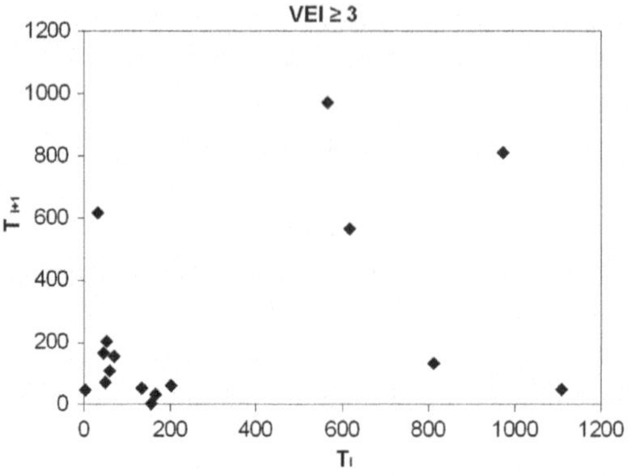

Fig. 3 Serial correlation diagram of successive repose intervals T_i and T_{i+1} (from Table 1, measured in months) between eruptions with magnitude VEI $\geq$ 3 of Volcán de Colima

6 Analysis of Repose Times Between Significant Explosive Eruptions

This non stationary component of the eruptive sequence of Volcán de Colima may be statistically represented by means of the shape parameter of the Weibull distribution with a value different than one. An alternative way to describe the distribution of repose intervals between events of an eruptive sequence, characterized by a succession of regimes with age-specific eruption rates, is through the use of the Mixture of Exponentials Distribution (MOED, Mendoza-Rosas and De la Cruz-Reyna 2009). The parameters of a MOED emphasize the effects of the time dependence of the eruptive process allowing clustering the eruption sequence in regimens as discussed below. A MOED is a mixture of m exponential distributions that has proved useful to describe the repose time distribution of eruptive sequences with non-stationary behaviour. The cumulative MOED function has the form:

$$F(t|\Lambda) = \sum_{j=1}^{m} w_j(1 - e^{-\lambda_j t}) \quad (1)$$

with a survival function:

$$S(t|\Lambda) = 1 - F(t|\Lambda) \quad (2)$$

and probability density function:

$$f(t|\Lambda) = \sum_{j=1}^{m} w_j \lambda_j e^{-\lambda_j t} \quad (3)$$

where $\Lambda = (\lambda_1, \ldots, \lambda_m, w_1, \ldots, w_m) > 0$ is the vector of parameters, and $\sum_{i=1}^{m} w_i = 1$.

The λ's represent the age-specific rates of eruption, namely the number of events per duration of each identified regime j. Of particular importance are the weighting factors w_j's, which for our purpose are calculated as the normalized complement of the corresponding proportions of the duration of regimes, considering that regimes of shorter duration tend to have higher rates of eruption.

$$w_i = \frac{D_t - D_i}{\sum_{i=1}^{m} (D_t - D_i)} \quad (4)$$

where D_t is the duration of the sampled interval (439 years for Colima), and D_i is the duration of the identified regime.

The hazard function, also known as the age-specific failure rate or as the force of mortality is a measure of the probability of an eruption occurring immediately after a time interval representing survival (i.e., no eruptions occurring in that interval), and it is defined as the ratio of the probability density function to the survival function:

$$h(t|\Lambda) = \frac{\sum_{j=1}^{m} w_j \lambda_j e^{-\lambda_j t}}{1 - \sum_{i=1}^{m} w_j(1 - e^{-\lambda_j t})} \quad (5)$$

The MOED has been successfully applied to eruptive series assumed complete for some volcanoes of México (Mendoza-Rosas and De la Cruz Reyna 2009, 2010) and to the Chilean Southern Volcanic Zone (Dzierma and Wehrmann 2010; Wehrmann and Dzierma 2011). For Volcán de Colima, Mendoza-Rosas and De la Cruz Reyna (2009) compared the probabilities of occurrence of at least one eruption calculated with MOED (with the parameters listed in Table 3) and with the exponential and Weibull distributions. Although the Weibull distribution showed a better fit to the repose-time data than the simple exponential function, the MOED improved that fit, having the additional advantage that its parameters may be easily obtained from the inspection of the cumulative number of eruptions plot when distinct rates of eruptions may be recognized as age-specific regimes, i.e., as a succession of periods, each with a constant rate, and clearly defined regime transition points, as has been the case of other Mexican and Chilean volcanoes (Mendoza-Rosas and De la Cruz-Reyna 2009, 2010; Dzierma and

Table 3 Observed eruptive regimes and calculated parameters of the MOED for Volcán de Colima (VEI > 2)

Regime	Time period	Number of eruptions	Duration of regime (yr)	Annual rate λ	Weighting factor w
1	1576–1622	6	47	0.127660	0.297646
2	1623–1868	3	246	0.012195	0.146545
3	1869–1913	8	45	0.177778	0.299165
4	1914–2014	1	101	0.009901	0.256644

Wehrmann 2010). This makes its use more physically intuitive. However, when the specific-age changes cannot be clearly identified as points between piece-wise linear regimes in the cumulative number of eruptions plot, specific statistical methods should be used to allow for a mathematically objective determination of the change points, using for example procedures of change-point identification by data partitioning based on two-sample Kolmogorv-Smirnov statistics (Mulargia and Tinti 1985; Mulargia et al. 1987), or applying hidden Markov models methods (Bebbington 2007).

One important feature of the MOED hazard function (5) is its time-dependence, as it decays after a previous event, as shown in Fig. 4 for Colima. This is a consequence of the Bayesian character of the MOED that uses a prior belief derived from the relevant prevalence of regimes in the past. As more time passes since the last eruption, the belief that we are in a high-rate regime decreases. This is revealed by the Weibull hazard function as well, also included in Fig. 4. Finally, it is important to be aware that the number of degrees of freedom of the MOED increases with the number of identified regimes, thus raising the number of required parameters. This may demerit its measure of goodness of fit when compared with other distributions, according to the Akaike Information Criterion (Akaike 1973; Bebbington 2007; Turner et al. 2008; Dzierma and Wehrmann 2010).

7 Assessing Eruption Hazard Through the Linking of Historical and Geological Eruptive Sequences

To deal with the difficulties derived from using an extended eruptive sequence that includes records in the historical and geological time

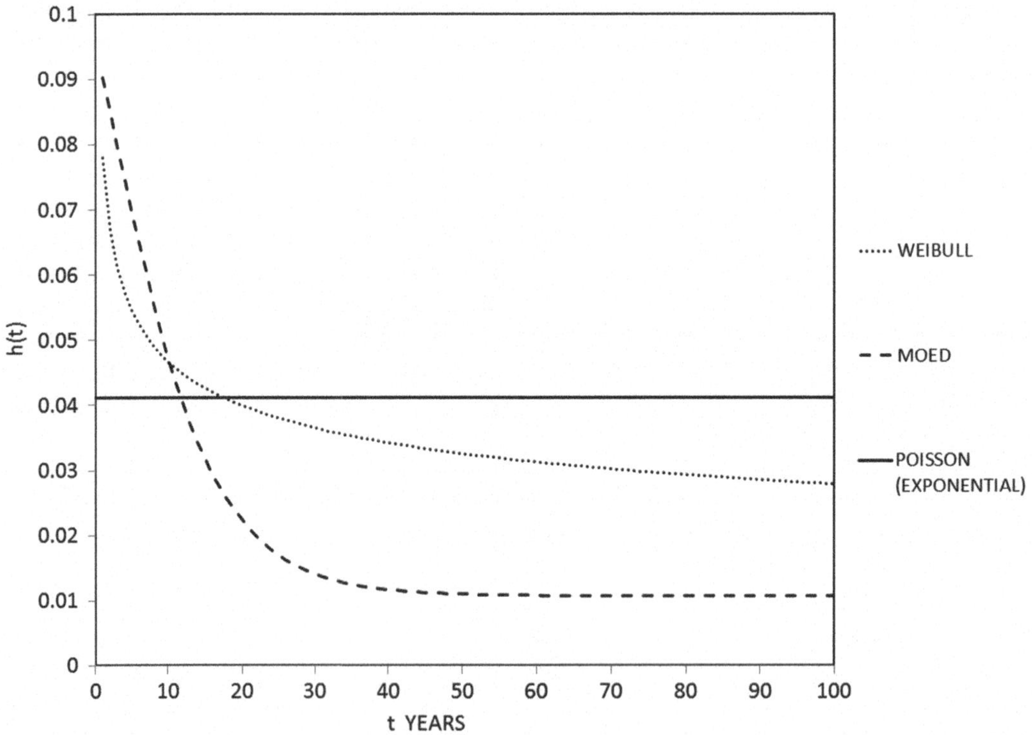

Fig. 4 The time dependent Weibull and MOED hazard functions, and the time-independent Poisson hazard function of Volcán de Colima

scales, and the possible incompleteness of the geological record, we apply a scaling law that relates the eruption size *Mvei* with the logarithmic eruption rate of that magnitude class λvei.

$$\log(\lambda vei) = a\,Mvei + c \tag{6}$$

Such a scaling law holds rather well for aggregations of volcanoes, and has proved useful for linking historical and geological time scales in specific volcanoes, allowing the expansion of the database in time and in missing information on eruption magnitudes (De la Cruz-Reyna 1991, 1993; Pyle 1995; De la Cruz-Reyna and Carrasco-Núñez 2002; De la Cruz-Reyna and Tilling 2008; Mendoza-Rosas and De la Cruz-Reyna 2008, 2010). However it must be used with care when attempting to transfer this type of analysis to other volcanoes, as the scaling may not hold for the available databases, or for volcanoes behaving differently from the average. Furthermore, in some cases, the scaling may hold for a magnitude-related parameter other than the VEI, as the tephra volume (Bebbington et al. 2008; Hurst and Smith 2010).

Using the scaling (6) on the Volcán de Colima eruption data, we may construct models of the distribution of large events using the available geological and historical time series in Tables 1 and 2, assigning likely VEI values to the uncertain data, and then select the model or models which best fit the rates of eruption occurrences, as shown in Fig. 5. The best fitting model in the figure is the "Case 2", and the most likely rates of

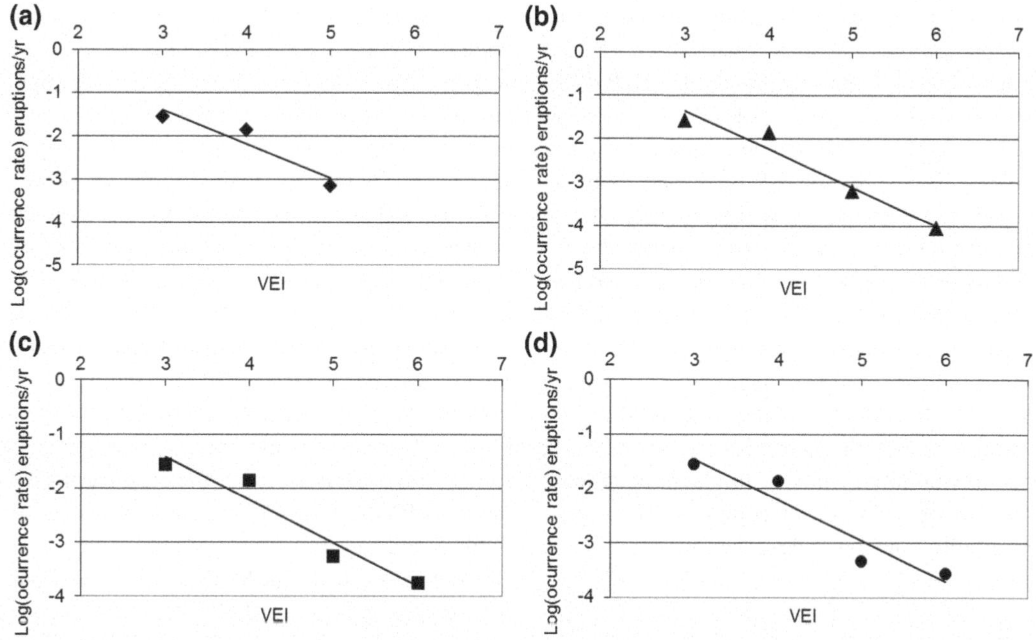

Fig. 5 Four possible models of the VEI magnitude versus rate of eruption distributions for major Holocene eruptions of Volcán de Colima. From the available VEI values and from the scaling law (5) we searched and tested for the most likely values of the eruption size distribution: **a** Case 1 (diamonds), we assume that the eight major geological eruptions listed in Table 1 were of about the same size (VEI 5). **b** Case 2 (triangles). We assume that one of the eruptions reached a VEI 6, while the others had VEI 5. **c** Case 3 (squares). This model assumes that six eruptions reached VEI 6 and two had VEI 5. **d** Case 4 (circles). This model assumes that five eruptions reached a VEI 6 and three had VEI 5. The best fittings to those distributions are the lines log (λvei) = −0.790 *Mvei* + 0.969 for case 1; log (λvei) = −0.879 *Mvei* + 1.285 for case 2; log (λvei) = −0.795 *Mvei* + 0.967 for case 3, and log (λvei) = −0.750 *Mvei* + 0.789 for case 4, with Pearson correlation coefficients 0.887, 0.953, 0.937 and 0.904 respectively. Case 2 has the best fit. However, the other models yield comparable results for the log (λvei) distribution

Table 4 Expected occurrence rates in the VEI range 3–6, and mean recurrence times obtained from the best fit "Case 2" model in Fig. 4

VEI	λvei Annual occurrence rate (er/yr)	$1/\lambda vei$ Mean recurrence time (yr)
3	0.04457	22
4	0.00589	170
5	0.00078	1283
6	$\sim 10^{-4}$	$\sim$10,000

eruption λvei for each VEI category obtained from the model are listed in Table 4.

Although more precise estimates of the distributions and probabilities of future eruptions may be obtained from the eruption rates calculated using the linked database, it is necessary to keep in mind that this does not guarantee the completeness of the whole database, and may underrate the hazard related to very large, rare events. To deal with this problem and obtain a more accurate mathematical quantification of the volcanic hazard we assume a Non Homogeneous Poisson process with a Generalized Pareto Distribution as intensity function (NHGPPP) (Davison and Smith 1990; Coles 2001; Reiss and Thomas 2001) to describe an eruptive sequence (Coles 2001; Mendoza-Rosas and De la Cruz-Reyna 2008, 2010; Sobradelo et al. 2011). This method belongs to the extreme value theory, in which the extreme values are atypical and rare events located at the tail of the distribution. The Generalized Pareto Distribution (GPD) is a robust tool that allows modeling of extreme values (Coles 2001; Davison and Smith 1990; Beguería 2005; Mendoza-Rosas and De la Cruz-Reyna 2008), such as the large-magnitude eruptions or major sector collapses. The cumulative GPD function is described by

$$G_{k,a}(y) = \begin{cases} 1 - \left(1 - \frac{ky}{\sigma}\right)^{1/k} & \text{for } k \neq 0 \\ 1 - e^{-y/\sigma} & \text{for } k = 0 \end{cases} \quad (7)$$

where k is a shape parameter, σ is a scale parameter, $y = x - u$ is the realization of an excess over a threshold u (also called the location parameter) and x is the exceedance over the established threshold. The GPD thus considers that an exceedance requires that an eruption exceeds a given VEI (threshold) of a time-series of magnitudes that may be stationary or not.

The NHGPPP depends only on the number of very large events exceeding a certain threshold, and not on the times of their occurrences, nor on their precise magnitudes, thus reducing the influence of time dependence and incompleteness. Still, it is very important to consider in this analysis that the VEI is not an open scale, ending in 8. Therefore the GDP probabilities of eruptions exceeding VEI 8 must be subtracted from the calculated probabilities to avoid an overestimation of the exceedances. In an eruptive sequence these data are represented by the right tail of the repose-time distribution and indeed influence any hazard estimation, as they represent events with very high destructive power. This method has been applied to calculate the probabilities of major eruptions in databases containing few and incomplete data in volcanoes of México (Mendoza-Rosas and De la Cruz-Reyna 2008, 2010) and the Canary Islands, Spain (Sobradelo et al. 2011). Homogeneous Poisson, Weibull, MOED and NHGPPP probabilities of occurrence of at least one eruption in specific time intervals using the eruptive history models of Volcán de Colima are summarized in Table 5 and Fig. 6.

Contrasting with the MOED, both, the Poisson process and the NHGPPP have the memoryless property. In both the magnitudes are independent of each other, and of the eruption timings. The distinction is that in the former the VEI is taken from the empirical distribution of the observed eruptive time series, whereas in the latter it is smoothed using an extreme value model which trades off some of the preservation bias against a frequency-magnitude assumption.

Table 5 Volcanic hazard as the probabilities of occurrence of at least one eruption with VEI $\geq$ 3 over different time intervals for Volcán de Colima

T YEARS	MOED	NHGPPP	WEIBULL	POISSON
10	0.5042	0.3787	0.4516	0.3364
20	0.6429	0.6140	0.6427	0.5596
50	0.7634	0.9075	0.8772	0.8713
100	0.8614	0.9914	0.9725	0.9834
200	0.9518	0.9999	0.9979	0.9997
500	0.9979	0.9999	0.9999	0.9999

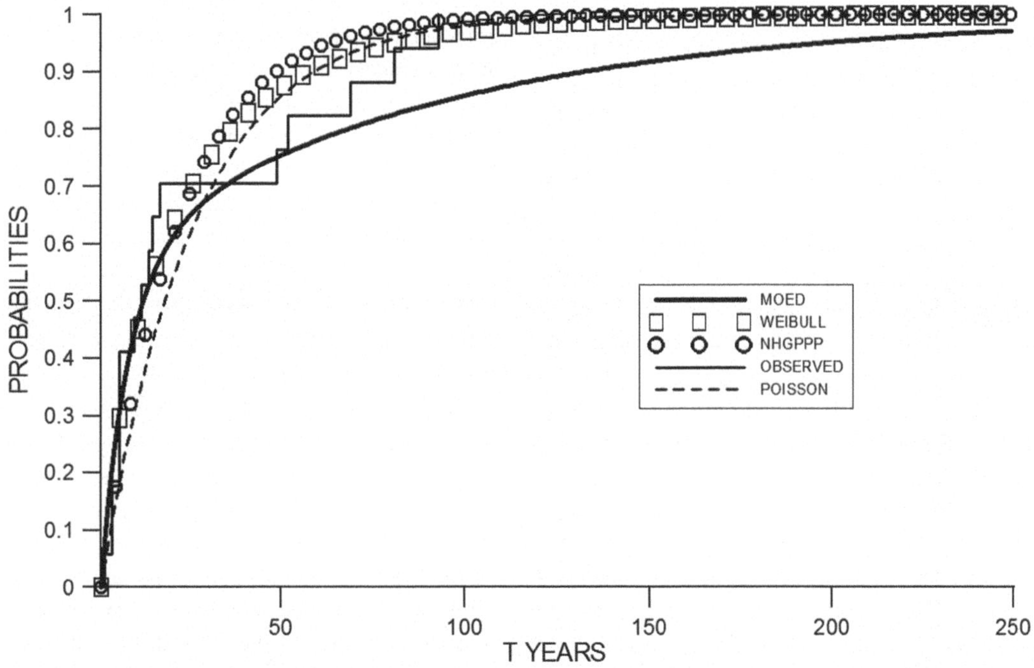

Fig. 6 Probabilities calculated by MOED, NHGPPP, homogeneous-Poisson and Weibull distributions of at least one eruption, with VEI $\geq$ 3 in a given time interval for "Case 2" in Fig. 5 and Table 4

8 Estimates of Debris Avalanche Hazards from Flank Collapses of Volcán de Colima

In the previous sections, the stochastic process of eruption occurrences has been addressed by different statistical methods with a common factor: all statistical distributions of occurrences have as a main parameter the mean eruption occurrence rate. Such parameter may be perceived as a quantity that reflects the physical process controlling the volcanic activity, i.e. the rate at which the complex volcanic system is supplied with magma and tries to reach equilibrium releasing it through eruptions.

In contrast, occurrences of debris avalanche events are conditioned by the instability of the volcanic edifice derived from its growth until it reaches a certain level of gravitational potential energy. The edifice growth depends on the magma production and accumulation rate, a feature that has been rather stable through the existence of Volcán de Colima (Luhr and

Carmichael 1990), introducing a cyclic component to the DAE occurrences. Such behaviour has been observed in other volcanoes with stable magma production rates, such as Mt. Taranaki in New Zealand (Zernack et al. 2012). Departures from a periodic cyclicity may be attributed to the random character of DAE triggering events, such as major eruptions, large earthquakes, or climatic factors. Although the available evidence from the DAE deposits indicate that most of the major DAE's at Volcán de Colima during the Holocene may have been caused by large eruptions (Cortés et al. this volume), we address the hazard derived from DAE's in a fashion different from the eruption hazard assessment, using a technique of analysis recently applied to Volcán de Colima by Borselli et al. (2011), which offers new insights for assessing the degree of instability in volcanic edifices.

This approach combines three methodologies (Borselli et al. 2011): (1) the analysis of the relative mass/volume deficit of the volcano structure, which was made using the VOLCANOFIT 2.0 software; (2) a slope stability assessment by limit equilibrium analysis (LEM) of multiple sectors on the volcano, considering fluid internal overpressure and dissipation; (3) a separate statistical analysis of major flank debris avalanche ages in the last 10,000 BP, adapted from stochastic arithmetic methods (Vignes 1993; Markov and Alt 2004) to the estimated ages of the DAE's and their error ranges.

Here, we revisit this statistical analysis of Borselli et al. (2011), now using radiocarbon calibrated ages, and considering an extended database of DAE's. The columns to the left in Table 6 shows the calibrated dates (and error ranges) of major DAE's, corresponding to dates of the major volcanic events listed in Table 1. In this approach, adding the mean interval between DAE's $\overline{\Delta Te}$ to the last event, including error propagation of the occurrence time yields a stochastic number (Vignes 1993; Markov and Alt 2004) that may be regarded as the forecast for the time of a future collapse. The method accounts for the error and uncertainty propagation among intervals between collapses, resembling the method proposed by Akçiz et al. (2010) for the assessment of large earthquakes recurrence times at the San Andreas Fault.

To estimate the most probable interval after the last DAE, the intervals between DAE's $\Delta Te_i = Te_{i-1} - Te_i$ are first calculated (third column in Table 6), where Te_i are the DAE's averaged calibrated ages listed in the left column. The uncertainty propagation between events is then obtained as $\varepsilon\Delta Te_i = \pm\sqrt{(\varepsilon Te_{i-1})^2 + (\varepsilon Te_i)^2}$, where $\pm\,\varepsilon\Delta Te_i$ is the propagated band of uncertainty calculated from the individual dating error bands $\pm\varepsilon Te_i$. The expected value of the DAE next to the last is calculated as the mean of the individual intervals (bottom line of the third column in Table 6), and its uncertainty or error is estimated as the standard deviation of the propagated errors $\varepsilon\Delta Te = \pm\sqrt{\frac{\sum_i (\varepsilon\Delta Te_i)^2}{n}}$, listed at the bottom line of the fourth column in Table 6. According to these results, an avalanche generating event may be expected in the time $(Te_{last} - \overline{\Delta Te}) \pm \varepsilon\Delta Te_{\text{next}}$ = (2864 yr BP(2014) – 2686)± 383 yr. Translating this into calendar dates, the interval is centred on 1836 AD, and ranges from to 1450 to 2222 AD. These results, presented in a graphic form in Fig. 7 are remarkably consistent to those based on uncalibrated radiometric ages reported by Borselli et al. (2011). It is striking that the effect of the added data points and calibrated ages does not significantly change the forecast made. These results may thus indicate that Volcán de Colima is, and has been mature for a major flank collapse for centuries.

9 Discussion and Conclusions on the Hazard Assessment of Volcán de Colima

Volcán de Colima is one of the most active volcanoes in the North American continent. Its wide range of eruptive styles, its propensity to flank collapse (Fig. 1), and the growing

Table 6 Ages of major debris avalanches in the Holocene

Calibrated age of DAE-generating collapse event (years BP-2014)	Uncertainty in calibrated age of DAE (yr)	Time interval between DAE's (yr) ΔTe_i	Dating uncertainty of interval from previous DAE (yr) $\varepsilon\Delta Te_i$	References
2864	52	1096	178	Luhr and Carmichael (1982), Luhr and Navarro (2002), Navarro and Luhr (2000), Siebe et al. (1992), Komorowski et al. (1993, 1997a, b), Cortés et al. (2005, 2010a, b), Cortés et al., this volume
3960	170	1071	349	Navarro and Luhr (2000), Kiryanov and Kolosov (1986), Komorowski et al. (1993, 1997a, b), Cortés et al. (2002, 2005, 2010a, b)
5031	305	2897	336	Luhr and Prestegaard (1988), Luhr and Navarro (2002), Robin et al. (1984), Cortés et al., this volume, Capra and Macías (2002), Capra (2007), Luhr et al. (2010)
7928	142	3197	232	Luhr and Carmichael (1982), Luhr and Navarro (2002), Komorowski et al. (1993, 1997a, b), Cortés et al. (2005, 2010b), Cortés et al., this volume, Luhr et al. (2010)
11125	184	5167	642	Navarro and Luhr (2000), Robin et al. (1984, 1987, 1991), Komorowski et al. (1993, 1997a, b), Cortés et al. (2005, 2010b)
16292	615	–	–	Robin et al. (1987), Cortés et al. (2005, 2010a), Capra (2007)
		Mean interval and standard deviation between DAE's $\overline{\Delta Te} \pm \mathrm{StDev}(\Delta Te) = 2686 \pm 1703\,\mathrm{yr}$	Uncertainty associated to the mean DAE interval $\varepsilon\Delta Te = \pm\ 383$ yr	

The uncertainties of intervals and their standard deviation were calculated with the method of Borselli et al. (2011). A pre-Holocene event is added in the last row to estimate the first Holocene time interval

population dwelling in areas that may be affected by volcanic phenomena justifies the need for a precise evaluation of the hazards it poses. The analyses described in this chapter reveal some features of the volcano's past that may prove insightful for more precise estimates of the probabilities associated with different hazards and for the design of preventive measures.

First in the list of such features are the stochastic character of the eruption occurrences, and the log-linear distribution of eruption rates as a function of the VEI magnitudes according to the scaling law described by Eq. (6). The combination of these features with an eruptive history made as complete as possible linking the historical and geological databases has allowed us to

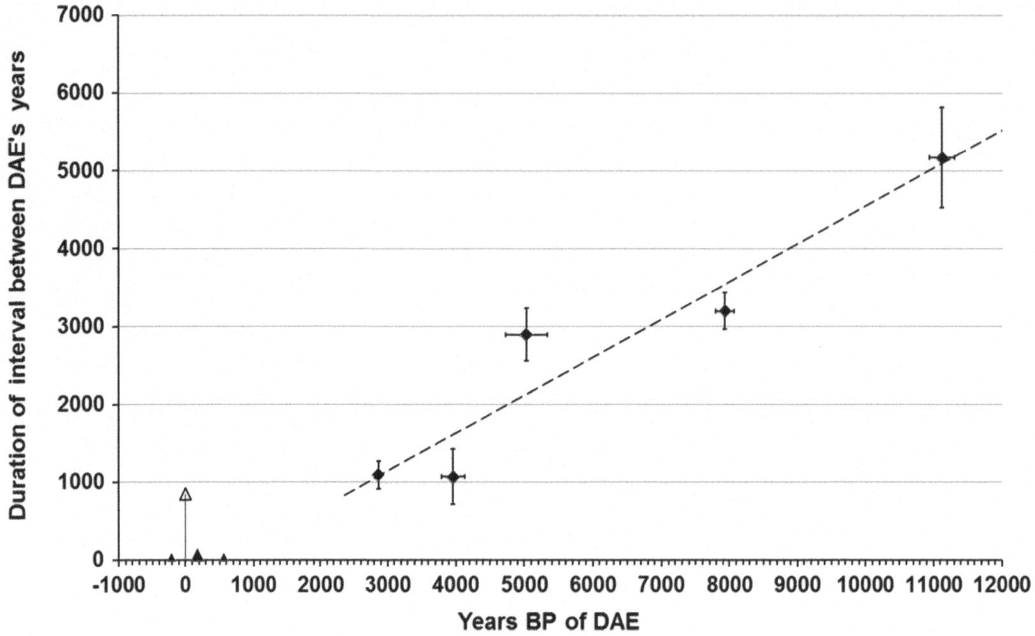

Fig. 7 DAE's as a function of the time since the previous event at Volcán de Colima (diamonds). Horizontal error bars represent the reported calibrated dating errors. Vertical error bars represent the uncertainty in the intervals calculated by the method of stochastic arithmetic used by Borselli et al., (2011). The forecast for the next DAE (largest triangle on the horizontal axis on the Gregorian year 1836) result from the estimation of the mean interval between DAE's counted from the last event. The uncertainty band (smaller triangles on the horizontal axis covering the range ±383 yr) of the forecast is also estimated using the stochastic arithmetic method. The inclined dashed line represents the trend from the best fit with the data points. The vertical arrow marks the present time

estimate the probabilities of eruptions considering the complete potential range of VEI's, as summarized in Table 4. Of particular significance is the increase in the probabilities of major eruptions (VEI $\geq$ 3) occurring in the intermediate time scale revealed by the NHGPPP method, which emphasizes the influence of very large events, despite the fact that very few are documented, or if only one has occurred far in the past.

Second, and referring to the hazard associated to DAE's, is the combined effect of the persistent magma feeding of the volcanic system, which translates into a magma production rate with a well-defined mean value with some fluctuations around it (see for example Fig. 2), combined with highly stochastic triggering phenomena of different nature (eruptive, seismic or climatic).

Referring to volcanic eruption triggers, Table 2 shows that at least 6 VEI 4 eruptions have been recorded in the historical timescale (from the 16th century at Colima), and no significant DAE has been reported in that period. On the other hand, inspection of Tables 1 and 6 reveals that volcanic eruptions as triggers of DAE's have a threshold at VEI 5, meaning that only eruptions with that index or larger are capable of causing flank collapses at this volcanic edifice.

The critical levels that climatic factors such as precipitation must reach have not been documented for Volcán de Colima, and little may be said about it, except that the recurrence of such factors should exceed any historical climatic extreme event, since no significant DAE's have occurred in the historical times scale even though the volcano may have been mature for a flank collapse since the fifteenth century, according to the arguments discussed in the previous section.

Earthquakes seem to have a better defined level of triggering in terms of the acceleration they may cause on an unstable volcanic edifice. The flank stability analysis of Borselli et al. (2011) shows that triggering a flank collapse in the volcano current configuration would require an earthquake with a magnitude 7.5 or higher, occurring at a hypocentral distance of 50 km or less. This type of event has a mean recurrence time of about 1200 yr (Mario Ordaz, Instituto de Ingeniería UNAM. Personal Comm. 2011). Curiously enough, this is about the same as the mean recurrence time of eruptions with VEI 5 expected from the "Case 2" model, as shown in Table 4. These values are within the standard deviation of the intervals between DAE's (2686 ± 1703 yr), and may support the model of a cyclic edifice building and a random DAE triggering process.

One final consideration relates to the current condition of the volcano. According to the stochastic arithmetic analysis of the previous section, Volcán de Colima should have reached a condition of instability since the 15th century. Although the time window of that forecasting extends to the 23th century, one may wonder why a significant sector collapse has not yet occurred. A speculative, yet possible explanation may be the relative abundance of VEI 4 eruptions, particularly in the 19th century. It is conceivable that such eruptions may have a stabilizing effect on the volcanic edifice since they have enough power to remove the upper parts of the volcanic cone, opening sizable craters, but are not strong enough to generate a flank collapse. If this were the case, the relatively long time since the previous VEI 4 eruption in 1913 may be considered as an additional hazard factor. In this circumstance, the scenario of a future VEI $\geq$ 5 must be very carefully considered, as an eruption of such dimension would probably lead to collapse.

Acknowledgements We are indebted to the reviewers, Mark Bebbington and David Pyle, and to Nick Varley, editor of this volume, whose comments and suggestion greatly helped to improve this manuscript. This research has been partially financed by DGAPA-PAPIIT-UNAM project IN106312. We are grateful for this support.

References

Akaike, H.: Information theory as an extension of the maximum likelihood principle. In: Petrov, B.N., Csaki, F. (eds.) Second International Symposium on Information Theory, pp. 267–281. Akademiai Kiado, Budapest (1973)

Akçiz, S.O., Grant Ludwig, L., Arrowsmith, J.R., Zielke, O.: Century-long average time intervals between earthquake ruptures of the San Andreas fault in the Carrizo Plain. Calif. Geol. **38**(9), 787–790 (2010). https://doi.org/10.1130/G30995.1

Arreola, J.M.: The recent eruptions of Colima Volcano (Edited and translated by F.Starr). J. Geol., 749–761 (1903)

Arreola, J.M.: Catalogo de las erupciones antiguas del volcán de Colima. Mem. Rev. Soc. Cient. Antonio Alzate, Mexico 32(11–12), 443–481 (1915)

Aspinall, W.P., Carniel, R., Jaquet, O., Woo, G., Hincks, T.: Using hidden multi-state Markov models with multi-parameter volcanic data to provide empirical evidence for alert level decision-support. J. Volcanol. Geotherm. Res. **153**, 112–124 (2006)

Bárcena, M.: Informe sobre el estado actual del Estado de Colima. In: El Estado de Colima. Periódico Oficial del Gobierno **XXI**(2), (January) (1887)

Bebbington, M.S., Lai, C.D.: Statistical analysis of New Zealand volcanic occurrence data. J. Volcanol. Geotherm. Res. **74**, 101–110 (1996a)

Bebbington, M.S., Lai, C.D.: On nonhomogeneous models for volcanic eruptions. Math. Geol. **28**, 585–600 (1996b)

Bebbington, M.S.: Identifying volcanic regimes using Hidden Markov models. Geophys. J. Int. **171**, 921–942 (2007)

Bebbington, M., Cronin, S., Chapman, I., Turner, M.: Quantifying volcanic ash fall hazard to electricity infrastructure. J. Volcanol. Geotherm. Res. **177**, 1055–1062 (2008)

Bebbington, M.: Trends and clustering in the onsets of volcanic eruptions. J. Geophys. Res. **115**, B01203 (2010). https://doi.org/10.1029/2009jb006581

Bebbington, M.: Models for temporal volcanic hazard. Stat. Volcanol. **1**, 1–24 (2013). https://doi.org/10.5038/2163-338X.1.1

Beguería, S.: Uncertainties in partial duration series modelling of extremes related of the choice of the threshold value. J. Hydrol. **303**, 215–230 (2005)

Borselli, L., Capra, L., Sarocchi, D., De la Cruz-Reyna, S.: Flank collapse scenarios at Volcán de Colima, Mexico: a relative instability analysis. J. Volcanol. Geotherm. Res. **208**, 51–65 (2011)

Bretón, M., Ramírez, J., Navarro, C.: Summary of the historical eruptive activity of Volcán de Colima, México 1519–2000. J. Volcanol. Geotherm. Res. **117**, 21–46 (2002)

Bronk Ramsey, C.: Dealing with outliers and offsets in radiocarbon dating. Radiocarbon **51**(3), 1023–1045 (2009)

Capra, L.: Abrupt climatic changes as triggering mechanisms of massive volcanic collapses. J. Volcanol. Geotherm. Res. **155**, 329–333 (2006)

Capra, L.: Volcanic natural dams: identification, stability and secondary effects. Nat. Hazards **43**, 45–61 (2007)

Capra, L., Macías, J.L.: The cohesive Naranjo debris-flow deposit (10 km^3). A dam breakout flow derived from the Pleistocene debris-avalanche deposit of Nevado de Colima Volcano (Mexico). J. Volcanol. Geotherm. Res. **117**, 213–235 (2002)

Capra, L., Borselli, L., Varley, N., Gavilanes-Ruiz, J.C., Norini, G., Sarocchi, D., Caballero, L., Cortes, A.: Rainfall-triggered lahars at Volcán de Colima, Mexico: surface hydro-repellency as initiation process. J. Volcanol. Geotherm. Res. **189**, 105–117 (2010)

Carey, S., Sigurdsson, H.: The intensity of plinian eruptions. Bull. Volcanol. **51**, 28–40 (1989)

Carta, S., Figari, R., Sartoris, G., Sassi, E., Scandone, R.: A statistical model for vesuvius and its volcanological implications. Bull. Volcanol. **44**, 129–151 (1981)

Coles, S.: An Introduction to Statistical Modeling of Extreme Values, 224 pp. Springer, London (2001)

Coles, S.G., Sparks, R.S.J.: Extreme value methods for modelling historical series of large volcanic magnitudes, In: Mader, H., Cole, S., Connor, C.B. (eds.), Statistics in Volcanology, Spec. Publ. of the Int. Assoc. of Volcanol. and Chem. of the Earth's Inter., Geol. Soc., London, vol. 1, pp. 47–56 (2006)

Cortés, A.: Depósitos de avalancha y flujos de escombros originados hace 3,600 años por el colapso del sector Suroeste del Volcán de Colima. M. Sc. Thesis, Universidad Nacional Autónoma de México. Mexico City. 121 pp. (2002)

Cortés, A., Garduño, V.H., Navarro, C., Komorowski, J. C., Saucedo, R., Macias, J.L., Gavilanes, J.C.: Carta Geológica del Complejo Volcánico de Colima, con Geología del Complejo Volcánico de Colima. Cartas Geológicas y Mineras 10, ISSN 0185–4798, 37 pp (2005)

Cortés, A., Garduño, V.H., Macias, J.L., Navarro, C., Komorowski, J.C., Saucedo, R., Gavilanes, J.C.: Geologic mapping of the Colima volcanic complex (Mexico) and implications for hazard assessment. Geol. Soc. Am. Spec. Pap. **464**, 249–264 (2010). https://doi.org/10.1130/2010.2464(12)

Cortés, A., Macias, J.L., Capra, L., Garduño, V.H.: Sector collapse of the SW flank of Volcán de Colima, México. The 3600 yr BP La Lumbre-Los Ganchos debris avalanche and associated debris flows. J. Volcanol. Geotherm. Res. **197**, 52–66 (2010b)

Cox, D.R., Lewis, P.A.W.: The Statistical Analysis of Series of Events, 285 pp. Methuen and Co., London (1966)

Davison, A.C., Smith, R.L.: Models for exceedances over high thresholds. J. Roy. Stat. Soc. **52**(B), 393–442 (1990)

Day, S.J.: Hydrothermal pore fluid pressure and the stability of porous permeable volcanoes. In: McGuire, W.J., Jones, A.P., Neuberg, J. (eds.) Volcano Instability on the Earth and Other Planets, Geological Society, London, Special Publications, vol. 110, pp. 77–93 (1996)

De la Cruz-Reyna, S.: Poisson-distributed patterns of explosive activity. Bull. Volcanol. **54**, 57–67 (1991)

De la Cruz-Reyna, S.: Random patterns of activity of Colima Volcano, Mexico. J. Volcanol. Geotherm. Res. **55**, 51–68 (1993)

De la Cruz-Reyna, S.: Long-Term probabilistic analysis of future explosive eruptions. In: Scarpa, R., Tilling, R.I. (eds.) Monitoring and Mitigation of Volcanic Hazards, pp. 599–629. Springer, Berlín (1996)

De la Cruz Reyna, S., Carrasco Núñez, G.: Probabilistic hazard analysis of Citlaltépetl (Pico de Orizaba) Volcano, Eastern Mexican volcanic belt. J. Volcanol. Geotherm. Res. **113**, 307–318 (2002)

De la Cruz Reyna, S., Tilling, R.: Scientific and public responses to the ongoing volcanic crisis at Popocatépetl Volcano, México: importance of an effective hazards warning system. J. Volcanol. Geoth. Res. **170**, 121–134 (2008)

De la Cruz-Reyna, S., Mendoza-Rosas, A.T.: An extreme event approach to volcanic hazard assessment. In: Chavez, M., Ghil, M., Urrutia-Fucugauchi, J. (eds.) Extreme Events: Observations, Modeling and Economics, ISBN: 978-1-119-15701-4, Wiley-Blackwell_AGU (2015)

Deligne, N.I., Coles, S.G., Sparks, R.S.J.: Recurrence rates of large explosive volcanic eruptions. J. Geophys. Res. **115**, B06203 (2010). https://doi.org/10.1029/2009JB006554

Díaz, S.: Efemerides del Volcán de Colima según las observaciones practicadas en los Observatorios de Zapotlan y Colima de 1893 a 1905. Sria. de Fomento, México (1906)

Dzierma, Y., Wehrmann, H.: Eruption time series statistically examined: probabilities of future eruptions at Villarrica and Llaima volcanoes, Southern Volcanic Zone. Chile. J. Volcanol. Geotherm. Res. **193**, 82–92 (2010)

Dzurisin, D.: Geodetic Detection of Inflating Stratovolcanoes: A Potencial Breakthrough for Mitigating Volcanic Hazards in the 21st Century. Eos 79, 973 pp (1998)

Elsworth, D., Voight, B.: Dike intrusion as a trigger for large earthquakes and the failure of volcano flanks. J. Geophys. Res. **100**, 6005–6024 (1996)

Fedotov, S.A.: Estimates of heat and pyroclast discharge by volcanic eruptions based upon the eruption cloud and steady plume observations. J. Geodyn. **3**, 275–302 (1985)

Gorshkov, G.S.: Directed volcanic blasts. Bull. Volcanol. **26**, 83–88 (1963)

Ho, C.H.: Bayesian analysis of volcanic eruptions. J. Volcanol. Geotherm. Res. **43**, 91–98 (1990)

Ho, C.H.: Time trend analysis of basaltic volcanism for the Yucca Mountain site. J. Volcanol. Geotherm. Res. **46**, 61–72 (1991)

Ho, C.H., Smith, E.I., Keenan, D.L.: Hazard area and probability of volcanic disruption of the proposed high-level radioactive waste repository at Yucca Mountain, Nevada, USA. Bull. Volcanol. **69**, 117–123 (2006)

Hurst, T., Smith, W.: Volcanic ashfall in New Zealand—probabilistic modelling for multiple sources. N. Z. J. Geol. Geophys. **53**, 1–14 (2010)

Jaquet, O., Löw, S., Martinelli, B., Dietrich, V., Gilby, D.: Estimation of volcanic hazards based on cox stochastic processes. Phys. Chem. Earth **25**, 571–579 (2000)

Jaquet, O., Carniel, R.: Estimation of volcanic hazard using geostatistical models. In: Mader, H.M., Coles, S. G., Connor, C.B., Connor, L.J. (eds.) Statistics in Volcanology. IAVCEI Publications n. 1, pp. 89–103. Geological Society, London (2006)

Keefer, D.K.: Landslides caused by earthquakes. Geol. Soc. Am. Bull. **95**, 406–421 (1984)

Kiryanov, V.Y., Koloskov, A.V.: Radiocarbon Dating of Volcanic Formations and the Properties of the Chemical, Quantitative-Mineral and Granulometric Composition of Rocks from the Chichinautzin Zone and Colima Volcano (based on the results of study of volcanic products sampled in Mexico in Summer 1984). Academy of Sciences of the USSR, Far East Scientific Centre, Institute of Volcanology, Petropavlovsk-Kamchatsky, pp. 1–71 (1986)

Klein, F.W.: Patterns of historical eruptions at Hawaiian volcanoes. J. Volcanol. Geotherm. Res. **12**, 1–35 (1982)

Komorowski, J.-C., Navarro, C., Cortés, A., Siebe, C., Rodriguez, S.: Multiple collapse of Volcán Colima, Mexico, since 10,000 y. BP: Implications for Eruptive Style, Magma Yield, Edifice Stability, and Volcanic Risk. (Abstract), International Association of Volcanology and Chemistry of the Earth's Interior, General Assembly, Canberra, Australia, 25 Sept–1 Oct 1993, Abstract Volume, p. 60 (1993)

Komorowski, J.-C., Cortés, A., Gavilanes, J.C., Navarro, C.: Relationships Between Repetitive Edifice Collapse, Emplacement of Debris Avalanche Deposits, and Volcanogenic Sedimentation at Colima Volcano. IAVCEI General Assembly, Puerto Vallarta, Mexico, 19–25 Jan 1997, 107 (1997a)

Komorowski, J.-C., Navarro, C., Cortés, A., Saucedo Girón, R., Gavilanes, J.C., Siebe, C., Espindola, J.-M., Rodriguez, S.: The Colima Volcanic Complex: Part I: Quaternary Multiple Debris-Avalanche Deposits; Part II: Historical Pyroclastic Sequences (1913, 1991, 1994). International Association of Volcanology and Chemistry of the Earth's Interior, General Assembly, Puerto Vallarta, Mexico, 19–25 Jan 1997, Fieldtrip Guidebook No. 3, 37 pp. 42 figs, appendix (1997b)

Luhr, J.F., Carmichael, I.S.E.: The Colima Volcanic Complex, Mexico; Part III. Ash and scoria fall deposits from the upper slopes of Volcán Colima. Contrib. Mineral. Petrol. **80**, 262–275 (1982)

Luhr, J.F., Carmichael, I.S.E.: Petrological monitoring of cyclical eruptive activity at Volcan Colima. Mexico. J. Volcanol. Geotherm. Res. **42**, 235–260 (1990)

Luhr, J.F., Carmichael, I.S.E.: Geology of Volcan de Colima. Boletín Instituto Geología, Univ. Nac. Auton. Mex. 107, 101 pp. (1990b)

Luhr, J.F., Prestegaard, K.L.: Caldera formation at Volcán Colima, Mexico by a large Holocene debris avalanche deposit. J. Volcanol. Geotherm. Res. **35**, 335–348 (1988)

Luhr, J.F., Navarro, C.: Guía de Excursión al Volcán Nevado de Colima, Octava Reunión Internacional, Volcán de Colima, pp. 1–50 (2002)

Luhr, J.F., Navarro-Ochoa, C., Savov, I.P.: Tephrochronology, petrology and geochemistry of Late-Holocene pyroclastic deposits from Volcan de Colima, Mexico. J. Volcanol. Geotherm. Res. **197**, 1–32 (2010). https://doi.org/10.1016/j.jvolgeores.2009.11.007

Marzocchi, W., Sandri, L., Gasparini, P., Newhall, C., Boschi, E.: Quantifying probabilities of volcanic events: the example of volcanic hazards at Mount Vesuvius. J. Geophys. Res. **109**, B11201 (2004). https://doi.org/10.1029/2004JB003155

Marzocchi, W., Sandri, L., Selva, J.: BET_EF: a probabilistic tool for long- and short-term eruption forecasting. Bull. Volcanol. **70**, 623–632 (2008)

Marzocchi, W., Bebbington, M.S.: Probabilistic eruption forecasting at short and long time scales. Bull. Volcanol. **74**, 1777–1805 (2012). https://doi.org/10.1007/s00445-012-0633-x

Markov, S., Alt, R.: Stochastic arithmetic: addition and multiplication by scalars. Appl. Numer. Math. **50**, 475–488 (2004)

McGuire, W.J.: Volcano instability: a review of contemporary themes. In: McGuire, W.J., Jones, A.P., Neuberg, J. (eds.) Volcano Instability on the Earth and Other Planets: Geological Society Special Publication, vol. 110, pp. 1–23 (1996)

Medina-Martínez, F.: Analysis of the eruptive history of the Volcán de Colima, Mexico (1560–1980). Geofís. Inter. **22**, 157–178 (1983)

Mendoza-Rosas, A.T., De la Cruz-Reyna, S.: A statistical method linking geological and historical eruption time series for volcanic hazard estimations: applications to active polygenetic volcanoes. J. Volcanol. Geotherm. Res. **176**, 277–290 (2008). https://doi.org/10.1016/j.jvolgeores.2008.04.005

Mendoza-Rosas, A.T., De la Cruz-Reyna, S.: A mixture of exponentials distribution for simple and precise assessment of the volcanic hazard. Nat. Hazards Earth Syst. Sci. **9**, 425–431 (2009). https://doi.org/10.5194/nhess-9-425-2009

Mendoza-Rosas, A.T., De la Cruz-Reyna, S.: Hazard estimates for El Chichón volcano, Chiapas, Mexico: a statistical approach for complex eruptive histories. Nat. Hazards Earth Syst. Sci. **10**, 1159–1170 (2010). https://doi.org/10.5194/nhess-10-1159-2010

Mooser, F.: Los Volcanes de Colima. Boletín Instituto Geología, Univ. Nac. Auton. Mex. **61**, 49–71 (1961)

Moriya, I.: "Bandaian eruption" and landforms associated with it. In: Collection of articles in memory of retirement of Prof. K. Nishimura from Tohoku University. Fac. Sci. Tohoku Univ., Sendai, pp. 214–219 (in Japanese with English abstract) (1980)

Mulargia, F., Tinti, S.: Seismic sample areas defined from incomplete catalogues: an application to the Italian territory. Phys. Earth Planet. Inter. **40**, 273–300 (1985)

Mulargia, F., Gasperini, P., Tinti, S.: Identifying different regimes in eruptive activity: an application to Etna Volcano. J. Volcanol. Geotherm. Res. **34**, 89–106 (1987)

Navarro, C., Luhr, J.: Late-Holocene tephrochronology at the Colima Volcanic Complex, México, Séptima Reunión Internacional Volcán de Colima, (abstract, p. 44). Marzo, 6–10 (2000)

Newhall, C.G., Self, S.: The volcanic explosivíty index (VEI): an estimate of explosive magnitude for historical volcanism. J. Geophys. Res. **87C2**, 1231–1238 (1982)

Newhall, C.G., Hoblitt, R.P.: Constructing event trees for volcanic crises. Bull. Volcanol. **64**, 3–20 (2002)

Ordóñez, E.: Les derniéres eruptions du Volcán de Colima, pp. 99–103. Mem. Soc. Cient. Antonio Alzate, Mexico (1903)

Pyle, D.M.: Mass and energy budgets of explosive volcanic eruptions. Geophys. Res. Lett. **22**, 563–566 (1995)

Pyle, D.M.: Forecasting sizes and repose times of future extreme volcanic events. Geology **26**, 367–370 (1998)

Pyle, D.M.: Sizes of volcanic eruptions. In: Sigurdsson, H., Hughton, B., McNutt, S.R., Rymer, H., Stix, J. (eds.) Encyclopedia of Volcanoes, pp. 263–269. Academic Press, San Diego (2000)

Reiss, R.D., Thomas, M.: Statistical Analysis of Extreme Values, 2nd edn, 316 pp. Birkhauser, Basel (2001)

Reyment, R.A.: Statistical analysis of some volcanologic data. Regarded as series of point events. PAGEOPH **74**(3), 57–77 (1969)

Robin, C., Camus, G., Cantagrel, J.M., Gourgaud, A., Mossand, P., Vincent, P.M., Aubert, M., Dorel, J., Murray, J.B.: Les Volcans de Colima (Mexique). Bulletin P.I.R.P.S.E.V.-C.N.R.S.-I.N.A.G. **87**, 1–98 (1984)

Robin, C., Mossand, P., Camus, G., Cantagrel, J.M., Gourgaud, A., Vincent, P.: Eruptive history of the Colima Volcanic Complex (México). J. Volcanol. Geotherm. Res. **31**, 99–113 (1987)

Robin, C., Camus, G., Gourgaud, A.: Eruptive and magmatic cycles at Fuego de Colima volcano (Mexico). J. Volcanol. Geotherm. Res. **45**, 209–225 (1991)

Rodríguez-Elizarrarás, S., Siebe, C., Komorowski, J.-C., Espíndola, J.M., Saucedo, R.: Field observations of pristine block- and ash-flow deposits emplaced April 16–17, 1991 at Volcán de Colima, Mexico. J. Volcanol. Geotherm. Res. **48**, 40–56 (1991)

Roverato, M., Capra, L., Sulpizio, R., Norini, G.: Stratigraphic reconstruction of two debris avalanche deposits at Colima Volcano (Mexico): insights into pre-failure conditions and climate influence. J. Volcanol. Geotherm. Res. **207**(1–2), 33–46 (2011)

Sarocchi, D., Sulpizio, R., Macías, J.L., Saucedo, R.: The 17 July 1999 block-and-ash flow at Colima volcano: new insights on volcanic granular flows from textural analysis. J. Volcanol. Geotherm. Res. **204**, 40–56 (2011)

Saucedo, R., Macías, J.L., Sheridan, M.F., Bursik, M.I., Komorowski, J.C.: Modeling of pyroclastic flows of Colima Volcano, Mexico: implications for hazard assessment. J. Volcanol. Geotherm. Res. **139**(1–2), 103–115 (2005)

Saucedo, R., Macías, J.L., Gavilanes, J.C., Arce, J.L., Komorowski, J.C., Gardner, J.E., Valdez-Moreno, G.: Eyewitness, stratigraphy, chemistry, and eruptive dynamics of the 1913 Plinian eruption of Volcán de Colima, México. J. Volcanol. Geotherm. Res. **191**, 149–166 (2010)

Saucedo, R., Macías, J.L., Gavilanes, J.C., Arce, J.L., Komorowski, J.C., Gardner, J.E., Valdez-Moreno, G.: Corrigendum to eyewitness, stratigraphy, chemistry, and 1913 Plinian eruption of Volcán de Colima, Mexico. J. Volcanol. Geotherm. Res. **207**, 67 (2011)

Savov, I.P., Luhr, J.F., Navarro-Ochoa, C.: Petrology and geochemistry of lava and ash erupted from Volcán de Colima, Mexico, during 1998–2005. J. Volcanol. Geotherm. Res. **174**, 241–256 (2008)

Settle, M., McGetchin, T.R.: Statistical analysis of persistent explosive activity at Stromboli, 1971: implications for eruption prediction. J. Volcanol. Geotherm. Res. **8**, 45–58 (1980)

Sheridan, M.F., Bonnard, C., Carreno, C., Siebe, C., Strauch, W., Navarro, M., Calero, J.C., Trujilo, N.B.: Report on the 30 October 1998 rock fall/ avalanche and breakout flow of Casita Volcano, Nicaragua, triggered by Hurricane Mitch. Lanslide News **12**, 2–4 (1999)

Siebe, C., Rodríguez-Elizarrarás, S., Stoopes, G., Komorowski, J.C., Sheridan, M.F.: How Many Debris Avalanche Deposits at the Colima volcanix complex or: Quo Vadimus. (abstract), Universidad de Colima, Tercera Reunión Nacional "Volcán de Colima y Segunda Reunión Internacional de Vulcanología. Colima, 15 (1992)

Siebert, L., Glicken, H., Ui, T.: Volcanic hazards from Bezimianny- and Bandai-type eruptions. Bull. Volcanol. **49**, 435–459 (1987)

Siebert, L., Simkin, T.: Volcanoes of the World: an Illustrated Catalog of Holocene Volcanoes and their Eruptions. Smithsonian Institution, Global Volcanism Program Digital Information Series, GVP-3, (http://www.volcano.si.edu/world/) (2002)

Simkin, T., Siebert, L.: Volcanoes of the World, 2nd edn. Geoscience Press, Tucson, Ariz (1994)

Simkin, T., Siebert, L.: Earth's volcanoes and eruptions: an overview. In: Sigurdsson, H. (ed.) Encyclopedia of Volcanoes, pp. 249–261. Academic Press, San Diego (2000)

Smithsonian Institution. Colima. Bull. Glob. Volcanism Netw. **30**(6) (2005)

Smithsonian Institution. Colima. Bull. Glob. Volcanism Netw. **31**(3) (2006)

Sobradelo, R., Martí, J., Mendoza-Rosas, A.T., Gómez, G.: Volcanic hazard assessment for the Canary Islands (Spain) using Extreme value theory. Nat. Hazards Earth Syst. Sci. **11**, 2741–2753 (2011)

Solow, A.R.: An empirical Bayes analysis of volcanic eruptions. Math. Geol. **33**(1), 95–102 (2001)

Sulpizio, R., Capra, L., Sarocchi, D., Saucedo, R., Gavilanes, J.C., Varley, N.: Predicting the block and ash flow inundation areas at Fuego de Colima volcano (Colima, Mexico) based on the present day status. J. Volcanol. Geotherm. Res. **193**, 49–66 (2010)

Tello, A. (Fray).: Cronica Miscelánea y Conquista espiritual y Temporal de la Santa Provincia de Jalisco en el Nuevo Reino de la Galicia y Nueva Vizcaína y Descubrimiento del Nuevo México. López Portillo y Rojas, J. (ed.) Guadalajara Jal, México (1650)

Turner, M.B., Cronin, S.J., Bebbington, M.S., Platz, T.: Developing probabilistic eruption forecasts for dormant volcanoes: a case study from Mt Taranaki. N. Z. Bull. Volcanol. **70**, 507–515 (2008). https://doi.org/10.1007/s00445-007-0151-4

Varley, N.R., Johnson, J., Ruiz, M., Reyes-Dávila, G.A., Martin, K.: Applying statistical analysis to understanding the dynamics of volcanic explosions. In: Mader, H.M., Coles, S.G., Connor, C.B., Connor, L.J. (eds.) Statistics in Volcanology. Special publication of IAVCEI, pp. 57–76 (2006)

Vignes, J.: A stochastic arithmetic for reliable scientific computation. Math. Comp. Sim. **35**, 233–261 (1993)

Waitz, P.: Datos históricos y bibliográficos acerca del Volcán de Colima. Mem. Rev. Soc. Cient. Antonio Alzate, Mexico **53**, 349–384 (1932)

Walker, G.P.L.: The Taupo pumice: product of the most powerful known (Ultraplinian) eruption? J. Volcanol. Geotherm. Res. **8**, 69–94 (1980)

Wehrmann, H., Dzierma, Y.: Applicability of statistical eruption analysis to the geologic record of Villarrica and Lanín volcanoes, Southern Volcanic Zone, Chile. J. Volcanol. Geotherm. Res. **200**, 99–115 (2011). https://doi.org/10.1016/j.jvolgeores.2010.11.009

Wickman, F.E.: Repose period patterns of volcanoes, 5. General discussion and a tentative stochastic model. Ark. Mineral. Geol. **4**, 351–367 (1965)

Wickman, F.E.: Markov models of repose—period patterns of volcanoes. In: Merriam, D.F. (ed.) Random Processes in Geology, pp. 135–161. Springer, Berlín (1976)

Zernack, A.V., Cronin, S.J., Bebbington, M.S., Price, R.C., Smith, I.E.M., Stewart, R.B., Procter, J.N.: Forecasting catastrophic stratovolcano collapse, a model based on Mt. Taranaki, New Zealand. Geology (published online 2012). https://doi.org/10.1130/g33277.1

Civil Protection and Volcanic Risk Management in Colima

Hugo Ignacio Rodríguez García

Abstract

Volcanic risk depends not only on the probability of impact by one of the volcanic hazards or by events derived from them. Paradoxically, the risk can be increased if it is evaluated, communicated and managed inappropriately, so the risk reduction strategy must be constructed in full interaction between at-risk populations, scientists and civil protection authorities. In this work, a review of the strategies to manage the volcanic risk by Civil Protection of the state of Colima is made; the analysis is based on the concept of Integral Risk Management proposed by Mexican legislation. The evidence suggests that it should perform a more social volcanic risk management, this risk reduction must be multidisciplinary so that the decisions taken are more effective and adequate, therefore the opinions of social scientists will have to be considered and included in the design and implementation throughout the risk management process. Finally, it is suggested that a risk communication program be developed, including an early warning system, updating hazard maps, and operational and emergency plans, in which the roles and responsibilities of each party interested in volcanic risk management are considered, and recommends the socialization of such maps and plans, as well as free access to them by society.

Keywords

Volcán de colima · Civil protection
Volcanic risk · Integral risk management

1 Introduction

Risk is generated by the interaction between the eruptive products of Volcán de Colima and nearby human settlements; however, it is important to remember that its origin is not limited to the probability of physical damage. Volcanic risk can imply other dimensions that are often ignored or disparaged by those who study it, produced and intensified both by those who manage it and those people living nearby the volcano, which has also been called by several local journalists The "Giant of Fire (El Coloso de Fuego)".

These dimensions are related to actions or omissions attributable to different social actors involved in the process of reducing the impact of volcanic activity, and direct or indirect

H. I. R. García (✉)
Departamento de Ciencias de la Naturaleza, Universidad de Guadalajara, Ciudad Guzmán, Jalisco, México
e-mail: hrodriguez382@gmail.com

N. Varley et al. (eds.), *Volcán de Colima*, Active Volcanoes of the World,
https://doi.org/10.1007/978-3-642-25911-1_10

consequences of those actions. This chapter considers that volcanic risk reduction goes beyond studying and monitoring the physical phenomena, meaning that geological and geophysical research does not provide a sufficient basis for appropriate decision making regarding risk management. The social, cultural, economic, and political dimensions of the lives of the people involved are crucial in enabling any institutional risk management efforts to be effective.

These considerations will be fundamental to the success of any institutional effort, since there is considerable evidence of the necessity of a multidisciplinary approach within physical science, including sociological knowledge and techniques to achieve a broader scope for reducing volcanic risk. It is necessary that the risk of volcanic eruptions be considered in a broader context alongside other risks to social and economic development, rather than as an isolated phenomenon (Barclay et al. 2008).

Ironically, the actions implemented during risk and disaster management can sometimes be even more disastrous than the impact from the natural phenomenon itself. In other words, the social, economic, and political costs can be higher for populations subjected to governmental intervention than the direct physical impact caused by the natural phenomena associated with the disaster. This chapter critically analyses the way in which volcanic risk management has been implemented in Colima, during recent years. The reduction of volcanic risk is a great challenge due to the complexity of forecasting and evaluating volcanic phenomena, as well as transmitting these assessments in the form of scenarios to the civil servants in charge of civil protection (Marzocchi et al. 2012; Barclay et al. 2008).

It is necessary to establish the basic theoretical elements needed to sustain risk-related research at Colima. The chapter presents details of how Civil Protection in the Mexican state of Colima handles risk management. For this purpose, reference is made to the different stages of the so-called Integrated Risk Management (Gestión Integral de Riesgos, locally abbreviated as GIR) which is defined in the 2012 law *Ley General de Protección Civil.*

It is important to point out that Integrated Risk Management (GIR), within the framework of the Mexican Civil Protection, is derived partially from the model of Emergency Management constituted in the United States. The difference is that the Emergency Management model covers the whole process, including activities defined as disaster phases (including prevention and preparation), whereas in the case of Mexico, GIR is restricted to a greater or lesser extent to interventions during the emergency or immediately after the manifestation of the disaster (Macías 2016). That is, the approach by the Mexican Government is reactive rather than preventive.

Macías (2016) concludes that this GIR approach was established without the consideration of an organizational change in the structure of Civil Protection. Instead it demands certain additional activities to those already defined. The challenge now is how to integrate all the organizations that participate in the process of civil protection (Macías 2016). Currently the structure defined in the Mexican legislation results in an inefficient treatment of risk.

Until now, research dedicated to volcanic hazards in Mexico has not analysed the specific role of Civil Protection in reducing risk or the organizational plans available to the authorities. Usually relationships between the scientists and the civil protection authorities are assumed to be optimal. Although in the last two decades in the case of Colima, some efforts to analyse this problem in a critical way have been made (Macías 1999, 2001; Gavilanes 2004; Cuevas 2005; Macías and Aguirre 2006; Gavilanes et al. 2009). This chapter further extends this critique by scrutinizing volcanic risk management during the recent eruptive period.

2 Volcanic Activity

2.1 The Context of Volcanic Risk

One of the most important characteristics of Volcán de Colima in terms of its hazard is its cyclicity for different types of activity (Varley, this volume). An approximate cycle of 100 years for Plinian or sub-Plinian eruptions has been described (Luhr 2002; see Crummy et al., this volume). It could be more frequent, however, with Mendoza-Rosas and De la Cruz-Reyna (2008) identifying eruptions with a Volcanic Explosivity Index 4 (VEI) in 1585, 1606, 1622, 1818, 1890 and 1913. Volcanologists have recently expressed an increasing concern considering that over 100 years have now passed since the last major eruption, combined with the increased intensity of activity witnessed since 1998 (Varley 2015; Varley, this volume).

The eruption of 1913 began on 17 January and culminated on the 20th of that same month through a series of eruptive phases that intensified in a period of 24 h (Saucedo et al. 2010; Saucedo et al., this volume). The first phase, started with Vulcanian explosions that emplaced small pyroclastic flows. In phase 2, the explosions increased in intensity and frequency, generating large volume pyroclastic flows. In Phase 3, a Plinian column of $\sim$23 km in height was generated which distributed ash to the NE. The Plinian column collapsed generating pyroclastic flows radially from the summit to a distance of 15 km (Saucedo et al. 2010; Saucedo et al., this volume).

Saucedo et al. (2010) indicate that a Plinian eruption is considered the most significant historical eruption scenario that could affect the population in the short and medium term, so it has been taken as a reference for the discussion of volcanic risks and risk scenarios, including the development of hazard and risk maps.

An upcoming VEI 4 or greater event would not have great significance, and would be observed only as a natural phenomenon if there were no population that could be directly or indirectly affected. At present, within a radius of less than 15 km from the top of the volcano, approximately 15,000 people are exposed to volcanic activity (Saucedo et al. 2005) and in a radius greater than 30 km more than 500,000 (INEGI 2010).

In the state of Colima, La Yerbabuena and La Becerrera, in the municipality of Comala are the communities closest to Volcán de Colima at 8 and 12 km, respectively. The two localities have been catalogued as high risk areas by government authorities and volcanologists (Martin del Pozzo et al. 1995; Sheridan and Macías 1995; Navarro and Cortés 2003; Gavilanes 2004; Macías et al. 2006).

These two communities have their origin in the agrarian reform. In the case of La Yerbabuena, the *ejido* (a system of communal land tenure in México) that gave rise to the community is officially termed as ejido of ex-hacienda of "San Antonio". It was founded in 1968 (González 2000). As for La Becerrera, officially known as "La Becerrera Ex-hacienda de San Antonio", it was constituted by a presidential resolution in 1975 (File 23/61, File 1, File General Agrarian). Currently, 47 people live in La Yerbabuena (resettled community, which will be discussed in Sect. 3.3.1) and 283 people in La Becerrera. It is important to consider that of these two populations, the oldest one is La Yerbabuena at 48 years, whilst La Becerrera has officially existed for 41 years. This makes them very recent, especially when compared with the eruptive cycles of Volcán de Colima. This is significant in terms of the relationship between the population and the volcanic threat, as well as its incidence in the perception of volcanic risk.

This interaction of volcanic phenomena with vulnerable populations defines the local risk. It can be said that the risk can be either generated, eliminated, intensified or diminished as a result of governmental actions related to volcanic risk management.

3 Volcanic Risk Management in Colima

In order to understand the way in which the state of Colima's Civil Protection authorities handle volcanic risk, it is necessary to take into consideration the concept of Integrated Risk Management (*Gestión Integral de Riesgos* GIR) presented in the *Ley General de Protección Civil.* GIR is defined therein as the set of actions directed at identifying, analysing, evaluating, controlling, and reducing risk, where the latter is considered as being multi-faceted and in a state of continual development. All three levels of government (federal, state and municipal) are responsible for these actions in conjunction with different sectors of civil society. They are classified in different stages: identification of risk and/or its process of formation, prevision, prevention, mitigation, preparedness, aid, recovery, and reconstruction. The *Ley General de Protección Civil* also includes references to those actions that the GIR direct towards the creation and implementation of public policies, strategies, and standardized procedures aimed at achieving sustainable development that considers the causes of disasters and their relationship with social structure, and at the same time strengthens the resilience of society.

In this context it becomes pertinent to question how risk management in Colima is being undertaken. In order to answer such a question, only the stages of risk identification, prevention, mitigation, preparedness, and aid will be considered. "Prevision" is defined separately in the *Ley General de Protección Civil*, but will be omitted because the distinction between this concept and identification is unclear. Reconstruction will also be left out because to date no disaster caused by volcanic activity has taken place in Colima.

Macías (2016) points out that risk identification is not a phase or stage of the disaster process as such, but a sub-stage, because this activity is included within prevention. He considers that its introduction as a phase within the GIR has been driven more by an interest in increasing the insurance market, that is, to move the emphasis from "risk determination" to "risk transfer" (Macías 2016).

However, in spite of this criticism, this study includes both risk identification as a phase, and GIR as a disaster management approach given their presence in the legislation. One objective is to analyse the preparation of the institutions regarding volcanic risk management in Colima, and their adoption of the strategies defined in relevant documents.

It is important to point out that not all actions taken by Colima's Civil Protection authorities to manage risk have been included in this chapter. Some topics, such as risk communication, alert systems, and evacuation drills, among others, have been excluded. The following sections are structured according to the identification, prevention, mitigation, preparedness, and aid phases of the GIR, using information from the "Volcán de Colima" Organization and Operation Manual and the Manual of Operations and Procedures in Case of Emergency, which are not accessible to the interested public or researchers, with no indication of author or publication date.

3.1 Identification of Risk

It is stated in the aforementioned *Ley General de Protección Civil* that the identification of risk consists of "recognizing and assessing probable loss and damage of affected parties and defining their geographical distribution, through risk and vulnerability analysis". In the case of Colima, so far only information regarding the identification of volcanic threats has been officially released, but an evaluation of probable losses or damage has yet to be published, although one scientific effort that estimates the former is available: the *Atlas Estatal de Riesgos de Colima*, specifically in the section dedicated to volcanic hazards. Here some of the potential costs of a volcano-related disastrous event have been calculated (Varley et al. 2017).

3.1.1 Zoning and Risk Maps

In the *Manual de Organización y Operación "Volcán de Colima"* (from now on the Manual OOVC) a classification of human settlements at risk is defined according to the degree of volcanic activity and the distance between these populations and Volcán de Colima. Risk zones were determined by concentric rings traced a certain distance from the crater. In the Manual OOVC, classification and zoning is very simplistic, due to the fact that only geographical locations of human settlements affected by volcanic risk are considered, without consideration of populations' forms of social organization, productive activities, or cultural or political elements of their lives.

Among the populations classified as being at risk of contact with pyroclastic material are listed the villages of La Yerbabuena (8.2 km), La

María (9.7 km), and La Becerrera (12 km) in the municipality of Comala. At risk from ashfall are, beside the aforementioned, Montitlán (12 km) and Quesería (14.5 km) within the municipality of Cuauhtémoc, and El Naranjal (12 km) within the municipality of Villa de Álvarez. La Becerrera is also threatened by the formation of lahars.

The Manual OOVC also states that, in the event of an eruption like the one that occurred in 1913, the following infrastructure could be affected: the bridge over La Lumbre ravine; the road that connects La Becerrera in Colima and San José del Carmen in Jalisco; La Becerrera's Secondary School (which is beside La Lumbre ravine); and the high-tension pylons of the *Comisión Federal de Electricidad* located in the Montegrande ravine.

Risk zones are classified as high, average, and low. Those settlements located a distance of up to 8 km in a straight line from the crater are considered to be at a high level of risk, like La Yerbabuena. Included in the medium risk zone are those settlements located a distance of between 8 and 15 km, like La Becerrera and Quesería, La Lima, El Naranjal, and Nuevo Naranjal. The category of low risk includes communities located further than 15 km from the crater, like Suchitlán, Agosto, La Caja, El Chivato, El Carrizal, Buena Vista, Chiapa, Ocotillo, Cerro Colorado, Palmillas and Alzada. The effects on these villages would depend specifically on the direction of the wind (see Fig. 1).

It is important to outline that the *Manual de Organización y Operación "Volcán de Colima"* is not available to the general public, and it is accessible for use only to Colima's Civil Protection authorities. Also this manual does not indicate the date when it was written and published, nor who participated in its elaboration.

On the other hand, there is currently a publically available map that has served both as a reference and scientific foundation for decision making in terms of risk management. That map is the geological map mistakenly labelled as "Mapa de Peligros Volcán de Colima", prepared by Navarro Ochoa and Cortés Cortés, and published in 2003 by the Universidad de Colima, in collaboration with the government of the State of Colima. In this map the geological fieldwork carried out by the authors during a three-year period with 525 field observation points is reflected. It includes zones where pyroclastic flows, debris avalanches, lava flows, lahars, and ballistic projectiles were mapped, as well as ashfall. Human settlements, water bodies, and roads are also indicated on the map (see Fig. 2).

The cartographical information contained in this map denotes it a geological map. In order to be considered a hazard map, it would have to provide information regarding probability of recurrence and impact of each episode of volcanic activity. To be considered a risk map it would have to feature information on conditions of vulnerability for human settlements with regards to volcanic activity.

As previously mentioned, the map illustrated as Fig. 2 has served as a scientific reference for decision making regarding volcanic risk management. An example of the latter is the decree that declares La Yerbabuena as a settlement at high volcanic risk, made acknowledgeable to the general public on 24 May 2003 by the state government of Colima. The first article of this decree states the following:

> The settlement of La Yerbabuena...is declared as being at high risk, for being within a range of 8 kilometres from the summit of the volcano, a distance at which that settlement is vulnerable to at least 3 types of volcanic hazards, which have been present in previous eruptions in the area, as shown in the hazard map of the Volcán de Fuego de Colima.

Despite being published a year after relocation of the settlement's inhabitants, it has provided legal support during a series of social-political and legal conflicts, generated and aggravated by the relocation project. This will be explained later in the mitigation phase section.

Let's keep in mind that human relocations have to be considered as the very last choice when it comes to mitigation and prevention of risk, due to the considerable social, economic, cultural, and political costs that those processes entail. For further information on the

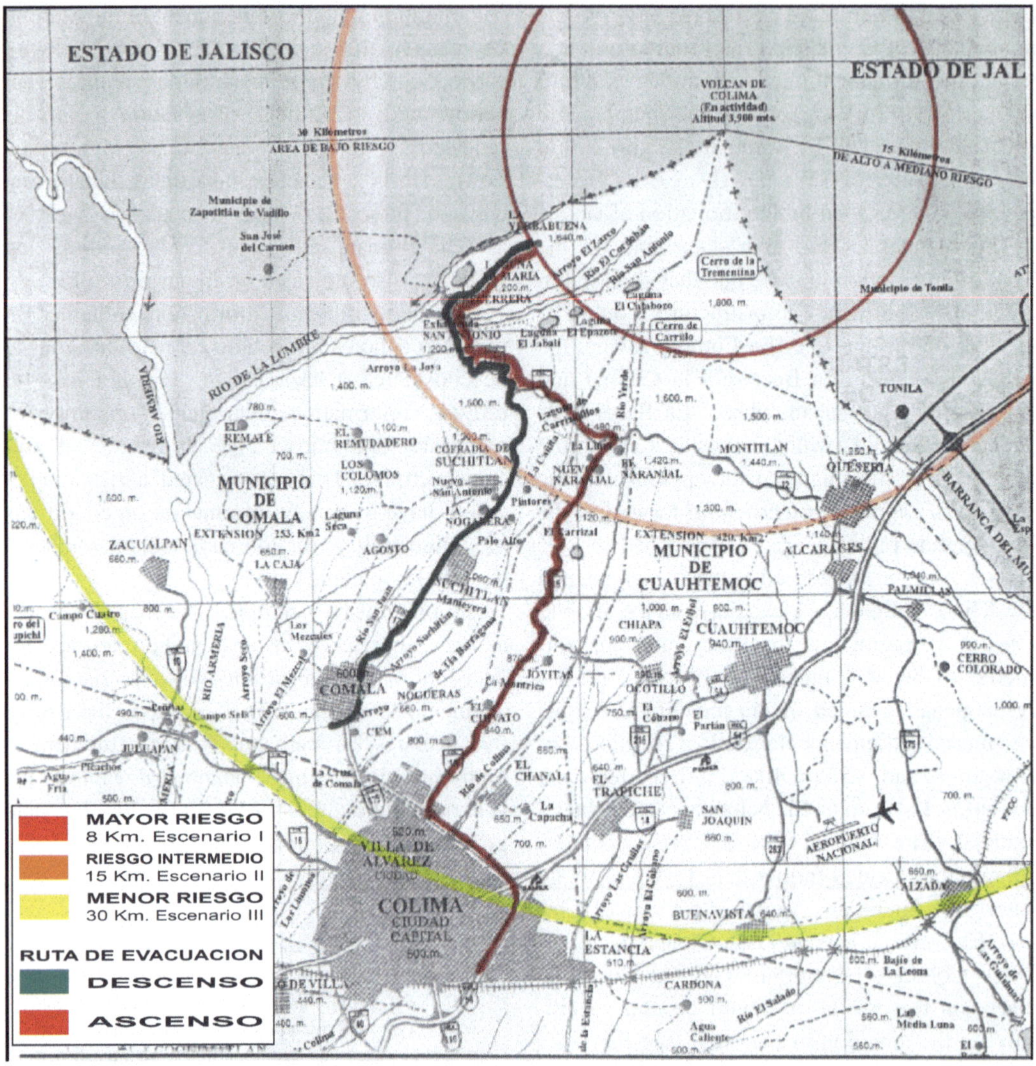

Fig. 1 Zoning of "risk" by means of concentric rings. This is the base map used by State Civil Protection to determine the delimitation of evacuation and exclusion zones in case of volcanic crisis and resultant scenarios. Likewise, the routes of evacuation for ascent and descent are shown, although in the volcanic crisis of July of 2015 these functioned as routes of evacuation in both directions, that is to say, each of them could be ascended or descended. The map has not been published and is without a date. *Source* Manual of organization and operation 'Volcán de Colima'

consequences of the relocation project, please review Cuevas and Seefoó (2005) and Macías (2001).

It has to be mentioned that some research has been dedicated towards producing volcanic hazard maps, and such maps were published prior to the one produced by Navarro and Cortés. Nevertheless, none of those other publications have been regarded as official pieces of cartography that can serve as a replacement for, or complement to the map produced by Navarro and Cortés (i.e. Martin del Pozzo et al. 1996; Capra et al. 2015). On the other hand, it is well known that risk and hazard maps have to be updated frequently, or every time that there is an eruption that modifies the relief around the volcano

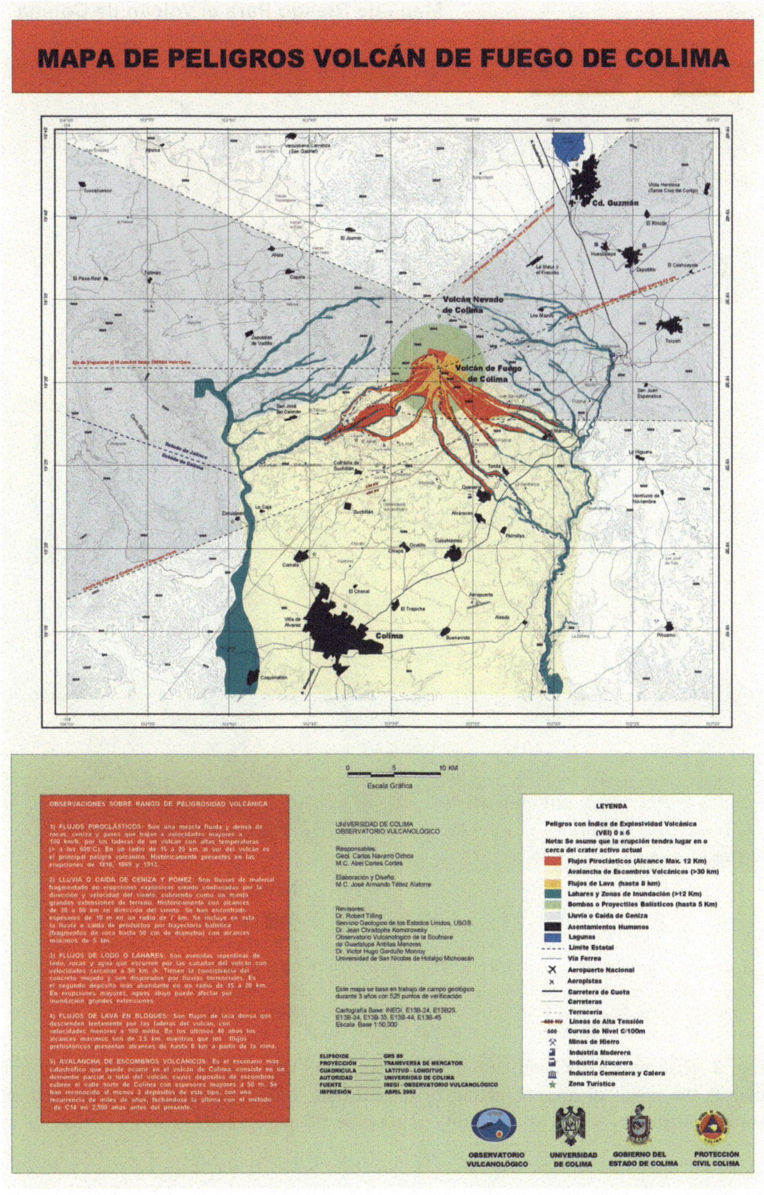

Fig. 2 Hazards map of Volcán de Colima. It is a map that summarizes the geological work carried out by researchers from the University of Colima. Although it has no data on impact probability and vulnerability factors, the map has been used for government decision-making. It possesses several problems, some of them technical, for example there are areas where the pyroclastic flows and lahares overlap, which does not allow clear distinction between them. Another problem is that the map is outdated, since its publication in 2003, it has not been updated, despite growth both in the population and agro-industrial activities in the vicinity of Volcán de Colima. *Source* Navarro and Cortés (2003)

(Guevara et al. 2004). In the case of Volcán de Colima, this happened in July 2015 (Capra et al. 2016; Reyes-Dávila et al. 2016). Therefore, it is vital that recent hazards zoning works (i.e. Capra et al. 2015) be considered by the civil protection authorities, in order to update volcanic hazards maps. Meanwhile the official release of the *Atlas Estatal de Riesgos* is still awaited. It includes a series of maps on different volcanic hazards and risks by taking into account the probability of impact derived from different eruption scenarios as well as social, economic, and natural variables and indicators related to the vulnerability of several settlements within the state of Colima (see Fig. 3). Without a doubt, such an Atlas is going to be very helpful for the local authorities, because it dynamically integrates updated and relevant information on the conditions of risk of Colima's risk-affected population, all of which will allow better decision making to be carried

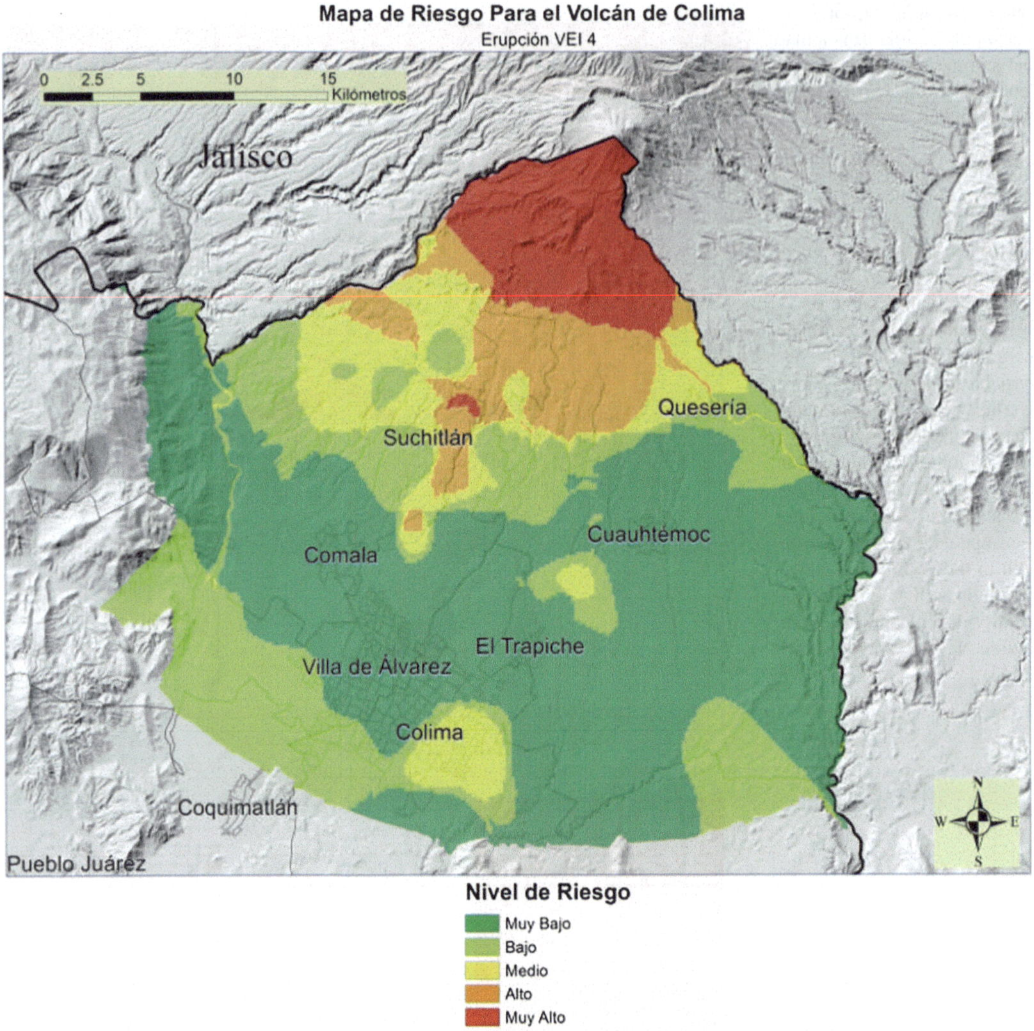

Fig. 3 Risk map associated with a VEI 4 eruption. This map is the result of a combination of hazard data with some elements of social vulnerability. This type of map allows better decisions to be made in the development of a volcanic risk management strategy. *Source* Varley et al. (2017)

out towards mitigation of volcanic risk in the short, mid, and long term.

Finally, it is worth pointing out that several aspects of the atlas risk analysis still have to be improved, because only a few vulnerability components, indicators and variables were studied for a few settlements, due to insufficient data and time to do fieldwork (Varley, personal communication). These limitations of the atlas risk analysis, constitute a barrier preventing an appropriate critical understanding of the real dimension of vulnerability (considering that alternative approaches are possible, other than numerical statistics and indicators). However, the work done on the atlas can be considered a big step forward for risk management. A start has been made to the study of risk, with the publication of an official risk map that includes important indicators of vulnerability, and not only geological and natural hazard cartography.

3.1.2 Monitoring the Volcano

Volcano monitoring is generally done by means of various disciplines, of which the most common are seismology, ground deformation and gas geochemistry (Donovan et al. 2012). Volcano monitoring in Colima is conducted mainly by an entity commonly and incorrectly known as "Observatorio Vulcanológico" (it cannot be considered as an observatory because the monitoring is not continuous, 24 h per day) officially named *Centro Universitario de Estudios e Investigaciones de Vulcanología*, CUEIV or University Centre for Study and Research in Volcanology, and the Faculty of Science of the University of Colima. Most seismic records from the seismologic telemetric network of the state of Colima, or RESCO, are received in the facilities of the civil protection authorities. The seismic data is obtained and processed, however, by the staff of RESCO with the goal of maintaining a monitoring capability (see Fig. 4).

Seismic monitoring of activity can therefore be carried out from Civil Protection's offices in Colima, however, the training that operatives have received is minimal so any interpretation will highly limited, making critical decision making difficult, if not impossible. This came to light during the second evacuation caused by a volcanic crisis in July 2015. The director of CUEIV and RESCO was unavailable and since that monitoring group has no continuous coverage or established communication protocol, the General Director of the state unit of Civil Protection was requested to provide an interpretation of the seismicity by the government communication centre (CECOM). The urgency of the situation resulted in Civil Protection having to make decisions regarding evacuation, without a high level of certainty concerning the current state of volcanic activity.

The aim of the CUEIV is to monitor Volcán de Colima's eruptive activity, to conduct scientific research within the state's volcanic

Fig. 4 State unit for Civil Protection's communication centre. Real time seismic volcanic data can be observed on the lower and central screens. In theory, this centre functions as a monitoring and communication station for natural phenomena and the associated risk that occurs in the state of Colima. However, like the CUEIV, there are only staff during office hours. During times of crisis additional volunteer staff are brought into handle telephone and radio calls outside regular hours

region, to coordinate the various groups of researchers working on aspects of volcanology, produce volcanic risk maps and strategize on disaster prevention in cooperation with both the state's Civil Protection system and with CENAPRED (Centro Nacional de Prevención de Disastres or National Centre for the Prevention of Disasters). In addition, they prepare documents with which to publicize technological and scientific development in Earth Sciences including in social media, organize local conferences and debates, etc. Monitoring is carried out with the participation of volcanologists from the Faculty of Science (see Varley, this volume for details on monitoring instrumentation).

Regarding the first goal, monitoring volcanic eruptions, it is important to point out that, despite actions taken and data produced since 1994, the centre has been faced with a series of problems that go beyond its scientific responsibility. Political changes have had an impact, as well as the economic consequences of the constant reduction or cancelation of financial resources destined to develop and maintain consistent and precise volcanic monitoring. The scarcity of funds means fewer staff available to achieve constant monitoring 24/7. Therefore, staff work only during normal business hours, although during critical times they have had to organize 24-h vigilance and continued monitoring from the university's facilities or from their own homes. Concerning the goal of consolidating a homogeneous group, with a clear objective of performing monitoring tasks, it has become necessary to strengthen bonds within CUEIV and being more open to welcoming researchers hailing from outside. In the case of the previous Scientific Assessment Committee and the current so-called Subcommittee of Geological Phenomena (see Sects. 3.2.1 and 3.2.2), less biased and more inclusive coordination is necessary on the side of Colima state's Civil Protection authorities to include, for example, social and behavioural science disciplines. From the viewpoint of this research, to successfully reduce volcanic risk, more synergy is required.

3.2 Prevention

Prevention is defined in the *Ley General de Protección Civil* as the set of actions and mechanisms implemented prior to the onset of a hazardous phenomenon, with the objective of obtaining knowledge on the hazards or risk: identifying them, subduing them, or eliminating them; to prevent or mitigate their destructive influence on people, economic goods, or infrastructure, and to be able to foresee and anticipate the processes of social construction of risk. It is important to outline that risk and disaster, being social processes, imply a complexity that is inherent to the classification of courses of action within the process itself. Limits need to be clearly defined so that the risk zoning, mapping, and monitoring mentioned above also form a part of prevention actions being undertaken in Colima to achieve successful volcanic risk management.

One of the most important steps taken towards volcanic risk management has been the joint work of scientists and the official authorities. Relationships between these two social actors, for instance, resulted in the creation of agreements with the University of Colima's scientific community to form a commission of expert volcanologists, replacing the now defunct Scientific Assessment Committee, with the Subcommittee of Geological Phenomena.

3.2.1 The Scientific Assessment Committee

The Scientific Committee for Risk Assessment was formed in 1998, starting with the signing of an agreement for cooperation between scientists and Colima's state civil protection system, in order to consolidate seismological and volcanic monitoring (Manual of Operational Proceedings in Case of Emergency, section dedicated to mitigation of volcanic risk). Scientific Assessment Committee refers to a group of professionals committed to the study of any kind of hazardous natural phenomena, and who possess the technical and scientific skills to be able to understand the source, evolution, and potential consequences of such phenomena (2nd Article, agreement by which Scientific Assessment Committees of the National System for Civil Protection are defined as technical consulting bodies in the matter of the prevention of disasters originated by geological, meteorological, chemical, sanitary, or social-organizational phenomena; *Diario Oficial de la Federación* the official Government document, published on 6 June 1995).

In the case of the state of Colima, the technical scientific committee is defined by the *Law of Civil Protection*, decree no. 376, as the "technical consultant group for Civil Protection for integrated risk management at state level" (4th Article, Agreement by which the Scientific Assessment Committees of the National System for Civil Protection are established for the prevention of disasters, DOF, 6 June 1995). This committee consisted mainly of earth science experts and for a limited period, a few social scientists, the latter component setting it apart from the Scientific Committee from CENAPRED, which was constituted after the volcanic crisis caused by Popocatépetl in 1994 (Macías 1999).

The Committee's functions were to establish a typological classification of potential crisis-causing phenomena for Volcán de Colima; evaluation of the data obtained from the volcano's monitoring networks in order to carry out a forecast of the most probable eruptive scenarios, and their consequences for Civil Protection; establish a methodology to maintain observation and follow up on eventual volcanic eruptive events; to assess the information obtained in the case of the occurrence of these kinds of events, and finally, formulate advice to the authorities about how to intervene in order to reduce risk linked to volcanic eruptions (Manual of Operational Proceedings in Case of Emergency (section dedicated to mitigation of volcanic risk), State System for Civil Protection, Colima).

The Committee would convene at least once monthly, or more frequently whenever existing volcanic risk conditions warranted. Each session of the Committee was organized in the following manner: list of attendees, analysis on monitoring parameters, consensus, recommendations, general affairs, moderators' signatures, meeting minutes, and bulletin.

Starting in July 2011, a new body was created to provide technical support to the State's Civil Protection System: the Scientific Assessment Committee was terminated and replaced by a new Technical Scientific Assessment Committee on Natural and Man-made Hazardous Phenomena. This new committee was formed by specialists in all five major categories of risk that exist in the state of Colima: geological, sanitary, chemical, social-organizational, and meteorological. In the case of geological phenomena, a geologist was appointed, who in turn gathered other specialists to form the subcommittee for geological phenomena.

It is important to indicate that the Operational Director of the State Unit of Civil Protection and the spokesman for the subcommittee for geological phenomena stated that the emergence of this new technical body was the result of a change in the law on civil protection (personal communication on 21 November 2014 and 4 August 2015, respectively). However, it has not been possible to track that particular reform in the context of the *Ley General de Protección Civil*, the *Ley Estatal de Protección Civil*, the *Manual of Organizational Operations* within the Sistema Nacional de Protección Civil, or in the Decree by which the Scientific Assessment Committees of the Sistema Nacional de Protección Civil were created.

3.2.2 Subcommittee of Geological Phenomena

It is worth outlining that the now-defunct Scientific Assessment Committee for Volcán de Colima was a large group of specialists in Earth Sciences from both Colima and Jalisco, with frequent participation from external researchers and advisors at the national and international level. The group who took up the task of evaluating the development of volcanic activity sometimes numbered more than 20 people. It was a group specifically formed to conduct analysis on the volcano. In contrast, the current Scientific Assessment Committee for Natural and Man-made Hazardous Phenomena is made up of a number of different scientists specializing in each individual hazard area. The Subcommittee for Geological Phenomena is formed by a group of only six specialists in Earth Sciences and volcanologists from the University of Colima, and excludes specialists located in Jalisco, though they have continued to participate in sessions held with Civil Protection authorities at the national level to evaluate activity during times of crisis.

This new technical Subcommittee for Geological Phenomena, although different from its predecessor in terms of structure and number of members, is essentially in charge of the same functions as was the former Scientific Assessment Committee. The main difference in structure of the subcommittee is its tendency to being more exclusive towards other researchers, whether from the University of Colima or external, for example from the University of Guadalajara (capital of Jalisco).

Until now, it is unknown whether there is document where the functions and attributions of the old Scientific Assessment Committee were described in detail. Written protocols are increasingly necessary in order to avoid poor decisions that can result from a lack of experience, for example, decisions on social matters being made by volcanologists in the midst of a future volcanic crisis (they may find themselves in a position where they are compromised to make such decisions). This has been the case when populations have to be evacuated, however, that is an action that the Civil Protection authorities are responsible for taking.

In this regard, Donovan et al. (2012) point out that the gap between risk assessment and risk management is generally complex, especially when the boundaries of the various actors are unclear. This can be dangerous especially during volcanic crisis situations, where scientists, officials and the public are not properly prepared and decisions have to be made under high levels of uncertainty in both volcanic risk assessment and decision-making. Therefore, it is suggested that social scientific analysis within the generation of scientific advice could facilitate its development by framing it within the political context in which it is carried out.

3.3 Mitigation

Mitigation refers to actions whose goal is to diminish the impact or damage caused by the presence or occurrence of hazards. In this sense, many stages of the disaster process and hence risk are directly related to effective mitigation.

3.3.1 Relocation of People as a Means of Risk Mitigation and Disaster Prevention

The relocation of human settlements involves extremely complex processes that go far beyond the mere displacement and reestablishment of human populations and the solution of their housing needs, which is why it must always be planned and executed with the active participation of the affected population involved, and according to their actual needs. Otherwise the social, economic, cultural, political, and environmental costs may be unnecessarily high (Barabás and Bartolomé 1992; Cernea 1995, 1997; Macías 2001, 2009; Oliver-Smith 2001). There are five main factors to be taken into account to ensure the success of relocation projects. These include the participation of the population that is to be relocated during the entirety of the process (even for making the decision of whether or not to relocate), their involvement in designing new housing (Macías 2001), physical and geographical aspects of the new settlement, and the possibilities of achieving the anticipated levels of social and economic development (Coburn et al. 2009; Macías 2009).

The case of relocation of the settlement of La Yerbabuena was a crucial action for risk management in terms of social, political, and scientific issues. In 1999, the Scientific Assessment Committee declared the need for La Yerbabuena to relocate after a series of evacuations in the prior months (Cuevas and Seefoó 2005; Macías 2001; decree by which the settlement of La Yerbabuena is declared as a high-risk area). The relocation of many of the people of La Yerbabuena generated conflicts and aggravated pre-existing ones. Some people chose to remain in the established settlement rather than relocate, and the resulting conflicts of interest produced disruption of the community's inner organization. The disruption was perceived by all the people involved (i.e. both the relocated people and those who chose to stay). For example, relationships between the inhabitants who refused to be relocated and the local government, especially the Civil Protection authorities, became problematic once cooperation on risk management between the two parties broke down.

The 57 relocated families were forced to adapt to a new environment, which was very different from the traditional lifestyles that they had in La Yerbabuena, and the 7 families that decided to resist relocation were forced by circumstance to commence new cultural and productive activities. They also had to implement a series of resistance strategies to make a living and fend for themselves in the context of new social dynamics, and on the other hand, to resist what they considered a flagrant violation of their human and individual rights. They believed that the claims by the authorities of the existence of volcanic risk were merely an excuse for the evacuation and relocation rather than a justification for it. To learn more on the complexity of this process, review the 2005 article (Cuevas and Seefoó).

It is important to mention that some volcanologists of the former Scientific Assessment Committee have insisted that they never participated in the formulation of the advice to relocate the settlement, which is at odds with the official version, indicated in the above-mentioned risk zone decree. Such ambiguity concerning an action of the magnitude of the relocation of a human settlement highlights the need for adequate and detailed ruling on the functions and responsibilities that scientific committees and subcommittees have in terms of risk management. As mentioned, they are mostly or entirely formed by earth-scientists with little or a total absence of both knowledge and skills in social research and hence, a very poor understanding of the causes and consequences of a variety of social phenomena relevant in actions such as evacuations and permanent relocations.

3.3.2 Volcanic Alert Systems

The objective of volcanic risk communication is to educate and to promote behaviour aimed at mitigating the risk. To do this it is necessary to link risk management and risk communication with a more inclusive approach, involving all stakeholders with robust methods that allow dialogue between the community, authorities and scientists at various levels (Barclay et al. 2008).

In Mexico, an instrument of communication and alertness, generically known as the Volcanic Warning Semaphore for Popocatépetl, has been developed. In the case of Colima, the *Manual of Operational Proceedings in case of Emergency*, in its section regarding volcanic risk mitigation, states that the traffic light warning system of Volcán de Colima is the mechanism of the state organization of Civil Protection, that keeps the people informed about the different levels of threat posed by volcanic activity. It also establishes that the alert system serves as a means to protect the families and the community as a whole. Each of its colours serves as an indicator of the particular level of danger present in a certain moment, and it is a reminder of things to be done as a consequence. The current Subcommittee for Geological Phenomena and Civil Protection authorities are the entities in charge of determining what colour the traffic lights should display within each community.

No physical device, such as a traffic light, has been installed in front of each vulnerable community to display colour signals in the state of Colima. Instead, the current alert or colour status has appeared in some press reports. The traffic light is an analogy that represents the level of danger and explains the actions that should be taken. In contrast to the situation at Colima, the volcanic traffic light of Jalisco does exist in each community as an actual device, and displays coloured signals to the local population (see Fig. 5). However, it is known that scientists from the University of Colima are currently collaborating with the Civil Protection authorities, in the design of a new volcanic alert system.

The aforementioned manual states that the volcanic alert system is the last link in the information chain (but not the communication chain) between the authorities and the people. In other words, out of an entire chain of processes, the traffic light's red light ends up being the means by which information resulting from the risk analysis, previously performed by the government and the scientific community, is given to the public, as well as the actions that must be taken as indicated in Fig. 6.

Figure 6 shows the simple and linear model of decision-making and risk management with which Civil Protection in the state of Jalisco acts. The population is considered as the last "link" in the "chain" of command, whereby they are represented as a passive group that only receives information, orders and is subject to government intervention.

As for the colours shown by the traffic light, the colour green indicates that normal, everyday activities can continue. Nevertheless, they ask the population to learn the different routes for emergency evacuation, to be aware of where meeting points determined by the authorities are located, attend training courses offered by Civil Protection units, partake in drills practiced in their communities, and remember that housing built on slopes and riverbanks is much more prone to be damaged, due to the fact that they represent the most common paths for volcanic material. Lastly, people have been asked to be aware of the benefits of building their houses with strong and preferably steep sloping roofs, in order to make them less vulnerable to ashfall

Fig. 5 Volcanic alert system or "traffic light" in the community of Juan Barragán, Jalisco. The device indicates the state of volcanic activity and courses of action associated with the three colours: green, yellow and red. They also present a sketch of the evacuation routes to be followed by the various communities

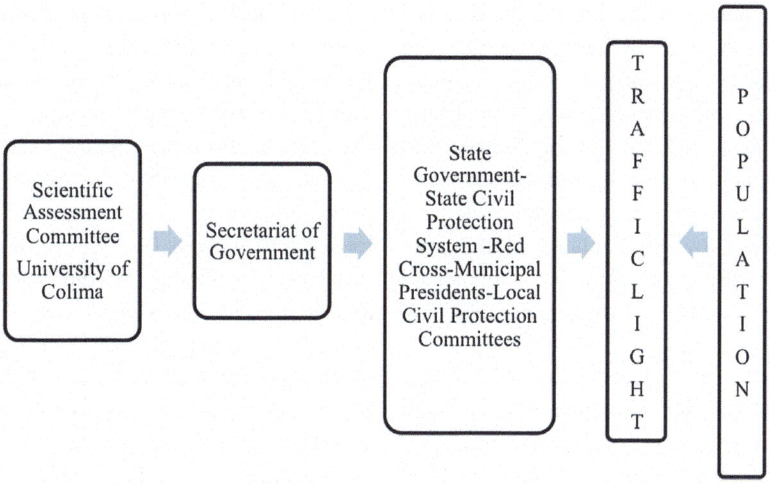

Fig. 6 Information flow and decision making for the volcanic alert system. In the scheme it can be seen that the alert system is considered the only mechanism for informing the population, that is to say, there is no further interaction between scientists, government authorities and Civil Protection and the population. The problem is that up to now in Colima, this alert system has not existed physically as an information mechanism. It is only referenced in the procedural manuals, which are not accessible to the public. *Source* Reproduction of the outline contained in the "Manual of Operational Proceedings in the Case of Emergency". *Source* Manual of operational proceedings in the case of emergency

(*Manual of Operational Proceedings in case of Emergency*).

The yellow colour means that people should be in a state of alert, pay close attention to what authorities declare through the media, and to actually follow the directions given. The authorities also recommend that families be prepared with individual cards indicating their names and addresses; close and shut down electricity, water and gas services in homes; to store drinkable water, food and medicine, as well as the most important documents such as property titles, birth and marriage certificates, military service carnets, etc.; to have a radio with charged batteries, a lantern, and house keys; and to cover water and food storage areas to prevent contamination by ash. Additionally, if they own cattle or other animals, they are asked to make an enquiry to Civil Protection officials on how to best protect them.

The red colour indicates alarm. When the signal is red, people must follow directions provided by authorities as a mandatory procedure. If they order an evacuation, people must gather together with their family members, have everyone properly identified, make sure that all doors and windows are closed, and before leaving, extend a white sheet or blanket in front of their houses to signal that their home is evacuated. They are then asked to immediately evacuate to meeting centres, taking with them only the essential items as listed in the previous paragraph. If possible, they are asked to evacuate diligently and without waiting for orders to be given; to reach the temporary shelter and take the place indicated to them. If medical attention is required, they must attend the closest healthcare facility. Once in the temporary shelter, they are asked to cooperate with whatever tasks they are assigned, to keep open communication with the authorities during their occupancy of the shelter and to pay attention regarding any developments related to the emergency until its end. If they are unable to locate the meeting centre, or if a means of transportation out of the danger zone is unavailable, they are asked to walk away from the volcano using the highest paths available to a safer place.

The intention is that both the current volcanic activity and the related risk assessments made by scientists are communicated through the analogy with the traffic lights, the aim being that the population take the appropriate protection measures. However, with a device such as the traffic light, at best, it serves to inform people but it is not a system that allows communication as such. It has been documented that increased awareness of hazards and knowledge related to the phenomena within the population does not necessarily lead to actions to reduce risks (Barclay et al. 2008).

Therefore, the incorporation of social scientists in the design and implementation of risk communication programs is necessary, to ensure that it is both appropriate and efficient. It is important that any alert system considers the communication and information needs of the populations at risk. Effective communication of volcanic risk needs to account for the socioeconomic situation of the population (Barclay et al. 2008), but must also include the socioeconomic and political conditions of the governmental authorities responsible for risk management.

3.3.3 Evacuation Routes

Considering the evacuation routes, distances between settlements and the most important road junctions have been calculated. Distances from La Yerbabuena on the evacuation route to La María is 3 km; and to the crossroads with La Becerrera it is 4 km; from there to El Cordobán bridge it is 4 km, and so on.

Some sections of the main evacuation route get blocked frequently, especially during the rainy season, and two particularly critical locations for hazards lie between El Cordobán bridge and La Becerrera and the junction to Carrizalillos, due to rockfalls and landslides in the area. This was clear during the passing of hurricane Jova in 2011, when La Becerrera and La Yerbabuena remained isolated for almost a week due to the roads being blocked. More recently, in 2015, after the impact of hurricane Patricia, the evacuation route that crosses El Zarco and El Cordobán bridges sustained damage and became unusable, as was the case with the road from La Becerrera to San José del Carmen, Jalisco, due to lahars and the rising of the river level (see Fig. 7).

Lastly, it is important to point out the fact that there are alternative routes of evacuation out of La Yerbabuena that have been defined by those villagers resisting the relocations. Those routes are different from the officially established ones in that, according to those people, the official routes pass closer to areas that could be affected by volcanic hazards (see Fig. 8). It is clear that a constant re-evaluation of the evacuation routes and their condition is necessary. This should be carried out jointly between scientists, civil protection and the population at risk. The improvement of these emergency routes, as well as the involvement of the population in their design and maintenance, would be a good mechanism to ensure their official recognition and appropriate use.

3.4 Preparation and Aid

3.4.1 Preparation

Preparation is defined as all the actions and measures implemented in advance to ensure a more effective response to the impact of a hazard in the short-, mid-, and long-term. As was already mentioned, most actions undertaken by the Civil Protection authorities can be classified in more than one stage or phase within the disaster process. In this sense, it can also be said that all of the aforementioned actions such as establishing evacuation routes, volcano alert system and relocation are part of the stage of preparation.

A clear example is the creation of manuals and emergency protocols described in previous sections of this chapter. Their contents and organizational charts are described below.

Manual of Procedures and Protocols in Case of Volcanic Emergency

As it has been explained, in the particular case of the state of Colima there is a series of official measures intended to achieve efficient management of volcanic risk. Large part of such measures are detailed in the *Manual of Organization and Operation 'Volcán de Colima*, as well as in the *Manual of Operational Proceedings in the*

Fig. 7 Evacuation route damaged by a lahar. The community of La Becerrera, Colima, is surrounded by La Lumbre, El Zarco and El Cordobán rivers, which experience lahars during almost every intense rainfall event, with some of them affecting the communication routes. The image shows a road that serves as an evacuation route and the damage sustained due to a series of lahars produced in El Zarco ravine during the impact of Hurricane Patricia in October 2015. The lahars within La Lumbre also damaged the road considerably, leaving La Becerrera cut-off from both Colima and Jalisco

Case of Emergency, the latter including a section that addresses mitigation of volcanic risk.

In the *Manual of Organization and Operation 'Colima Volcano*, a twelve-page booklet, a variety of topics are briefly addressed such as: general aspects of the volcano (mainly its physical and geographical features), the populations at risk, risk zoning and classification, parameters for volcano monitoring, the evacuation route map (for the settlements of Colima, Comala and Cuauhtémoc), details of communication options (radio and telephone lines), an information bulletin (resulting from sessions held by the now defunct Scientific Assessment Committee), an estimation on the number of inhabitants at risk due to volcanic activity, the location of a service centre for the people affected, and a definition of the governmental dependencies that participate in the Operational Plan for evacuation of high risk settlements. Included are the activities that each of those entities is expected to perform.

The *Manual of Operational Proceedings in the Case of Emergency*, which includes a section dedicated to the mitigation of volcanic risk, contains a subsection which includes a definition of risk and ways to reduce it, through what is conceptually referred to as active protection. It also contains a subsection dedicated to explaining how volcano monitoring in Colima is performed. Other subsections feature digital maps that present potential risk scenarios; one presents the map of volcanic hazard by Navarro and Cortés (2003). A description of the "volcano observatory" (CUEIV) and the Scientific Assessment Committee and its role within the State System of Civil Protection are also described. Lastly, a subsection summarises how ordinary and extraordinary sessions of the committee

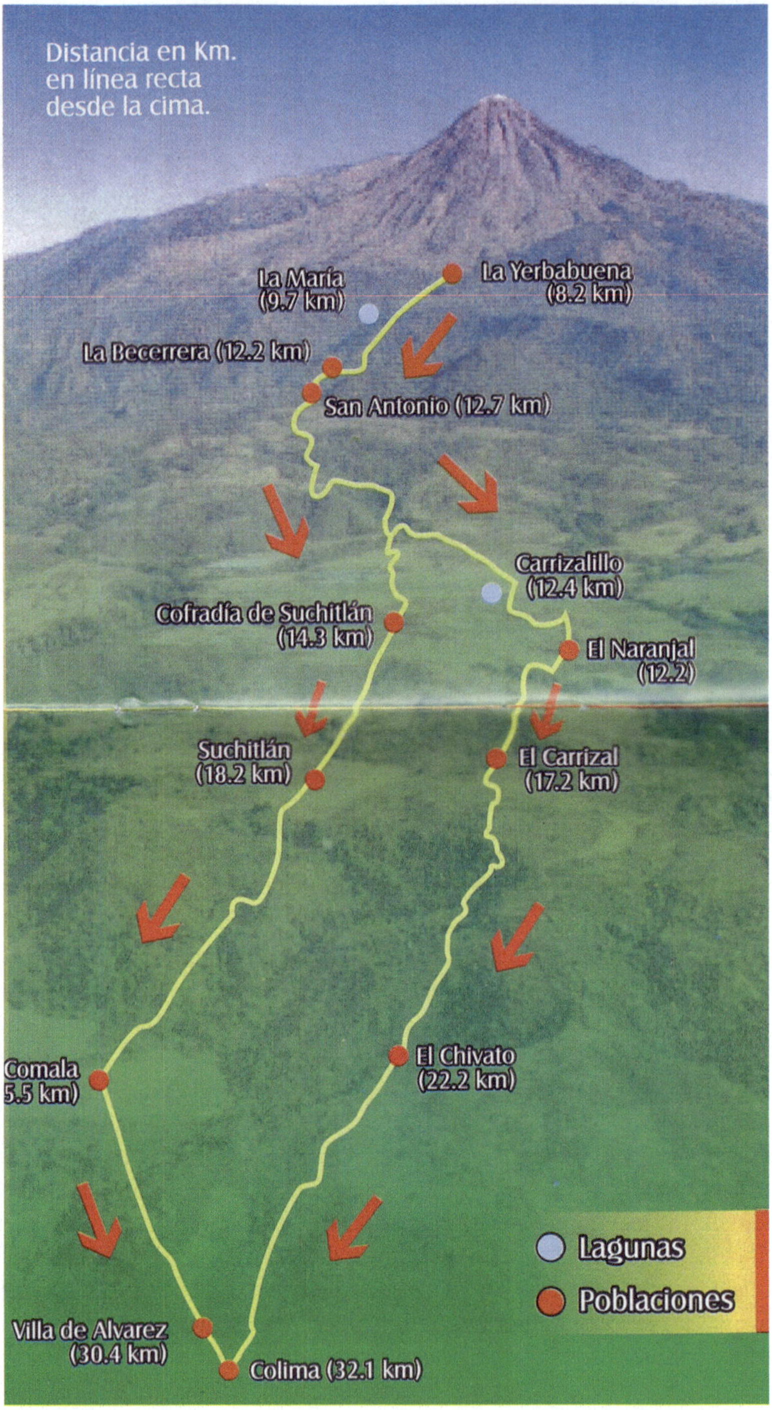

Fig. 8 Evacuation routes in the state of Colima. This leaflet was prepared by the University of Colima and Civil Protection Colima, and shows the evacuation routes in case of a volcanic emergency. Also the most relevant communities and the distances between them and the crater of Volcán de Colima are shown. *Source* Colima State Civil Protection

were held, the flow of information in the case of an emergency related to the volcano, the volcanic alert system, and evacuation routes.

It is necessary to emphasize that these documents are not available for the general public, they are kept in the offices of Civil Protection.

These documents are not updated, and have not yet been approved by Colima's elected governor, who according to the law, is the President of the State System for Civil Protection. These documents were developed based on the general outlines of the Volcanic Emergency Management manual of the United Nations Disaster Relief Office (UNDRO 1985), which gives us an idea of the implications of these international schemes, for example, schemes were created based on the experience gained in developed countries and therefore are not fully applicable in so-called third world countries. In that sense, successful transfer of technology proposed by the UNDRO model requires special attention to community livelihoods, especially economic, family, social and cultural subsystems. Likewise, the UNDRO model needs to evolve to make disaster mitigation the responsibility of the people of these communities working hand-in-hand with local emergency management agencies, informed public officials and scientific groups (Macías and Aguirre 2006).

It is important to note that neither of the two manuals of operational procedures (OOVC Manual or OPCE) establish the criteria or the basis on which a decision is taken to move from one alert phase to another, nor do they establish the parameters and criteria for deciding on an evacuation or relocation of a population at risk, nor who determines such decisions. These criteria should be developed and established with the full participation of the at-risk population.

3.4.2 Aid

The stage of aid is defined as the response directed to the people at risk, or affected by a emergency or disaster, and provided by public or private specialized working teams, or by units of Civil Protection, as well as all actions intended to defend any other risk-prone populations.

In Colima, in the case of volcanic risk, the aid phase is implemented when the Subcommittee of Geological Phenomena identifies a significant increase in volcanic activity that, because of its nature and dimensions, could affect certain human settlements; after that, they discuss with the State unit of Civil Protection the evaluation of the situation and any decision on which actions shall be taken. In the case of evacuations or any other specific orders, the Civil Protection authorities at the state level report to upper governmental entities via radio and telephone. The municipal commissioner in charge of the settlement to be evacuated is advised via radio, so he and his support team can distribute the information using loud speakers or by going door-to-door. Transport is prepared to assist in the evacuation. The inhabitants are placed together in public parks and then transferred with use of public transportation (see Fig. 9). A preliminary count of the evacuated population is done during their transportation, and then again when they arrive at the shelters.

It is important to note that the decision-making process is inherently multidisciplinary, for example a particular expertise is required so that consideration can be given to the way in which people are informed about volcanic hazards. This is an aspect in which social scientists and communication experts can play an extremely important role (Marzocchi et al. 2012).

During the volcanic eruptions of July 2015, evacuations were made once the pyroclastic flows were generated from the eruption, and once volcanic ash began to cover the populations of La Yerbabuena and La Becerrera. The evacuation of 10 July 2015 was made based on the decision taken only by the Operational Director of the State Unit of Civil Protection. The next day, the evacuation of 11 July was decided collectively between the Subcommittee on Geological Phenomena, Civil Protection officials and other government authorities. It is important to note that the two evacuations were reactive rather than preventive.

The *Secretaría de la Defensa National* (Ministry of Armed Forces, SEDENA) and local councils take care of preparing, managing, and operating the temporary shelters. Public security personnel at both the state and local level support the evacuation processes and guard the perimeters of the evacuated areas. The *Secretaría de Salud* (Ministry of Health) is responsible for performing check-ups of the evacuated people. There is a long list of other governmental

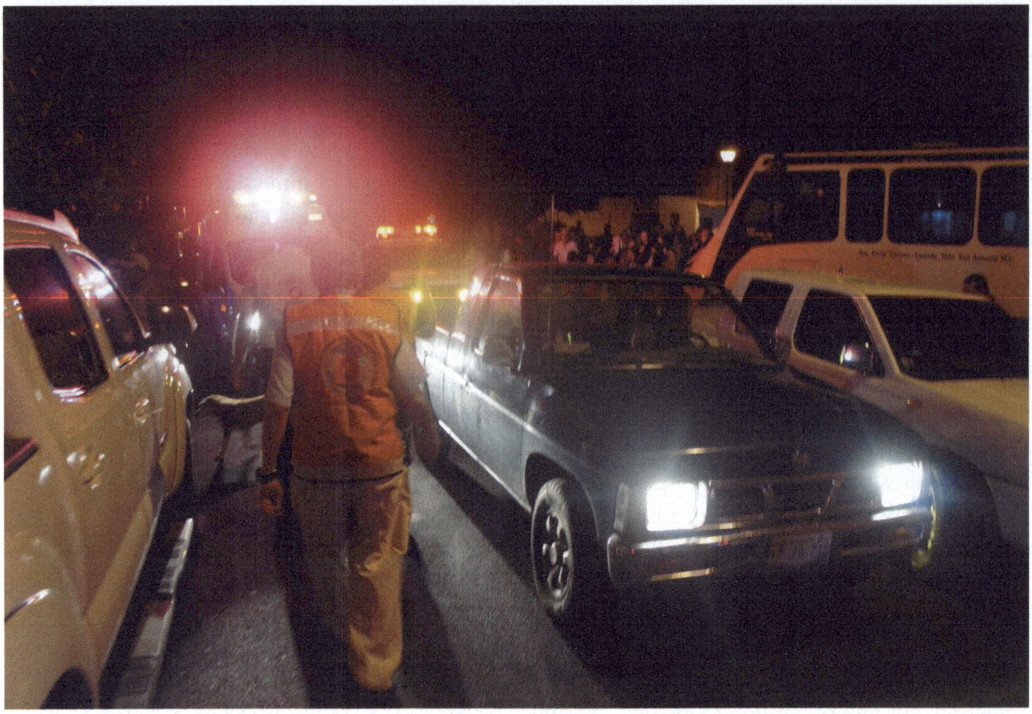

Fig. 9 People assembled in a public park in La Becerrera, before being evacuated in September 2016. The photograph shows the moment in which the people of the community were gathered in the main garden to receive indications from the Director General of Colima State Civil Protection (person with orange vest), before being evacuated in anticipation of the increase in volcanic activity

institutions that partake in this operational plan, which includes the *Dirección de Transportes* (Department of Transport), *Comisión Nacional del Agua* (National Water Commission), *Desarrollo Integral Familiar* (Department for Family Development), the Fire department, the Mexican Red Cross, *Secretarìa de Administración del Gobierno del Estado* (Department of Administration of the State Government), *Procuraduría General de Justicia* (Justice Department), *Teléfonos de México* (State Telephone Company), *Secretaría de Comunicaciones y Transportes* (Secretary of Communications and Transport), *Desarrollo Urbano* (Department of Urban Development), Rancho "El Jabalí" (a private entity), and groups of volunteers.

Although in Colima there is a list of participating institutions and government agencies in the event of a volcanic emergency, there is no established protocol to follow, i.e. the mechanisms by which the operational plan is implemented and the procedures by which each government agency is notified.

4 Conclusion

The understanding of volcanic risk must go beyond the probability of impact of volcanic hazards, and actions to reduce volcanic risk should not be based solely on the monitoring of the volcanic phenomenon. Many social, economic, political and cultural factors related to the populations at risk, civil protection personnel, scientists and other authorities should be considered in the design of strategies to manage and reduce volcanic risk. Otherwise, one only addresses part of the problem, rather than its complex entirety: in its construction as a process. To do this, the risk management strategy must be more socially engaging, through the integration of social scientists in its design and implementation, and with the total participation of the population in the process. In that sense, the relationship and interaction between scientists, Civil Protection and population must be much closer and more active. These aspects are detailed below with respect to each phase of Integral Risk Management.

For risk identification, maps of volcanic hazards should be regularly updated due to the constant changes in the environment, such as the volcano drainage pattern, due to the continuous eruptions and often intense rainy seasons, as well as the increase of agro-industrial activities in the volcanic zone and the increasing construction of ranches and country houses. Also worthy of note is the release of the State Risk Atlas, in the section on volcanic risk, where vulnerability maps were constructed by integrating some elements of social vulnerability with volcanic hazards. It is therefore suggested that such information be integrated into the design of the risk management strategy. With respect to monitoring, it is necessary to employ more personnel to enable 24 h a day, 365 days a year monitoring, or at the very least during moments of increased activity. The incorporation of university students as volunteers could be an alternative to achieve continuous monitoring.

With regard to prevention, the functions of the Subcommittee of Geological Phenomena should be established in detail, and social scientists should be involved in risk assessment and decision-making. It is important that actions to reduce risks should be carried out mainly in moments of volcanic calm, so that when the activity increases, populations, scientists and authorities are prepared.

In the mitigation phase, it is necessary to consider the political, scientific and social experiences acquired during the process of relocation of La Yerbabuena, since criticism has arisen due to the different perspectives of Civil Protection and social scientists, but relocation could be the only option for certain communities. In addition, new mechanisms for the reduction of volcanic risk must be sought with the full participation of the target populations. With regard to the volcanic alert system, a broader and more permanent volcanic risk communication program should be developed to increase the efficiency of the system currently under development. Alone it is not enough to ensure an appropriate self-protection reaction within the communities. The participation of social scientists will be of great importance to make any headway. Likewise, in this communication program, social mechanisms must be established so that communication is not unilateral: from Civil Protection to the population, but is multidirectional, in which the various social actors can communicate and express their needs and concerns, and that these are heard, understood and answered. Finally, evacuation routes should be re-evaluated and conditioned to make them more effective and to reduce the risk from other hazards during an eruption crisis.

For the preparation and assistance stages, the existing operational and manual plans with definitions of procedures must first be re-evaluated and updated. These documents are based on the general schemes of the Volcanic Emergency Manual prepared by the United Nations Disaster Relief Office (UNDRO), and they require further adaptation to the political, economic, scientific and social context of Mexico and in particular Colima. They should detail the roles and responsibilities of each social actor involved in risk management, and the capacities and resources available to deal with volcanic hazards. An adequate decision-making mechanism corresponding to the roles and responsibilities of each participant should be detailed. This will avoid conflicting situations, such as those experienced in the relocation of La Yerbabuena, and will allow actions to be more appropriate with respect to the target population. Later, the documents should be distributed to the population, with continued availability to the public and the research community.

These recommendations have the sole objective of creating a more social and therefore more adequate and effective volcanic risk management strategy.

Acknowledgements I thank the reviewers Jesús Manuel Macías, Luisa Macedo Franco and Chris Gregg for their valuable contributions to the work. I also thank Nick Varley and Juan Carlos Gavilanes for their help in improving the document, and Graeme Alexander William Sinclair for his help in translation.

References

Barabás, A., Bartolomé, A.: Antropología y relocalizaciones. Alteridades. Revista de Ciencias Sociales y Humanidades **2**(4), 5–15 (1992)

Barclay, J., Haynes, K., Mitchell, T., Solana, C., Teeuw, R., Darnell, A., Crosweller, H., Cole, P., Pyle, D., Lowe, C., Fearnley, C., Kelman, I.: In Liverman, D., Pereira, C., Marker, B. (eds.) Communicating Environmental Geoscience. Geological Society, London, Publications, 305, pp. 163–177 (2008)

Capra, L., Gavilanes-Ruiz, J.C., Bonasia, R., Saucedo-Giron, R., Sulpizio, R.: Re-assessing volcanic hazard zonation of Volcán de Colima, México. Nat. Hazards **76**, 1–21 (2015)

Capra, L., Macías, J., Cortés, A., Dávila, N., Saucedo, R., Osorio, S., Arce, J., Gavilanes, J., Corona, P., García, L., Sosa, G., Vázquez, R.: Preliminary report on the July 10–11, 2015 eruption at Volcán de Colima: pyroclastic density currents with exceptional runouts and volume. J. Volcanol. Geoth. Res. **310**, 39–49 (2016)

Centro Universitario de Estudios e Investigaciones de Vulcanología (CUEIV). http://portal.ucol.mx/cueiv/. Consulted 15 Dec 2016

Cernea, M.: Primero la gente. Variables sociológicas en el desarrollo rural. Fondo de Cultura Económica, Mexico (1995)

Cernea, M.: The Risk and Reconstruction Model for the Resettling of Displaced Populations. World Bank, Washington DC (1997)

Coburn, A., Leslie, J., Tabban, A.: In: Macías, J.M., Investigación evaluativa de reubicaciones humanas por desastres en México, pp. 49–97. Centro de Investigaciones en Estudios Superiores en Antropología Social, Mexico (2009)

Cuevas, A.: (2005) Dinámicas de interfaces entre diversos actores sociales: Reubicación de una población campesina por un posible riesgo volcánico. Ph.D. thesis, El Colegio de Michoacán, Mexico

Cuevas, A., Seefoó, J.: Reubicación y desarticulación de La Yerbabuena; entre el riesgo volcánico y la vulnerabilidad política. Desacatos, Centro de Investigaciones en Estudios Superiores en Antropología Social, pp. 41–70, Sept–Dec, number 019 (2005)

Donovan, A., Oppenheimer, C., Bravo, M.: Science at the policy interface: volcano-monitoring technologies and volcanic hazard management. Bull. Volcanol. **74**, 1005–1022 (2012)

Gavilanes, J.C.: Simulación de escenarios eruptivos del Volcán de Colima y aportaciones al plan de contingencias del estado de Colima. Masters thesis, Posgrado en Geografía de la Universidad Autónoma de México (2004)

Gavilanes, J., Cuevas, A., Varley, N., Gwynne, G., Stevenson, J., Saucedo, R., Pérez, A., Aboukhalil, M., Cortés, A.: Exploring the factors that influence the perception of risk: the case of Volcán de Colima, México. J. Volacanol. Geoth. Res. **186**, 238–252 (2009)

González, L.: Evaluación de la vulnerabilidad social en la comunidad Ex-hacienda San Antonio, La Yerbabuena, Edo. de Colima. Undergraduate thesis, Facultad de Filosofía y Letras, Colegio de Geografía, Universidad Nacional Autónoma de México (2000)

Guevara, E., Quass, R., Fernández, G.: Guía Básica para la Elaboración de Atlas Estatales y Municipales de Peligros y Riesgos, p. 155. Centro Nacional de Prevención de Desastres, Mexico (2004)

INEGI: Censo de Población y Vivienda 2010. www.inegi.org.mx (2010)

Ley General de Protección Civil: Diario Oficial de la Federación 6 June 2012, with last amendments on 3 June 2014. https://www.gob.mx/indesol/documentos/ley-general-de-proteccion-civil-60762 (2012)

Luhr, J.F.: Petrology and geochemistry of the 1991 and 1998–1999 lava flows from Volcán de Colima, Mexico: implications for the end of the current eruptive cycle. J. Volcanol. Geoth. Res. **117**(1–2), 169–194 (2002)

Macías, J.: Riesgo volcánico y evacuación como respuesta social en el Volcán de Fuego de Colima. Textos urgentes, Centro de Investigaciones y Estudios Superiores en Antropología Social y Universidad de Colima, México, pp. 87 (1999)

Macías, J.: Reubicación de comunidades humanas. Entre la producción y la reducción de desastres, p. 280. Universidad de Colima, Mexico (2001)

Macías, J.: Investigación evaluativa de reubicaciones humanas por desastre en México. Centro de Investigaciones y Estudios Superiores en Antropología Social, p. 614. México (2009)

Macías, J.: Los Modelos gubernamentales para enfrentar los problemas del riesgo-desastre (Defensa Civil, Protección Civil, Manejo de Emergencias). Texto elaborado para la Especialidad de Gestión Integral de Riesgo de Desastre en la Escuela de Administración Pública de la Ciudad de México: 1–71 (2016)

Macías, J., Aguirre, B.: A critical evaluation of the United Nations volcanic emergency management system: evidence from latin America. J. Int. Aff. Spring/Summer **59**, 43–61 (2006)

Macías, J., Saucedo, R., Gavilanes, J., Varley, N., Velasco, S., Bursik, M., Vargas, V., Cortés, A.: Flujos piroclásticos asociados a la actividad explosiva del Volcán de Colima y perspectivas futuras. GEOS **25** (3), 340–351 (2006)

Martin del Pozzo, A., Sheridan, M., Barrera, D., Lugo, J., Vázquez, L.: Potential hazards from Colima Volcano,

Mexico. Geofísica Internacional **34**(4), 363–376 (1995)

Martin del Pozzo, A., Sheridan, M., Barrera, D., Lugo, J., Vázquez, L.: Mapa de Peligros Volcán de Colima. Instituto de Geofísica, Universidad Autónoma de México, Mexico (1996)

Marzocchi, W., Newhall, C., Woo, G.: The scientific management of volcanic crises. J. Volcanol. Geoth. Res. **247–248**, 181–189 (2012)

Mendoza-Rosas, A.T., De la Cruz-Reyna, S.: A statistical method linking geological and historical eruption time series for volcanic hazard estimations: applications to active polygenetic volcanoes. J. Volcanol. Geoth. Res. **176**(2), 277–290 (2008)

Navarro, C., Cortés, A.: Mapa de Peligros Volcán de Colima, Universidad de Colima, Mexico. http://portal.ucol.mx/content/micrositios/169/image/mapa.gif (2003)

Oliver-Smith, A.: Consideraciones teóricas y modelos del reasentamiento de comunidades. In: Macías, Reubicación de comunidades humanas. Entre la producción y la reducción de desastres. Universidad de Colima, Mexico (2001)

Reyes-Dávila, G.A., Arámbula-Mendoza, R., Espinasa-Pereña, R., Pankhurst, M.J., Navarro-Ochoa, C., Savov, I., Vargas-Bracamontes, D.M., Cortés-Cortés, A., Gutiérrez-Martínez, C., Valdés-González, C., Domínguez-Reyes, T., González-Amezcua, M., Martínez-Fierros, A., Ramírez-Vázquez, C.A., Cárdenas-González, L., Castañeda-Bastida, E., Vázquez Espinoza de los Monteros, D.M., Nieto-Torres, A., Campion, R., Courtois, L. Lee, P.D.: Volcán de Colima dome collapse of July, 2015 and associated pyroclastic density currents. J. Volcanol. Geoth. Res. **320**, 100–106 (2016)

Saucedo, R., Macías, J., Sheridan, M., Bursik, M., Kmorowski, J.: Modeling of pyroclastic flows of Colima Volcano, Mexico: implications for hazard assessment. J. Volcanol. Geoth. Res. **139**, 103–115 (2005)

Saucedo, R., Macías, J., Gavilanes, J., Arce, J., Komorowski, J., Gardner, J., Valdez, G.: Eyewitness, stratigraphy, chemistry, and eruptive dynamics of the 1913 Plinian eruption of Volcán de Colima, México. J. Volcanol. Geoth. Res. **191**, 149–166 (2010)

Sheridan, M.F., Macías, J.-L.: Estimation of risk probability for gravity-driven pyroclastics flows at Volcan Colima, Mexico. J. Volcanol. Geoth. Res. **66**(1–4), 251–256 (1995)

United Nations Disasters Relief: Volcanic Emergency Management, pp. 86. United Nations (1985)

Varley, N.: Evaluación del riesgo volcánico del Volcán de Colima, México. Presentado en el Foro Internacional sobre Gestión del Riesgo Geológico, Arequipa, Perú (Oct 2015)

Varley, N.R., Cernas, J.A., García, A., López, Z., Meza, M.I.: Riesgos Volcanológicos, Atlas de Riesgos Estatal de Colima, Universidad de Colima, Protección Civil del Estado de Colima, Colima (2017)

Suárez, C., Saavedra, G.: Análisis y mapa de riesgo del Volcán Colima, México. Centro de investigaciones de la Facultad de Geografía, Universidad de Guadalajara

The manufacturer's authorised representative in the EU is Springer Nature Customer Service Centre GmbH, Europaplatz 3, 69115 Heidelberg, Germany. If you have any concerns regarding our products, please contact ProductSafety@springernature.com

Printed and bound by CPI Group (UK) Ltd, Croydon, CR0 4YY
30/09/2026
02225591-0019